SELECTION OF MATERIALS FOR COMPONENT DESIGN

Source Book

A collection of outstanding articles from the technical literature

SELECTION OF MATERIALS FOR COMPONENT DESIGN

Source Book

A collection of outstanding articles from the technical literature

Compiled by
Consulting Editor

Howard E. Boyer
Technical Consultant
Former Editor, Metals Handbook

American Society for Metals
Metals Park, Ohio 44073

Library of Congress Catalog Card No.: 86-70248
ISBN: 0-87170-256-8
SAN 204-7586

PRINTED IN THE UNITED STATES OF AMERICA

Contributors to This Source Book

KENNETH J. ALTORFER
Kidd Creek Mines Ltd.

A. A. ANCTIL
Army Materials and Mechanics
    Research Center

M. AZRIN
Army Materials and Mechanics
    Research Center

R. G. BAKER
Brunel University

W. T. BAKKER
Electric Power Research Institute

WILLIAM BALLANTINE
ITT Harper

H. J. BOVING
Laboratoire Suisse de
    Recherches Horlogères
    (LSRH)

WALTER E. BURD
Carpenter Technology Corp.

TONY CALAYAG
Kidd Creek Mines Ltd.

J. E. CHAFEY
General Atomic Co.

NEIL J. CULP
Carpenter Technology Corp.

LOUIS R. DePATTO
Lehigh University

B. R. DRENTH
Metal Research Institute TNO

A. M. EDWARDS
Cabot Corp.

W. M. FARRELL
Fern Engineering

DEAN FERRES
Kidd Creek Mines Ltd.

H. E. FEWTRELL
Lockheed-California Co.

TATSUHIKO FUKUOKA
Taiho Kogyo Co., Ltd.

R. J. HENRY
University of Pittsburgh,
    Johnstown

H. E. HINTERMANN
Laboratoire Suisse de
    Recherches Horlogères
    (LSRH)

H. C. HOFFMAN
American Iron and Steel
    Institute

DENNIS D. HUFFMAN
Timken Research

A. G. IMGRAM
Gulf Research and Development

C. F. JATCZAK
The Timken Co.

JOHN JOHNSON
SKF Industries, Inc.

W. R. JOHNSON
G. A. Technologies, Inc.

SOJI KAMIYA
Taiho Kogyo Co., Ltd.

HIROMITSU KATO
Taiho Kogyo Co., Ltd.

PETER W. KELLY
Barber-Colman Co.

K. KONDO
Toyota Motor Corp.

D. KOTCHICK
AiResearch Co. of California

R. KUMAR
National Metallurgical
    Laboratory
Council of Scientific and
    Industrial Research (India)

D. F. LAURENCE
Petrolite Corp.

K. J. McCARTHY
Petrolite Corp.

ARTHUR F. McLEAN
Ford Motor Co.

R. K. MAHANTI
National Metallurgical
    Laboratory
Council of Scientific and
    Industrial Research (India)

JAMES H. MAKER
Associated Spring

JOHN L. MASON
The Garrett Corp.

NOTE: Affiliations given were applicable at date of contribution.

PREFACE

Selection of materials for component design is a complex task requiring consideration of many interrelated factors, not all of which are necessarily compatible. Thus, compromises and tradeoffs among various design factors are routinely made. A designer must know the relative importance of these factors and how they interact before he can make intelligent choices between compatible requirements. Currently, the designer's decision-making process is endless. If he thinks he has made the right decision one day, the events of the next day or the next week—specifically, product misuse, government regulations on materials or inconsistent manufacture of the material—may prove him wrong. Fortunately, however, the complexity of the decision-making process applies equally to all materials.

Materials selection should be approached as part of the overall design process, which also includes determination of the configuration of the product and its various component parts, and processes by which it is to be made. In its early stages, the process consists of evaluating various combinations of preliminary configuration, candidate material and potential method of manufacture, and comparing these with the previously established performance specifications. The relationship between configuration and material (processed in some specific way) is that every configuration places certain demands on the material, and the material has certain capabilities to meet these demands. A common specific relationship is that between the stress imposed by the configuration and the strength of the material. Of course, there are other relationships as well. Changes in processing can change the properties of a material, and certain combinations of configuration and material cannot be made by some manufacturing processes.

The size and weight of a part can affect the choice of both material and manufacturing process. Small parts often can be economically machined from solid bar stock, even in fairly large quantities. The material cost of a small part may be far less than the cost of manufacturing it, perhaps making the use of relatively expensive materials feasible. Large parts may be difficult or impossible to heat treat to high strength levels. There are also size limits on parts that can be formed by various manufacturing processes. Die castings, investment castings and powder-metallurgy parts are generally limited to a few kilograms. When weight is a critical factor, parts often are made from materials having high strength-to-weight ratios.

The quantity of parts to be made can affect all aspects of the engineering design process. Low-quantity production runs can seldom justify the investment in tooling required by production processes such as forging or die casting, and may limit the choice of materials to those already in the designer's factory or those stocked by service centers. High-quantity production runs may be affected by the capability of materials producers to supply the required quantity. Mass-produced parts sometimes can be designed and redesigned, requiring a large expenditure for engineering and evaluation, but providing enough savings, considering the large quantity involved, to make the effort worthwhile. Design for low-quantity parts may be limited to finding the first design and material that serve the required purpose.

Products that may be manufactured in several locations can present additional problems for designers because the cost and availability of materials can vary from place to place. If the product is to be made in different countries, the nearest equivalent grades of steel, for example, might be different enough to affect service performance. In some areas that have low prevailing labor costs, it may be desirable to design a labor-intensive product; in a high-cost labor market, the designer often attempts to design the product to fit the capabilities of automated manufacturing equipment.

It is relatively late in the design process before designers, materials engineers and manufacturing engineers, by working together, can establish detailed criteria for materials selection. Only then can the field of candidate materials and manufacturing processes be narrowed to a manageable number of alternatives. The implications of each of these alternative materials and manufacturing processes can then be evaluated, and any required changes in configuration can be made.

Two often-cited reasons for selecting a certain material for a particular application are that the material has always been used in that application, and that the material has the right properties. Neither is evidence of original thinking or even careful analysis of the application. The collective experiences gained from common usage of a material in a particular application are useful information, but not justification in themselves for selecting a material. The time has passed when each application has its preferred material and a particular material its secure market. The term "property" connotes something that a material inherently possesses. However, it should be regarded as the response of the material to a given set of imposed conditions. A property should be understood as the characteristic of the material in its final available and processed form, as a material's properties are almost always affected by manufacturing processes.

Regardless of specific expertise, every engineer concerned with hardware of any description (and this includes essentially all engineers) must deal constantly with selecting an appropriate material (or combination of materials) for his design. Except in trivial applications, it certainly is not sufficient to indicate that the components should be "steel" or "aluminum" or "plastic." Rather, the engineer must focus his attention, knowledge and skill on the general factors in materials selection: function requirements and constraints, mechanical properties, design configuration, available and alternative materials, fabricability, corrosion and degradation resistance, stability, properties of unique interest and cost.

One of the complicating factors in materials selection is that virtually all materials properties, including fabricability, are interrelated. Substituting one material for another or changing some aspect of processing in order to effect a change in one particular property generally affects other properties simultaneously. Similar interrelations that are more difficult to characterize exist among the various mechanical and physical properties and variables asso-

ciated with manufacturing processes. For example, cold drawing a wire to increase its strength also increases its electrical resistivity. Steels that have high carbon and alloy contents for high hardenability and strength generally are difficult to machine and weld. Additions of alloying elements such as lead to enhance machinability generally lower long-life fatigue strength and make welding and cold forming difficult. The list of these relationships is nearly limitless.

In principle, one could write a mathematical expression describing the merit of an engineering design as a function of all these variables, differentiate it with respect to each of the criteria for evaluation and solve the resulting differential equations to obtain the ideal solution. No one has any illusion that we are in a position to do this. The principle is valid, however, and should provide a basis for action by the designer. In some instances, a standard, readily available component may be much less costly than, and yet nearly as effective as, a component of optimized nonstandard design.

This Source Book deals with materials selection in a practical manner. Instead of considering a material and its possible applications, this volume is organized by component, approaching materials selection in much the same way as the design engineer. Most of the major groupings of structural components are included: bearings, gas turbines, gears, generating plant components, heat exchangers, pressure vessels, shafts, springs, fasteners and tools. In each section, a wide variety of materials is discussed, from steels and alloys to nonferrous metals and ceramics. The text is preceded by a concise introduction that clearly explains and outlines the component approach to materials selection. It is hoped that this Source Book will serve as a guide and an aid to designers, engineers and technicians who select materials based on design specifications.

*   *   *   *   *

The American Society for Metals extends most grateful acknowledgment to the many authors whose work is presented in this Source Book, and to their publishers.

HOWARD E. BOYER
Technical Consultant
Former Editor, *Metals Handbook*

CONTENTS

Introduction: Selection Criteria

by Howard E. Boyer

Selection of the ideal material for a given application would be a relatively simple matter if a perfect or near-perfect material were available. Such a material would have high strength, high toughness, good ductility and good fabricability; that is, machinability, weldability and formability.

In addition, an ideal material would be impervious to environmental attack, would retain its properties at elevated as well as cryogenic temperatures, and would be readily available at a reasonable cost.

Obviously, no such material exists, so most materials are selected for specific applications through a system of compromises.

The Philosophy of Tradeoff

The term "tradeoff," which apparently originated in the aerospace industry, has come into common use and generally is considered synonymous with the term "compromise." Tradeoff means to sacrifice in one property area in order to gain in another. This approach can be extremely complex, and thus the designer must be acquainted with all the requirements — especially the service requirements.

In some instances, a simple graph may be used to indicate how one property degrades as another increases. Figure 1 shows how the strength and/or hardness of a quenched-and-tempered steel decreases as shock resistance (toughness) increases, as controlled by tempering temperature. Unfortunately, in most instances tradeoff is more complex than is illustrated in this figure. Often, several properties must be considered before a final selection is made. The degree to which a given property must be sacrificed in order to gain the required amount of another often cannot be tolerated. The designer then must proceed on a different path entirely.

As an example, a component must have high strength and high resistance to stress-corrosion cracking. Therefore, if too much strength must be sacrificed to comply with the maximum allowable for stress-corrosion cracking, a different material must be sought — a material that has good resistance to stress-corrosion cracking at high strength levels.

Weight and strength are common tradeoff properties and may involve consideration of several materials. For example, in certain applications, notably aerospace, minimum weight is of maximum importance. In other industries, for

example locomotive construction, it was necessary to increase weight to obtain acceptable operation. In one instance, a fine diesel-electric locomotive was designed and put into production. This new locomotive was substantially lighter than previous models. It met all the service requirements but one: it could not move the train because of its light weight, and its wheels merely spun on the tracks. This condition was remedied by the addition of some 20 tons of concrete poured into the center frame of the unit.

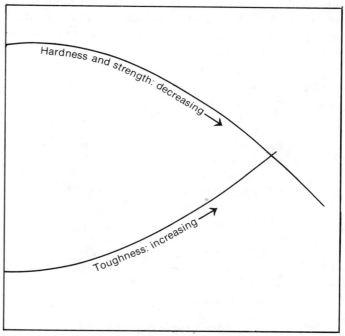

Tempering temperature: increasing ⟶

Fig. 1. Typical tradeoff of strength for toughness or vice versa in quenched-and-tempered steel, as controlled by the tempering temperature.

Quantity of parts to be manufactured plays a major role in selection of material as well as method of fabrication. For example, the extra strength or other specific properties that can be provided by a closed-die forging — that is, properties that are difficult or impossible to obtain from castings — are often desirable. However, quantity requirements must be considered, as well as whether the cost of expensive dies can be economically justified. In some rare instances, dies have been constructed for production of a few pieces wherein it was determined that a forged shape was required regardless of the enormous die write-off per forging. Even though such a practice has sometimes been used — notably in the aerospace industry — any of three alternative approaches is more common: (1) select a stronger material so that another product form could be tolerated; (2) make a very rough shape by open-die forging, accepting the attendant loss of metal and greatly increased machining required to produce the final shape; and (3) use the blocker-type forging approach, which would entail far less die cost than for fully closed-die practice but would require far less metal and machining than in the open-die practice.

In the sections that follow, other examples will be cited that illustrate the philosophy of tradeoff in selecting materials as well as processes for specific applications.

Specific Criteria

It is virtually impossible to list all the criteria that could apply in all cases. The 16 considerations that follow cover those criteria that most often apply. For any individual application several of the items in the list usually apply and are generally interrelated:

- Size considerations
- Shape considerations
- Weight considerations
- Strength considerations
- Wear resistance
- Knowledge of operating environment
- Fabricability
- Type of loading
- Life requirements
- Quantity requirements
- Availability
- Cost
- Existing specifications and/or codes
- Feasibility of recycling
- Scrap value
- Standardization

Size Considerations

In most situations, size is fixed and there may be no leeway or relationship to material selection. On the other hand, the massiveness of a component may sharply restrict not only material composition but also fabrication methods, so that options on material become highly desirable — for economy if not for other reasons.

For large components, assuming that requirements appear to be rigorous and that use of a relatively costly material is indicated, the most logical approach is to review the possibilities of a multi-part assembly. Components of such an assembly might be joined by some mechanical means or by welding. For example, a part might demand high resistance

to wear in one or more areas, which could be accommodated by inserts of a wear-resisting material or even by hardfacing these areas. In any event, the multi-component possibility should always be explored where large parts are involved.

Shape Considerations

Shape and size are so closely related that considerations are generally similar. Shape of the part is usually related to fabrication method — forging, casting, weldment or hogout — and may also influence the material selected.

Shape becomes a major factor in material selection when the part is to be heat treated. Long, unwieldy parts or other complex shapes may warp severely when heat treated. One approach to avoiding this problem is to select a material that will exhibit little deformation in heat treatment — for example, an air-hardening steel instead of one that must be severely quenched to achieve full hardness. However, air-hardening steels are much costlier and generally more difficult to machine than carbon or low-alloy steels, so that low deformation is traded for higher cost. Another approach (provided a machinable hardness or strength range can be tolerated) is to rough machine, heat treat and then finish machine. Here the tradeoff is to gain minimum deformation at the sacrifice of high finishing cost.

In some instances, the trade may be nearly even. For example, corrosion resistance and strength requirements indicate the use of a high-strength stainless steel, such as quenched-and-tempered type 410 or 420. Part shape may dictate that the deformation that would accompany these grades could not be tolerated. Under such conditions, a precipitation-hardening grade might be considered. With this approach, the part would be solution treated, finish machined, and then aged to attain the required mechanical properties. Cost would probably not be a factor.

Weight Considerations

Myriads of data are available for strength-to-weight ratios for hundreds of alloys. Iron-base alloys (carbon steels, alloy steels, stainless steels) weigh approximately the same per unit volume. Iron-base and copper-base alloys are generally not chosen for a part for weight per unit volume. As a rule, the iron-base alloys are a great deal stronger than most copper-base alloys because the former are more amenable to heat treatment.

Titanium alloys have presented enormous possibilities because they can be heat treated to high strength levels and they are only about two-thirds the weight of steel for a given volume.

Aluminum alloys are approximately one-third the weight of steel. Many aluminum alloys can also be heat treated to increase their strength level, but this level is far from what can be accomplished when steel is heat treated.

Ever since aluminum alloys became competitively low in

cost, they have replaced steel in thousands of applications where strength could be traded for lighter weight and increased corrosion resistance. On the other hand, there have been many notable applications where such a trade did not bring about the desired results. For example, in certain aerospace applications it was found that impractically thick sections were required when aluminum alloys were used, and that the use of steel (stainless steel when corrosion resistance was required) actually resulted in a weight saving because of the strength advantage of the heat treated steel.

Magnesium alloys are only about one-fourth the weight of steel for a given volume. In addition, many of the magnesium alloys can be heat treated to relatively high strengths — comparable in some instances to those of the aluminum alloys. The greatest drawback in the use of magnesium alloys is their lack of resistance to corrosion in almost any environment.

A magnesium alloy should be considered only if it is well understood that the magnesium parts must have their surfaces protected at all times — even during fabrication, when periods of time between operations may be more than a few days. Sufficient protection can usually be maintained during fabrication by simple conversion coating, such as dichromate dipping.

This plus paint will usually suffice to provide good, permanent protection for unlimited lengths of time if service conditions are such that the painted surfaces do not become marred. In one case, the use of magnesium was discontinued for this reason. A manufacturer of chain saws wanted to keep the total weight of the machine to a minimum. The housings were made from magnesium alloy castings painted for protection. However, due to the rigorous service to which the chain saw was subjected, much of the painted surface became marred. Wherever the paint chipped, corrosion immediately began. The tradeoff (sacrifice of corrosion resistance to reduce weight) failed, resulting in a change to aluminum alloy housings.

Strength Considerations

With the tremendous array of materials and all the strength data available, selection for strength might appear simple, were it not for other considerations. For example, selecting high-strength alloy steels for maximum strength invariably results in sacrifice of toughness (Fig. 1).

It is often necessary to sacrifice some strength to gain resistance to stress-corrosion cracking or corrosion-fatigue cracking. In some environments, such as ammonia, many high-strength alloys are susceptible to stress-corrosion cracking. Lowering the strength requirement lessens this susceptibility. Of course, in some cases a change of material will serve without sacrificing strength.

For many materials, notably for heat treated steels, fatigue strength is proportional to tensile strength. Commonly, fatigue strength is about one half the tensile strength.

Wear Resistance

In selecting a metal for wear resistance, the designer must answer four major questions before an acceptable selection can be made:

1. Will there be any form of lubrication?
2. If it is nonlubricated abrasive wear, what is the form of abrasive to which the metal parts will be exposed?
3. What is the operating temperature?
4. Is the abrading material corrosive to any appreciable degree?

Case hardened steels are most widely used for combating wear in lubricated components such as gears and pinions. They are also used extensively for nonlubricated applications — notably, where elevated temperatures and/or corrosive environments are not severe.

For applications involving severe wear at elevated temperatures, high-carbon stainless steels and sometimes high-alloy tool steels are potential selections. However, these materials are expensive, and fabrication cost is greatly increased. These materials also have a measure of resistance to corrosion.

Where corrosion and/or heat is severe, the most common approach for gaining maximum wear resistance is to use overlays (generally hardfacing) with high-carbon alloy steels, nickel-base alloys, cobalt-nickel-chromium alloys or other materials. The choice of overlay depends on the service environment. Therefore, the designer must be fully aware of the service conditions to which the parts will be subjected.

Knowledge of Operating Environment

In many cases, parts have failed to meet service requirements simply because the operating environment was not known or not properly considered.

A material may function satisfactorily under normal conditions, but would deteriorate under conditions of elevated temperature or in an aggressive atmosphere. Although exposure to elevated temperatures and aggressive atmospheres should be considered separately, these conditions often coexist, presenting an environment of "hot corrosion" that may require further consideration and tradeoffs before an optimum material can be selected. Data are generally available that will indicate the best material when conditions are clearly known.

Fabricability

Under any conditions, material selection is related to the initial method of making the shape (that is, forging, casting, weldment or hogout) and also to the fabricating operations necessary to complete the part.

For example, if welding is required for fabrication, weldability must be considered. Although almost all alloys can be welded, they are by no means equal in weldability. Steels are often rated for their weldability by the carbon equivalent (CE):

$$CE = C + \frac{Mn}{6} + \frac{Cr + Mo + V}{5} + \frac{Ni + Cu}{15}$$

Working this formula for a specific steel will provide a number. If this number exceeds 0.40%, welding may cause difficulty principally in terms of weld cracking. Special precautions such as low-hydrogen electrodes, preheating or postheating must be used. Naturally when one or more extra procedures must be used, a tradeoff between cost and quality begins.

Type of Loading

In almost any instance, the designer must be aware of the type of loading to which the part will be subjected independent of temperature and environment. Eventually these factors must also be considered.

Types of loading include static tension, static compression, tension-compression and other varieties of alternating stress loading. The designer should also know, when the part is under conditions of alternating stress, whether the stress is low- or high-cycle fatigue, and whether stresses are constant. There have been many notable examples of component failure in fatigue when loads were cycled, whereas a steady maximum load did not result in failure. Components such as bolts and studs should always be loaded under stress, or premature fatigue failure may result.

An aggressive atmosphere may cause failure regardless of whether the part is loaded statically or dynamically — via the mechanisms of stress-corrosion cracking or corrosion-fatigue cracking. Thus, the atmosphere to which the component will be subjected may influence the final choice of material. Operating temperature may also influence final selection. In any case, the operating temperature must be known and carefully considered.

Life Requirements

The life requirements of a component can vary from one cycle (as in a rocket) to repeated loading, as in a dynamic machine expected to operate continuously for years without failure. In a dynamic machine that operates at or near room temperature, it is essential that the designer thoroughly understand fatigue strengths and fatigue limits for the candidate materials. With this understanding, selection should be relatively easy because fatigue data for structural metals are readily available.

Selection of materials for elevated-temperature life requirements is usually based on creep and creep-rupture properties. If aggressive environments are involved along with elevated temperature, the problem of making an opti-

mum selection becomes more complex, as there are fewer applicable data. Under these conditions, selection is largely based on service data for similar environments. Under conditions of fluctuating stress, elevated temperature and an aggressive environment, stress-corrosion cracking becomes an important consideration. Some alloys are far more resistant to stress-corrosion cracking than others. Data are available on a wide variety of metals for resistance to stress-corrosion cracking.

Corrosion rates are usually factors that must be reckoned with in final selection for the service life requirements. First, the designer must be familiar with the operating environment. Corrosion rates are vastly different in a low-humidity nonindustrial region from those in a region along a seacoast where both saline and industrial atmospheres prevail.

When the conditions are known and service life is established, the selection process begins. For almost any condition or set of conditions, there exists a material, or materials, that will resist corrosive attack. However, in many instances the ideal material for the prevailing conditions may be prohibitively expensive or not readily available in the desired product form. Thus an alternative material must be sought. Often there is more than one approach for obtaining the required service life at minimum cost.

For example, many materials corrode rapidly at first but soon form an impervious layer of corrosion products that are strongly resistant to further attack. Examples of such metals include several aluminum alloys, certain copper alloys, a few low-alloy steels, and stainless steels that are resistant to corrosion because of the very thin oxide layer that forms in an oxidizing atmosphere.

The application of overlays to non-corrosion-resisting material, by electroplating or by any of several other techniques, is commonly used and should always be considered in the selection process.

Painting is probably the most frequently used and least costly approach for protection against corrosion. There are, of course, many applications in which painting cannot be used, especially those involving elevated temperatures. But painting should always be considered when one is selecting a material for finite life.

Quantity Requirements

There are notable instances in which certain requirements are a must, and there are simply no tolerances for changes regardless of whether the quantity is one or one million. As a rule, although there are always exceptions, the influence of quantity on selection is directly related to cost.

Very often the influence of quantity on selection is related to the product form. For instance, the total order may be for five pieces with no indication of future orders. A die forging is desired, but obviously the cost of dies could not be justified for such a small quantity. One approach would be to select a grade of material that could be partly forged in open dies and then completed by machining. Another approach would be to machine the part from a solid — one of the tradeoffs being an excessive loss of material, perhaps very costly. Depending on the application and the material, both the open die practice and the total hogout must be viewed critically because of the possibility of end grain exposure, which can be highly detrimental for some high-stress applications.

The possibility of using castings under such conditions should be considered. The obvious questions then follow: Is the selected material castable? If not, how much alteration is required in the composition to attain the required degree of castability, and can such a composition change be tolerated? And will the mechanical properties of a casting meet the requirements?

When larger quantities are anticipated, studies will be made to determine the most economical production. For instance, if the part will require a considerable amount of machining, it must be determined if the free-machining version (if there is one) of the selected material could be used, considering the possible sacrifice of some mechanical properties and corrosion resistance. If such a change can be tolerated, the next question involves another tradeoff — will more be spent for the free-machining material than is saved in machine time? Though each part must be studied for an accurate evaluation, a rule of thumb is that unless about 20% or more of the total envelope for one part will be turned into generated scrap, the extra cost for free-machining material cannot be justified.

Quantity requirements may also be related to other fabricating operations, such as welding. When a material is chosen and the component requires welding, can a similar material with better weldability be compositionally altered and used, without sacrifice of the required properties?

The same may sometimes apply to a part that will require one or more cold forming operations. Can minor changes be made in the specified compositions to increase formability?

Availability

In the final selection process, availability must always be considered, especially when the parts will be manufactured in substantial quantities. There is obviously no use in specifying a material or a product form if it is unavailable.

Although degrees of availability may be interpreted somewhat differently in the various plants where material selections are made, there is some general agreement, and suggested degrees of availability are broken down into five categories:

1. Available at nearly all metal service centers
2. Available at some service centers and readily available from several mills
3. Not available in most service centers and only from mills on delayed delivery
4. Not readily available and, when available, only from one or two sources

5. Not really produced except for experimental or development work

Therefore, it behooves the designer to evaluate the sources of supply and to determine whether the availability, or lack of availability, can be tolerated.

Cost

It would seem that all criteria of metal selection are related to the "bottom line," or cost. As a rule, products are designed first to do the job, and then to be manufactured and sold in the marketplace, which means that they must compete in cost.

There may be several tradeoffs possible in the total selection process. In some instances it is good economy to pay more for a specific material and save in fabrication. For example, it is well known that heat treating operations are costly. Perhaps in one instance parts could be machined from previously quenched and tempered or hard drawn steel bars, thereby saving the cost of heat treating and the subsequent cleaning. Of course, here again one encounters a tradeoff. Heat treated bars cost more than annealed bars, in addition to which machining of heat treated or otherwise strengthened bars increases cost compared with machining of annealed bars.

To arrive at the most favorable cost for any given part, there must be close cooperation among the designer, the materials engineer and the manufacturing engineer. To be competitive, the part must not only comply with mechanical and metallurgical specifications, but it must also lend itself to being manufactured at minimum cost.

Existing Specifications and/or Codes

In making any final selection of material, it is essential to establish whether or not the proposed component is already covered by a specification or code or possibly both.

Specifications and codes often overlap in their coverage, although their intended coverage is generally somewhat different. For the present purpose, a specification is a detailed description of a metal part as a whole — a statement or an enumeration of particulars of size, composition, quality level and performance. A code is a body of laws or rules arranged systematically for easy reference. One of the most commonly known codes is the ASME Code for Boilers and Pressure Vessels. When designing any component related to boilers, pressures vessels and heat exchangers, one may usually be assured that it will be covered by a code, which should be carefully followed.

Among the thousands of specifications (or standards) that have been established for making, processing and testing of metal parts are those of the Society of Automotive Engineers and the American Society of Testing and Materials, Aeronautical Material Specifications, and Military Specifications.

Feasibility of Recycling

Recycling has become one of the favorite practices of the decade, and there is no doubt that it offers savings all along the line that can be passed on to the customer. However, much thought must be given to material selection, as well as to design, to permit the rebuilding of a single part or an assembly of parts to make it feasible for recycling.

If, for instance, the part will be reclaimed by welding, it should initially be made from a weldable material. High-wear areas of parts (press dies, for example) are very often built up and refinished by welding with large savings, compared to the cost of making new dies. However, not all press dies are made of materials that can be readily welded — with a reasonable degree of safety, that is. To be safely weldable, dies must be made from steel with a very low hardenability. This is one factor that accounts for the extensive use of low-hardenability water-hardening steels such as W1 for many types of press dies: They can be recycled.

Scrap Value

The scrap value of the more common materials — carbon steel, alloy steel, brasses, aluminum alloys and so on — is not usually great enough to become an important factor in initial material selection, although it is always prudent practice to consider scrap value. It might be a factor once the choice is down to two or more materials considered acceptable for a specific part.

As a rule, scrap value becomes more important as cost of the material increases. Beginning with stainless steels, then tool steels, nickel alloys, and the nickel- and cobalt-base heat-resisting alloys, salvage becomes an important consideration in the material selection process.

As an example, consider selection of alloys for heat processing furnace parts and fixtures. These parts are very expensive castings, often costing thousands of dollars. A number of different alloys are used, depending largely on the furnace temperature. One popular alloy is HT, which contains a nominal 15% Cr and 35% Ni. Obviously, such an alloy would have a high scrap value, but not nearly as high as that of alloy HX, which contains 17% Cr and 66% Ni. Naturally, a given cast fixture of HX would cost substantially more than the same fixture made from HT, but at the same time much longer life would be expected from HX for the same operating conditions. In addition, HX would have a far higher scrap value than HT.

The above is given merely as an example of the factors that must be considered in making a material selection. Clearly, one should have operating data on life to make the best decision, but there have been notable cases in which it was far less costly to use HX than to use a cheaper alloy, partly because of the difference in scrap value.

Standardization

Most manufacturing plants (or their governing corporations) have developed their own systems of material and part standardizations or specifications, and standard-writing societies such as SAE have developed standards for a variety of parts, such as threaded fasteners, which are denoted by an assigned number.

Standardization is a great help in material selection because it actually registers thousands of successful selections and minimizes the constant influx of new materials, new sizes and new designs that occurs when standardization is not enforced.

Introducing new materials into a manufacturing plant is costly and upsets the system. Of course, it is at times necessary to introduce a new material, but this should be done only after much consideration. Buying, identifying and storing additional material compositions can be rather expensive, so it is to the advantage of the material selector to attempt to make final selections without increasing the number of compositions already in inventory.

The articles that follow contain some case histories involving the selection criteria discussed in this introduction.

SECTION I
Bearings

BEARING STEELS OF THE 52100 TYPE WITH REDUCED CHROMIUM

by C. F. Jatczak, The Timken Company

The Timken Company is primarily a manufacturer of tapered roller bearings which are made from carburizing steels that generally contain low amounts of chromium - typically between 0.25 and 0.50%. At the same time, however, the Company is a major supplier of 52100 high carbon steels to various ball bearing manufacturers. By contrast with the carburizing grades, the 52100 steel contains 1.5% Cr. This makes our job of supplying it without interruption very vulnerable to chromium shortages.

In anticipation of a possible Cr shortage in the early 1970's (which later did occur), we set out to apply our combination of steel and bearing knowledge to develop a suitable low Cr alternate which would satisfy all of the technical and bearing performance requirements of standard 52100 ball bearing components. A secondary aim was to make the new composition so economically attractive that our customers would be willing to evaluate its performance as soon as possible.

The resulting composition, designated TBS-9, is shown in Figure 1. A savings of 1% Cr was achieved. Test results to date indicate that TBS-9 offers processing characteristics and service performance which are in fact superior to those of 52100 at a lower cost. For example, the metallurgical characteristics of the steel are such that its annealing times are much shorter, thus lowering our costs. We were able to pass these savings on to our customers, which in 1971 amounted to $60/70 per ton for steel in tubular form furnished with a spheroidized microstructure. Today this savings is not as lucrative as it was in 1971, due largely to a large increase in the price of Mo; nevertheless it is still significant.

From the Cr conservation standpoint, the 1% lower Cr content in TBS-9 amounts to a rather modest but important savings of 20#/ton. Based on AISI data for 1981, in which approximately 91,000 tons of high carbon bearing steels were sold domestically, the bulk savings in Cr could amount to 910 tons annually if TBS-9 were substituted across the board.

Since 1971, TBS-9 has received wide acceptance as an alternate for 52100. About one-third of Timken's high carbon steel business is now TBS-9 steel.

Hardenability played an important role in the development of this grade. The steel was designed to meet the hardenability specifications of the 52100 grade. This was easily accomplished

Reprinted from Technical Aspects of Critical Materials Used by the Steel Industry, Vol. II-B, 1982, 34.1-34.5, National Bureau of Standards

3

as shown in Figure 2, because of extensive prior work conducted by Timken on the influence of the various alloying elements on hardenability of high carbon steels.* The process was simply one of substitution of Mn and Mo for Cr based on the data shown in Figure 3.

Metallurgical and processing characteristics were also designed to be at least equivalent. The results were equivalent forging and tube piercing characteristics and as already noted superior annealing behaviour, because less time is required to produce the spheroidized structure which provides ease of machining. The maximum as annealed hardness is BHN 207; however, the typical range is 179/192 Brinell.

The effects of various austenitizing and tempering temperatures on hardness of TBS-9 are illustrated in Table 1.

The ultimate criterion for a ball bearing steel is its resistance to rolling contact fatigue. The fatigue resistance of TBS-9 was compared with 52100 and lower carbon version 5160 in our laboratories - our fatigue machines provide a rolling contact test that controls all variables except the material itself. Contact geometry is very similar to a ball bearing. The test conditions employed were: specimen, 4 by 5/8 in. diameter cylinder; load rings, 1 in. crown radius on an 8½ in. diameter; speed, 9000 rpm; load, 1680 lb.; maximum Hertz stress, 700 ksi; lubricant, SAE 20 oil; temperature, 175°F.

The results are shown in Figure 4. The contact resistance of TBS-9 is superior to that of 5160 and similar to that of 52100. Actual bearing field tests conducted by our customers since these RCF tests have confirmed the equality of TBS-9 with 52100 with respect to bearing life. As was noted earlier, TBS-9 has replaced at least 1/3 of our high carbon bearing steel orders. This is the best confirmation of its viability as a ball bearing steel.

To summarize, it is clear that a reduction in Cr in high carbon bearing steels need not cause any consternation in the industry at all. Simple substitution of the TBS-9 should provide uninterrupted production of ball bearing races even during drastic shortages of the element chromium.

---

* See "Determining Hardenability from Composition," by C. .F. Jatczak, Metal Progress, September, 1971, p. 60.

FIGURE 1

COMPARISON OF 52100 AND TBS-9 CHEMISTRY

Type	C	Mn	Si	Cr	Mo
52100	.95/1.10	.25/.45	.20/.35	1.30/1.60	–
TBS-9	.89/1.01	.50/.80	.20/.35	.40/.60	.08/.15

Table I — Effects of Austenitizing and Tempering
Temperatures on Hardness of TBS-9

Temperature, F	Hardness, Rc	
Austenitizing	As Quenched	After 300 F Temper
1,475	66.0	63.5
1,500	65.0	64.0
1,525	66.0	63.5
1,550	66.0	63.5
1,575	65.0	63.0
Tempering (1,550 F Quench)	After 1 Hr Temper	After 2 Hr Temper
200	65.0	65.0
300	64.0	63.5
400	60.5	60.0
500	58.5	58.0
600	55.5	55.5
700	52.5	52.0
800	49.0	49.0
900	45.0	45.0
1,000	40.0	39.0
1,100	36.0	35.0
1,200	32.0	30.0
1,300	26.0	24.0

Typical Rockwell C hardness (normalize 1,650 F, anneal, quench 1,550 F): J1 position (1/16 in. Jominy distance), 65; J2, 65; J3, 65; J4, 61; J5, 47; J6, 43; ⁷. 43; J8, 43; J10, 43; J12, 41.

Source: Technical Aspects of Critical Materials Used by the Steel Industry, Vol. II-B, 34.1-34.5

TYPE	HEAT NO.	C	Mn	Si	Cr	Mo	NORM. TEMP. °F	QUENCH TEMP. °F
52100	Avg 16 hts.	.98/1.10	.25/.45	.20/.35	1.30/1.60		1650	1500
TBS-9	Avg 35 hts.	.89/1.01	.50/.80	.15/.35	.40/.60	.08/.15	1650	1500

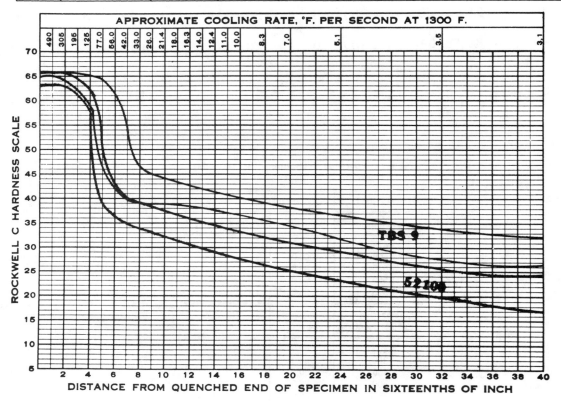

Figure 2.   Comparison of Hardenability of TBS-9 versus 52100.

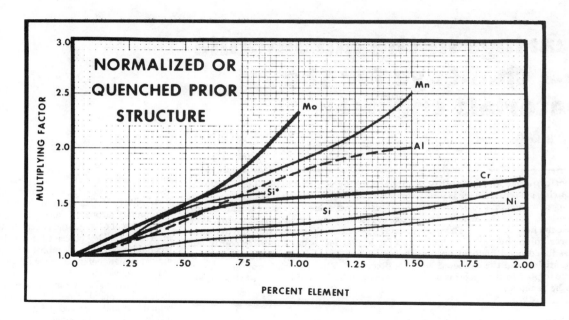

Figure 3. Multiplying factors for the calculation of hardenability in high carbon quenched from 1525°F (830C), see reference indicated below.

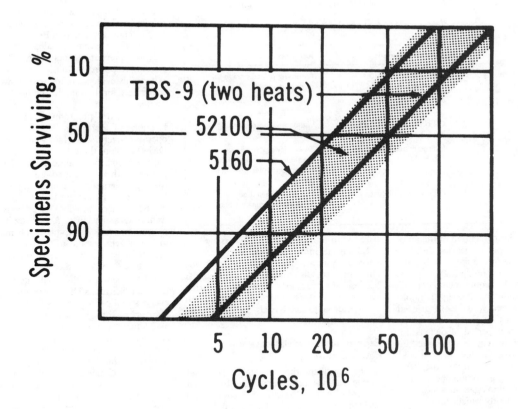

Figure 4. Comparison of rolling contact fatigue life.

Source: Technical Aspects of Critical Materials Used by the Steel Industry, Vol. II-B, 34.1-34.5

Bearings in the automobile — a challenge for the materials engineer

by R G Baker *

The performance and reliability of modern motor vehicles depend critically on the bearing systems. There is now a serious challenge to the materials engineer to develop new systems to meet both economic and social pressures. The major economic factor is the expected progressive rise in energy cost in real terms, over at least the next 20 years. The major social pressures will come from the demand for lower noise and pollution levels and for higher levels of reliability combined with reduced need for routine maintenance. The required development targets are identified and the expected benefits discussed. The materials engineer has a key role to play in the systems approach that will be needed if most of the targets are to be reached.

A bearing can be regarded as a mechanism for transmitting a load between two surfaces in relative motion. Thus, ever since the invention of the wheel, bearings have played a key role in transport. The wide spectrum of loads and speeds and the high reliability which is demanded from modern motor vehicles make bearing performance a key factor in the overall behaviour of the engine or system.

The widespread introduction of the automobile provided a spur to bearing development and led to the introduction of bearing systems capable of sustaining higher speeds and loads to meet the requirements of modern engines. Reliability has increased dramatically; most car owners would feel aggrieved if their car bearings or pistons had to be replaced during their period of ownership, whilst running-in procedures demanded by modern engines are far less critical and irksome to the driver than in the past. Why then, is there a challenge for anyone, least of all the materials technologist, in a situation which sounds eminently satisfactory?

The explanation is that there is a major need for change in current technology in response to both economic and social pressures. The major economic factor is the expected progressive rise in energy cost in real terms, over at least the next 20 years. The major social pressures will come from the demand for lower noise and pollution levels and for higher levels of reliability, combined with reduced need for routine maintenance. Considerable technological changes in car engines and systems may be expected, and at an accelerated innovation rate, compared with that of the last 20 years.

Significant innovation will be required in bearing systems. This poses a searching challenge to the materials technologist because the way in which most bearing systems work is not understood well enough to provide a rational basis for developing improved systems.

Economic and social case for development

If one considers the very large potential energy savings possible in the European Community as a result of technological development of engines, the case for change seems overwhelming. Some of the best available estimates of the potential savings have been made by an Energy Committee of the American Society of Mechanical Engineers (ASME), which has issued two reports[1,2] and by Jost and Schofield[3]. From this Baker[4] estimated a possible ultimate saving of up to 8–10 billion ECU/annum in the European Community at present value. An estimated breakdown of the potential benefits is shown in Table 1. The perceived priorities for technological change are probably greater to individual national governments in the EEC than to individual owners of vehicles.

The individual owner may regard savings in fuel as marginal compared with the large capital cost and overheads with which he is involved. Indeed he may give priority to higher reliability and a reduction in maintenance charges rather than to a reduction by marginal increments in the amount of fuel he uses.

There is evidence that both individual governments and the European Community are aware of the needs, and regard the necessary development as a high priority. One might expect, following experience in the USA where new cars have to satisfy minimum Federal government requirements for fuel consumption, that European governments will be tempted to adopt a 'stick-and-carrot' approach through legislation and funding respectively. In the meantime, there is certainly evidence of increasing awareness of the need for funding the required developments.

In the BMFT programme[5], in the Federal Republic of Germany, which has been in operation over the last four years at a level of DM40 million, there has been considerable emphasis on solving some of the tribological problems which will otherwise be inherent in the new generation of engines. In the current BRITE initiative of the EEC[6], a high priority has been placed on tribology amongst other technological objectives. A forthcoming EEC programme is being discussed on materials in motor vehicles and in the past considerable emphasis has been placed on the potential future use of ceramics in diesel engines[7].

What then are the challenges for the materials scientists? The problem in nearly every specific bearing application is that there are no firm grounds on which to design a significantly improved system. Most past research work has been carried out on tribological systems in the dry, unlubricated state whilst most engine bearing systems are lubricated. Generally, tribological models have been based on the supposed interaction of asperities on the bearing surfaces in relative motion, and formalised models have been evolved to account for different types of wear and surface damage. These permit observed behaviour to be rationalised, but improved physical models are required to provide a basis for prediction.

Table 1: *Estimates of potential benefits of bearing-related developments in automobiles in the European Community (per annum)*

Item	Potential benefit	Remarks
Continuously variable transmission drives	Speculative, estimates between 1 & 10 b ECU	Assuming full advantage taken of opportunities
Adiabatic diesel development	4 b ECU ultimately	After existing engines have been replaced.
Piston assembly development	0.2 b ECU perhaps rising to over 2 b ECU	Ultimately and depending on success.
Engine bearings development	Indirectly perhaps up to 2 b ECU	Direct savings small but development necessary to implement other potential benefits.
Cam — follower development	0.95 b ECU ultimately	Through reduced cost, reduced maintenance and longer life.

*Professor Baker CEng, FIM is a consultant and Professor Associate at Brunel University in the Department of Materials Technology.

Reprinted with permission from Metals and Materials, January 1985, 45-52, © 1985 Penton/IPC

In addition, the relationships between material characteristics of the bearing materials and the design criteria for the engine are very imperfectly defined. Where they exist at all the basis is empirical and the data is usually of a comparative nature. Many times in the *Tribology Handbook*[8] and other definitive publications, one sees a qualification to the effect that recommendations are indicative only and must be validated by type- or service-tests. For this reason the cost of testing, which accompanies bearing innovation, is high even for small changes in the system.

Three challenges are therefore posed to the materials technologist in addition to that of providing new materials; to define more realistic physical models of bearing behaviour; to use these as a basis for defining constitutive relationships between bearing materials and system characteristics so as to provide development targets; and to evolve a reliability methodology to minimise the testing costs which presently accompany innovation.

Mechanism of bearing operation

If there is lubricant between the two bearing surfaces, a number of different modes of operation are possible as shown in fig 1, and the effective friction depends on the mode of operation. If the geometry of the two surfaces is compatible a stable film can be formed as shown in fig 2 a & b, which transmit very heavy loads. In this regime one of the major sources of the apparent 'frictional losses' is the result of shear in the lubricant. Minimum friction may be experienced at the transition to boundary lubrication illustrated in fig 1.

So what actually happens in engines? Clearly, when an engine starts for the first time any oil film must be so thin conditions are well to the left of the boundary lubrication area in fig 1. In piston rings and in cam/tappet interactions apart from ball and roller bearings, the loads are so high that any simple view of bearing operation must lead to the assumption of metal to metal contact. However, the pressures transmitted by the oil film are so great that local deformation of the two contact surfaces greatly increases the effective surface area and reduces the peak pressure. Clearly the extent to which this elastohydrodynamic (EHD) effect occurs depends on the modulus of the two surfaces in contact, and upon the lubricant characteristics. This regime is shown in fig 2 c.

The degree of mechanical interaction will depend on the surface roughness, since the greater this is, the more interactions there will be at any mean distance of separation. Foreign particles will also increase mechanical interaction as will any dynamic distortions of the surfaces in contact and factors such as misalignment and variance in machining tolerances in the engine. Some degree of mechanical interaction is clearly inevitable, even where 'pure' hydrodynamic or EHD lubrication are involved.

What form does the mechanical interaction take? Recent research has shown that in the very early stages of dry surface interaction an adherent transfer layer is built up on

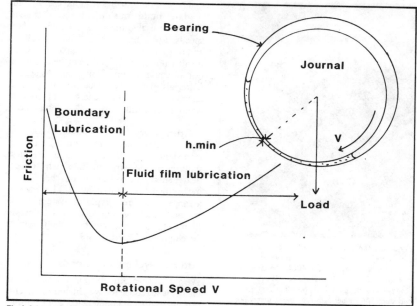

Fig 1 *Schematic diagram showing the different operating regimes for a journal bearing with fluid film lubrication*

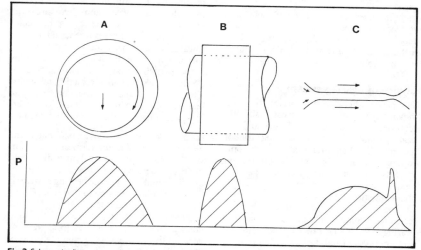

Fig 2 *Schematic diagrams to show the pressure profiles in the fluid films in lubricated bearings. Diagrams (a) and (b) show the profiles through the film perpendicular and parallel to the journal axis respectively. (c) shows schematically the local deformation and pressure profile in elastohydrodynamic conditions.*

both the surfaces in contact and there is very considerable local hardening. In experiments in which a hardened steel ring was moved in contact with a copper disc, Rigney and Heilmann[9] showed that the transfer layer consisted of very finely divided, virtually amorphous phase mixtures of copper and steel as shown in fig 3. Beneath this layer, the surface of the copper had received a very high degree of local deformation and had a very fine grain size.

This may explain why the friction between many metallic pairs studied in tribology experiments rapidly reaches a constant value. After a very short initial period, friction is determined by the characteristics of the 'third body' — the transfer layers on both the surfaces in contact. Probably, analogous layers are built up in lubricated systems, almost certainly incorporating some of the lubricant constituents and, perhaps, com-

pounds formed between the metallic constituents of the transfer layer and constituents of the lubricant.

Workers at INSA have recognised the importance of such third-body layers and have studied some of their properties both directly and in model systems[10].

It is interesting that the concept of transfer layers has been recognised for some time as of importance in the case of dry bearings such as Glacier DU, but is only beginning to be recognised as applicable to lubricated metal systems. Perhaps one of the reasons is that an opportunity for studying characteristics of these third bodies has only comparatively recently been provided in the shape of the advanced metallographic and analytical techniques which are now available. The implications are considerable, because it means that the initiation of seizure and scuffing may depend critically on the

Fig 3 *Thin film transmission micrograph through a copper surface after it had been in rubbing contact with a hardened steel ring (440C). The direction of motion is indicated. There is a fine, almost amorphous surface layer, a phase mixture of copper and steel. The copper substrate shows a very fine grain size, as a result of the intense deformation. (Courtesy Rigney and Heilmann)*

transfer layer or 'third body' which is developed between them. This in turn will depend both on the characteristics of the surfaces and lubricant and might well be sensitive to minor constituents in either.

Choice of bearing system

The preferred choice is invariably the most inexpensive system which will fulfil the requirement of the engine. Roller and ball bearings have very low starting friction which makes them preferred for such applications as wheel bearings. Plain bearing systems in the hydrodynamic operation mode have high stiffness and damping characteristics and low friction at high surface velocities, which makes them preferred for engine main and large small-end bearings. These and other applications, the preferred system and the factors determining the choice are summarised in Table 2.

The engine bearings also have to accommodate build misalignments, dirt and dynamic misalignments, all at high surface speeds. Local breakdown of both lubricant and third-body films is inevitable. However, it must not result in sufficient heat generation to cause seizure. This demands special materials with low surface modulus and characterised by low energies of rupture when local surface to surface interactions occur.

Other highly loaded systems such as piston ring/liners and cam/tappets are loaded in the EHD regime and have less requirement for low modulus, whilst local heat generation is less serious because of more efficient heat dissipation. This leads to a range of different types of system in the engine, which will be discussed below in more detail in the light of the probable engine developments of the future.

Future development trends

Weight saving: An obvious way of saving energy is to reduce the weight of the car. This process is already taking place. For example, the progressive replacement of metals by polymers in the automobile has been taking place for some years and there are even experimental all-polymeric internal combustion engines. There is a general tendency towards lower density metals and to reduce the weight of individual components by reducing their size and thickness as much as possible consistent with mechanical reliability. This tends to increase the dynamic distortion of engine components and thus affects bearing system loading throughout the engine.

Frictional losses — Pistons: About half of the energy consumed in the form of fuel is converted into motive power. This is dissipated mainly by aerodynamic drag, braking losses and frictional losses in the bearing system. It is perhaps not surprising to find that piston/cylinder losses account for about half the mechanical losses at about 8% of the total energy used. The other bearing systems account for a further 4%. From the point of view of governments, if not of individuals, the immense amount of money this involves in terms of a nation's fuel bill makes improvements in even plain bearing systems worthwhile. This is leading to work to reduce piston assembly weight and to reduce frictional losses between rings and skirts and the liners.

One of the best ways of achieving increased fuel efficiency is by raising the temperature of combustion. This increases the efficiency of the combustion cycle, and presents the opportunity for energy to be extracted from the higher temperature exhaust gases through a secondary drive mechanism. The optimum temperature is beyond the operating range for existing piston/liner systems and also of existing lubricants. However, the principles involved in both progressively raising the temperature with existing materials and then changing it to ceramic materials in the future need considerable development. There have already been initiatives to develop wholly-ceramic and ceramic-surfaced combustion chambers and pistons. The significant potential benefits provide a strong incentive for such work to continue.

Frictional losses — Bearings: Even though the losses in the engine bearings are a relatively small proportion of the total, the large potential benefits provide an incentive to achieve even small savings. The best way of achieving reduction in energy loss is far from clear, although there are many ideas and strongly held theories. In principle, reducing the bearing diameter will be beneficial in reducing energy loss and this will tend to increase the loading on bearings and reduce the minimum oil film thickness as result both of the reduced load bearing area and of the increased dynamic distortion of the crankshaft.

Noise reduction: This may become of increasing importance in the future. Engine bearings are a primary noise generator. Generally, reducing the clearance reduces the noise. This demands closer tolerances, and may lead to thinner oil films.

Reliability: For a popular model of motor car,

Table 2: *Current systems and the associated selection criteria*

Application	Material	Criteria	Remarks
Crankshaft bearings	Al-20Sn, unplated Al-6Sn & Al-Si plated; Cu-Pb & Cu-Pb-Sn plated.	High stiffness and damping characteristics, seizure resistance. Up to 40 Pa nominal loads and 25 m/s. Low surface modulus, dirt tolerance	Material chosen of lowest hardness to give required fatigue strength. Crankshaft may need surface hardening for higher loads.
Pistons	Cast iron or aluminium alloy	Easily cast, scuff resistant on cast iron, light, strong.	May need cooling and high temp inserts in diesel engines
Piston rings	Cast iron — grey, carbide & nodular. Sintered irons. May be plated with Cr or other matls.	Good wear behaviour of both ring & liner or bore. Good resistance to scuffing. Good fatigue resistance.	Ring design is critical to performance. Surface finish also critical.
Cylinders & liners	Cast irons. Aluminium alloys	Compatibility with piston & ring materials. Low cost & weight & good wear resistance.	Liner and bore finish is critical.
Wheel bearings	Tapered roller UHT steel	Low starting & running friction. Fatigue resistance.	Must be well sealed against dirt & moisture.
Cam followers	Chilled Cl. Surface hardened steels	Resistance to indentation, pitting, wear scuffing.	High stresses require good surface finish.
Dry & suspension bearings	Filled PTFE (eg Glacier DU)	Good wear resistance, no added lubricant, low loads.	Eg windscreen wipers, steering system bushes suspension struts.

there might be 100,000 engines in service at any particular time. It would be an embarrassment to have more than one or two per year of the 20 or so half-bearings which would be used in the engine, returned for a specific type of defect. This implies a reliability of about 10^{-7} failures per bearing per engine type per annum, comparable with what is expected for highly critical engineering components. Other engine components including bearing systems are expected to have comparable reliability and the future requirements are likely to become even more severe.

Fuel substitution and improved efficiency: Fuel additives such as ethanol will become more common. Some authorities predict that methanol derived from coal will be a preferred liquid fuel substitute in the future unless and until power derived from nuclear sources becomes available. Meanwhile, the trend towards diesel fuel in cars may continue and there will be some substitution of natural gas for liquid fuels. Electricity would be of considerable potential benefit, if the problems of adequate and efficient storage could be solved. However, this seems unlikely in the immediate future and will not be considered further in this article.

The major effects on bearing systems derive from the contamination of the lubricant with corrosive combustion products. Efficient diesel engines are characterised by lower engine speeds and higher firing loads than in the case of gasoline engines. The trend towards pressure injection of the gas mixture will continue because of the improved efficiency possible particularly in diesel engines.

Continuously-variable transmission (CVT) systems: If engines could operate at a constant speed, a considerable improvement in fuel efficiency would be obtained because all of the elements of the combustion system could be optimised for that particular engine speed. To make this possibility practical would require the introduction of drive systems in which the effective reduction ratio was continuously variable (CVT systems). Such

systems would also be compatible with regenerative braking and other energy conserving measures. ASME[1] originally estimated potential savings which would amount to more than 10 billion ECU per annum in the EEC. Jost and Schofield[3] regarded the potential savings as speculative, as did ASME[2] in their second report. In Table 1 they are estimated at 1–10 billion ECU pa in the European Community. Whichever estimate is taken, the potential benefit is so large that the topic will inevitably command a high priority as long as there is a reasonable chance of success.

ENGINE BEARING SYSTEMS
Plain crankshaft bearings

The major modes of failure for plain bearings are fatigue and seizure. Fatigue loads are derived from both cylinder firing loads and from inertia. The number of cycles experienced in life usually exceeds 10^8. The fatigue life is certainly related to the loading which is experienced by the bearing, but most existing theory assumes, in calculation, that the load is distributed more or less evenly over the projected bearing 'contact' area. This is certainly not so; the load can only be transmitted through the oil film between the journal and the bearing surface, which is by no means continuous over the whole bearing surface. In any case it contains steep pressure gradients as shown schematically in fig 2. Thus, whilst the calculated loading on bearings may be of the order of about 50 Pa, peak loads on local surface regions may be an order of magnitude higher, assuming a rigid system. As the system is not rigid, and local deformation of the bearing surface will occur, not to speak of the possible dynamic distortions in the crankshaft and bearing housing, the true peak stress and stress distribution is usually unknown.

As there is no basis for direct design of the bearing and system, the only alternative is to carry out comparative standard tests for fatigue on formed bearings in laboratory rigs, and then rely on engine experience to validate the tentative conclusions reached. This procedure is extremely expensive because large engine testing programmes may be

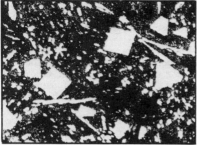

Fig 5 *Microstructure of whitemetal bearing (courtesy Glacier Metal Co) shows a matrix of tin reinforced with harder compounds of tin with copper and antimony. Magnification × 150*

necessary even for quite minor changes in bearing design, because of the lack of reliable design models. This undoubtedly inhibits innovation and will continue to do so unless a significantly improved understanding of the relationship between the structure and characteristics of the bearing on the one hand, and the characteristics of the engine on the other, can be established.

It is easy to see how changes in engine characteristics can affect bearing loading, but perhaps less obvious is that the nature of the lubricant may also be critical. However, the dynamic distortion and inertial effects in an engine are very strongly affected by the characteristics of the oil. It has also been stressed that the load is transmitted between the crankshaft and the bearing housing through an oil film, the extent of which also depends on the lubricant characteristics.

Engines have to start and they have to stop, so that hydrodynamic conditions are not invariably maintained in life. Furthermore, the trend to increasing flexibility of engine housings and crankshafts, and to higher loads, is expected to lead to thinner oil films at the point of closest approach of bearing and crankshaft. The gap between the two surfaces can certainly be bridged by particles in the oil too small to be removed by present day filtration systems. It will also be small enough to be bridged by the hills and valleys inherent in the surface roughness.

Perhaps the greatest danger of substantial interaction arises if the point of closest approach occurs outside the limit of the oil film, which is possible in some engines under some conditions. Thus, even in normal 'hydrodynamic' operation, the bearing has to be able to survive at least for short periods without seizure if 'contact' occurs between the journal and bearing surfaces.

Need for optimisation in design

The demands made on bearing materials at present are conflicting in their implications. A high fatigue resistance is required, which tends to favour a hard material. Resistance to wear is necessary, bearing in mind that the oil film thickness is likely to be bridged at the point of closest approach, and this also favours a hard material.

Sufficient conformability to deal with imperfections of alignment is required. No engines are built perfectly and, in any case, flexure in the crankshaft and other parts of

Fig 4 *Schematic diagram showing the requirements for a plain bearing material, as a function of hardness, illustrating the need for optimisation.*

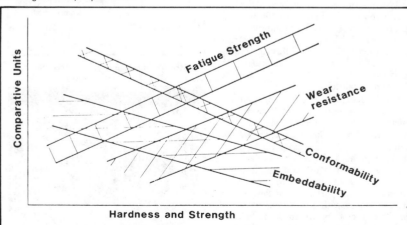

the engine produces dynamic elastic misalignment. This requires a low modulus and maybe a soft material to provide the necessary behaviour.

Finally, seizure resistance, (a low energy generation rate on surface contact coupled with the ability to embed foreign particles) is required and this implies a soft matrix. The requirements are summarised schematically in fig 4. The history of modern bearing development shows that these conflicting requirements have been solved in a very pragmatic spirit of compromise. White or Babbit metal consists of a tin or lead based matrix containing harder compounds produced by adding copper and antimony to produce a 'reinforced' matrix.

Whitemetal, illustrated in fig 5, even if it owes its existence to empiricism or serendipity, is one of the major successes of 19th century engineering innovation. It virtually provided all the car engine bearing requirements for the first half of the 20th century and continued until very recently to provide most of the engine bearings for the American market. The advent of the thinwall bearing in the 1930s, which rapidly replaced its cast-in predecessor, for good economic reasons, made possible a new lease of life for whitemetal as progressive increase in engine load led to premature fatigue failure. This was brought about by deliberately reducing the thickness of the bearing metal layer so that the steel backing could increase the apparent strength, undoubtedly by reducing the local strains occurring in the whitemetal.

Modes of failure

It is perhaps worthwhile examining the characteristic mode which fatigue failure takes in a bearing. A system of cracks; some perpendicular to the bearing surface, some

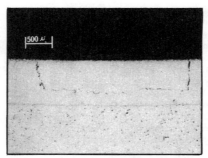

Fig 7(a) *Mode of fatigue crack propagation in an aluminium-tin bearing. (Courtesy Glacier Metal Co)*

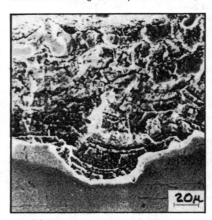

Fig 7(b) *SEM photograph of bearing fatigue damage illustrating fatigue striations. (Photograph courtesy Dr T S Eyre)*

parallel to it, and near the junction with the steel, spread out either from a free edge or perhaps from some imperfection in the bearing itself. Dirt between the housing and the bearing wall may slightly raise the bearing surface locally and provide an initiating point for fatigue. If adhesion between the bearing lining and the steel is inadequate, this may also trigger off a premature failure. Fig 6 and fig 7(a) and (b) show the way in which fatigue cracking occurs in engine bearings.

Eventually, increasing engine loads led to the substitution of stronger materials for whitemetal. Development tended to follow two separate paths. In each case, however, the prescription has either been to produce hard particles in a softer matrix, as in the case of whitemetal and aluminium-silicon bearings, or the reverse, as with aluminium-tin and copper-lead bearings. The microstructures of these materials are illustrated in fig 8(a) and 8(b), and show phase mixtures with a uniform and relatively fine dispersion of the minor phase particles.

As the strength of plain bearing materials increases, it is perhaps not surprising that resistance to seizure progressively decreases. The minimum oil film thicknesses are less than the diameter of foreign particles in the oil. This, perhaps, needs some explanation. Although modern engines contain a filtration system, it is usually designed to remove only the larger foreign particles. The designers of filters are faced with a dilemma, because if the filter interstices are made sufficiently small to remove even fine particles, there is a

tendency for the filter to clog. If the filter is in-series, this either leads to rupture or to oil starvation. If the filter is in by-pass, then as it clogs, smaller proportions of oil actually pass through it. Thus element filters tend to be designed so they remove the bulk of the larger and more harmful particles throughout their life, but at the expense of reducing their efficiency in removing the smaller particles below about 20 microns in diameter.

By-pass centrifugal filters remove smaller particles more efficiently, but currently they are usually only fitted in heavy duty high speed diesel engines, say for long distance trucks. For the bulk of motor vehicles, therefore, the bearings are required to deal with foreign particles at least up to about 15 to 20 microns in diameter. These can be wear particles from other parts of the engine, dirt, grit, and occasionally particles which result from chemical reactions in the lubricating medium itself.

The margin against seizure in both copper-based and aluminium-based bearings is eroded by both increase in strength and increase in diameter, to a point where it is unacceptable. However, an extension to the effective stress range is provided in both cases by plating a thin overlay onto the base material. Overlays are typically either lead-tin, lead-tin-copper, or lead-indium. The fatigue strength, whilst intrinsically low, depends on the thickness of the overlay plated layer which is typically about 20 to 25 microns. This is a compromise between the need to achieve improved resistance to seizure throughout life, (which would demand the thickest layer possible), and the need to achieve reasonable fatigue properties, which would demand the thinnest layer possible. It could be thought that the properties of the underlying material were no longer impor-

Fig 8(a) *Microstructure of an aluminium – 20% tin alloy. Mag × 75. (Photograph courtesy Glacier Metal Co)*

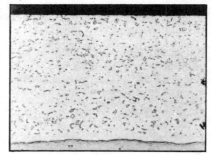

Fig 8(b) *Microstructure of a cast copper-lead bearing. Mag × 75. (Courtesy Glacier Metal Co)*

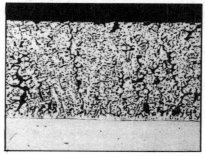

Fig 6 *Fatigue cracking in engine bearings. Macro photograph of fatigue damage in an aluminium-tin bearing. (Courtesy of Glacier Metal Co)*

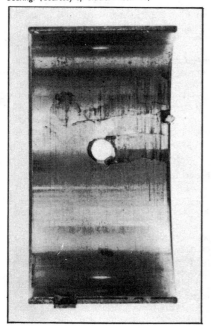

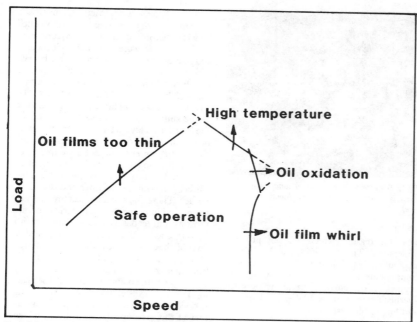

Fig 9 *Schematic diagram showing the criteria limiting the region of safe operation of an engine bearing material (after Martin and Garner, Glacier Metal Co)*

tant in an overlay-plated bearing. However, they affect both the resilience of the bearing surface and also the behaviour of the bearing system in the event that wear takes place completely through the overlay. This is likely to happen during the expected life of the engine bearing.

Scope for further development

There is little margin, if any, between the present performance requirements for motor vehicles and plain bearing capability as shown schematically in fig 9. If, as seems likely, the trends are towards thinner oil film and higher loads it is not easy to see how these requirements can be met by the further development of metallic bearings along conventional lines.

Pistons and liners

Frictional losses in piston assemblies are a high proportion of the mechanical losses in an engine. Phosphorus cast irons, for piston rings and bores, were for many years extremely successful. The requirements are for good wear and scuff resistance combined with adequate sealing behaviour. Scuffing is the term given to excessive interaction between the ring and the bore of the piston leading to significant local damage. Fig 10(a) shows a section through a piston ring, and a liner in fig 10(b). Fig 11 shows an example of scuffing on the inner wall of a liner.

The mode of operation is elastohydrodynamic (EHD) and a high standard of surface finish is required on both ring and bore. There is an optimum finish which permits the retention of sufficient lubricant without causing deleterious interaction, and the detailed profile of the rings is of great importance. With the advent of higher performance diesel engines and of demands for longer engine life, cast irons containing car-

bides have been introduced to reduce wear. More recently, hard, wear resistant surface coatings have been applied to the top compression ring or even to the whole ring pack.

During the operation of the piston, contact takes place between both the ring and the skirt of the piston and the liner. The ASME committee estimated that potentially about 25% of the losses in current systems could be saved by improved piston cylinder systems. Over the EEC this could exceed two billion ECU per annum ultimately, after all engines have been replaced with improved systems. J & S took a less ambitious and shorter term view, but nevertheless estimated savings which could amount to a 225 million ECU per annum.

Opportunities for development

One obvious method of saving energy is to reduce the weight of the piston assembly, and many objectives for doing this are being pursued by component manufacturers involved in engine production. There is, however, a problem. The overall objective must be to evolve piston cylinder systems that can consistently operate with lower energy loss, and to define the boundary conditions for such systems so that they can be operated successfully in engines. The problem is complicated by the fact that if the mode of operation is changed, for example by altering the lubricant, its temperature, or the number of cylinders in an engine, it is possible for the energy loss characteristics in the engine as a whole to be altered dramatically. There is accordingly limited value in studying part of the overall system in isolation.

Part of the reason for this complication are the dynamic distortions which take place in the cylinder lining and in the engine generally, including the crankshaft. These things all

have an effect upon the mean oil film thickness which is one of the major factors controlling the frictional loss. Evidently, a systems approach is necessary if efforts to reduce energy loss are to be attended by a reasonable chance of success. Nevertheless, as a contribution to this, better models of the interaction between rings and skirts and liners are necessary. Currently the prediction of oil film conditions involved in piston ring interactions is limited in accuracy by inability to take account of dynamic elastic effects principally in the liner. Work in the USA by Rhode[11] has permitted roughness to be included in oil film thickness calculations through a modification of the Reynolds equation. Following that, Dowson *et al*[12] have presented a quantitative method of dealing with EHD effects taking roughness in to account. At the same time workers at INSA[13] and the University of Poitiers have been working on models permitting dynamic elastic effects to be modelled. Thus the tools now exist for a systematic attack to develop the necessary design basis and which have a high probability of success.

Cams and followers

Savings in the EEC from a 10% increase in life of the cam follower system could ultimately be about 50 million ECU per annum across the whole vehicle park.

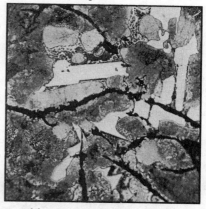

Fig 10(a) *Microstructure of a piston ring from a diesel engine. The ring consists of carbide particles in a grey iron matrix. Mag × 75*

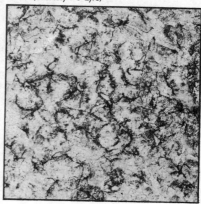

Fig 10(b) *Cylinder liner in phosphorus grey iron. Mag × 75. (Courtesy T S Eyre)*

Source: Metals and Materials, January 1985, 45-52

Fig 11 *Damage caused by scuffing on the liner of a cylinder of grey cast iron. Mag × 500. (Courtesy T S Eyre)*

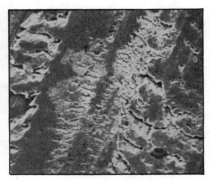

Fig 12 *Damage on a cam follower surface due to severe wear (scuffing) Mag × 75. (Courtesy T S Eyre)*

Savings in maintenance costs of only 0.5% could ultimately total about 500 million ECU per annum. A reduction in cost of 10% in the cam/tappet system would save about 400 million ECU per annum with immediate effect. All of these targets should be feasible. Savings on fuel and maintenance costs would only apply to new vehicles and would, therefore, build up over 10 or 12 years. There would be further knock-on benefits which are less tangible and the results would have a modest catalytic effect on engine development. Cams and followers operate under high local stresses and low speeds, and thus are in the boundary lubrication mode where EHD and dynamic elastic effects can be significant. Locally severe attrition can take place due to pitting or scuffing such as is shown in fig 12, but even minor wear can lead to non-optimum operation of an engine.

This leads to a serious increase in fuel consumption and even a reduction in the life of other parts of the engine. The effect of minor wear can be partially offset by modern engine control devices which can monitor and adjust the engine tuning automatically. However, frequently there is a sharp transition between minor and unacceptably heavy wear for reasons which are not properly understood. This transition may be difficult to predict when different low cost or density components are introduced into automotive engines. The current trend is towards overhead cam systems, higher local stresses and longer life systems.

Dry bearings and suspension bearings

Most metal bearing systems require external lubrication by oil, grease or some other fluid with lubricating properties. Porous bronze bearings can be made which, when impregnated with oil, remain lubricated throughout life, but at the expense of load carrying capacity and reliability. Certain types of polymer, in contrast, have inherently good frictional characteristics. PTFE is perhaps the best known example. PTFE can be used to great effect as a low friction surface against other materials. It does tend to creep under quite modest loads, and modest rises in temperature make it soften to a point where pick-up can occur on an opposing metal surface. Its use is therefore limited to modest loads and speeds.

The obvious low friction potential of PTFE led the Glacier Metal Company to introduce a dry bearing material called Glacier DU, in which the low friction properties of PTFE were utilised in an ingenious and interesting way. The bearings are backed with steel strip to provide rigidity and the PTFE rich lining is bonded to the steel by a sintered layer of tin-bronze particles. The structure is shown in fig 13. The open nature of the sinter provides a mechanical key for the PTFE rich layer. This contains additives of which an important constituent is lead. The bronze sinter particles extend to the lining surface and thus provide an excellent heat-conducting path away from the contact surface between the polymer and the counterface. These and the filler also provide stiffening and improve the load bearing characteristics. The lead is oxidised on contact with the opposing surface and forms a thin surface layer on the counterface with some transferred PTFE. The subsequent operation of the bearing in the dry condition involves interaction between the polymer surface of the bearing and the transfer layer on the opposing surface. The general characteristics of bearings of this type are shown in fig 14 which shows the different regions of behaviour within which different bearing types can operate. The presence of a transferred layer on the counterface is necessary for the efficient dry operation of PTFE and the nature of the layer is of great importance.

Major future opportunities

One of the major contributions which it is possible for materials scientists and engineers to make is in the improvement of the physical models describing interactions between lubricated surfaces making full use of modern equipment. This is one of the most important aspects of the systems approach which will probably be needed to overcome the more important problems facing the industry. There are some important corollaries.

Seizure is the result of a catastrophe occurring at some point between two interacting surfaces. It is easy to see that, once a significant interaction has occurred, the local rise in temperature can reduce oil film thickness, increase the local temperature and provide an autocatalytic mechanism. It is also clear that an acceptable risk of seizure exists in engines even when the chance of the event occurring is extremely rare, about 10^{-7} per annum. Investigating seizure resistance under realistic conditions entails being sure the load can be increased and the speed can be reduced. The problem, however, is that under these circumstances if an incident initiating seizure occurs, the resulting catastrophic interaction may destroy all evidence of what caused the initial breakdown in the system.

What seems to be necessary is a high resolution method of detecting local frictional changes so that surface interactions can be detected when they first occur and the position on the surface at which they first occur can be defined precisely. Some work along these lines has been carried out at the National Bureau of Standards by Blau[14] and Bhansali[15] and may point the way for future

Fig 13 *Structure of a filled PTFE dry bearing (Glacier DU). The filled PTFE is bonded to the steel backing by a layer of sintered tin-bronze powder. Mag × 100 (Courtesy Glacier Metal Co)*

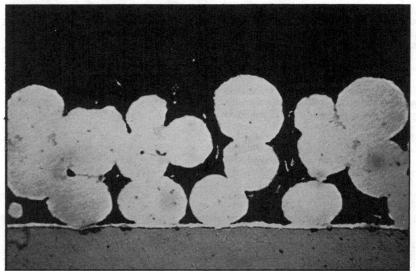

work of the same type. Effectively, what was done was to use a computer to measure local friction and temperature at very rapid successive intervals as a ring was drawn progressively across a flat specimen, some 14,000 measurements per inch being made. In this way very precise resolution of surface incidents could be made for subsequent metallographic study.

NEW MATERIALS

Coatings: A big opportunity has been provided for the materials and automotive engineer by the rapid development in coating techniques which has taken place in recent years. Coatings are already being extensively applied to wear surfaces such as gear teeth, tappets, cams, piston rings, etc. Many different coatings methods are available, but the quality obtainable and, perhaps, more important, maintainable, is somewhat variable. The advantages are that many of the methods have low cost, great flexibility, and can be made automatic or semi-automatic. The many challenges include the need to model coating behaviour in a realistic way and to define realistic quality control criteria and production practices which satisfy them.

Bearing materials: The extension of polymeric materials to plain bearings would seem to have great future potential. The lower surface modulus would be expected to be beneficial since it would tend to increase both the minimum oil film thickness and the area over which the maximum pressure was transmitted. It would thus effectively smooth out the peak stresses to a greater extent than with current systems. Should significant interactions still occur between the bearing material and the opposing metal surface, it would be possible to reduce the risk of seizure by incorporating constituents deliberately designed to lower the friction. In work in the USA on polymeric bearing systems such as PPS (polyphenylene sulphide) and PEEK (poly ether ether ketone), PTFE has been found to be effective in this respect in unlubricated experiments. Experiments carried out by the Glacier Metal Company have shown that both PPS and PEEK possibly with certain additives can function well under lubricated conditions, although their fatigue properties are somewhat limited. Glacier HX consists of PEEK[16], but with a patented additive package including PTFE. This shows fatigue properties comparable with aluminium-20% tin bearings under the test conditions used in Glacier laboratories.

SUMMARY

Existing materials and their capability limits are summarised in fig 13. The diagram is schematic and is certainly oversimplified. Limiting behaviour depends on many parameters in addition to load and speed. Reference to load instead of stress has been used deliberately, to denote the total load divided by the total projected area. The future development targets are summarised in Table 3.

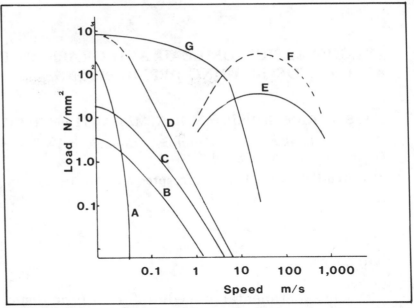

Fig 14 *Diagram showing the limiting performance of different types of bearing system as a function of load and speed.* **A** *is hard steel on hard steel;* **B** *filled PTFE;* **C** *porous metal;* **D** *rubbing bearing;* **E** *plain journal nominal stress;* **F** *plain journal probable peak stress;* **G** *EHD operation.*

REFERENCES

[1] Strategy for Energy Conservation Through Tribology, ASME, 1977

[2] Strategy for Energy Conservation Through Tribology, Second Edition, ASME, 1981

[3] Jost, H P and Schofield, J. Energy Saving Through Tribology: A Techno-Economic Study *Proc. IMechE*, 1981, **195**, No 16, 151–173

[4] Baker, R G. Basic Technological Research in Tribology. Study Prepared for EC DG XII, 1983.

[5] Tribologie; 1976, Report BMFT-FB T 76–38, Bundesministerium für Forschung und Technologie.

[6] Proposal for a Council Decision adopting a multiannual research and development programme of the EEC in the field of basic technological research; COM(83) 350, Brussels 1983

[7] Timoney, S and Flynn, G. For example: A low friction unlubricated SiC diesel engine, 1983. Paper 830313 presented to the SAE conference on the Adiabatic Diesel Engine, Detroit, March 1983. Preprints No SP 543. 11–19

[8] *Tribology Handbook;* Edited M J Neale. 1973, Newnes – Butterworth, London.

[9] Rigney, D A & Heilmann. Wear and Erosion of Metals. Metallurgical Treatises, 1982, Access No: 50608, Class 58–1, 621–641.

[10] For example; Extrapolation in Tribology: Godet, M; Wear, 77(1982), 29–44

[11] Rhode, S M. A Mixed friction model for dynamically loaded contacts with application to piston ring lubrication, 1981, Proc. 7th Leeds-Lyon Symp IPC PB Ltd, London, 262–278

[12] Dowson, D, Ruddy, B L & Economou, P N; The elasto hydrodynamic lubrication of piston rings, 1983, to be published, Proc Roy Soc, London

[13] Godet, M, Frene, J, Fantino, B and others; INSA and University of Poitiers; 1983, Private Communications

[14] A study of the mechanisms involved in the unlubricated break-in behaviour of 52100 steel on Al-Si-Cu Alloy; Blau, P J & Whitenton, F P; submitted to Trib Int

[15] Wear of Materials; Bhansali K J & Miller A E; 1981, NY

[16] Glacier HX; Data Sheet R&D/PDD/GJD/913, 1984, Glacier Metal Company ∎

Table 3: *Summary of Development Objectives*

Application	Development Objectives
Crankshaft bearings	Capability of dealing with higher stresses and thinner oil films in more dynamically flexible engines. Increase in temperature and corrosion resistance.
Pistons & assemblies	Lighter pistons and assemblies. Lower friction systems. Higher temperature systems leading to adiabatic engine concept.
Wheel bearings	Lower weight, improved sealing reliability.
Cam followers	Lower cost & weight, improved reliability, higher load capability
Dry & Suspension	Lower cost, improved wear performance, improved reliability.
CVT traction drives	Improved power transmission/weight ratio.

THE DISPERSION OF LEAD AND GRAPHITE IN ALUMINIUM ALLOYS FOR BEARING APPLICATIONS

C. S. SIVARAMAKRISHNAN, R. K. MAHANTI and R. KUMAR

National Metallurgical Laboratory, Council of Scientific and Industrial Research, Jamshedpur 831007 (India)

(Received February 21, 1983; accepted May 18, 1984)

Summary

The experimental procedure for uniform dispersion of lead and graphite in some aluminium alloys which are particularly suitable for bearing purposes is described in this paper. The results of thermal expansion studies and wear testing of the lead- and graphite-dispersed Al–Si systems have been included to demonstrate their good bearing properties.

1. Introduction

Aluminium-base bearing alloys are being used extensively because of their good bearing and mechanical properties [1 - 3] such as the wear resistance, antiseizure properties, fatigue limit, yield and tensile strengths, elongation and hardness. The embeddability of aluminium alloys is poor, however, compared with that of tin Babbitt metals and conventional bronze and other alloys but this handicap has been overcome successfully through the introduction of 6% - 10% Sn [4] into the aluminium alloy composition. The addition of tin has been found to have the further advantage of improving the scuffing resistance and antiseizure property of the material. Since tin is expensive and has to be imported into India, its substitution by indigenously available and/or less costly metals is desirable. The development of a new series of aluminium-base bearing alloys dispersed with lead and graphite has opened up a promising area. Whereas tin is soluble in aluminium alloys in the liquid state [5] and its introduction into the alloy could be achieved through conventional casting and solidification techniques, the Al–Pb system has an immiscibility gap and graphite is completely immiscible in aluminium. In other words conventional means of melting and casting are inadequate for the Al–Pb or Al–graphite systems. Therefore alternative casting techniques have to be evolved not only to retain these materials in

Reprinted with permission from Wear, Vol. 96, 1984, 121-134, © 1984 Elsevier Sequoia/Printed in the Netherlands

the alloy but also to have them suitably distributed in the matrix of the aluminium or aluminium-rich solid solution. It may be pointed out that graphite particles have been successfully dispersed in aluminium alloys to improve the wear behaviour [6] by coating the particles with copper or nickel to match the densities and to impart some wettability [7 - 9].

In this work an attempt has been made to introduce lead and graphite into partially crystallized alloys of aluminium containing copper, nickel and iron through the vortex generated with the help of a graphite stirrer. The distribution of the lead and graphite was studied metallographically. The thermal expansion behaviour and wear resistance were also evaluated in lead- and graphite-containing alloys to determine their potential for bearing applications.

2. Experimental details

2.1. Dispersion of Pb–graphite in aluminium alloys

A specially designed arrangement was fabricated to disperse lead and graphite in aluminium alloys. It essentially consisted of a butterfly graphite stirrer 35 mm in diameter firmly attached to a $\frac{1}{4}$ h.p. electric motor through a mild steel spindle. The stirrer, together with the motor, was kept inclined at 30° with respect to the vertical axis to give better dispersion. The studies were carried out in three stages with this set-up.

In the first stage binary aluminium alloys containing 4, 8 and 12 wt.% Si and a commercial Y alloy, designated henceforth as the A series in this paper, were melted in a clay-bonded graphite crucible in a resistance furnace. The crucible together with the melt were taken out of the furnace after the alloys had been completely melted and the preheated graphite stirrer was inserted into the melt. When crystallization commenced the solid–liquid slurry was agitated at 1440 rev min^{-1} with the graphite stirrer to form a vortex into which weighed quantities of Pb–graphite were added. This procedure enabled the dispersion and retention of lead in the bulk of the solid–liquid slurry. The stirred alloys were then cast while still in the solid-liquid range into plate or cylindrical moulds.

Alloys for the second stage of the work were chosen mainly from Al–Si-base compositions which are suitable for bearing applications (B series). Thus alloys of aluminium containing silicon in the hypoeutectic, eutectic and hypereutectic range were prepared in an electrical resistance furnace to which nickel, iron and copper were introduced as alloying elements with a view to imparting strength to the matrix and forming interlocking phases during solidification which would enable the retention of a greater proportion of the lead that is dispersed by the rotating stirrer.

In the third stage of the work, attempts were made to disperse uncoated graphite with a view to studying the effect of its distribution on solidification. These alloys were designated as the C series. Various sizes of the graphite particles, namely mesh sizes of +52, −52 to +120, and −120, were intentionally introduced into the vortex of the slurry to study the effect of size on the extent of retention in the solidified ingot.

The graphite contents in the alloys were determined by analysis of the residues left after dissolution of the metallic part in each of the representative slices cut from the ingots.

In all the above three sets of experiments the melts, after vigorous stirring and the introduction of lead or graphite, were cast into 10 mm × 100 mm chill plate moulds or cylindrical graphite moulds 25 mm in diameter where the solidification took place immediately because the melts were already in a partially crystallized condition during pouring.

2.2. Thermal expansion and wear behaviour of Pb–graphite-dispersed aluminium alloys

As a low coefficient of thermal expansion is an important requirement for alloys meant for bearing applications, it was decided to determine these values for the alloys belonging to the B and C series. For this, specimens 6 mm in diameter and 25 mm in length were prepared and thermal expansion data were determined in a dilatometer at 20 °C intervals from room temperature to 240 °C. The specimens were maintained at each temperature for 15 - 20 min in the uniform temperature zone of the furnace to ensure a uniform temperature throughout the specimen.

The wear characteristics of the B and C series alloys, of which the former corresponded to conventional bearing compositions, were determined in an Amsler wear testing machine in lubricated conditions. Standard wear test specimens were prepared and the mating surfaces were matched by trial runs extending over 10 - 12 h against a standard steel base of hardness 252 HV rotating at 200 rev min^{-1}. The actual tests were carried out at an applied pressure of 3.43 MN m^{-2} only after the surfaces matched perfectly.

As the B and C series alloys have potential applications in bearings their mechanical properties such as the tensile strength and Young's modulus were determined in a Hounsfield tensometer. Grain size measurements were also made to study the relationship between grain size and wear characteristics.

3. Results and discussion

The compositions and mechanical properties of the A, B and C series alloys are reported in Tables 1 - 5. It can be seen from Table 1 that although 6% Pb was added during stirring of the solid–liquid slurry the ingots actually retained only 20% - 50% of the addition, the rest of the lead settling at the bottom of the crucible during the period of transfer of the material into the mould. The binary Al–Pb diagram [5] shows that lead has a solubility of less than 1% at 650 °C but the present attempt to disperse it in the two-phase

TABLE 1

Details of the composition of the Al–Si and Al–Si–Pb alloys

Alloy	Compositions (wt.%)				Pb distribution density (particles mm^{-2})
	Nominal		Actual		
	Si	Pb	Si	Pb	
A2	4.0	6	3.92	3.94	5
A3	8.0	6	7.28	3.73	12
A6	12.0	6	9.52	3.08	2
A7a	12.0	—	12	—	
A8	12.0	—	12	—	
A12	Y alloy	6	Y alloy	3.16	2
MH-B6	15	1.0 (Fe)	14.29	1.0 (Fe)	

aNormal casting.

TABLE 2

Composition, grain size and mechanical properties of aluminium alloys

Alloy	Composition (wt.%)						Pouring temperature (°C)	Grain size (μm)	Ultimate tensile strength (MN m^{-2})	Young's modulus (MN m^{-2})
	Si	Mg	Cu	Fe	Ni	Pb				
B1	8.12	—	1.48	—	—	1.71	550	1118 ± 126	—	—
B2	10.51	0.23	1.03	0.96	—	0.95	—	1089 ± 235	117.6	5586
B3	7.0	1.0	1.0	0.78	—	1.2	—	1050 ± 185	156.8	9996
B4	5.95	2.10	0.94	1.16	0.12	0.78	—	845 ± 141	147.0	28469
B5	14.29	1.3	—	—	—	1.06	—	1403 ± 188	117.6	7546
B6	15.0	1.3	1.3	1.0	1.3	—	650	1259 ± 136	92.1	17444

Source: Wear, 1984, 121-134

TABLE 3

Wear testing data and thermal expansion coefficient

Alloy	Mating area (×10⁻⁴ m²)	Pressure on the surface (MN m⁻²)	Mass before testing (g)	Mass after testing for 5.5 h (g)	Mass loss (g)	Coefficient of friction	Hardness (HV 5)	Bearing parameter[a] ZN/P	Coefficient of thermal expansion (×10⁻⁶ °C⁻¹)
B1	2.80	3.49	24.0788	24.0676	0.0112	0.061	108	0.64	18.27
B2	2.80	3.49	24.2382	24.2380	0.0002	0.067	88	0.64	13.92
B3	2.70	3.63	25.1990	25.1982	0.0008	0.078	121	0.53	17.21
B4	2.75	3.57	24.5680	24.5676	0.0004	0.058	128	0.52	16.55
B5	2.65	3.69	24.4340	24.4320	0.0020	0.067	109	0.54	13.44
B6	2.70	3.63	24.7240	24.7220	0.0020	0.067	102	0.53	12.74

[a]Z (N s m⁻²), viscosity; N (rev min⁻¹), rotational frequency; P (MN m⁻²), pressure.

TABLE 4

Analysis of graphite retained in aluminium alloys (C series)

Sample	Composition	Amount of graphite added (wt.%)	Size of graphite particles (mesh size range)	Graphite distribution (particles mm^{-2})	Amount of C in alloys (wt.%)
C1	1 wt.% Cu	1.0	−52 to +70	4	Not significant
C2	1 wt.% Cu	1.0	−100 to +120	—	Not significant
C3	1 wt.% Cu	1.0	−120 to +200	—	Not significant
C4	4.2 wt.% Cu	5.0	−52 to +120	10	0.52
C5	7.8 wt.% Cu	5.0	−52 to +120	8	0.47
C6	Y alloy	5.0	−52 to +120	—	2.06
C7	Y alloy	5.0	−120	—	0.21
C8	Y alloy	5.0	−52	—	0.31
C9	Y alloy	5.0	−52 to +120	—	0.20
C10	5 wt.% Mg	5.0	−52 to +120	—	2.91
C11	5 wt.% Mg	5.0	−120	—	0.37
C12	5 wt.% Mg	5.0	−52	—	0.64
C13	4wt.%Zn−2wt.%Mg	5.0	−52 to +120	—	3.50
C14	4wt.%Zn−2wt.%Mg	5.0	−120	—	0.57
C15	4wt.%Zn−2wt.%Mg	5.0	−52	—	0.15
C16	12wt.%Si−(1 - 3)wt.%Cu−(1 - 3)wt.%Ni−1.0wt.%Fe	1	−52 to +70	6	0.20
C17	12wt.%Si−(1 - 3)wt.%Cu−(1 - 3)wt.%Ni−1.0wt.%Fe	1.1	−70 to +100	—	0.12
C18	12wt.%Si−(1 - 3)wt.%Cu−(1 - 3)wt.%Ni−1.0wt.%Fe	0.85	−100 to +120	—	0.32
C19	12wt.%Si−(1 - 3)wt.%Cu−(1 - 3)wt.%Ni−1.0wt.%Fe	0.85	−120 to +200	—	0.09
C20	12wt.%Si−(1 - 3)wt.%Cu−(1 - 3)wt.%Ni−1.0wt.%Fe	0.88	−200	—	0.08
C21	Al−8wt.%Si	1	−52 to +120	8	—
C22	Al−12wt.%Si	1	−52 to +120	4	—
C23	Al	1 (SiC)	—	1	—

Source: Wear, 1984, 121-134

TABLE 5

Wear, mechanical and thermal properties of some graphite-containing aluminium alloys

Alloy	Mating area ($\times 10^{-4}$ m²)	Pressure (MN m⁻²)	Specimen masses (g)		Mass loss (g)	Coefficient of friction	Bearing parameter[a] ZN/P	Ultimate tensile strength (MN m⁻²)	Hardness (HV 5)	Thermal expansion coefficient ($\times 10^{-6}$ °C⁻¹)
			Before testing	After testing for 5.5 h						
C4	2.60	3.77	25.016	25.000	0.016	0.062	0.59	168	64.2	15.5
C5	2.70	3.63	25.550	25.548	0.002	0.032	0.61	175	73.6	18.5
C10	2.68	3.66	24.240	24.2316	0.0104	0.036	0.60	221	74.1	21.2
C13	2.60	3.77	24.9685	26.9635	0.0050	0.021	0.59	345	139	20.3
C22	2.65	3.70	24.3610	25.3540	0.0070	0.042	0.59	164	54.1	13.6

[a]$Z = 0.01$ N s m⁻², viscosity; N (rev min⁻¹), rotational frequency; P (MN m⁻²), pressure.

field has enabled retention of about 3% Pb. This is possibly due to the formation of interlocking intermetallic compounds which minimize the segregation of lead globules dispersed by the rotating stirrer. Some typical microstructures of lead-dispersed alloys are shown in Figs. 1 - 4. Microstructures from different sections of the ingot (25 mm in diameter × 200 mm in length) are also shown in some figures to demonstrate the uniformity of the distribution of lead in the interdendritic spaces.

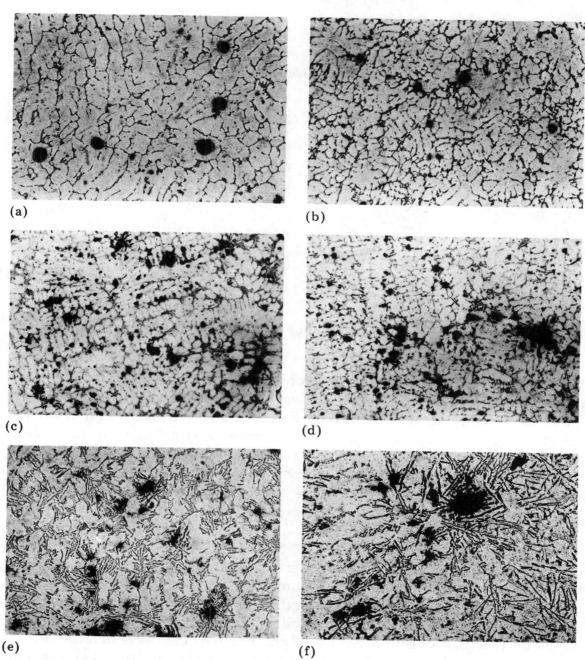

Fig. 1. Lead-dispersed Al–Si alloys: (a) Al–4wt.%Si (bottom of the ingot);(b) Al–4wt.%Si (middle of the ingot); (c) Al–8wt.%Si (bottom of the ingot); (d) Al–8wt.%Si (top of the ingot); (e) Al–12wt.%Si (bottom of the ingot); (f) Al–12wt.%Si (top of the ingot). (Magnifications, 152×.)

Source: Wear, 1984, 121-134

23

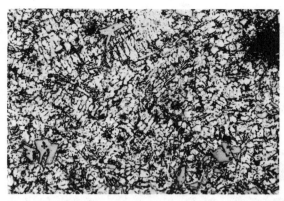

Fig. 2. Lead-dispersed Al–14wt.%Si–1.3wt.%Mg–0.12wt.%Ni alloy. (Magnification, 76×.)

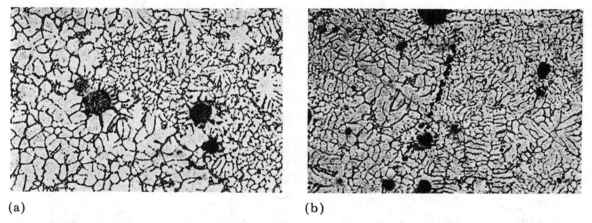

(a)   (b)

Fig. 3. Lead-dispersed Y alloy: (a) bottom of the ingot; (b) top of the ingot. (Magnifications, 152×.)

It can be observed from the microstructures that (i) lead particles are coarse and widely distributed in Al–4wt.%Si alloys and (ii) lead particles are finer and closely distributed in Al–8wt.%Si alloy. This is supported by statistical counting of the number of dispersed lead globules in one square millimetre of the matrix in each of the samples, as recorded in Table 1. In all these cases it can be seen that the dispersed lead has occupied interdendritic spaces and assumed the shape of the space left by the already crystallized silicon needles and primary phases. No lead is observed in the primary phase in any case.

Apart from the composition of the alloys, Table 4 also records the graphite particle sizes, distribution density and the percentage of graphite retained on solidification in some aluminium alloys. The table shows that only about 10% of the particles added could be retained on solidification. The rest was partly thrown off during addition and/or rejected by the solidifying melt. The distribution densities of graphite particles were determined by counting the number of particles per unit area and the statistical averages of counts are reported. It can be noticed that graphite in the mesh range from −52 to 120 is retained in greater quantities than in the other ranges investigated.

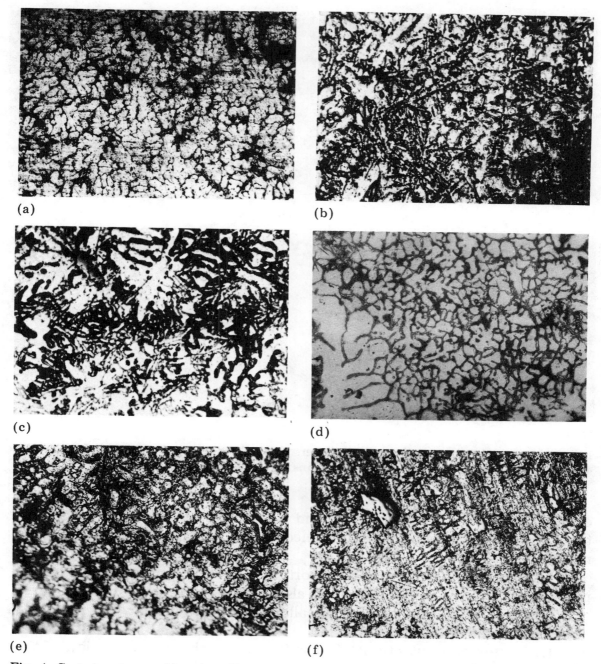

Fig. 4. Cast structures of bearing alloys: alloy B1 (Al–8.12wt.%Si–1.48wt.%Cu–Pb alloy); (b) alloy B2 (Al–10.5wt.%Si–0.23wt.%Mg–1.03wt.%Cu–0.96wt.%Fe–Pb alloy); (c) alloy B3 (Al–7wt.%Si–1wt.%Mg–1wt.%Cu–0.78wt.%Fe–1.2wt.%Pb alloy); (d) alloy B4 (Al–5.95wt.%Si–2.1wt.%Mg–0.94wt.%Cu–1.16wt.%Fe–0.78wt.%Pb alloy); (e) alloy B6 (Al–15.0wt.%Si–1.3wt.%Mg–1.3wt.%Cu–1.0wt.%Fe–1.3wt.%Ni alloy); (f) alloy B5 (Al–14.29wt.%Si–1.3wt.%Mg–0.12wt.%Ni–1.06wt.%Pb alloy). (Magnifications: (a) - (c) 152×: (d) 70×; (e), (f) 152×.)

Figure 5 shows a few typical micrographs of unetched sections of graphite-dispersed Al–Cu alloys. It can be seen that a uniform and fine distribution of graphite particles is achieved in the alloys corresponding to

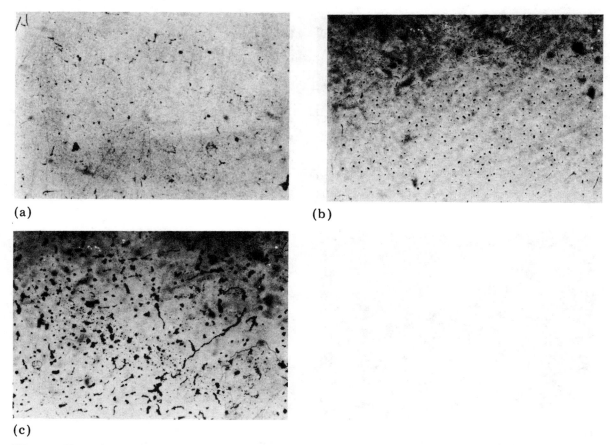

(a)

(b)

(c)

Fig. 5. Unetched sections of graphite-dispersed Al–Cu alloys: (a) Al–1wt.%Cu; (b) Al–4.2wt.%Cu; (c) Al–7.2wt.%Cu. (Magnifications: (a) 152×; (b), (c) 76×.)

compositions within the primary solid solubility range (4.2 wt.% Cu) and that, unlike lead, graphite has occupied both the grain boundaries and the matrix.

The thermal expansion data of vortex-cast aluminium bearing alloys are plotted in Figs. 6 and 7. Figure 6 also includes the data obtained for a conventionally cast eutectic Al–Si alloy for comparison. The following observations can be made.

(i) Vortex-cast eutectic Al–Si has the lowest thermal expansion of the alloys examined.

(ii) Lead does not seem to influence the expansion behaviour of the hypoeutectic alloy, but it increases the thermal expansion coefficients in the eutectic alloy, the effect being more pronounced at higher temperatures.

(iii) The thermal expansion behaviour does not change significantly in the alloy containing silicon, copper, iron and nickel in the composition ranges investigated.

The details of friction tests such as mass loss and coefficient of friction are incorporated in Tables 3 and 5. The tables also contain a column giving the values of bearing parameters as calculated from the expression ZN/P where Z (N s m^{-2}) is the viscosity of the oil, N (rev min^{-1}) is the rotational

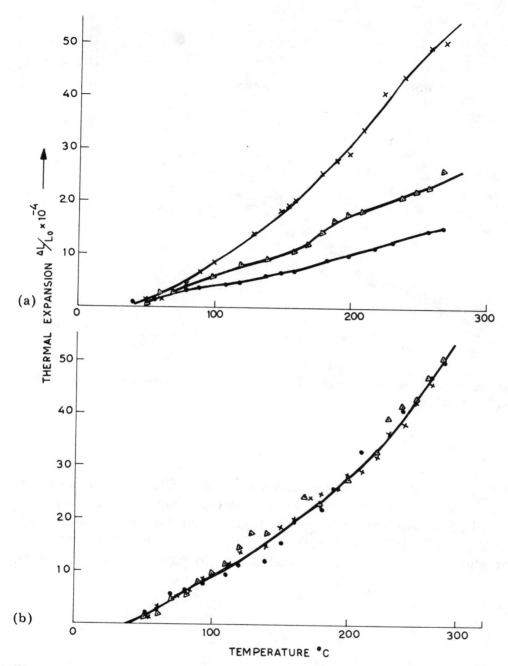

Fig. 6. Thermal expansion of some stir-cast Al–Si and Al–Si–Pb alloys: (a) Al–12wt.%Si (△, conventional; ●, present work) and Al–12wt.%Si–Pb (×); (b) Al–4wt.%Si (△), Al–4wt.%Si–Pb (●) and Al–8wt.%Si–Pb (×).

frequency and P (MN m^{-2}) is the load. These values give an idea of the index of lubrication. The present values of the coefficient of friction and the bearing parameter value of about 0.58 are in agreement with those of the earlier work [10] and correspond to boundary-lubricated conditions. Since the materials did not seize under the conditions of the present investigation, their galling resistance appears to be good.

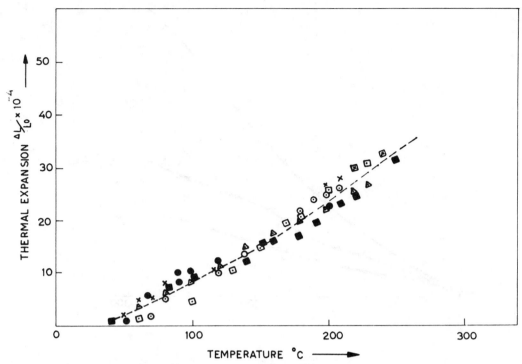

Fig. 7. Thermal expansion plots of some aluminium bearing alloys containing lead: ⊙, B1; △, B2; ⊡, B3; ×, B4; ●, B5; ■, B6.

A cursory glance at the average grain size data in Table 2 and the wear data in Table 3 shows that the mass losses corresponding to grain sizes of 1118, 1403 and 1259 μm are significantly higher than those corresponding to grain sizes of 1089, 1050 and 850 μm. This only shows that there is a tendency for alloys with relatively smaller grains to have better wear resistance.

4. Conclusions

The present investigation has shown the following.

(i) A uniform dispersion of about 3 wt.% Pb in aluminium alloys can be obtained by introducing the lead into the vortex formed by a stirrer in the partially crystallized melt.

(ii) The expansion behaviour of aluminium alloys containing silicon, copper, iron and nickel is not basically changed in the presence of lead or graphite.

(iii) Friction tests have shown that vortex-cast alloys containing lead and graphite are suitable for bearing applications.

(iv) The wear resistance is improved in alloys with a relatively small grain size.

Acknowledgment

The authors record their thanks to the Director, National Metallurgical Laboratory, Jamshedpur.

References

1 H. Y. Hunsicker, *Sleeve Bearing Materials*, American Society for Metals, Metals Park, OH, 1949.
2 F. T. Barwell, *Met. Rev.*, 4 (1949) 141.
3 H. Y. Hunsicker and L. W. Kenip, *SAE Q. Trans.*, 1 (1) (1947) 6.
4 H. N. Bassett, *Met. Ind.*, 52 (1938) 25.
5 P. M. Hansen, *Constitution of Binary Alloys*, McGraw-Hill, New York, 1958.
6 V. G. Gorbunov, V. D. Parshin and V. V. Panin, *Russ. Cast. Prod.*, (August 1974) 348.
7 F. A. Badia and P. K. Rohatgi, *Trans. Am. Foundrymen's Soc.*, 77 (1969) 402.
8 F. A. Badia, D. F. McDonald and J. R. Pearson, *Trans. Am. Foundrymen's Soc.*, 79 (1971) 265.
9 B. C. Pai and P. K. Rohatgi, *Trans. Indian Inst. Met.*, 27 (2) (1974) 97.
10 P. K. Rohatgi and B. C. Pai, *Wear*, 59 (1980) 323.

Zinc + aluminum =

new bearing options

The addition of large amounts of aluminum to zinc can provide high-performance low-speed bushings and bearings.

A group of high-performance, high-aluminum zinc alloys have been shown to have excellent bearing characteristics. They should be considered for low speed applications in such areas as construction equipment where bronzes and leaded bronzes have been used extensively for bearings and bushings.

For the foundry, the new materials offer energy savings and pollution-free processing. For the user, they provide cost savings and performance benefits in selected applications.

Alloying zinc with aluminum, copper, and other metals began early in this century, though some early examples lost strength and distorted. These problems resulted from intercrystalline corrosion and overaging due to impurities and high copper content, respectively.

Availability of high-purity (99.99%) zinc and technological advances provided bases for introduction of die casting alloys in the 1930s. When properly alloyed, cast, and applied they provided useful engineering materials with reliable properties. Germany suffered a copper shortage before and during World War II, and developed gravity-cast zinc-based alloys as substitutes for copper and its alloys. Among those alloys were some containing 10-30% aluminum, which rivaled the performance of the bronze bearing materials they were intended to replace.

These materials were dubbed "white bronze," but most applications returned to the use of bronze after the war. Some firms, however, continued development of the high-aluminum zinc alloys as bearing metals. One Austrian firm developed two proprietary bearing alloys from the Zn-Al-Cu system. Both are used in Europe for plain bearings, sliding elements, hydraulic components, worm wheels, spindle nuts, and other parts generally made of bronze in North America. The new materials were found to have:

• Price advantages and lower shipping costs
• Shorter machining times
• Lower frictional coefficients
• Better emergency running properties (no shaft seizing after lube failure)
• Better embedding properties
• Higher internal damping characteristics (thus quieter running)
• Longer life in many applications.

With such promise, documented by field experience, why were such materials not used here? There seem to be three reasons. First, though the die casting alloys gained a respected position as engineering materials, most gravity casting of zinc was confined to non-structural (i.e., decorative) applications. From this, it gained a poor-performing "white metal" image in designers' eyes. Second, virtually all non-ferrous foundries were deeply involved with bronze and aluminum. Zinc provided no obvious incentive to these producers to substitute it for either material. Third, the car manufacturing

Reprinted with permission from Automotive Engineering, Vol. 90, No. 9, 1982, 40-44, © 1982 Society of Automotive Engineers, Inc.

boom may have made the zinc industry complacent via its use in die casting. Only when weight saving became important, and injection molded plastics and aluminum replaced it in many areas, did the zinc industry seek new markets and new products.

Complacency is gone, and with it came renewed emphasis on older applications such as galvanizing or other forms of zinc coating for corrosion protection of mobile and other structures. However, the product development expected to broaden the zinc market is a new family of alloys.

High-aluminum zinc alloys

These zinc alloys have been developed for gravity casting in sand or permanent molds and are thus often referred to as "foundry" or "gravity casting" alloys, though at least one of them is proving itself as a die casting alloy. They might thus become best known as high-performance, high-aluminum zinc casting alloys.

Their high-aluminum description is a comparison with the original die casting alloys noted. Those alloys contained about 4% aluminum vs the 8-27% range of the present range.

The "high-performance" description is also comparative. With the higher aluminum content and controlled amounts of copper and magnesium these alloys have mechanical properties which are significantly better than those of conventional zinc die casting alloys.

Table 1
Zinc Alloy Compositions, wt-%

Metal	ZA-8*	ZA-12*	ZA-27*
Al	8.0-8.8	10.5-11.5	25-28
Cu	0.8-1.3	0.5-1.25	2.0-2.5
Mg	0.015-0.030	0.015-0.030	0.01-0.02
Fe	0.10	0.075	0.10

*Pb 0.004, Cd 0.003, Sn 0.002, Zn bal.

Of more practical importance is the comparison of the properties of these three alloys with cast iron (GR 30), aluminum (356-T6), and bronze (SAE 660) shown in Table 2. With advantages over one or more of them in tensile strengths, yield strengths, and hardness (not shown here), they obviously can compete in performance for selected applications. When performance is appropriate, they are likely to have cost advantages, particularly over cast iron and bronze. They can be cast to closer tolerances and machined far more readily than iron, so total parts cost can be much less, though the material cost of cast iron is much lower. All other factors being equal, the Zn-Al alloy costs are about a third that of bronze.

Zinc/aluminum alloy advantages

Broadly speaking, the following are the advantages of these zinc casting alloys:
• Low energy requirements for melting — lower than for any other casting metal
• Cleanliness — they satisfy OSHA and EPA requirements wthout need for pollution control equipment
• Excellent castability — better than iron, and generally equal to or better than aluminum and bronze
• Excellent wear properties — important for bearings and bushings
• Good mechanical properties — particularly strength and toughness
• Fine finishing versatility — better than iron and aluminum, equal to bronze
• Lower material cost than bronze, and
• Usually lower finished cost than cast iron when machining and tooling are considered.

A rapidly growing area of application for these alloys is that of replacing iron, bronze, aluminum, plastics, and other materials for parts which have bearing inserts such as gear boxes and connecting rods or links. Because these alloys are bearing materials, costs of the bearing insert and its assembly can be eliminated. Their main applications, however, lie in substituting for selected applications traditionally held by bronze.

Sourcing and cost

If the zinc alloys had equal or marginally improved performance, there would be no incentive to replace bronze, unless sourcing and cost were important factors.

Sourcing — It was noted that no capital investment in air pollution equipment was needed for casting zinc alloys. However, foundries handling the leaded bronzes common to bearings face significant expenses in meeting air quality standards for lead. This has lowered the number of available suppliers and perhaps made availability less dependable, particularly at times of high demand.

An inherent problem for bearing bronzes is their dependency on tin, a strategic metal not mined in the U.S. and subject periodically, to cartel-like market behavior. Zinc is readily available, domestically, and is thus subject mainly to domestic market forces.

Cost — These pollution and availability factors also play a part in costs, whose trends almost have to favor zinc vs tin and copper. Similarly, the cost of melting bronze is about double that for zinc, so fuel price increases will also probably favor zinc.

Material cost for a part is a function of density and raw material cost. On a cost per unit volume basis, the ZA-12 and ZA-27 materials should be about 35-40% as expensive as bronze, while on a per unit mass basis it would approximate 55-65% vs SAE 660 bronze.

Of key importance in heavy machinery fields — as bearing size increases, the material cost becomes a more significant portion of delivered cost. Thus, the larger the bearing, the more significant

Table 2
Property Comparisons of High-performance Zinc Alloys with Cast Iron, Aluminum and Bronze

	ZA-8	ZA-12 (ASTM B-669-80)	ZA-27	Cast Iron (Class 30)	Aluminum (AA356T6)	Bronze (SAE-660)
Density (lb/in³) at 70°F	0.227	0.218	0.181	0.26	0.097	0.322
Specific gravity	6.30	6.03	5.00	7.22	2.68	8.83
Casting temp. range (°F)	800-900	850-1000	950-1100	2550-2750	1350-1400	1950-2150
Solidification shrinkage (%)	1.0	1.2	1.3	1	--	--
Coefficient of expansion (in/in/°F×10⁻⁶) at 68-212°F	12.9	15.5	14.4	5.8	11.9	10.0
Thermal conductivity (Btu × ft/ft² × hr×°F) at 75°F	66.3	67.1	72.5	26.6	87	34.0
Electrical conductivity (%IACS) at 68°F	27.7	28.3	29.7	2.6	39	12.0
Pattern makers shrinkage (in/ft)	1/8	5/32	5/32	1/8	5/32	7/32

the saving for the Zn-Al alloys.

Performance

Cost benefits are, necessarily, meaningless without corresponding performance evaluations. These can be indicated by physical properties, laboratory evaluations, and field performance. European experience with zinc alloy bearings cannot, however, be directly transferred due to compositional differences.

Comparisons of the ZA alloys with bronze indicate that the tensile strengths of ZA-12 (40-50 ksi) and ZA-27 (58-64 ksi), depending on their casting practices, exceed those of SAE 660 bronze (30-38 ksi). Yield strengths vary even more widely at 30-32 ksi, 53-64 ksi, and 17-21 ksi, respectively.

The zinc alloys have natural lubricity — a low coefficient of friction. When antifrictional properties are favorable, hardness is a direct indicator of bearing life. Here, again, the zinc alloy hardnesses exceed those of the bronze by up to 65 BHN. Impact strengths of the zinc alloys are adequate for most applications, but bronze is superior and would be preferred for high-impact bearing applications.

ZA-27 is a heat treatable alloy, and treatments can be selected to accelerate aging and stabilize dimensions or to provide higher ductility and impact strengths. Since tensile and yield strengths as well as hardness are reduced, such treatment should be a consideration, rather than a routine procedure.

Emergency properties — Excellent behavior under emergency conditions, such as lubrication failure or extreme overloads, is a tangible property for high-aluminum alloys vs bronze. The usual behavior of bronze under these conditions is high heat generation, intermittent shaft seizures, and finally bearing failure with seizure of the shaft. The bronze is likely to fuse to the shaft, which may be scored or worn beyond recovery.

In contrast, the ZA-alloy bearings have a greater affinity for lubricants and thus retain their bearing properties somewhat longer and cause less shaft wear under lubricant deprivation. They are unlikely to seize or damage the shaft and, if fusing occurs, the shaft can readily be scraped clean.

Laboratory comparison—Design and performance data for typical bronze bearing alloys have been developed over the years, though gathering the ZA-alloy data is in its early stages. The zinc industry, working through ILZRO — International Lead Zinc Research In-

stitute — is conducting extensive research on the absolute and comparative bearing properties of zinc alloys at the Battelle Columbus Laboratories.

An early program compared ZA-12 and another material with SAE 660 and SAE 40 bronzes. The report stated, "It is apparent that the range of loads and speeds imposed during the program very adequately covered the capability range indicated for (ZA-12 . . . but) many of the load/speed combinations were too severe for the SAE 660 and SAE 40 bronzes." The report concluded that "High temperature rather than high friction or wear appears to be the limiting factor for (ZA-12.) It outperformed the bronzes in all comparisons."

Today, ILZRO is sponsoring further research on bearing properties of zinc gravity casting alloys (Fig. 1). In this test, specimens (Fig. 2) are placed under increasing loads and the shaft surface sliding speed is increased to a 140°C operating temperature limit. ZA-27 outperformed SAE 660 bronze over the entire range, as did ZA-12 at the lower sliding speeds.

A coefficient of friction for bronze is about 0.1 at maximum load capacities. That of ZA-27 ranged from 0.03 to 0.07, and it appeared to maintain this lower coefficient at high load ranges. ZA-27 also seems capable of maintaining its low coefficient at greater loads than ZA-12, an indication favoring its selection for tougher bearing applications.

In Fig. 3, a bearing wear comparison, a uniform stress of 1000 psi was placed on each test sample. Sliding speeds were varied, with wear measured at various distances. The results are encouraging: for example, at the point that each material had reached 0.002 inch wear, ZA-12 had gone about 15,000 feet whereas the bronze had traveled only 4000 feet — even though the ZA-12 was moving 50% more rapidly. It appears that the ZA-27, at an indicated 83 feet/minute, might move 100,000 feet or more before reaching the same amount of wear. These data represent 24-hour break-in periods, whereas longer term properties after wear is established would be of greater importance, and are being explored.

Limitations — While there are indications that these zinc alloys can outperform bronze for bearings, they should not be considered to be universal replacements for several reasons. Both tensile strength and hardness decrease with rising temperature. Thus maximum operating temperatures for ZA-27 have been stated as 250-

300°F; those for ZA-12 should be about 50°F less. A conservative rule of thumb would be to hold to about 200°F for both alloys to protect against the occasional installation in which environmental factors push normal operating temperatures upward. This 200-250°F maximum range is suggested for all zinc alloy applications, not those for bushings and bearings alone.

Zinc corrosion rates are least where pH lies in the 6.5-12.5 range. Given this range and the additional protection provided by most lubricants, corrosion is not a problem for most earthmoving applications. If a zinc alloy is in direct contact with another metal, normal galvanic corrosion can occur. Most common lubricants are excellent insulators, however, so bearing/shaft galvanic corrosion is not a problem.

Application guidelines

Generally, if a bronze bearing already exists, straight physical substitution of a zinc alloy version is often possible. In a raised temperature application, it may be desirable to add clearance. If the design is new, the following parameters should be reviewed before its completion:
• There is no clear demarcation for selecting between ZA-12 and ZA-27. ZA-27 is frequently used without heat treating, but such

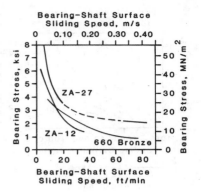

Fig. 1 — **Operating limits** of ZA alloys and SAE 660 bronze indicate expected benefits, particularly at higher stresses, of change to zinc alloys.

treatment improves its ductility, impact strength, and low temperature properties (while reducing its hardness). The heat treated product should be considered as a separate material.
• Bearing length should not exceed shaft diameter by a factor of more than 1.5 to prevent excessive edge pressures. (For bronze, a factor of 2.0 is normal.) ZA alloys are harder, less ductile, and have a lower elastic modulus than bronze, lessening the factor. Length-to-bore ratios can be greater for heat treated ZA-27.
• Wall thickness is typically 5% or more of shaft diameter, with a

Fig. 2 — **Specimens** two inches long, with two inch I.D. and half inch wall, were used in obtaining Fig. 1.

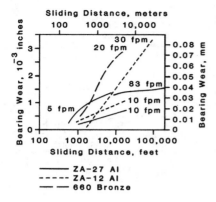

Fig. 3 — Wear comparisons for bearings stressed at 1000 psi show marked differences (note differences at 0.002 in. wear).

minimum thickness of (0.045 x D) + 0.020 inches.

• Clearances for bronze are suitable for zinc alloys at normal temperatures. With expected temperatures of 100-250°F, the zincs will require about 30% greater assembly clearance (in the bearing, not the shaft) due to their greater expansion coefficient).

• Surface speeds should be kept below 1400 feet/minute, though virtually all oscillating applications are acceptable.

• Bearing surface finishes normal for bronze bearings are acceptable. Better finishes, however, will provide longer bearing life, just as they would with bronze.

• Shaft hardness should be ~100 BHN greater than that of the zinc alloys. Due to ZA-27's greater hardness, it is better for harder shafts, while ZA-12 is better for softer ones.

• Shaft surfaces for bronze applications are quite satisfactory, though improved surfaces improve bearing lives.

• Maximum running temperatures are ~200°F for ZA-12, ~250°F for ZA-27.

• For bearings in steady contact with aggressive liquids, pH should be between 6.5 and 12.5.

• A wide range of neutral oils and greases such as mineral lubricants are suitable. Avoid those containing animal or vegetable constituents as well as those which are unstable or acid in nature. Keep bearings well lubricated.

• Zinc alloy bushings can be cold shrunk or press-fitted.

Though not yet readily available from distributors who normally carry full lines of bronze fittings, many foundries are aware of both varieties and are familiar with their processing needs. Avoidance of segregation and of contamination is important; the former eliminating centrifugal casting as a suitable procedure for ZA-27. Continuous casting is feasible, and the availability of such shapes frequently assists in areas such as maintenance for which time delays in normal procurement may be unacceptable. □

Based on SAE paper 820643, "High-Performance, High-Aluminum Zinc Alloys for Low-Speed Bearings and Bushings," by **Tony Calayag** and **Dean Ferres,** Kidd Creek Mines Ltd.

Silicon-nitride ceramics for bearings — a progress report

Fully dense silicon nitride has high-temperature strength, light weight, high hardness, low friction, and good fatigue life — all important qualities in a bearing material. Unlike other ceramic materials, which can be crushed in a bearing, this material fails by fatigue spalling, the common and less catastrophic mode of failure for bearing steels. Hot-pressed silicon nitride, the most promising for rolling elements and races, has high fracture toughness and fine grain size. Parts also have extremely good surface finish.

The results of SKF research and development programs to date indicate that:

- Silicon nitride in rolling contact forms elastohydrodynamic films similar to steel.
- Dry traction forces of silicon nitride against itself or steel are lower than steel against steel.
- Rolling traction forces of silicon nitride in lubricated contact are identical to that of steel.
- Wettability of silicon nitride with oils is comparable to steel.
- Silicon-nitride components have an endurance life greater than steel components.
- Bearings containing silicon-nitride balls run significantly cooler than all-steel bearings when lubrication is limited.
- Smaller centrifugal forces of the low-density silicon-nitride balls increase high-speed bearing life.
- Conventional grinding machinery can be used to finish ceramic bearings.
- Silicon-nitride bearings will be as reliable as high-quality steel bearings within a few years.

These results indicate that silicon nitride is an excellent candidate for rolling elements in high-speed bearings, where the centrifugal force generated by rolling elements at the outer race becomes greater than the application loads. In addition, because of the ceramic's hardness, silicon-nitride bearings can be used in highly abrasive applications, such as slurry pumps and rotary rock bits. Silicon nitride resists scuffing or seizing, so it can be used in unlubricated bearings.

In the past, ceramics had poor rolling-contact fatigue life. Recent improvements in silicon-nitride purity and grain structure have improved high-stress fatigue life. Under high test loads, rolling-contact fatigue life of NC-132 grade silicon nitride bearings is now equal to or better than that of M-50 steel. Usable strength also is highly dependent on quality, so process control of powder purity, density uniformity, and surface flaws is extremely important.

The internal geometry of rings and rolling elements must be designed specifically for ceramic characteristics. For

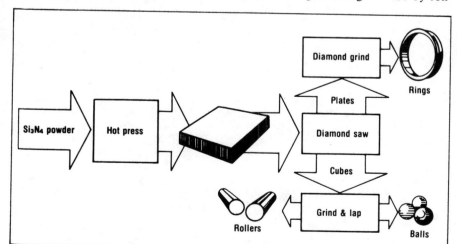

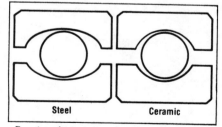

Bearing design must be altered for ceramics. Primarily, greater contact area is required to distribute stresses more uniformly throughout the balls and races.

The manufacture of silicon-nitride bearings begins with a hot-pressed and sintered slab from which the races, balls and rollers are sawed, ground, and lapped.

An all-ceramic ball bearing made from silicon nitride. Resting on a hot-pressed silicon-nitride slab, from which they were made, are balls and rollers in various stages of manufacture.

Bearing materials today and tomorrow

	Silicon nitride	M50 tool steel
Density, lb/in.³ (g/cm³)	0.115 (3.19)	0.283 (7.87)
Coefficient of thermal expansion, °F (°K) × 10^{-6}	1.83 (3.2)	7.39 (13.3)
Modulus of elasticity, psi (GPa)	45 (310)	30 (207)
Poisson's ratio	0.26	0.29
Hardness, Rockwell C	75-80	58-62
Strength* psi (MPa)		
Room temperature	120,000 (850)	350,000 (2400)
570 F (573 K)	—	300,000 (2100)
1100 F (873 K)	—	150,000 (1000)
1650 F (1173 K)	110,000 (750)	< 15,000 (<100)
2550 F (1673 K)	50,000 (350)	—

*Modulus of rupture for silicon nitride and tensile strength for steel

example, greater surface conformity (larger contact area) helps to reduce stresses in the ceramic. Designs should include means to sweep away wear particles and to help prevent contamination of the bearing from external sources. Also, bearing installation will have to be free of excessive vibration and shock loading, or include damping mechanisms to keep such loads from reaching the silicon-nitride elements. Also, special care must be exercised in assembly and in bearing maintenance to avoid chipping ceramic surfaces. Silicon-nitride bearing components — balls, rollers, and rings — can be made on production equipment using commercially available grinding wheels. Manufacturing processes are capable of holding the extreme dimensional accuracy required. To detect inclusions and voids responsible for premature spalling, fluorescent dye penetrant inspection is used. ∎

By John Johnson, Director, Product & process development, SKF Industries, Inc.

Hard particles dispersed in bearing metals may reduce wear

The addition of silicon to tin and aluminum alloy bearings may lessen bearing wear and increase critical seizure resistance.

Copper and lead alloys are now used for the sliding bearings in automotive engines. However, these alloys are being replaced by aluminum alloys because of aluminum's greater resistance to corrosion and fatigue. To lessen the possibility of bearing seizure in some applications, hard particles of silicon have been dispersed in the tin and aluminum alloy bearing material. This "new alloy" shows much less surface wear than the copper and lead alloy bearing material and exhibits sufficient extra load capacity without the use of lead-based overlays when used as crankshaft bearings.

There are two basic types of aluminum alloy bearings: one is made from high tin and aluminum alloys and the other is an aluminum alloy with a lead-based overlay. Passenger car engines are generally equipped with the high tin and aluminum alloy type bearings. High tin and aluminum alloy bearings are more easily affected by shaft roughness and foreign particles, and more often cause seizures when used with nodular cast iron shafts. Special attention is needed to reduce the bearing's load, prevent roughness on the shaft which interfaces the bearings, and eliminate the presence of foreign particles between the bearing and shaft interface.

More and more new engines are designed for higher speed operation and increased output and are lubricated by lower viscosity oil to save energy. The proneness of high tin and aluminum alloy bearings to seizures has become a greater problem. A study was conducted to analyze problems of seizure between high tin (20%) and aluminum alloy bearings and nodular cast iron shafts.

The results indicated that the high tin and aluminum alloy bearings had inferior seizure resistance as compared to the copper and lead alloy type. The critical seizure load was extremely low when the high tin and aluminum alloy bearings were used with nodular cast iron shafts. The seizure load became even lower as the engine oil temperature increased. This tendency of being adversely affected by high oil temperature was also found with the copper and lead alloy bearings. A danger of similar degree is anticipated in actual production when oil temperatures reach 140°C. On the other hand, the critical seizure point when using the bearings with forged steel shafts decreased only slightly when the oil temperature increased. No danger is anticipated with the forged steel shafts since the critical seizure point remained high.

The critical seizure point of the bearings always went down when they were used with the nodular cast iron shafts. In this regard, it would be correct to say that the high tin and aluminum alloy bearings should not be used with nodular cast iron shafts. A question arises: why is it that high tin and aluminum alloy bearings cause seizure more easi-

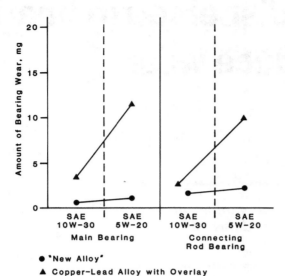

Fig. 1 — Amount of bearing wear using silicon-tin-aluminum alloy bearings and copper and lead alloy bearings with lead overlay during engine endurance test.

ly when they are used with nodular cast iron shafts?

Assumed seizure process

The assumed mechanism of seizure of high tin and aluminum alloy bearings was determined. Graphite grains are always found at the surface of the nodular cast iron shaft and burrs created by grinding are found around them. In an initial friction stage of fluid-lubricated sliding bearings, the bearings inevitably contact the shafts directly although such metal-to-metal contact may be minimized. It is therefore considered that the burrs around the graphite first wear the bearings through abrasion. Aluminum has a peculiar tendency to adhere to other materials and its abrasive debris easily adheres to the burrs on the nodular cast iron shaft. Once they start adhering, they will gradually grow in size. Then, same-metal friction is caused between the adhering aluminum and the lining aluminum of the

bearings which will then lead to seizures and other problems. From this assumed process of wear and seizure, it appears that it is very important to remove the burrs on the graphite and the adhering aluminum to improve seizure resistance. However, it is extremely difficult to remove the fine burrs through machining and it is impossible to remove aluminum once it adheres to the burrs.

New alloy

In an attempt to solve the seizure problem, a hard material was added to the bearing material to make it a new alloy. Hard particles (Si, W, Ti, or Cr) could be added to the bearing material which would remove the burrs on the shaft surface of the nodular cast iron and the adhering aluminum alloy.

The chosen "new alloy" uses three percent silicon with only a 10% tin content. The new alloy contains aluminum with 10% Sn, 1.8% Pb, 3% Si, 0.4% Cu,

and 0.3% Cr. Testing showed that the silicon-tin-aluminum alloy bearings had a higher critical seizure load than the previous 20% tin-aluminum alloy bearings. It was also found that these new bearings worked well with the low viscosity oils.

There was concern about the fatigue resistance of the new alloy used for the bearings. Fatigue strength was measured by the 10^7 load cycle method. Test results showed that the fatigue resistance of the new alloy was greater (100 MPa vs. 80 MPa) than that of the 20% tin and aluminum bearings.

Engine tests

Engine tests using the new alloy were performed using a four-cylinder, 1.8-L gasoline engine. The crankshaft was made of nodular cast iron (hardness-Hv of 230). The engine was lubricated with SAE 10W-30 oil with the oil temperature ranging from 130-135°C. The test was conducted at wide open throttle (100 hp) at a maximum engine speed of 5700 rpm. The results of the endurance test showed that contact between shaft and bearing was excellent. The contact distribution of the connecting rod bearing was slightly more dense at both sides than at the center. There were no shiny areas on the bearing surface, as often associated with high tin and aluminum alloy bearings.

Another engine test was run using SAE 5W-20 engine oil. The new alloy bearings performed extremely well, despite a slightly uneven contact distribution. Fig. 1 shows the results of a comparative test of bearing wear between the new alloy bearings and copper and lead alloy bearings (with SAE 49 and SAE 19 equivalent overlays). Granting the difference in specific gravity of the alloys, the new alloy showed much less surface wear than the copper and lead alloy bearing. The surface of the copper and lead alloy bearing became extremely worn and came close to seizure.

Uneven bearing contact can be caused by a misalignment between the bearings and crankshaft, distortion in the crankshaft and connecting rod due to load, and other factors which cannot be completely controlled. Uneven bearing contact cannot be eliminated and therefore the bearings must be manufactured with extra capacity against such uneven contact. Generally, such extra load capacity is provided by use of a lead-based overlay. The new alloy bearing does not have such an overlay but tests have shown that it does have sufficient extra load capacity without a lead-based overlay. □

Based on SAE paper 830308, "Aluminum Alloy Bearings Containing Hard Particles for Use with Nodular Cast Iron Shaft," by **Tatsuhiko Fukuoka, Hiromitsu Kato, Soji Kamiya,** Taiho Kogyo Co., Ltd. and **Norimune Soda,** The Institute of Physical and Chemical Research.

TiC-Coated Cemented-Carbide Balls in Gyro-Application Ball Bearings

BOVING, H. J. and HINTERMANN, H. E.
Laboratoire Suisse de Recherches Horlogères (LSRH)
Neuchâtel, Switzerland
STEHLE, G.
Roulements Miniatures SA (RMB), Biel, Switzerland

Stainless steel (oil-grease) lubricated ball bearings show wear tracks on the races after 3000 hours at 24 000 rpm. The lubricant becomes opaque and loses its qualities.

When TiC-coated cemented carbide balls are used, the bearings perform much better. After 20 000 hours at 24 000 rpm, the bearings are still running smooth and the lubricant remains clear. The uncoated steel races as well as the coated cemented carbide balls have suffered some damage.

The bearing characteristics with coated and uncoated balls are determined. After the test, the bearing components are analyzed by visual examination and by SEM.

INTRODUCTION

Ball bearings with TiC-coated races have been used successfully in hostile environments, such as vacuum or inert gases, at room temperature or at more elevated temperatures (200–300°C) (1), (2), (3). The very hard refractory coating acts as a diffusion barrier between the metallic components in sliding, rolling, and stationary contact with one another.

The low coefficient of friction between TiC and steel results in a very low wear rate of the contact partners. The fatigue resistance of coating-substrate assemblies has also been studied and has been reported elsewhere (4).

Standard stainless steel bearings give signs of deterioration after 3000 hours. The lubricant becomes opaque and loses its qualities. It is believed that the blackening and deterioration of the lubricant is due to the small wear particles created by pitting and fretting corrosion. This behavior has been studied and reported by B. Baxter (5). Bearings with TiC-coated stainless steel races still operated perfectly after 25 000 hours. At the ASME-ASLE Joint Lubrication Conference in San Francisco, in 1980, the authors presented a paper on tests with ball bearings having TiC-coated races (6). At that same meeting were presented test results obtained with low speed, large, outside diameter (80 mm) bearings using TiC-coated balls (7).

The work reported in this paper concerns lubricated miniature ball bearings with coated cemented-carbide balls tested at high speed; good performance was obtained during the 20 000 hours of the test.

Two types of tests are reported:

1. *Functional comparative tests,* aimed at comparing characteristics of bearings with TiC-coated balls with uncoated standard bearings, and a
2. *lifetime test,* aimed at evaluating the performance of bearings with TiC coated balls.

The cemented carbide balls underwent high-quality and precision polishing before and after the TiC-CVD treatment (8).

THE TiC COATING

The TiC coating is obtained by *chemical vapor deposition* (CVD). This method is detailed in the literature (3) and will only be briefly described here. CVD consists of chemical reactions occuring in the gas phase under controlled temperature and pressure, energized by radiation or heat, to form solid products for industrial applications and volatile byproducts.

The typical reaction for the production of TiC is as follows:

$$TiCl_4(g) + CH_4(g) \xrightarrow[H_2]{T} TiC\,(s) + 4HCl\,(g)$$

The coating treatment itself takes place in an apparatus which is presented schematically in Fig. 1. The different gases are mixed, passed through the $TiCl_4$ evaporator, introduced into the alloy reactor containing the workpieces to be coated, and heated to the required temperature by means of an electric resistance furnace. Data on the physical and tribological properties of TiC can be found in (3), (6). In addition to their favorable friction coefficients against steel, TiC coatings, due to their low solubility in steel, can be used as diffusion barriers between metallic components, thus drastically reducing adhesive wear.

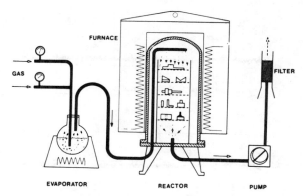

Fig. 1—Schematic view of a CVD apparatus

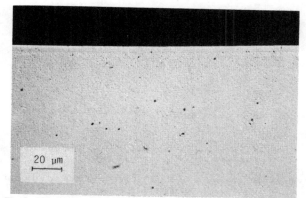

Fig. 2—Micrograph of the TiC coating on the WC-Co cemented carbide substrate.

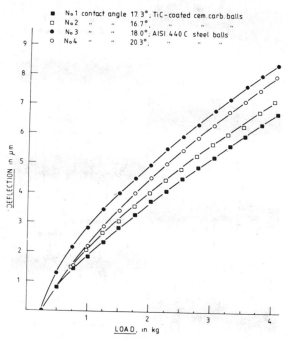

- No 1 contact angle 17.3°, TiC-coated cem. carb. balls
- No 2   "   "   16.7°,   "   "   "   "
- No 3   "   "   18.0°, AISI 440 C steel balls
- No 4   "   "   20.3°,   "   "   "   "

Fig. 3—Deflection curves of RA 6016X bearings (R-3)

Due to the elevated temperature (900–1000°C) and the duration (ca 5 hours) of the coating treatment, a certain amount of diffusion, from the coating to the substrate and from the substrate to the coating, takes place. This contributes greatly to the adherence of the coating to the substrate. Figure 2 shows a metallographic section through a coated cemented-carbide ball: the TiC coating thickness is ca 2 μm.

FUNCTIONAL TEST

By replacing the steel balls with coated cemented-carbide balls, some characteristics of the bearing are modified. In order to evaluate the importance of these modifications, a series of functional measurements were made on ball bearings with steel and coated cemented carbide balls, respectively. These measurements included axial deflection, static torque, noise at constant speed, and dynamic vibrational behavior. The bearings used for these comparisons were: RA 6016X-257 BB – outer ⌀ 12.7 mm; inner ⌀ 4.7 mm; width 3.9 mm; balls ⌀ 2.381 mm (R-3 bearing).

Figure 3 shows the deflection curves of two bearings with steel balls and two with coated cemented-carbide balls. The bearing with coated cemented-carbide balls has a smaller deflection than the bearing with steel balls. This is due to the higher modulus of elasticity of cemented carbide. This characteristic decreases the load-bearing capacity of the ce-

mented-carbide balled bearing. However, for the lifetime test reported here, the loads are low and far away from the load limit.

Figure 4 shows the torque readings according to MIL Std 206 A. The values measured at 2 rpm are practically equivalent for the two bearing types.

Figure 5 shows that the vibration levels at 1000 rpm under axial load of 4 N are practically equivalent for the two bearing types. For the determination of the dynamic vibrational behavior of the bearings, a gyro mounted on two bearings was used. The axial load was 8 N, the rotor weight 3.5 N and the speed decreasing from 25 000 to 5000 rpm. The different noise levels obtained are shown in Fig. 6. The gyro equipped with the cemented carbide balls behaves better in this dynamic vibration test. The deceleration time from 24 000 to 0 rpm was 350 s for the cemented carbide balled bearing and 220 s for the steel balled bearing. These functional comparative tests show that the two types of ball bearings have similar characteristics.

LIFETIME TEST

The lifetime test was performed on the test rig described hereafter, using a rotor mounted on bearings of the following type:

—Bearings: RA 6190X-257BB — 20° — J613/61
outer: ⌀: 19 mm; inner ⌀: 6 mm; width: 6 mm
races in AISI 440 C steel
balls: ⌀ 3.175 mm
   a. AISI 440 C steel
   b. cemented-carbide (WC + 6 percent Co)/TiC-coated
contact angle: 20°

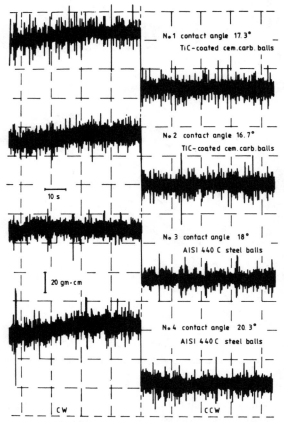

Fig. 4—Torque recordings of RA 6016X (R-3) bearings according to MIL Std 206 A, under 400 g and 2 rpm.

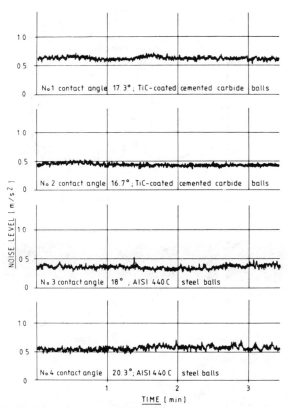

Fig. 5—Vibration levels of RA 6016X (R-3) bearings under an axial load of 400 g and at 1000 rpm.

—Lubrication: cage impregnated with oil MIL-L-00831768 (phenol 258 BB); 4–5 mg Andok C grease; same quantities in coated as in uncoated standard bearings.

The apparatus used to test two bearings at a time is shown schematically in Fig. 7. The weight of the rotor is 4 N, which provides both bearings with a radial load of 2 N. In the axial direction, the load of 1 daN is provided by means of a spring. The electrical energy is provided through a ternary-phased generator of 400 Hz. The rotation speed is 24 000 rpm.

The lifetime test consists of a continuous rotation at 24 000 rpm at ambient temperature. At regular intervals, the deceleration time has been recorded. The lubricant loss has been determined by weight loss of the bearings.

Performance of the bearings

The bearings with TiC-coated cemented-carbide balls reached 20 000 hours without incident, and could have gone on; the test with standard stainless steel bearings were not continued after 3000 hours because of the elevated torque noise and vibration levels of these bearings.

During the first 80 hours, the 24 000–0 rpm deceleration time was around 130 seconds for both types of bearings. Later, the TiC-coated cemented-carbide balled bearings had a deceleration time that oscillated between 165 and 195 seconds, all the way through the 20 000 hours. For the steel balled bearings, the deceleration time decreased notably

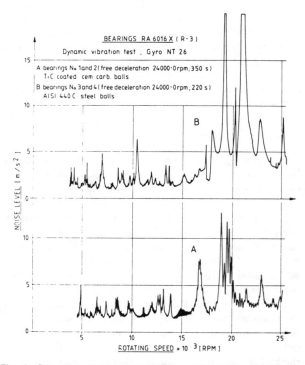

Fig. 6—Dynamic vibrational behavior, under decreasing speed (25 000– 5000 rpm), of NT 26 gyro mounted on RA 6016X (R-3) bearings.

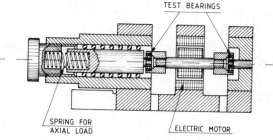

Fig. 7—Schematic view of the test equipment

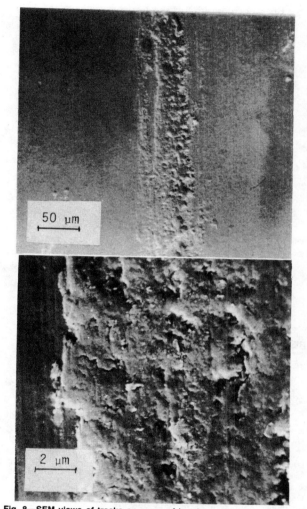

Fig. 8—SEM views of tracks on races of bearings with uncoated steel balls after 3000 hours at 24 000 rpm.

Fig. 9—SEM views of the racetrack on one of the two inner rings of bearings with TiC coated cemented carbide balls, after the 2000 hours test.

after a few hundred hours; after 3000 hours, the deceleration time was of only a few seconds, indicating bearing deterioration.

The loss of lubricant amounted to 60 to 70 percent after 3000 hours, for both steel balled and TiC-coated cemented-carbide balled bearings. For the latter, the further lubrication loss from 3000 to 20 000 hours was negligible; the remaining 30 to 40 percent "lubricant" which remains in the TiC-coated cemented-carbide balled bearing is sufficient to lubricate the friction couple "steel against TiC." In the case of the steel balled bearing, the remaining 30 to 40 percent "lubricant" lost at least part of its lubricating properties, resulting in cage wear and increased torque.

Examination of bearing components

After 3000 hours, the rings, cages, and balls of the steel balled bearings are covered with a dried-out, well-adherent black material with apparently poor lubricating properties; as can be seen from Fig. 8, there is an important track of these races, due most probably to transferred cage-wear debris.

After 20 000 hours, the components of the TiC-coated cemented-carbide balled bearings are covered with grease which appears dried out but in good condition; the cages have no sign of wear; as can be seen from Fig. 9, the races after 20 000 hours at 24 000 rpm, show some damage: the right-hand-side view has been made on the site where the damage was the greatest. Spotted areas amounting to roughly 5 percent of the total surface can be observed on the used TiC-coated cemented-carbide balls. Figure 10 shows a SEM view of a spot.

Final viewing of the ball surfaces suggested that the spots

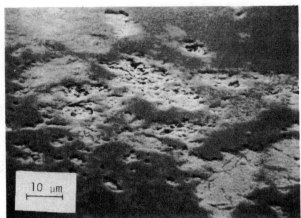

Fig. 10—SEM view of a spot on a TiC-coated cemented carbide ball after the 20 000 h test.

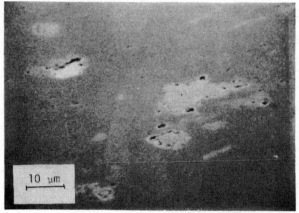

Fig. 11—SEM view of a spot on an unused TiC-coated cemented carbide ball.

correspond to areas where the coating has either partially or totally disappeared. Unused coated balls belonging to the same batch as the ones used for the present tests, were examined. Figure 11 shows that untested balls also have defects on the surface. It can be seen that the coating has spalled locally, even on untested coated balls. The portion of the surface affected by this defect is well below 5 percent.

It appears thus that the balls used for the present lifetime test had defects at the start. These defects, however, allowed the test to be run without difficulties for 20 000 hours. Finally, viewing of the ball surfaces indicates that the spalled regions have increased during the lifetime test. This can be seen from the surface roughness measurements made on both untested and tested balls.

DISCUSSION

H. E. SLINEY (Fellow, ASLE)
NASA-Lewis Research Center
Cleveland, Ohio 44135

The chemical vapor deposition (CVD) process is an effective method of achieving well-bonded carbide coatings.

Untested: cla 0.011, 0.012, 0.013, 0.019 μm
Tested : cla 0.029, 0.035, 0.038, 0.047, 0.058 μm

The cages of both tested bearings did not give any sign of wear.

CONCLUSIONS

1. It has been shown that ball bearings with TiC-coated cemented-carbide balls operated satisfactorily at 24 000 rpm for at least 20 000 hours.
2. Compared to standard uncoated bearings, no measurable cage-wear can be observed and the lubricant does not deteriorate; the bearings with coated balls run smoother.
3. The damage observed on the races must be due to small TiC particles spalled off the coated cemented-carbide balls.
4. The TiC-coated cemented-carbide balls show wear, which is mostly due to the reduced quality of the coating; since the start of the reported lifetime test, progress has been made and good quality coated balls are now being produced.
5. It has to be stressed, however, that even with low-quality coated balls, the bearings performed much better than standard uncoated bearings under the same conditions.
6. Basically, similar results are obtained whether the TiC coating is on the races or on the balls.
7. Finally, TiC coatings can be used to increase the life time and performance of high-speed ball bearings; potential application fields include: guidance systems, tape recorders, turbomolecular pumps, turbine engines, etc.

REFERENCES

(1) Boving, H. et al, "Wear Resistant Ball Bearings for Space Applications," *Proc. 11th Aerospace Mechanisms Symp.* NASA Conf. Publication 2038, NASA, GSFC, 1977.
(2) Hintermann, H. E. et al, "Wear Resistant Coatings for Bearing Applications," *Wear,* **48,** pp 225–236 (1978).
(3) Hintermann, H. E., "Exploitation of Wear- and Corrosion-Resistant CVD-Coatings," *Tribology Intl.,* **13,** pp 267–277 (1980).
(4) Maillat, M. et al, "Load Bearing Capacities of TiC Layers on Steel and Cemented Carbide," *Thin Solid Films,* **64,** pp 243–248 (1979).
(5) Baxter, B., "Chemical Processes Occuring in the Region of Rubbing Contact in S 166 Bearings for Satellite Tape Recorders Use," Int. Ball Bearing Symposium. Charles Drapers Lab., 1973.
(6) Boving, H. et al, "TiC-Coated Ball Bearings for Spin-Axis Gyro Application," ASLE Preprint No. 80-LC-3B-2.
(7) Lammer, J. et al, "TiC Coatings for Bearings in SPACELAB-Instrument-Pointing System," ASME/ASLE Intl. Lubr. Conf., San Francisco, August 18–21, 1980.
(8) Saphirwerk Industrieprodukte AG, Nidau, Switzerland, Manufacture of Precision Parts of Superhard Materials, Precision, October (1980).

As the authors indicate, good bond is achieved because of interdiffusion of the coating material and the substrate material during the CVD process. This can be attributed to the promotion of diffusion by the high temperatures (900-1000°C) that are used in the process. Diffusion results in a relatively gradual change in properties from the coating to the substrate compared to the step changes that occur at the bond line of a nondiffused coating. However, the photomicrographs of Fig. 2 in this paper and in the authors' previous paper (A1) almost seems to contradict this line of reasoning. The bond lines for CVD/TiC on stainless steel and on cemented carbides are quite sharply delineated. Does the demonstration of diffusion effects require higher magnification?

There are other more traditional methods of achieving diffusion-modified surfaces such as case carburizing, molten salt immersion, and hot nitriding. Have these processes been evaluated and compared to CVD for surface modification of rolling element bearing materials? These processes all involve high temperature and long durations. The authors stated that the CVD application of TiC requires about five hours at 900–1000°C. What is the effect of this treatment on surface finish and on the retention of the dimensional accuracy of the treated parts? Since the excellent bond appears to be attributable to diffusion effects, the bond strength of coatings applied by low-temperature processes might be considerably enhanced by post-heat-treating. It would be interesting to compare the advantages and disadvantages (including cost) for these various approaches.

The authors indicated that the oil had degraded in uncoated stainless steel gyro bearings after 3000 hours of operation. They believe the degraded lubricant contained metallic wear debris. This discusser suggests that the generation of metallic wear debris may have accelerated the oil degradation. It has been shown that steel is catalytic to the degradation reactions of oils (A2). Wear particles with their large specific contact area would be more efficient catalysts than the polished surfaces of the steel raceways in the bearings equipped with TiC-coated cemented carbide balls. Some coatings may be as valuable in reducing the catalytic effect of bearing surfaces on oil degradation as they are in reducing wear. In fact, the effects are obviously coupled. Certain soluble additives in formulated oils are considered to be "metal coaters" or deactivators to poison the catalytic effect of metals on oil degradation. It would be interesting to investigate the effectiveness of pre-applied coatings to perform the same function.

In Ref. (A1), the data show that gyro bearings with CVD/TiC-coated stainless steel balls survived about the same duration of bearing operation as bearings with CVD/TiC-coated cemented carbide balls. Therefore, any advantage of one over the other must be in some area other than extended bearing life. For example, which substrate provides the higher yield of usable balls after the high-temperature CVD treatment? Are there any other differences?

Did the authors test any bearings with uncoated cemented carbide balls? Why should carbide balls with carbide coatings be any better than uncoated carbide balls? It would have been instructive to use uncoated carbide balls as a second reference material along with the uncoated steel balls.

REFERENCES

(A1) Boving, H. J., Hintermann, H. E., and Stehle, G., "TiC-Coated Ball Bearings for Spin-Axis Gyro Application," *Lubr. Eng.*, **37**, 9, pp 534–537 (1981).

(A2) Lahijani, J., Lockwood, F. E., and Klaus, E. E. The Influence of Metals on Sludge Formation, *ASLE Trans.*, **25**, 1, pp 25–32 (1982).

AUTHORS' CLOSURE

It is true that the photomicrographs of TiC on AISI 440C and on cemented carbide are sharply delineated. There is, indeed, a sudden change in phase going from the substrate into the coating; in the case of the AISI 440C steel, the substrate consists of a martensitic phase and mixed iron chromium carbides, while, in the cemented carbide case, we have WC grains in a cobalt binder phase. The TiC layer, on the contrary, is single phased but takes several elements in solution. Solid solution promoted by the high temperatures used in the process has been observed by analytical methods. With the steel substrate, Fe and Cr were found present on the surface of the TiC (ca 2 percent each); amounts of about 10 percent were found at about one-third of the coating thickness away from the substrate. In the case of the cemented carbide substrate, notable amounts of Co and W were found in the coating [see Ref. (B1)].

Case carburizing, molten salt immersion, and hot nitriding can be used to diffuse elements like C and N into steel-type substrates. After such treatments, however, the surface consists mostly of two phases, e.g. a metallic (for example, martensite) and a ceramic type (carbide or nitride) phase. Such a surface is still liable to be subjected to adhesive wear when in contact with steel. The CVD-TiC is single phased and has, in addition, good tribological properties against steel [see Ref. (B2)]. After the CVD treatment, the surfaces have to be polished even when the substrates were polished prior to the treatment. The discusser is right when he worries about the retention of the dimensional accuracy of the treated parts; problems of this type do exist when steel races are being coated but do not exist with cemented carbide substrates.

The authors believe, indeed, that the diffusion leads to a good bond strength since the post-heat-treating of coated steel parts does not affect the coating. The bond strength of low-temperature process coatings might, indeed, not be sufficient for post-heat-treatment; however, with these low-temperature processes, there is usually no need for post-heat-treating.

The authors agree with the discusser when he assumes that the lubricant degradation is due to the metallic wear particles with their large specific contact area [see Ref. (B3)]. The "metal coaters" mentioned by the discusser might, indeed, slow down the degradation of the lubricant but do not necessarily decrease the amount of wear particles present in the lubricant. What the authors claim is that the use of TiC on races or on the balls may practically eliminate the production of wear particles.

It is true that the performances concerning bearing life were similar when the TiC coating was on the races or on the balls. The reasons for choosing one type of bearing over the other are, at the present state, only technical. As the authors mentioned earlier, with steel components there is a problem of dimensional stability; with cemented carbide such a problem does not exist. In general, it can be said that it is easier to polish perfect geometrical bodies like balls than bearing races.

No bearings with uncoated cemented carbide balls were tested. The discusser is right when he mentions that for complete comparison such tests would be necessary. However, the aim of the authors was to test steel-TiC contact compared to steel-steel contact, regardless of the substrate being used.

REFERENCES

(*B1*) Chollet, L. and Beguin, J., "Electron Microprobe Analysis of TiN Coating," *Electron Microscopy*, **3,** pp 164–165 (1980).

(*B2*) Habig, K. -H, Evers, W., and Hintermann, H. E., "Tribological Tests on Surface layers of Titanium Carbide, Titanium Nitride, Chromium Carbide and Iron Boride," *Z. Werkstofftech*, **11,** pp 182–190 (1980).

(*B3*) Hintermann, H. E., "Exploitation of wear- and corrosion-resistant cvd-coatings," *Tribology Intl.*, p. 274 (December, 1980).

Consideration of Mechanical, Physical, and Chemical Properties in Bearing Selection for Landing Gear of Large Transport Aircraft

H. E. FEWTRELL
Lockheed-California Company
Burbank, California 91520

Airline experience has shown that, under certain load conditions, there is a basic incompatibility between chrome-plated high-strength steel surfaces and aluminum nickel bronze bearing material. This incompatibility has resulted in the generation of severe thermal spikes over small discrete areas which have caused cracking of the high-strength steel substrate with, in some cases, resulting catastrophic failures.

This paper compares the performance of aluminum nickel bronze and beryllium copper under test and service conditions. It shows that mechanical, physical, and chemical properties must be considered in the selection of a viable bearing material. Each of these properties is related to specific performance requirements of pin joints lubricated with diester grease and shock strut pistons lubricated with MIL 5606 hydraulic oil.

Test data are shown demonstrating how mechanical properties can affect life in one case and, in another, how the greater thermal diffusivity of beryllium copper can effectively prevent "hot-spotting" under marginal lubrication conditions and relatively high transient PVs. Also, data are presented showing how the chemical reactivity of beryllium copper in the presence of the tricresyle phosphate (TCP) of MIL 5606 hydraulic oil can effectively preclude the generation of "ladder cracks" in the chrome-plated piston of large transport landing gears.

INTRODUCTION

The use of aluminum-nickel bronze bushings and bearings working with chrome-plated steel pins and surfaces in the landing gear of large jet aircraft is a well-recognized industry design practice. Their high bearing strength and excellent wear characteristics enable operators to realize extended times between overhaul, typically 20 000 flight hours.

Their use has provided a marked improvement over materials used in the first generation of jet transports, which required complete gear overhaul at 7000 flight hours and before. These first-generation jet aircraft used fixed, bare pins and steel bushings with either cadmium or silver plating for both rotating and fixed joints. Lubrication was provided only for the rotating joints. As a result, the pins in the rotating joints were severely pitted and galled after only 5000 hours of operation, while pins used in fixed, unlubricated joints displayed severe fretting and corrosion.

These design features on the first generation of jet transports were handed down from the old propeller-driven aircraft which were so slow in comparison to the jets that the flight hours per landing was about three times as great. Landing gear maintenance, when related to airframe and engine maintenance, was not considered burdensome for these aircraft. With the more frequent landings per flight hour of the jets, landing gear maintenance frequency assumed a far greater significance.

In 1967, one of the large landing gear manufacturers, in preparation for the advent of the second-generation wide-bodied jet transports, surveyed all of the major airlines and airframe manufacturers with respect to maintenance problems and their individual actions taken with respect to problem areas. The purpose of this survey was to gather information useful in the future design of landing gears which would ultimately improve the life and safety and extend the TBOs and also reduce the cost of overhaul.

One fallout of the survey is the almost universal use of aluminum-nickel bronze bearings working in concert with chrome-plated steel surfaces. The consequence has been a dramatic improvement in the life and performance of landing gear on the current generation of wide bodies.

Their use, though, has not been without problems. There has been demonstrated a basic incompatibility between chrome-plated surfaces and aluminum-nickel bronze under certain dynamic load conditions where there exists a high-relative-velocity component. This incompatability has been demonstrated in the presence of both AF-7 diester based grease and MIL-H-5606 hydraulic oil.

PROBLEM DEFINITION

Aircraft industry experience has shown that aluminum-nickel bronze has a strong propensity to form junctions with a chrome-plated surface. The resultant formation of metal-to-metal junctions is not self-limiting. Instead, it becomes regenerative and self-generating under conditions of high-translational velocity. The physical manifestation is bronze transfer to the chrome surface and cracking of the steel substrate. The cracks are always associated with a zone of overtempered martensite which, in a majority of cases, has an overlay of reformed untempered martensite. These metallurgical changes in the steel are the result of a phase transformation which, for the steel involved, occurs above 788°C. The hardness of the untempered martensite, Rc 60, indicates a very rapid quench from the austenitizing temperature.

There exists several diverse manifestations of this type of cracking in landing gear components. One is associated with landing gear shock strut pistons and is characterized by very long streaks (0.46m) of transformed material whose width is typically less than 2.5 mm. Associated with the transformed zone is a network of cracks aligned much like the rungs of a ladder, henceforth referred to as "ladder cracks." The fracture faces of these cracks are always inter-granular and are characteristic of quench-induced cracking. Another manifestation is a dime-sized semicircular crack on the surface of a journal pin which is subjected to high PVs for only 0.5 seconds during the touchdown condition. As in the previous case, there was a zone of transformed material relating to the crack. Fractographic analysis again indicated that the crack was intergranular, indicative of a quench crack.

In the above case of ladder cracks, one manufacturer redesigned the lower bearing of the shock strut such that piston lubrication was not dependent on seepage past the seals for lubrication by MIL-H-5606 hydraulic oil. The joint became fully wetted by immersion. This was to no avail, "ladder cracks" were just as prevalent as before. The design was then modified to provide lubrication to the lower bearing through a complex labrinyth of grease grooves with external lube fittings. Various greases were used without success. The "ladder crack" phenomenon persisted.

TEST PROGRAMS

Recognition of an industry-wide problem of the basic incompatability between aluminum-nickel bronze and chrome-plated high-strength steel under high-relative-velocity conditions prompted a series of test programs to characterize the problem and seek a viable solution. It was the intent of the test program to evaluate alternate bearing materials lubricated with current technology greases and hydraulic oil.

Friction Burn

An in-house test program was initiated to characterize alternate bearing materials to preclude the generation of dime-sized cracks on the aforementioned journal bearing. Four bronzes plus beryllium copper were subjected to the pressure-velocity-time environment known to exist in the troublesome joint. A typical test load spectrum is shown in Fig. 1. Of the five materials tested, only beryllium copper failed to produce transformation and associated cracking. Each of the bronze specimens generated localized "hot-spots." Cracking of the steel in discrete areas was in response to these "hot-spots." The beryllium copper test specimens demonstrated no tendency to "hot-spot." For the more severe loads of the test, metal transfer occurred for all test specimens, including the beryllium copper. Metallurgical examination of the steel in zone of metal transfer showed that for the bronzes, metallurgical transformation had taken place with concomitant cracking. The same examination of the steel member used in the test of the beryllium copper show no metallurgical transformation had occurred. Thus, there was no mechanism present to cause thermal or transformation cracking.

The room temperature strength of the beryllium copper is higher than any of the bronzes and, yet, under the most severe test environment, transferred to the chrome surface much the same as the bronzes.

The lack of transformation of the steel substrate in the beryllium copper test showed that the interfacial temperatures during the test were markedly less than those generated using any of the bronzes. A review of the physical properties showed that they were substantially different. These differences can provide a rationale for performance improvement of beryllium copper. The melting temperature of beryllium copper is approximately 177°C lower than any of the bronzes. The thermal diffusivity is almost three times as great. It became apparent that these physical properties were causal to the dramatically improved performance of beryllium copper, due to reduced interfacial temperatures. As the duration of the transient is only 0.5

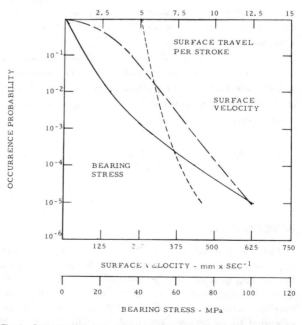

Fig. 1—Stress-velocity-travel probability for a dynamically loaded journal bearing.

seconds, the sleeve bearing was able to supply an adequate heat sink such that the increased thermal diffusivity could be an effective mechanism in the control of interfacial temperature. In addition, the lower melting temperature of beryllium copper results in lower shearing stresses in the bearing material, thus, also lowering the interfacial temperatures.

Adhesive Wear

Previously, a series of journal bearing wear tests had been performed where the mechanical properties and, to a lesser extent, the chemical properties of beryllium copper versus aluminum-nickel bronze were the predominant factors. These data are presented in Fig. 2.

These data show the effects of adhesive wear. As the test was performed in a clean laboratory environment and the load direction reversed every 100 cycles, as well as being lubricated coincident with the load reversal, there is very little chance that abrasive wear was a factor in the differential wear rates of the two materials. The low wear rate of BeCu versus aluminum nickel bronze becomes significant in showing that the mechanical properties of the two materials is the dominant factor under these test conditions. The following test shows how the mechanical properties have virtually no impact on wear life.

Chemical Wear

In another test run with MIL-H-5606 hydraulic oil, the wear rates of aluminum-nickel bronze and beryllium copper were compared with the results proving to be dramatically different than for the aforementioned journal bearing test using grease. The wear data are presented in Fig. 3. These data show that, at 172 MPa bearing stress and a vector velocity of 0.48 m/s, the wear rate slope is approximately 50 times as great for beryllium copper than for aluminum-nickel bronze. It can be seen from Fig. 3 that the wear rate of aluminum-nickel bronze is very linear and is directly related to the bearing stress. These data also show that the wear rate of beryllium copper is exponential over the range of test bearing stresses.

The presence of tricresyl phosphate (TCP) in the MIL-H-5606 hydraulic oil is the operative factor in the disparate wear rates between the two tested materials. It is apparent that the activation energy in which the TCP reacts with the beryllium copper occurs at a bearing stress below 35 MPa.

At first glance, it would seem that the extreme wear rate of beryllium copper versus that of aluminum-nickel bronze would make it a totally unacceptable alternate to aluminum-nickel bronze even though its reactivity with TCP and formation of a very low shear strength protective coating can effectively preclude generation of the aforementioned "ladder cracks" on landing gear pistons.

Commercial transport aircraft are designed for a landing sink speed of 3.05 m/s as an ultimate condition, which theoretically occurs only once in the lifetime of the aircraft. The typical landing is at a sink speed of only 0.55 m/s. The energy dissipated by the gear at this sink speed is less than four percent of that of which it is capable. It is the occasional hard landing with its concomitant greater energy dissipation requirement that can cause "ladder cracks." The higher the sink speed, the higher the translation velocity of the piston with respect to its lower bearing.

Associated with high sink speeds are transient load spikes of high magnitude with durations in terms of milliseconds. These load spikes, when combined with the high-translation velocities, result in the formation of junctions at asperity

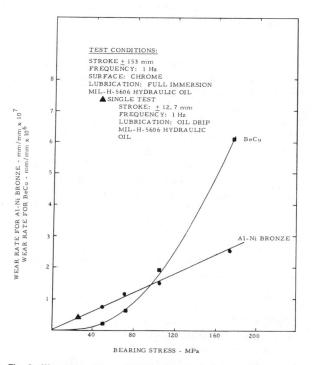

Fig. 2—Comparative wear of beryllium copper versus aluminum-nickel bronze.

Fig. 3—Wear rates versus bearing stress for BeCu and Al-Ni bronze bearing against chrome-plated steel heat-treated to 1792 MPa UTS

Source: Lubrication Engineering, March 1983, 153-157

			Wear Rate		Wear in 25K Flts	
Bearing Pressure	Piston Displacement Per Flight	Piston Displacement in 25K Flts	$mm \times mm^{-1} \times 10^6$		mm	
MPa	mm	m	Be Cu	Al-Ni Br	Be Cu	Al-Ni Br
23.00	142	3 550	0.0900	0.0380	0.319	0.134
15.03	670	16 500	0.0380	0.0250	0.627	0.412
9.20	615	15 375	0.0120	0.0150	0.184	0.230
5.60	203	5 075	0.0050	0.0100	0.025	0.051
4.55	309	7 600	0.0025	0.0090	0.019	0.068
3.46	203	5 075	0.0020	0.0065	0.010	0.033
2.96	879	21 975	0.0001	0.0050	0.002	0.044
1.10	391	9 775	0	0.0025	0	0.024
				Σ	1.185	0.995

contact points. Once formed, in the case of aluminum-nickel bronze bearing material, there is no self-limiting action. It becomes regenerative or self-generating. The translational velocity of the piston with respect to the lower bearing is such that healing is precluded and junctions are continually formed over virtually the total translational distance, up to 0.46 m, even though the sustained load is a mere fraction of the initiating load.

As shown in Fig. 3, the exponential increase in rate of wear versus bearing stress of beryllium copper in the presence of TCP in MIL-H-5606 hydraulic oil is highly suggestive of chemical wear. The continual formation and depletion of a low shear strength film as a function of the activation energy is the mechanism which effectively precludes the initial formation of metal-to-metal junction or if formed are healed immediately. It is through this mechanism that long streaks of transformed material in the chrome-plated high-strength steel piston are eliminated.

As mentioned previously and shown in Fig. 3, the wear rate of beryllium copper is so high in comparison to aluminum-nickel bronze at the higher bearing stresses, that it would appear to be a poor substitute based on wear life. For the particular application in question, this is not the case. A typical main landing gear wear spectra is shown in Table 1. From this table, it can be determined that 63 percent of the wear travel is at a bearing stress below 7.0 MPa while only 5 percent of the travel is at a stress where chemical wear could be significant. Integration of the wear rates of the two material versus travel distance at the bearing stresses indicated in Table 1 shows that either material will provide a wear life of 25 000 flights.

DISCUSSION

In the design of landing gears, primary emphasis is placed on the structural integrity of the gear as to both its ultimate capability and fatigue life. Testing is primarily to verify its shock-mitigating capability and fatigue life. To achieve the required wear life at the fixed and rotating joints, design criteria are established which dictate the maximum allowable bearing stresses. These cut-off stresses are typically 35 MPa or below. As part of the fatigue test, the wear life of the joints is verified. Missing from this testing is the velocity component which the gear sees in its real-world operation. It is this dynamic condition that has caused problems, two of which have been discussed in this paper.

In the solution of these problems, the industry has come to recognize that mechanical properties alone cannot always characterize the performance of bearing materials. It has been shown that the mechanical, physical, and chemical properties can each play a significant role in determining the absolute performance of a bearing material in a specific application.

CONCLUSION

The problems and solutions discussed in this paper are not unique to the lubrication engineer nor to automotive industry. The addition of extreme-pressure additives to petroleum base stocks has been practiced for over fifty years and has reached a high level of sophistication in our current fuel-efficient automobiles.

It has been just since the advent of the large, heavy commercial transports that the aforementioned problems have come to light. The solutions suggested here are unique to landing-gear design and have resulted in a broader characterization of bearing materials for landing gear application to enhance service life and reduce overhaul costs.

DISCUSSION

HAROLD E. SLINEY (Fellow, ASLE)
NASA-Lewis Research Center
Cleveland, Ohio 44135

This is an interesting paper on a practical study of lubrication problems in aircraft landing gears. The experimental program was straightforward and compared a limited number of bearing alloys for sliding against chromium-plated steel lubricated with diester grease or with MIL H 5606 hydraulic oil. The author did not provide alloy compositions nor criteria for their selection. This information could easily be provided and is important in trying to interpret the results.

The occurrence of ladder cracks and martensitic transformation in the steel underneath the chrome plate certainly suggests extremely high interfacial temperatures. This, in turn, suggests totally inadequate boundary lubrication under some operating conditions. The resulting high friction, in addition to generating high temperatures, also causes high biaxial stresses in the sliding contacts. High friction superimposes a large tangential stress component on the normal compressive stress. This discussion suggests that the resultant severe stress field may be contributing to the "ladder cracks" which propagate perpendicularly to the sliding direction. Reducing the friction (the source of added heat and stress) may be the key to a really satisfactory solution of the problem. Because friction is so important, it would have been helpful if friction coefficients had been reported by the author.

The author attributes the elimination of ladder cracks in the steel when beryllium copper was used to its superior thermal diffusivity. This discussion suggests that a lower friction coefficient may also have contributed to the improvement. [Heat input (Q in) via frictional heating is equally as important as heat dissipation (Q out) in determining the interfacial temperatures.]

The present study provided a useful fix to the problems addressed. However, a really satisfactory solution may require a more basic research approach to identify materials combinations that provide low friction and adhesion when lubricated with diester grease or MIL H 5606 hydraulic fluid under conditions relevant to landing gear application.

Source: Lubrication Engineering, March 1983, 153-157

Seawater-Lubricated Mechanical Seals and Bearings: Associated Materials Problems

FREDERICK J. TRIBE
Admiralty Marine Technology Establishment
Ministry of Defence, United Kingdom

The tribological performance of materials can seldom be predicted with confidence from consideration of bulk properties. Far less so can this be done when the lubricant is seawater for, in addition to the normal tribological problems, corrosion and/or dimensional instability can adversely affect component life. This paper presents results from an ongoing program of work to characterize and assess a wide variety of materials under consideration for either high-pressure mechanical seals or bearings operating in seawater. It discusses a number of failure mechanisms including corrosion of white metal-impregnated carbons, dimensional stability of resin-impregnated carbons, and the corrosion of metallic substrates supporting thermally sprayed coatings.

INTRODUCTION

The selection of materials for use in tribological applications is often a difficult problem for the designer because there is no sound basis for predicting surface behavior from a knowledge of bulk properties. Progress in this important technological area usually requires extensive empirical testing of candidate materials under conditions which simulate as closely as possible the actual working situation. The adverse effects of friction and wear can be minimized by suitable lubrication but care must be exercised in selecting a lubricant with acceptable viscosity characteristics, good chemical and thermal stability, and the capability of conferring some degree of protection to its system. In the marine world, many components have to operate in seawater and the use of seawater lubrication, particularly in warships, has certain advantages; notably, availability, excellent heat-transfer characteristics, and the elimination of expensive oil or grease systems requiring regular maintenance. The disadvantages of seawater in this context include its low viscosity resulting in thin lubricating films, the possibility of marine fouling causing abrasive wear or impeding water flow, and, most significantly, the associated corrosion problems. It is not uncommon for the lives of seawater-lubricated components to be determined by corrosion rather than wear.

The Admiralty Marine Technology Establishment at Holton Heath in Dorset has an ongoing program to characterize, assess, and improve materials for use in seawater-lubricated bearings and seals and this paper discusses some of the technical problems experienced in this work. The work is exclusively related to materials in rubbing-surface contact and primarily to the contacting faces of high-pressure mechanical seals.

COATINGS

General

Coatings are a highly effective method of conferring particular properties to the surfaces of suitable substrates and the useful lives of innumerable components operating in dry-, oil- or grease-lubricated conditions have been significantly extended. However, when used in a seawater environment, similarly coated components have failed to give the expected service life and many have failed in a matter of weeks or even days. This work is not concerned with the sacrificial corrosion of anodic coatings or, indeed, with the selective attack of specific phases in coatings. For example, the dissolution of cobalt in seawater leading to failure of cobalt bonded tungsten carbide coatings is well known and the problem can be avoided by the selection of nickel or duplex cobalt-chromium bonded cermet coatings. The work is concerned with the failure of otherwise sound coatings by corrosion of their substrates.

All thermally deposited coatings are porous to some degree and frequently contain microcracks or other microstructural defects which, more often than not, form continuous leak paths from the coating/seawater interface to the substrate metal. The life of the component will then almost certainly be determined by the corrosion resistance of the substrate under severe crevice conditions. The initial sign of substrate deterioration is "water marking" on the coating surface followed by blister formation. The blisters then swell as the volume of corrosion product under the coating increases, and ultimately crack, resulting in local coating detachment. Figure 1 shows an area of a plasma-sprayed chro-

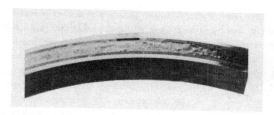

Fig. 1—Blister and crack formation in chromium oxide deposit on stainless steel substrate.

mium oxide coating deposited onto a stainless steel substrate and "water marking," blistering, and cracking can all be seen. Figure 2 shows a section through a chromium oxide coating which has been physically separated from its substrate by a "bridge" of corrosion products. Figure 3 is a section through a blister on a cobalt-chromium bonded tungsten carbide coating on a nickel aluminum bronze substrate showing selective attack of the Kappa phase. The problem of substrate corrosion has been experienced with a number of metals including mild steel, various stainless steels, phosphor bronze, aluminum bronze, and Monel. Paradoxically, it is the high-quality coatings that appear to be the most vulnerable to substrate attack since their finished thickness is usually small, between 50 and 100 μm, and the probability of continuous leak paths is correspondingly greater. Also, although their cumulative porosity is small (4 percent or less) the greater part of the porosity is submicron in size and virtually impossible to seal with resin impregnation. Some commercial organizations do offer an impregnation service but, in the author's experience, the

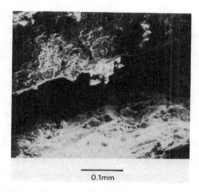

0.1mm

Fig. 2—Chromium oxide layer physically separated from stainless steel substrate by a bridge of corrosion products.

0.1 mm

Fig. 3—Corrosion of kappa phase in nickel aluminum bronze substrate beneath tungsten carbide coating.

depth of penetration achieved is minimal and the sealant is largely, if not totally, removed during subsequent grinding and lapping treatment.

Corrosion Program

In an attempt to establish whether continuous leak paths were an integral, albeit variable, feature of the deposition of thermally sprayed coatings or were perhaps introduced during surface finishing treatments, the following experimental program was carried out. Two types of coating were selected, a ceramic (chromium oxide) and a cermet (cobalt-chromium bonded tungsten carbide). The substrates chosen were mild steel, known to be vulnerable and to give early visual evidence of attack, and alloy 625 (ASTM B446), because of its high resistance to crevice corrosion. The specimens were cylinders of nominal diameter 12.5-mm by 60-mm long, reduced to 6-mm diameter for a 10-mm length at one end. This stub end was not coated. Half of the total specimens from each group were grit blasted before coating and the remainder solvent degreased. Half of the mild steel specimens were resin impregnated immediately after coating, the remainder and the alloy specimens were not. One third of the specimens from each particular category were exposed without any surface finishing and a further third diamond ground to roughness average (Ra) of 0.4/0.5 μm. The remainder were given a sanding treatment resulting in an intermediate surface finish of Ra 0.7/0.8 μm. The final depths of coating varied between 0.14 and 0.20 μm. The specimens were assembled into Perspex boxes and sealed with "O" rings such that a coated axial length of 25 mm was exposed to and aligned normal to circulating seawater. The arrangement is shown schematically in Fig. 4. Separate seawater circulating systems were used for the mild steel and alloy 625 specimens, the former being exposed at room temperature (16 − 22°C), but the alloy specimens were maintained at 30 ± 1°C. At weekly intervals, half the seawater in each system (12 liters) was replenished but at no time were the specimens uncovered. Evaporation losses were made good with deionized water to maintain normal saline concentrations. The electrical resistance R_b, between specimen substrate and a bare alloy 625 electrode in the seawater flow (i.e. between points A and B, Fig. 4) was measured at regular intervals by AC bridge (1 kHz) together with the potential developed between specimen and a silver/silver chloride electrode (Ag/Ag Cl). All measurements were made through the coating.

Mild Steel/Chromium Oxide

A total of 21 mild steel substrates coated with chromium oxide were immersed and within 24 hours 16 were recording R_b values of less than 100 ohms and had taken up potentials, relative to Ag/AgCl, more negative than −0.6 volts. This potential is, of course, the average value of uncoated mild steel in seawater and together with the low resistance path strongly suggests that seawater had completely penetrated the coating and that the substrates were under attack. Measurements of potential of a section of chromium oxide rod, supplied as a consumable for plasma spraying but not from the same source as the coatings above, gave repeatable values to Ag/AgCl in seawater of about −0.4 volts. The R_b

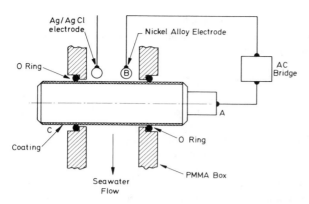

Fig. 4—Diagramatic illustration of spray-coated specimen exposed to circulating seawater.

resistance of 3 of the remaining 5 specimens fell steadily and their potentials moved more negative such that within 12 days exposure, they exhibited similar values to the main group above. Rust spots began to appear on the surface of some of the coatings after 24-hours immersion and the incidence and extent of rusting increased fairly rapidly until, after a few days, most specimens appeared to be affected. It was not clear at this stage whether the surface rust on any particular specimen had originated from its own substrate or had been deposited from other specimens under more advanced attack. Two specimens which had been grit blasted, sprayed, resin impregnated, and left in the "as deposited" condition, exhibited higher resistance readings and less negative potentials throughout the program, lending support to the contention that resin impregnation is simply a surface effect, which if left undisturbed by surface grinding could possibly extend component life. Seven specimens were removed after 28-days exposure and a further 13 after 61-days exposure, including one of the 2 specimens described previously. The sole remaining specimen was left immersed for 363 days during which time its resistance remained largely unchanged at about 1000 ohms and its potential drifted slightly either side of zero volts.

Mild Steel/Tungsten Carbide

Twenty-three specimens were immersed and within 24 hours all had shown R_b values of 12 ohms or less, reflecting the conductivity and continuous network of the cobalt-chromium phase of the coating. All specimen electrode potentials were negative, of which 14 were measured between -0.5 and -0.6 volts and the remaining 9 evenly distributed over the range -0.1 to -0.5 volts. The distinction between the coating and mild steel in terms of electrode potential to Ag/AgCl was not clearly defined since, although average values for both materials were -0.50 and -0.62 volts, respectively, scatter bands were sufficiently wide to give substantial overlap. The individual values continued to move more negative with time and, after 28 days, no less than 19 specimens were recording values more negative than -0.5 volts. The 4 remaining specimens, which had all been grit blasted, resin impregnated, and exposed in the "as deposited" condition, were still exhibiting potentials between -0.1 and -0.25 volts after 61 days. Three specimens were removed after 28 days and the remainder after 61 days,

except for one of the 4 above which was left exposed for 363 days, during which time its potential remained fairly constant at -0.1 volts. The first indications of rusting appeared within 4 hours of exposure and, after 24 hours, only the 4 specimens just mentioned showed no visual sign of attack. It seems likely that attack upon the mild steel substrates of this group of specimens was stimulated by the cathodic nature of the coating. Again, as in the previous category of chromium oxide coated specimens, some measure of protection had been conferred by resin impregnation, but only to those specimens left in the "as deposited" condition.

Alloy 625/Chromium Oxide

Twelve specimens of this type were exposed and within 24 hours the R_b values had fallen to below 600 ohms and 5 were actually less than 50 ohms. All specimen potentials were found to lie within the range 0.15 to -0.20 volts. The specimens have now been exposed for 26 months (April 1981) and during this time very little change has been seen. The resistance values have continued to fall overall and now range from 20 to 200 ohms, but potentials have scarcely altered and all lie within the range 0.12 to -0.18 volts. These values are slightly less negative than the normal value for uncoated alloy of -0.25 volts. There has been no visual evidence of corrosion.

Alloy 625/Tungsten Carbide

Again 12 specimens were exposed and, as expected with the metallic phase of the coating, the resistance R_b was measured at 5 ohms or less in every case. Specimen potentials fell within a very narrow band between -0.25 and -0.27 volts, the value that would be expected from uncoated alloy 625. During the 26 months' exposure, individual resistance measurements have not altered significantly, but there has been a slow and progressive movement of potentials to a less negative state and all potential readings are now located within the range -0.14 and -0.17 volts. No visual evidence of attack to either substrate or coating has been observed.

Metallography

Great care was exercised in the metallographic preparation of the specimens selected for microscopic examination since there is a very real danger of introducing polishing artifacts into the structure of ceramic and cermet coatings. Even so it was extremely difficult to quantify small differences in pore size and distribution, microcracking, and other defects. The 12 chromium oxide/mild steel specimens examined showed continuous or near-continuous bands of corrosion at substrate/coating interfaces, leading, in many cases, to significant coating detachment. Axial and/or radial cracking was observed in 9 of the 12, which in a few cases had resulted in coating loss. Twelve tungsten carbide/mild steel specimens were similarly examined and showed great similarity in the degree of substrate corrosion and coating damage, to that seen in the chromium oxide specimens. The few specimens which showed different electrical resistance and potential values from the rest of their respective groups showed, not unexpectedly, the least corrosion damage to their substrates, but no significant structural differences in these coatings were seen that could have accounted for their improved performance.

Confirmatory Work

In a further series of experiments, a number of mild steel specimens of similar geometry were grit blasted and plasma sprayed with chromium oxide to an average depth of 0.18 mm. Half were left in the "as deposited" condition and the remainder surface ground to R_a 0.4/0.5 μm with a final coating thickness of 0.15 mm. The specimens were assembled into a Perspex box as before and the resistance A to C (Fig. 4) measured, i.e. the resistance through the coating in the dry state. These resistances were generally in excess of 60 K ohms. Also included was an uncoated alloy 625 specimen used as a control. The box was filled with a non-conducting liquid (petroleum ether, boiling range 40°–60°C) and the resistances R_b confirmed as infinity. It was then drained, refilled with ethanol, and the resistances redetermined. Initially, the values ranged from 100 K to 125 K ohms but these fell quickly during the first few hours exposure, generally levelling out between 34 K and 39 K ohms after 142 hours. The values for the alloy control tended to fall within the range obtained from the coated specimens and all can be represented by a single curve. The box was again drained and refilled with boiled out deionized water. The R_b values rose to about 75 K ohms, virtually double the final value in ethanol, and then fell rapidly, but it was now possible to differentiate between the alloy control and the mean values of the coated specimens. Paradoxically, it was the coated specimens which gave the lower readings, levelling out at about 7.75 K ohms after 160 hours compared with the 9.25 K ohms obtained from the control. The reason for this effect remains obscure. The deionized water was then removed and the box refilled with seawater. Predictably, the resistances plummeted but, as expected, the uncoated control now showed the lowest value of 7 ohms, about one quarter of that of the coated specimens. When the seawater was replaced with boiled-out deionized water, the resistances rose immediately to the level recorded at the end of the first deionized water exposure but fell very quickly indeed, the control levelling out at about 850 ohms, some 500 ohms lower than the coated specimens. No further changes occurred and the test was terminated after 160 hours. The tests were carried out under an argon shield to prevent contamination of the deionized water with atmospheric carbon dioxide. All the results are shown graphically in Fig. 5.

The results of the work strongly suggest that continuous leak paths are formed in plasma and other high-quality thermally sprayed coatings during deposition, from interconnected porosity, microcracks, and similar structural defects. Their formation is not significantly influenced by either grit blasting of the substrate or surface grinding of the deposited coating. Resin impregnation appears to be entirely a surface effect and is unlikely to be beneficial unless the coating is left substantially undisturbed.

CARBONS

It is common practice to use a form of carbon-graphite

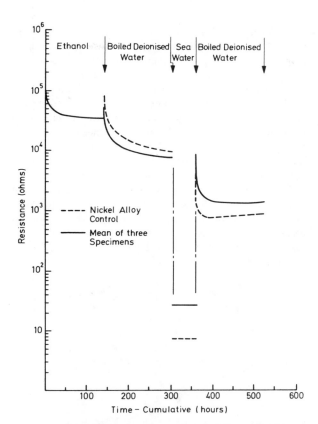

Fig. 5—Resistance vs time curves for chromium oxide coated mild steel and uncoated nickel alloy (ASTM B446) in various liquids under argon shielding.

material for the sealing ring of a mechanical seal and for the carbon to be impregnated during manufacture with metal, polymer, or resin. Impregnation prevents liquid permeation through the bulk of the carbon and can be used selectively to enhance mechanical strength, electrical and thermal conductivity, and reduce friction and wear. Two specific types of carbon graphite were included in the program, a lead-rich white metal impregnated material and four resin-impregnated materials. The specimens were cylindrical in form, counterbored on one face to present an annular rubbing track to the countersurface. They were fitted into alloy 625 holders using epoxy adhesive and both the annular carbon face and the reverse face of the holder were lapped until both faces were flat to within 2 light bands sodium light, and parallel to within 10 μm. Experience had shown that too great a departure from parallelism could significantly affect friction and wear characteristics by increasing hydrodynamic lift.

White Metal Impregnated Carbon-Graphite

This material performed well in short term wear tests of up to 3 days and, when rubbing against cobalt alloys or thermally sprayed coatings, consistently outperformed the resin-impregnated grades in terms of lower friction and wear. In seawater lubricated tests of longer duration (12 days) it was found that the white metal impregnant was attacked and leached out of the bulk material. Figure 6 shows a scanning electron micrograph of the surface of such a specimen in which needle-like crystals, subsequently iden-

Source: Lubrication Engineering, May 1983, 292-299

tified as lead chloride by x-ray analysis, can clearly be seen. Also visible are cracks developing from these corrosion sites. More prolonged exposure is known to result in complete exfoliation of the surface and a number of similarly defective carbon seal rings have been examined following failure of the units in service. Figure 7 shows a microsection through the nose of a large-diameter seal ring where a zone extending for about 1 mm below the surface of the carbon has been denuded of white metal. This condition ultimately leads to partial exfoliation of the surface layer as shown in Fig. 8 and it seems likely that once this has happened, the rate of failure is rapid as erosion becomes the dominant mechanism. The greatest depth of attack was generally found at the loaded face of the nose and, while unloaded surfaces of the carbon exposed to seawater were also corroded, the depth of attack was far less, probably reflecting the lower temperatures of these areas.

Fig. 6—Intergranular cracking and formation of lead chloride crystals in surface of white metal impregnated carbon.

Fig. 7—Showing absence (removal) of white metal impregnant from surface layers of carbon seal ring.

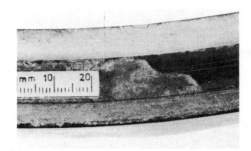

Fig. 8—Showing exfoliation and erosion of surface of white metal impregnated carbon seal ring.

It seems strange that, although this material has been used in mechanical seals in seawater pumps for many years, this type of failure mechanism has not been recorded before. It may be that failures of this type were simply not recognized by the maintenance authorities and were regarded as just another seal failure. Alternatively, since corrosion is time dependent, affected seals may have failed from other causes before the problem just described had developed.

Resin Impregnated Carbon Graphite

Interest in resin impregnated carbons for high-pressure mechanical seals operating in seawater was stimulated when it became apparent that the life of white metal impregnated carbons was limited to 12 months or less. The compatability of various grades of the material was assessed against a variety of countersurfaces including reaction-bonded, hot-pressed and sintered silicon carbides, cobalt alloys, and sprayed coatings. The test conditions attempted included rubbing speeds ranging from 0.1 to 3.5 m/s and pressures of 0.5 to 10 MPa under seawater lubrication. Additionally, some dry rubbing tests were carried out at the lower end of the speed and pressure ranges. Four grades were included in the program, two from each of two manufacturers.

Generally, all four carbons performed well against most of the silicon carbide variants but were very much less effective than the white metal impregnated carbon when run against cobalt alloys and sprayed coatings. The degree of wear suffered by the carbons against silicon carbide was invariably small and it became necessary to improve the accuracy of measurement. A custom-built gauging tool, shown in Fig. 9, was used for the purpose capable of accuracy to ± 2 μm. The gauge has a motorized turntable, two large drum micrometers and two Shneeberger precision bearing carriages. The vertical carriage carries a linear transducer and a spring-loaded probe tipped with a tungsten carbide ball of 0.8-mm diameter. The spring load is small, variable up to 20 gm, to avoid damage to the surfaces of specimens, and the transducer has linearity better than 2 percent over the range ± 500 μm. Vertical control of the probe is attained by the first of the micrometers while the second micrometer controls the movement of the horizontal carriage which, in turn, carries the whole of the vertical assembly. Specimens are located on the turntable and sit

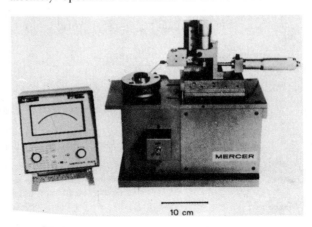

Fig. 9—Gauge for measuring height of carbon wear test specimens.

upon three equispaced tungsten carbide feet which are used as a measuring datum. The datum is used to indicate whether the probe has been displaced since previous measurements and what, if any, correction to apply. The transducer is connected to an analogue meter and the output of the meter is fed to the input of a digital multimeter and printer. Wear measurements have been automated by the addition of an indexing attachment and control circuit which causes the turntable to stop at 8 equispaced positions per complete revolution. Further sets of readings can be taken at differing diameters by repositioning of the probe using the horizontal micrometer. The presence of a datum mark on the specimen holder enables pre- and post-test measurements to be taken in the same places. The measurements are, in fact, deviations from a value obtained by preloading the probe on to the specimen and recording the vertical micrometer reading. This initial value is critical and a large illuminated magnifying screen is used to improve the accuracy of reading the micrometer scale. The value is independently verified before the specimen is tested. It is normal practice in this work to measure wear along 3 diameters of the annular track and to average the 24 individual readings thus obtained. This mean value is then used to calculate specific wear rate thus:

$$K \ mm^3/Nm = \frac{\bar{d}A}{Wl}$$

Where $\bar{d}$ is mean depth of wear (mm)
A is area of annulus (mm^2)
W is applied force (Newtons)
l is rubbed distance (meters)

Dimensional Stability

Post-wear test measurements carried out on certain carbon specimens revealed that a small increase in axial length had occurred and not the expected reduction due to wear. Once the obvious possibility of operator error had been eliminated, a possible explanation of this behavior was thought to be the transfer of material from the silicon carbide countersurface. However, repeated x-ray analysis of the affected carbons surfaces yielded no indication of silicon. Thermal expansion caused by variation in room temperature was also considered and temperature measurements were recorded whenever specimens were measured, but the variations were small and could not have accounted for the length discrepancies observed. Tests were then carried out to assess the dimensional stability of resin impregnated carbons in seawater and it was found that specimens could increase in length by up to 25 μm (0.1 percent of original length) after 7 days immersion. This value was highly significant since the majority of wear measurements recorded in the resin carbon/silicon carbide program were less than 50 μm. A further series of tests was carried out to try and establish the reproducibility of growth and the effect of pre- and post-exposure drying. The results, which are given in Table 1 are remarkable principally for their inconsistency. Some tenuous conclusions may be drawn however:

(a) dimensional changes do occur on exposure to seawater, generally represented by overall growth
(b) growth usually increases with time up to 14 days and possibly beyond
(c) there can be a marked variation between grades of material and even between samples of the same material taken from the same batch
(d) the result of post-exposure drying is unpredictable and specimens may increase or decrease in length
(e) the use of pre- and post-exposure drying treatments cannot be relied upon to eliminate or control the variable of dimensional instability in water.

The variation in growth between different materials or the same material taken from different batches could be explained in terms of different resin impregnants or variations in manufacturing techniques. However, variations in growth between specimens of the same material taken from the same batch can only be explained by the quantity and distribution of impregnant within the specimen, its alignment within the parent block, distance below the impregnation surface, and so on. While the variations in growth appear to be small, they are highly significant in two ways. First, in assessing wear in low-wear situations and second, but probably of far greater importance, in relation to the gaps between the faces in mechanical seals.

With regard to the former, a procedure has been adopted in this laboratory to minimize this source of error. Carbon specimens are mounted, lapped in their holders, and immersed in seawater for 7 days. They are then tissue dried and measured immediately before assembly into the wear test rigs. On completion of the tests, the specimens are immediately rinsed, tissue dried, and remeasured. With regard to the second category, a number of researchers have reported the face separation in correctly operating mechanical seals as ranging from 1 to 5 μm. Clearly any non-uniform swelling around the seal face equal to or exceeding these values could lead to the formation of artificially large separations, with resulting high leakage. Paradoxically the effect would be worst, and perhaps only significant, with low-wear rubbing pairs because it would not then be self-correcting.

Seal failures, defined in terms of unacceptable leakage, have been reported from service in units where a stationary cresylic resin (Material A, Table 1) impregnated carbon has been engaged with a chromium oxide sprayed countersurface. The general appearance of the carbon noses were highly polished except for areas where erosion of the carbon surface had occurred. The mechanism of failure would have been consistent with irregular swelling of the carbon ring, the creation of abnormally large gaps, and very high leakage causing the localized patches of erosion.

CONCLUSIONS

Bearings and mechanical seals operating in seawater are subject to the usual corrosion processes but the situation is complicated when corrosion and wear can proceed simultaneously and often synergistically.

The service life of spray-deposited wear-resistant coatings will more frequently be governed by the crevice corrosion

TABLE 1—Resin Impregnated Carbons. Dimensional Changes Following Immersion in Seawater and Post Exposure Drying

Material and Specimen Reference	Pre-Immersion Drying	Mean Length Before Immersion (mm)	Mean Growth (µm)		Mean Growth After Post-Immersion Drying (µm)				Nett Change After Immersion	Nett Change After Immersion and Drying
			7 Days Immersion	14 Days Immersion	2 Hours @ 120°C Cooled in Desiccator	+ 16 Hours @ 120°C Cooled in Desiccator	+ 3 Days in Desiccator at Room Temperature	+ 3 Days in Laboratory Atmosphere		
A (2a)	None	23.175	3	4	4	11	12	Not Applied	Growth	Growth
B (2b)	None	22.678	20	25	16	− 2	− 3	Not Applied	Growth	Shrinkage
A (3ai)	8 hours @ 120°C Cool in Desiccator	23.670	6	3	Not Applied	1	Not Applied	2	Growth	No Significant Change
A (3aiii)	8 hours @ 120°C Cool in Desiccator	23.190	6	1	Not Applied	15	Not Applied	12	No Significant Change	Growth
B (3aiii)	8 hours @ 120°C Cool in Desiccator	22.674	25	28	Not Applied	12	Not Applied	8	Growth	Growth
B (3aiv)	8 hours @ 120°C Cool in Desiccator	22.457	16	19	Not Applied	15	Not Applied	15	Growth	Growth
A (3bi)	8 hours @ 120°C Cool in Desiccator	15.648	5	9	Not Applied	− 4	Not Applied	Not Applied	Growth	Shrinkage
A (3bii)	8 hours @ 120°C Cool in Desiccator	15.717	9	9	Not Applied	<1	Not Applied	Not Applied	Growth	No Significant Change
B (3biii)	8 hours @ 120°C Cool in Desiccator	15.396	<1	9	Not Applied	− 7	Not Applied	Not Applied	Growth	Shrinkage
B (3biv)	8 hours @ 120°C Cool in Desiccator	15.557	− 2	4	Not Applied	− 14	Not Applied	Not Applied	Growth	Shrinkage

All growth values are relative to initial lengths before exposure

resistance of individual substrates rather than wear and will be related to the available leak paths through the coatings. Leak paths are formed as a variable function of the deposition process and do not appear to be significantly influenced by either grit blasting of the substrate or surface grinding of the coatings. Resin impregnation is largely a surface effect and unlikely to be beneficial unless the coating is employed in the "as deposited" condition—scarcely the preferred surface finish for either bearings or seals. For critical applications, a substrate of proven crevice corrosion resistance, such as alloy 625, would be cost effective.

White metal impregnated carbons are susceptible to failure by corrosion of the white metal and are generally unsuitable for seawater systems.

Resin impregnated carbons are dimensionally unstable in seawater and irregular swelling can occur within a single component. The total movement may be small but is significant in relation to face separations in correctly operating mechanical seals. It is worse in low-wear rubbing pairs, such as resin impregnated carbon/silicon carbide, where there is little chance of self-correction.

ACKNOWLEDGMENT

The assistance of Graham Green in carrying out much of the experimental work and of J. Manson, J. Cooper, and J. Jones in conducting scanning electron microscope surveys is most gratefully acknowledged. The author would also like to record his appreciation to J. Rowlands and Ann Edwards for advice and assistance in obtaining and assessing corrosion data.

DISCUSSION

IVARS FREIMANIS
Crane Packing Company
Morton Grove, Illinois 60053

The author, in his paper, has offered an enlightening discussion on several of the various available mechanical seal face material combination performance characteristics under seawater service conditions. The test results presented largely parallel the collective experience at the discusser's company.

The dissolution of cobalt binder from tungsten carbide materials, either coatings or solid parts, has been noted on numerous occasions even in demineralized water applications. However, when solid tungsten carbide face materials are selected, their inherently greater carbide-to-carbide bond strength and lower total cobalt content, allows them to operate even with considerable surface cobalt dissolution.

The presented evidence that "white metal"-impregnated carbon graphite, believed to be one of the commercial babbitt bearing alloys, is visibly attacked and leached after 3 to 12 days exposure to seawater is unexpected. The metal's chemical composition here may be a critical factor and possibly a major portion of the leaching in the seal contact area could be explained by metal flow and pull-out due to frictional heat.

Resin-impregnated carbon graphite material dimensional changes resulting from exposure to moisture and storage temperature variations has been noted to contribute to mechanical seal face leakage—especially in applications like compressors and deep wells where refrigerant fluid or oil leakage is closely scrutinized. Among the numerous resin-type carbon graphites available, some do provide greater stability but, unfortunately, their proper selection requires experimentation.

No mention is made of buffer layers underneath the coatings to protect substrate corrosion. This possibility might be explored for steel substrate considerations.

AUTHOR'S CLOSURE

I would like to thank the discusser for his interesting comments.

1. While I would agree with his remarks regarding the dissolution of cobalt from both solid and spray-coated tungsten carbide alloys, this feature was outside the scope of the paper.
2. It is quite possible that some feature specific to the test procedure was responsible for the early attack on the metal phase of the white metal-impregnated carbon specimen. Certainly similar material in the form of large-diameter sealing rings has performed at sea for periods of up to 12 months before failing in the manner described.
3. Experience of buffer or intermediate layers beneath wear resistant coatings has been disappointing. Preliminary layers of electroless nickel or ion-plated chromium, gold or platinum have been deposited onto stainless steel and nickel aluminum bronze substrates before the deposition of chromium oxide or tungsten carbide. In every case, the coatings exfoliated in seawater following substrate corrosion; the intermediate coatings were themselves porous to some degree.

Materials for Sliding Bearings

Condensed from Metals Handbook, Ninth Edition, Volume 3, pages 802 to 822.

A SLIDING BEARING (plain bearing) is a machine element designed to transmit loads or reaction forces to a shaft that rotates relative to the bearing. Journal bearings are cylindrical (full cylinders or segments of cylinders) and are used when the load or reaction force is essentially radial — that is, perpendicular to the axis of the shaft. Thrust bearings are ring-shape bearings (full rings or segments of rings) and are used when the load or reaction force is parallel to the direction of the shaft axis. Both radial and axial loads can be accommodated by flange bearings, which are journal bearings constructed with one or two integral thrust-bearing surfaces. The sliding movement of the shaft surface or thrust-collar surface relative to the bearing surface is characteristic of all plain bearings. In many applications, plain bearings offer advantages over rolling-contact bearings — advantages such as lower cost, smaller space requirements, ability to operate with marginal lubricants, resistance to corrosion and ability to sustain high specific loads.

CLASSIFICATIONS

Functions and Applications. When considered in terms of function or location in a machine, bearings are commonly called by the following names, which describe their use:

- Connecting-rod bearing
- Main bearing
- Piston-pin bushing
- Camshaft bearing or bushing
- Crankshaft thrust washer
- Leaf-spring bushing
- Front-axle bushing
- Idler-gear bushing
- Brake-pedal shaft bushing
- Rocker-arm bushing
- Transmission-gear bushing
- Reverse-pinion bushing
- Differential-trunnion bushing
- Intermediate gear thrust washer
- Countershaft thrust washer
- Electric-motor shaft bushing.

The terms "bearing" and "bushing" are used interchangeably, and do not have meanings that are significantly different in terms of function or location in a machine.

A major distinction can be drawn, however, between (a) connecting-rod and main bearings and (b) the remaining items in the list. In a reciprocating engine the connecting-rod and main bearings must support the entire firing load on the pistons, which is transmitted through the connecting rods to the crankshaft. These bearings therefore are subject to exceptionally severe cyclic loading, shaft-deflection effects, and variations in lubricant-film thickness. The conditions imposed on the other bearings listed differ mainly in type of load (steady or intermittent), magnitude of load, speed of operation, direction of operation (unidirectional or oscillating) and operating temperature.

Structural Characteristics. Bearings are also frequently classified according to material construction as solid (single-metal), bimetal (two-layer) or trimetal (three-layer) bearings.

In current practice, connecting-rod and main bearings of reciprocating engines are usually of either bimetal or trimetal construction, but single-metal bearings are employed occasionally. Bearings for other machinery applications are most often of single-metal or bimetal construction, although trimetal construction is sometimes required.

Size, Configuration and Manufacturing Method. With respect to size, bearings commonly are classified as either thin-wall or heavy-wall bearings; in general, bearings greater than about 5 mm (0.2 in.) in wall thickness and not less than 150 mm (6 in.) in diameter are considered heavy-wall bearings. Configuration may be further described as half round, full round, flanged or washer. SAE standards classify bearings used in mass-produced machinery (virtually all thin-wall bearings) into three groups: sleeve-type half bearings, split-type bushings and thrust washers.

OPERATING CONDITIONS

It is necessary to analyze both the mechanical design and the material requirements of a bearing in terms of the system in which it will be used. The important mechanical components of a system include the bearing housing, the lubricant film, the surface and subsurface of the bearing itself, and the surface of the journal or thrust collar. The interactions among these components in an operating machine can be exceedingly complex, and they are not subject to rigorous analysis on the basis of any established general theory.

However, a substantial body of empirical knowledge and a growing fund of theoretical understanding exist, permitting engineering decisions to be made on questions of bearing design and material selection with a minimum of trial-and-error testing.

The technical factors that are most important from the standpoint of material selection are discussed briefly below.

Loads and Speeds. The approximate magnitude of the specific load to which a bearing will be subjected should be known, so that materials of insufficient strength can be eliminated from consideration. However, precise values of actual bearing unit loads in operating machinery are not easily obtained. The most common approximation is the mean unit load, P, which is calculated from the equation

$$P = \frac{F}{L \times D} \qquad \text{(Eq 1)}$$

where F is maximum total force acting on the bearing, L is axial bearing length and D is bearing diameter.

Total force usually can be estimated with reasonable accuracy from known machine design parameters, and sometimes can be determined experimentally. Values of P for bearings in commercial machinery vary from nearly zero to 105 MPa (15 ksi). Where the loading on a bearing is steady or varies in a simple pattern, the P value is a satisfactory reference for selecting bearing materials of adequate strength.

Lubrication and Friction. In order to minimize the friction associated with the sliding movement of plain bearings, a fluid lubricant is almost invariably used. Oils and greases are the most commonly used lubricants; but other fluids, including hydraulic fluids, water and even air, may be used in special applications.

Every effort should be made to design any bearing so that a continuous film of lubricant will be maintained between the sliding surfaces during operation.

Heat and Temperature. Even in the absence of metallic contact, generation of some frictional heat is unavoidable in machinery bearings. In any heat

engine, additional heating occurs by conduction from the working fluid. Artificial cooling of bearings generally is accomplished by use of a heat exchanger installed in the lubricant system, by means of which excessively high oil and bearing temperatures can be avoided. In machinery other than heat engines, it is not usually necessary to employ artificial lubricant cooling.

In general, successful lubrication and bearing operation require that the inlet lubricant temperature be maintained below about 135 °C (275 °F) if the lubricant is a hydrocarbon oil. Under these conditions, bearing-back temperatures of 175 °C (350 °F) or less usually can be maintained. Temperatures as high as 260 °C (500 °F) can be tolerated by some synthetic lubricating fluids, with correspondingly higher bearing-back temperatures. Actual surface temperatures of bearing liners in operation are rarely known, but must be assumed to be higher than corresponding bearing-back temperatures.

Corrosion. Except in pumps that handle corrosive fluids, plain bearings usually are not required to operate in extremely hostile chemical environments. It is possible, however, for corrosive problems to develop in lubricating oils as a result of oxidation and/or by reaction with engine coolants and combustion products. Fatty acids, alcohols, aldehydes and ketones formed in this way can cause corrosion of bearing metals. Acidic sulfur compounds in lubricating oils, which may be initially present or which may result from oxidation or from contamination by combustion products, also can act as corrodents.

PROPERTIES OF BEARING MATERIALS

Surface and Bulk Properties. The nature of the conditions under which plain bearings must operate and the wide ranges over which these conditions can vary lead to concern for bearing-material properties of two kinds: (a) surface properties (those associated with the bearing surface and immediate subsurface layers) and (b) bulk properties.

Conventional engineering definitions of material properties are not sufficient to characterize all the essential attributes of bearing materials. Although there is no universally accepted system of nomenclature, measurement or testing for these properties, they can be defined and studied in terms of the following characteristics:

- *Compatibility:* the antiwelding and antiscoring characteristics of a bearing material when operated with a given mating material
- *Conformability:* the ability of a material to yield to and compensate for slight misalignment and to conform to variations in the shape of the shaft or of the bearing-housing bore
- *Embeddability:* the ability of a material to embed dirt or foreign particles and thus prevent them from scoring and wearing shaft and bearing surfaces
- *Load capacity:* the maximum unit pressure under which a material can operate without excessive friction, wear and fatigue damage
- *Fatigue strength:* the ability of a material to function under cyclic loading below its elastic limit without developing cracks or surface pits
- *Corrosion resistance:* the ability of a material to withstand chemical attack by uninhibited or contaminated lubricating oils
- *Hardness:* the ability to resist plastic flow under high unit compressive loads, convention-

ally measured by indentation hardness testing
- *Strength:* the ability to resist elastic and plastic deformation under load, conventionally measured by compression, shear and tensile testing.

Compatibility can be regarded as a purely surface characteristic. Conformability and embeddability involve the surface and immediate subsurface, but are strongly related to the bulk properties of strength and hardness. The other characteristics relate principally to bulk properties.

Measurement and Testing. Of all the characteristics in the list above, only hardness and strength can be measured satisfactorily by standard laboratory test methods. Many special dynamic test rigs and test methods have been developed in the plain-bearing industry to evaluate and measure the other characteristics and their interactions. Although much useful information has been developed through laboratory rig testing, it is still often necessary to test bearing materials and designs in full-size operating machines in order to clearly establish their over-all suitability. Such testing is necessarily expensive and time-consuming. It should be undertaken only after careful study of the conditions under which the bearing will operate and of prior experience in similar applications.

Bearing-Material Structures. All commercially significant bearing metals except silver are used as polyphase alloys. Typical microstructures are shown beginning on page 232 in Volume 7 of the 8th Edition of this Handbook. Four general microstructural types can be recognized:

- *Soft Matrix With Discrete Hard Particles.* Lead babbitts and tin babbitts are of this type.
- *Interlocked Soft and Hard Continuous Phases.* Many copper-lead alloys are of this type.
- *Strong Matrix With Discrete Soft-Phase Pockets.* Leaded bronzes, aluminum-tin alloys and aluminum-lead alloys are of this type.

Laminated Construction. One of the most useful concepts in bearing-material design came with the recognition in 1941 that the effective load capacities and fatigue strengths of lead and tin alloys were sharply increased when these alloys were used as thin layers intimately bonded to strong bearing backs of bronze or steel. Use is made of this principle (Fig. 1) in both two-layer and three-layer constructions, in which the surface layer is composed of a lead or tin alloy, usually no more than 0.13 mm (0.005 in.) thick.

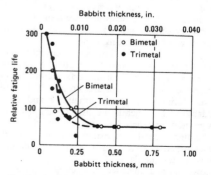

Bearing load: 14 MPa (2000 psi) for all tests.
Fig. 1. Variation of bearing life with babbitt thickness for lead or tin babbitt bearings

Source: Metals Handbook Desk Edition, 1985, 20.40-20.48

Unimpaired compatibility is provided by such a layer, together with reasonably high levels of conformability and embeddability. Other useful compromises can be effected between surface and bulk properties by employing an intermediate copper alloy or aluminum alloy layer between the surface alloy layer and a steel back. In these three-layer constructions, use of surface-layer thicknesses as low as 0.013 mm (0.0005 in.) offers even more favorable compromises between surface and bulk properties than are possible with two-layer constructions.

BEARING-MATERIAL SYSTEMS

Because of the widely varying conditions under which bearings must operate, commercial bearing materials have evolved as specialized engineering materials systems rather than as commodity products. They are used in relatively small tonnages and are produced by a relatively small number of manufacturers. Much proprietary technology is involved in alloy formulation and processing methods. Successful selection of a bearing material for a specific application often requires close technical cooperation between the user and the bearing producer.

Single-Metal Systems

Virtually all single-metal sliding bearings are made of either copper alloys or aluminum alloys. Considerable ranges of compositions and properties are available in the older copper group.

Single-metal systems do not exhibit outstandingly good surface properties, and their tolerance of boundary and thin-film lubrication conditions is limited. As a result, the load-capacity rating for a single-metal bearing usually is low relative to the fatigue strength of the material from which it was made. Because of their metallurgical simplicity, these materials are well suited to small-lot manufacturing from cast tubes or bars, using conventional machine-shop processes.

Copper Alloys. Except for commercial bronze and low-lead tin bronze, copper alloys in single-metal systems are almost always used in cast form. This provides thick bearing walls (3.20 mm, or 0.125 in.) strong enough so that the bearing is retained in place when press fitted into the housing.

Commercial bronze and low-lead tin bronze (alloys C22000 and C47600) are used extensively in the form of wrought strip for thin-wall bushings, which are made in large volumes by high-speed press forming. The poor compatibility of these alloys can be improved by embedding a graphite-resin paste in rolled or pressed-in indentations, so that the running surface of the bushing consists of interspersed areas of graphite and bronze. Such bushings are widely used in automotive engine starting motors.

The lead in leaded tin bronzes is present in the form of free lead, dispersed throughout a copper-tin matrix so that the bearing surface consists of interspersed areas of lead and bronze. This improves compatibility, conformability and embeddability. In general, the best selection from this group of materials for a given application will be the highest-lead composition that can be used without risking excessive wear, plastic deformation or fatigue damage.

Low-lead and unleaded bronzes are also used in porous bushings produced by powder metallurgy methods. The sintered bushings are impregnated with oil, which provides a built-in supply of lubricant. Such bushings are widely used in applications involving light loads and requiring self-lubricating properties.

Aluminum Alloys. Virtually all solid aluminum bearings used in the United States are made from alloys containing from 5.5 to 7% tin, plus smaller amounts of copper, nickel, silicon and magnesium. Starting forms for bearing fabrication include cast and wrought tubes as well as rolled plate and strip, which can be press formed into half-round shapes. As in the case of solid bronze bearings, relatively thick bearing walls are employed.

The tin in these alloys is present in the form of free tin, dispersed throughout an aluminum matrix so that the bearing surface consists of interspersed areas of aluminum and tin. Surface properties are enhanced by the free tin in much the same way that those of bronze are improved by the presence of free lead.

The high thermal expansion of aluminum poses special problems in maintaining press-fit and running clearances. Various methods are employed for increasing yield strength, through heat treatment and cold work, to overcome plastic flow and permanent deformation under service temperatures and loads.

Bimetal Systems

All bimetal systems employ a strong bearing back to which a softer, weaker, relatively thin layer of a bearing alloy is metallurgically bonded. Low-carbon steel is by far the most widely used bearing-back material, although alloy steels, bronzes, brasses and (to a limited extent) aluminum alloys are also used. When steel bearing backs are employed, load-capacity ratings for both copper and aluminum alloys are sharply increased above those of the corresponding single metals, without degrading any other properties.

Bronze-back bearings do not exhibit combinations of performance characteristics substantially different from those of steel-back bearings. The practical advantages of bronze as a bearing-back material lie partly in the economics of small-lot manufacturing and partly in the relative ease with which worn bronze-back bearings can be salvaged by rebabbitting and remachining. From the standpoint of performance, the advantage of bronze over steel as a bearing-back material is the protection bronze affords against catastrophic bearing seizure in case of severe liner wear or fatigue.

Although the surface properties of bronze bearing-back materials are not impressive, they are superior to those of steel, and these "reserve" bearing properties can be of considerable practical importance in large, expensive machinery used in certain critical applications.

Trimetal Systems

All trimetal systems employ a steel bearing back, an intermediate layer of relatively high fatigue strength, and a tin alloy or lead alloy surface layer. Most trimetal systems are derived from steel-back bimetal systems by addition of a surface layer.

The strengthening effects of thin-layer construction are notable in those systems that incorporate electroplated lead alloy surface layers approximately 0.025 mm (0.001 in.) thick. Comparison of fatigue-strength and load-capacity ratings of these systems with those of corresponding bimetal systems shows that the thin lead alloy surface layer upgrades not only surface properties but also fatigue strength. The increase in fatigue strength can be attributed at least in part to the elimination of surface stress raisers, from which fatigue cracks can propagate.

Trimetal systems with electroplated lead-base surface layers and copper or aluminum alloy intermediate layers provide the best available combinations of cost, fatigue strength and surface properties. Such bearings have high tolerances for boundary and thin-film lubrication conditions, and thus can be used under higher loads than can any of the bimetal systems.

TIN ALLOYS

Tin-base bearing materials (babbitts) are alloys of tin, antimony and copper that contain limited amounts of zinc, aluminum, arsenic, bismuth and iron. The compositions of tin-base bearing alloys, according to ASTM B23 and SAE specifications, are given in Table 1.

The presence of zinc in these bearing metals generally is not favored. Arsenic increases resistance to deformation at all temperatures; zinc has a similar effect at 38 °C (100 °F), but causes little or no change at room temperature. Zinc has a marked effect on the microstructures of some of these alloys. Small quantities of aluminum (even less than 1%) will modify their microstructures. Bismuth is objectionable because, in combination with tin, it forms a eutectic that melts at 137 °C (279 °F). At temperatures above this eutectic, alloy strength is decreased appreciably.

In high-tin alloys, such as ASTM grades 1, 2 and 3, and SAE 11 and 12, lead content is limited to 0.50% or less because of the deleterious effect of higher percentages on the strength of these alloys at temperatures of 149 °C (300 °F) or above. Lead and tin form a eutectic that melts at 183 °C (361 °F). At higher temperatures, bearings become fragile as a result of formation of a liquid phase within them. Mechanical properties of ASTM grades 1 to 3 are given in Table 2.

Table 1. Compositions of tin-base bearing alloys

Designation	Sn(a)	Sb	Pb (max)	Cu	Fe (max)	As (max)	Bi (max)	Zn (max)	Al (max)	Others (max total)
ASTM B23 alloys										
Alloy 1	91.0	4.5	0.35	4.5	0.08	0.10	0.08	0.005	0.005	...
Alloy 2	89.0	7.5	0.35	3.5	0.08	0.10	0.08	0.005	0.005	...
Alloy 3	84.0	8.0	0.35	8.0	0.08	0.10	0.08	0.005	0.005	...
Alloy 11 ...	87.5	6.8	0.50	5.8	0.08	0.10	0.08	0.005	0.005	...
SAE alloys										
SAE 11 ...	86.0	6.0–7.5	0.50	5.0–6.5	0.08	0.10	0.08	0.005	0.005	0.20
SAE 12 ...	88.0	7.0–8.0	0.50	3.0–4.0	0.08	0.10	0.08	0.005	0.005	0.20

(a) Desired in ASTM alloys; specified minimum in SAE alloys.

Table 2. Properties of selected ASTM B23 tin-base bearing alloys

Designation	Specific gravity	Compressive yield strength(a)(b) At 20 °C (68 °F) MPa	ksi	At 100 °C (212 °F) MPa	ksi	Compressive ultimate strength(a)(c) At 20 °C (68 °F) MPa	ksi	At 100 °C (212 °F) MPa	ksi	Hardness(d), HB At 20 °C	At 100 °C	Solidus temperature °C	°F	Liquidus temperature °C	°F	Pouring temperature °C	°F
Alloy 1	7.34	30.3	4.40	18.3	2.65	88.6	12.85	47.9	6.95	17.0	8.0	223	433	371	700	440	825
Alloy 2	7.39	42.1	6.10	20.7	3.00	102.7	14.90	60.0	8.70	24.5	12.0	241	466	354	669	425	795
Alloy 3	7.46	45.5	6.60	21.7	3.15	121.3	17.60	68.3	9.90	27.0	14.5	240	464	422	792	490	915

(a) The compression-test specimens were cylinders 1 1/2 in. long and 1/2 in. in diameter, machined from chill castings 2 in. long and 3/4 in. in diameter. (b) Values for yield point were taken from stress-strain curves at a deformation of 0.125% reduction of gage length. (c) Values for ultimate strength were taken as the unit load necessary to produce a deformation of 25% of the length of the specimen. (d) Tests were made on the bottom face of parallel machined specimens cast at room temperature in a steel mold 2 in. in diameter by 5/8 in. deep. The Brinell hardness values listed are the averages of three impressions on each alloy, using a 10-mm ball and applying a 500-kg load for 30 s.

The mechanical-property values obtained from massive cast specimens are dependent on temperature. Hardness and compression tests are sensitive also to duration of the load because of the plastic nature of these materials. Bulk properties may be of some value in initial screening of materials, but they do not accurately predict behavior of the material in the form of a thin layer bonded to a strong backing, which is the manner in which the babbitts are normally used. The relationship that exists between bearing life and thickness of babbitt is shown in Fig. 1, and the marked influence of operating temperature is shown in Fig. 2.

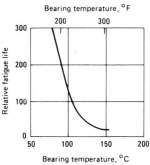

Fig. 2. Variation of bearing life with temperature for SAE 12 bimetal bearings

SAE 12 alloy lining, 0.05 to 0.13 mm (0.002 to 0.005 in.) thick, on steel backing. Bearing load: 14 MPa (2000 psi).

Compared with other bearing materials, tin alloys have low resistance to fatigue, but their strength is sufficient to warrant their use under low-load conditions. These alloys are commercially easy to bond and handle, and they have excellent antiseizure qualities. Furthermore, they are much more resistant to corrosion than lead-base bearing alloys.

LEAD ALLOYS

There are two types of lead babbitts: (a) alloys of lead, tin, antimony and in many instances arsenic; and (b) alloys of lead, calcium, tin and one or more of the alkaline earth metals. Many alloys

Table 3. Nominal compositions of lead-base bearing alloys

Designation	Pb	Sb	Sn	Cu (max)	Fe (max)	As	Bi (max)	Zn (max)	Al (max)	Cd (max)	Others
ASTM B23 alloys											
Alloy 7 (a) ......Rem		15.0	10.0	0.50	0.1	0.45	0.10	0.005	0.005	0.05	...
Alloy 8 ......Rem		15.0	5.0	0.50	0.1	0.45	0.10	0.005	0.005	0.05	...
Alloy 13 (b) .....Rem		10.0	6.0	0.50	0.1	0.25(a)	0.10	0.005	0.005	0.05	...
Alloy 15 (c) .....Rem		16.0	1.0	0.50	0.1	1.10	0.10	0.005	0.005	0.05	...
Other alloys											
SAE 16 ......Rem		3.5	4.5	0.10	...	0.05(a)	0.10	0.005	0.005	0.005	...
AAR M501 (d) ..Rem		8.75	3.5	0.50	...	0.20(a)	...	...	...	...	...
SAE 19 ......Rem		...	10.0	...	...	...	...	...	...	...	...
SAE 190 .....Rem		...	7.0	3.0	...	...	...	...	...	...	...
Proprietary alloys											
A 95.65		...	3.35	0.08	...	...	...	...	...	...	0.67 Ca
B 83.30		12.54	0.84	0.10	...	3.05	...	...	...	...	...
C ....Rem		10.0	3.0	0.20	...	...	...	...	...	...	2.0 Ag

(a) Also SAE 14. (b) Also SAE 13. (c) Also SAE 5. (d) Association of American Railroads, Specification M501; also ASTM B67.

of the first group have been used for centuries as type metals, and were probably employed as bearing materials because of the properties they were known to possess. The advantages of arsenic additions in this type of bearing alloy have been generally recognized since 1938. Alloys of the second type were developed early in the 20th century.

Nominal compositions of lead-base bearing alloys covered by ASTM specifications are listed in Table 3, along with compositions of other proprietary alloys. Additional information on the mechanical properties of some of these alloys is given in Table 4.

In the absence of arsenic, the microstructures of these alloys comprise cuboid primary crystals of SbSn, or of antimony embedded in a ternary mixture of Pb-Sb-SbSn in which lead forms the matrix. The number of these cuboids per unit volume of alloy increases as antimony content increases. If antimony content is more than about 15%, the total amount of the hard constituents increases to such an extent that the alloys become too brittle to be useful as bearing materials.

Arsenic is added to lead babbitts to improve their mechanical properties, particularly at elevated temperatures. All lead babbitts are subject to softening or loss of strength during prolonged exposure to the temperatures (95 to 150 °C; 200 to 300 °F) at which they serve as bearings in internal-combustion engines. Addition of arsenic minimizes such softening. Under suitable casting conditions, the arsenical lead babbitts—for example, SAE 15 (ASTM grade 15)—develop remarkably fine and uniform structures. They also have better fatigue strength than arsenic-free alloys.

Arsenical babbitts give satisfactory service in many applications. Use of these alloys increased greatly during the second world war, particularly in the automobile industry and in diesel engines. The alloy most widely used is SAE 15 (ASTM grade 15), which contains 1% arsenic. Automobile bearings of this alloy usually are made from continuously cast bimetal (steel/babbitt) strip. When properly handled, this alloy can withstand the considerable strain that results from forming the bimetal strip into bearings.

Diesel-engine bearings often are cast as individual bearing shells by either centrifugal or gravity methods. In applications where higher hardness is required and where formability requirements are less severe (rolling-mill bearings, for instance), an alloy that contains 3% arsenic

Table 4. Properties of selected ASTM B23 lead-base bearing alloys

Designation	Specific gravity	Compressive yield strength(a)(b) At 20 °C MPa	ksi	At 100 °C MPa	ksi	Compressive ultimate strength(a)(c) At 20 °C MPa	ksi	At 100 °C MPa	ksi	Hardness(d), HB At 20 °C	At 100 °C	Solidus temperature °C	°F	Liquidus temperature °C	°F	Pouring temperature °C	°F
Alloy 7	9.73	24.5	3.55	11.0	1.60	107.9	15.65	42.4	6.15	22.5	10.5	240	464	268	514	338	640
Alloy 8	10.04	23.4	3.40	12.1	1.75	107.6	15.60	42.4	6.15	20.0	9.5	237	459	272	522	340	645
Alloy 15	10.05	...	...	...	...	...	...	...	...	21.0	13.0	248	479	281	538	350	662

(a) The compression-test specimens were cylinders 1.5 in. long, 0.5 in. in diameter, machined from chill castings 2 in. long, 0.75 in. in diameter. (b) Values were taken from stress-strain curves at a deformation of 0.125% reduction of gage length. (c) Values were taken as the unit load necessary to produce a deformation of 25% of the length of the specimen. (d) Tests were made on the bottom face of parallel-machined specimens that had been cast at room temperature in a steel mold, 2 in. in diameter by 0.625 in. deep. Values listed are the averages of three impressions on each alloy, using a 10-mm ball and applying a 500-kg load for 30 s.

has been used successfully (alloy B in Table 3).

For many years, lead-base bearing alloys were considered to be only low-cost substitutes for tin alloys. However, the two groups of alloys do not differ greatly in antiseizure characteristics, and when lead-base alloys are used with steel backs and in thicknesses below 0.75 mm (0.03 in.), they have fatigue resistance that is equal to, if not better than, that of tin alloys. Bearings of any of these alloys remain serviceable longest when they are no more than 0.13 mm (0.005 in.) thick (see Fig. 1). The superiority of lead alloys over tin alloys becomes more marked as operating temperature increases. For this reason, automotive engineers generally favor lead-base alloys of compositions that approximate ASTM alloys 7 and 15 and SAE alloy 16. SAE alloy 16 is cast into and on a porous sintered matrix, usually copper-nickel, bonded to steel. The surface layer of babbitt is 0.025 to 0.13 mm (0.001 to 0.005 in.) thick.

The fatigue resistance of bearing materials depends to a great extent on the design of the bearing. The strength and rigidity of the supporting structure, the thickness of the backing metal (steel or bronze), the thickness of the bearing material and the character of the bond between the bearing material and the backing are all factors of consequence in bearings for use in high-speed reciprocating engines, such as the main and connecting-rod bearings of automobile and aircraft engines.

Resistance to fatigue is somewhat less important in bearings that operate under static load—for example, journal bearings in traction-motor supports for diesel locomotives and in railway freight cars. In such bearings, antiseizure characteristics, conformability, compressive strength, and resistance to abrasion and corrosion are of greater significance. The lining metal generally employed in such journal bearings is the low-arsenic AAR alloy (ASTM B67) cast onto a leaded bronze back.

Pouring temperature and rate of cooling markedly influence the microstructures and properties of lead alloys, particularly when they are used in the form of heavy liners for railway journals. High pouring temperatures and low cooling rates, such as result from use of overly hot mandrels, promote segregation and formation of a coarse structure. A coarse structure may cause brittleness, low compressive strength and low hardness. Therefore, low pouring temperatures (325 to 345 °C; 620 to 650 °F) usually are recommended. Because these alloys remain relatively fluid almost to the point of complete solidification (about 240 °C, or 465 °F, for most compositions), they are easy to manipulate and can be handled with no great loss of metal from drossing.

Use of lead babbitts containing calcium and alkaline earth metals is confined almost entirely to railway applications, although these babbitts also are employed to some extent in certain diesel-engine bearings. One of the more widely used alloys contains 1.0 to 1.5% tin, 0.50 to 0.75% calcium, and small amounts of various other elements. The strength of this alloy approximates that of a tin alloy containing 90% Sn, 8% Sb and 2% Cu. Hardness of this lead alloy is about 20 HB, and the solidus is 321 °C (610 °F). The liquidus is probably near 338 °C (640 °F). The pouring temperature, which varies from 500 to

Table 5. Designations and nominal compositions of copper-base bearing alloys

UNS number	SAE	Other	Former SAE	Cu	Sn	Pb	Zn	Other	Form	Use
Commercial bronze										
C22000	795	...	...	90	0.5	...	9.5	...	Wrought strip	Solid bronze bushings and washers
Unleaded tin bronzes										
C90300	C90300	...	620	88	8	0	4	...	Cast tubes	Solid bronze bearings
C90500	C90500	...	62	88	10	0	2	...	Cast tubes	Solid bronze bearings
C91100	...	...	...	84	16	0	0	...	Cast tubes	Solid bronze bearings
C91300	...	...	...	81	19	0	0	...	Cast tubes	Solid bronze bearings
Low-lead tin bronzes										
C92200	C92200	...	622	88.5	6	1.5	4	...	Cast tubes	Solid bronze bearings
C92300	C92300	...	621	87.0	8.5	0.5	4	...	Cast tubes	Solid bronze bearings
C92700	C92700	...	63	87.5	10	2	0.5	...	Cast tubes	Solid bronze bearings
Medium-lead tin bronzes										
C54400	791	...	...	88	4	4	4	...	Wrought strip	Solid bronze bushings and washers
...	...	F32/62	...	87	4	4	3	2 Fe	Cast on steel back	Bimetal bushings and washers; trimetal intermediate layer
C83600	C83600	...	40	85	5	5	5	...	Cast tubes	Solid bronze bearings and bronze bearing backs
C93200	C93200	...	660	83	7	7	3	...	Cast tubes	Solid bronze bearings
C93600	793	...	...	85	4	8	3	...	Cast on steel back	Bimetal surface layer
...	798	...	...	88	4	8	...	...	Sintered on steel back	Bimetal surface layer
C93700	C93700	...	64	80	10	10	...	...	Cast tubes	Solid bronze bearings and bronze bearing backs
...	792	...	...	80	10	10	...	...	Cast on steel back	Bimetal surface layer and trimetal intermediate layer
...	797	...	...	80	10	10	...	...	Sintered on steel back	Bimetal surface layer
High-lead tin bronze										
C93800	C93800	...	67	78	6	16	...	...	Cast tubes	Solid bronze bearings and bronze bearing backs
...	...	AMS 4825	...	74	10	16	...	...	Cast on steel back	Bimetal surface layer
...	794	...	...	71.5	3.5	23	2	...	Cast on steel back	Bimetal surface layer
...	799	...	...	74	3	23	...	...	Sintered on steel back	Bimetal surface layer
...	...	AMS 4824	...	75	1	24	...	...	Cast on steel back	Trimetal intermediate layer
...	...	F780	...	74	2.5	23.5	...	...	Sintered on steel back	Trimetal intermediate layer

Table 5 (continued)

UNS number	SAE	Other	Former SAE	Cu	Sn	Pb	Zn	Other	Form	Use
					Nominal composition, %					
High-lead tin bronze (continued)										
...	...	F15/112	...	72	3	25	...	...	Cast on steel back	Bimetal surface layer and trimetal intermediate layer
...	...	AMS 4840	...	70	5	25	...	...	Cast tubes	Solid bronze bearings
Copper-lead alloys										
...	49	...	...	75.5	0.5	24	...	...	Cast on steel back	Trimetal intermediate layer
...	49	...	...	75.5	0.5	24	...	...	Sintered on steel back	Trimetal intermediate layer
...	48	...	...	70	...	28	...	1.5 Ag	Cast on steel back	Bimetal surface layer and trimetal intermediate layer
...	482	...	...	67	5	28	...	...	Sintered on steel impregnated with Pb-Sn	Bimetal surface layer
...	480	...	...	65	...	35	...	...	Cast on steel back	Bimetal surface layer
...	480	...	...	65	...	35	...	...	Sintered on steel back	Bimetal surface layer
...	481	...	...	55	0.25	40	...	5 Ag	Cast on steel back	Bimetal surface layer
...	484	...	...	55	3	42	...	(a)	Sintered on steel back, infiltrated with Pb	Bimetal surface layer
...	485	...	...	48	1	51	...	(a)	Sintered on steel back, infiltrated with Pb	Bimetal surface layer
				98	2	...	...	(b)	Sintered on steel back, infiltrated with Pb	Bimetal surface layer
...		F510	...	41	2	48	...	7 Ni, 2 Sb(a)	Sintered on steel back, infiltrated with Pb-Sn-Sb alloy (SAE 16)	Bimetal surface layer and trimetal intermediate layer
				86	2	...	...	12 Ni(b)	Sintered on steel back, infiltrated with Pb-Sn-Sb alloy (SAE 16)	Bimetal surface layer and trimetal intermediate layer
...		M100A	...	40	1	48	...	9 Ni, 2 Sb(a)	Sintered on steel back, infiltrated with Pb-Sn-Sb alloy (SAE 16)	Bimetal surface layer and trimetal intermediate layer
				85	...	...	...	15 Ni(b)	Sintered on steel back, infiltrated with Pb-Sn-Sb alloy (SAE 16)	Bimetal surface layer and trimetal intermediate layer

(a) Composition of dense alloy after infiltration. (b) Composition of open grid before infiltration.

Source: Metals Handbook Desk Edition, 1985, 20.40-20.48

520 °C (930 to 970 °F), is relatively high and accounts for the formation of a larger volume of dross than that encountered in melting of Pb-Sb-Sn alloys. Care must be taken to avoid contamination of the alloy with antimonial lead babbitts, and vice versa. Deformability and resistance to wear are of the same order as those of the other lead babbitts. Most alloys of this type are subject to corrosion by acidic oils.

OVERLAYS

The improvement in fatigue life that can be achieved by decreasing babbitt-layer thickness has already been noted. Economically as well as mechanically, it is difficult to consistently achieve very thin uniform babbitt layers bonded to bimetal shells by casting techniques. Therefore, the process of electroplating a thin precision babbitt layer on a very accurately machined bimetal shell was perfected. A specially designed plating rack allows the thickness of the coplated babbitt layer to be regulated so accurately that machining usually is not required.

Coplated tin babbitts were found to be inferior in performance to lead babbitts. Plated babbitts are somewhat different in structure and composition from their cast counterparts. SAE alloy 190 is the most widely used overlay plate. The tin content of this alloy gives it better wear resistance than that of pure lead, and is necessary to protect the lead from corrosion; the copper content increases fatigue life.

When an SAE 190 overlay is plated directly onto a copper-lead bimetal surface, the tin has a tendency to migrate to the copper-lead interface, forming a brittle copper-tin intermetallic compound and/or diffusing into the lead phase. This decreases the corrosion resistance of the overlay and causes embrittlement along the bond line. To avoid this deterioration, a thin, continuous barrier layer, preferably nickel about 1.3 μm (0.05 mil) thick, is plated onto the copper-lead surface just prior to plating of the overlay. In addition to providing better surface behavior, overlays improve fatigue performance of the intermediate layer by preventing cracking in this layer. Plated overlays generally range in thickness from 0.013 to 0.05 mm (0.0005 to 0.002 in.), with fatigue life increasing markedly as overlay thickness decreases. In order to take full advantage of the improved fatigue life achieved with thin overlays, it is necessary to minimize assembly imperfections (such as misalignment) and to maintain close tolerances on machined shafts and bearing bores. Engine components must be thoroughly cleaned before assembly, and adequate air and lubricant filtration must be maintained if the overlay is to survive during the useful life of the bearing. Under adverse wear conditions, however, premature removal of the overlay will not necessarily impair operation of the bearing, because the exposed intermediate bearing alloy layer should continue to function satisfactorily.

COPPER ALLOYS

Copper-base bearing alloys comprise a large family of materials with a considerable range of properties. They include commercial bronze, copper-lead alloys, and leaded and unleaded tin bronzes. They are used alone in single-metal bearings, as bearing backs with babbitt surface layers, as bimetal layers bonded to steel backs, and as intermediate layers in steel-backed tri-metal bearings.

Pure copper is a relatively soft, weak metal. The principal alloying element used to harden and strengthen it is tin, with which it forms a solid solution. However, when tin content is higher than about 8%, a hard constituent (the alpha-delta eutectoid) develops in cast copper-tin alloys because of deviation from true equilibrium. This constituent is quite hard, and its presence causes a considerable improvement in wear resistance. Lead is present in all cast copper-base bearing alloys as a nearly pure, discrete phase, because its solid solubility in the matrix is practically nil. The lead phase, which is exposed on the running surface of a bearing, constitutes a site vulnerable to corrosive attack under certain operating conditions.

The antifriction behavior of copper-base bearing alloys improves as lead content increases, although at the same time strength is degraded because of increased interruption of the continuity of the copper alloy matrix by the soft, weak lead. Thus, through judicious control of tin content, lead content and microstructure, an entire family of bearing alloys has evolved to suit a wide variety of bearing applications.

Table 5 gives specification numbers and nominal compositions of copper-base bearing alloys, as well as the forms in which the alloys are used and general notations on typical applications.

Commercial Bronze. Lead-free copper alloys are characterized by poor antifriction properties but fairly good load-carrying ability. Wrought commercial bronze strip (SAE 795) with 10% zinc can be readily press formed into cylindrical bushings and thrust washers. Strength can be increased by cold working this inexpensive material.

Unleaded Tin Bronzes. The unleaded copper-tin alloys are known as phosphor bronzes because they are deoxidized with phosphorus. They are used principally in cast form as shapes for specific applications, or as rods or tubes from which solid bearings are machined. They have excellent strength and wear resistance, both of which improve with increasing tin content, but poor surface properties. They are used for bridge turntables and trunnions in contact with high-strength steel, and in other slow-moving applications.

Low-Lead Tin Bronzes. The inherently poor machinability of tin bronzes can be improved by adding small amounts of lead. Such additions do not significantly improve surface properties such as lubricity, however, and applications for these alloys are essentially the same as those for unleaded tin bronzes.

Medium-Lead Tin Bronzes. The only wrought strip material in this group of alloys is SAE 791, which is press formed into solid bushings and thrust washers. C83600 is used in cast form as bearing backs in bimetal bearings. SAE 793 and 798 are chemically similar low-tin, medium-lead materials that are cast or sintered on steel backs and used as surface layers for medium-load bimetal bushings. SAE 792 and 797 are higher in tin and slightly higher in lead, are cast or sintered on steel backs, and are used for heavy-duty applications such as wrist pin bushings and heavy-duty thrust surfaces.

High-lead tin bronzes contain medium to high amounts of tin, and relatively high lead contents to markedly improve antifriction characteristics. SAE 794 and 799 are widely used as bushings for rotating loads, and have the same chemical

matrix as 793 and 798 but with three times as much free lead. Both are generally cast or sintered on steel backs and are used for somewhat higher speeds and lower loads than alloys 793 and 798. The 3Sn-25Pb alloy cast on a steel back provides a much stronger bronze matrix than plain 75-25 copper-lead alloy, and is used with a plated overlay as the intermediate layer in heavy-duty trimetal bearing applications such as main and connecting-rod bearings in diesel truck engines. This construction provides the highest load-carrying ability available at the present time in copper alloy trimetal bearings.

Copper-lead alloys are used extensively in automotive, aircraft and general engineering applications. These alloys are usually cast or sintered to a steel backing strip from which parts are blanked and formed into full-round or half-round shapes depending on final application. Copper-lead alloys continuously cast on steel strip typically consist of copper dendrites perpendicular and securely anchored to the steel back, with an interdendritic lead phase. In contrast, sintered copper-lead alloys of similar composition are composed of more equiaxed copper grains with an intergranular lead phase.

The high-lead alloys (28 to 40% Pb) may be used bare on steel or cast iron journals as medium-duty automotive bearings. Tin content in these alloys is restricted to a low value to maintain a soft copper matrix, which along with the higher lead improves the antifriction/antiseizure properties. Bare bimetal copper-lead bearings are used less frequently today than they were some years ago because the lead phase, present as nearly pure unalloyed lead in all cast copper alloys, is susceptible to attack by corrosive products that can form in the crankcase lubricant during the longer oil-change periods now in use. Therefore, many of the copper-lead alloys with lead contents near 25% are used as bases for plated overlays in trimetal bearings for automotive and diesel engines.

Other alloys included in this group are the special sintered and infiltrated or impregnated materials SAE 482, 484, and 485. The last two items in Table 5 consist of an open copper-nickel or copper-nickel-tin grid, which is sintered onto a steel back, then infiltrated with a Pb-Sn-Sb alloy (SAE 16) to make bimetal grid bearings. Alternatively, the lead-base alloy may be overcast so that it completely covers the grid. Excess babbitt can then be machined off, leaving a very thin layer covering the grid, to make a medium-duty trimetal bearing.

ALUMINUM ALLOYS

Successful commercial use of aluminum alloys in plain bearings dates back to about 1940, when low-tin aluminum alloy castings were introduced to replace solid bronze bearings for heavy machinery. Production of steel-backed strip materials by roll bonding became commercially successful about 1950, permitting the development of practical bimetal and trimetal bearing-material systems using aluminum alloys in place of babbitts and copper alloys.

The ready availability of aluminum and its relatively stable cost have provided an incentive for continuing development of its use in plain bearings. Aluminum single-metal, bimetal and trimetal systems now can be used in the same load ranges as babbitts, copper-lead alloys and high-lead tin bronzes. Moreover, the outstanding cor-

rosion resistance of aluminum has become an increasingly important consideration in recent years, and has led to widespread use of aluminum alloy materials in automotive engine bearings in preference to copper-lead alloys and leaded bronzes.

Designations and Compositions. Alloy designations and nominal compositions of the commercial aluminum-base bearing alloys used most extensively in the United States are listed in Table 6. In these alloys, additions of silicon, copper, nickel, magnesium and manganese function to strengthen the aluminum through solid-solution and precipitation mechanisms. Fatigue strength and the opposing properties of conformability and embeddability are largely controlled by these elements. Tin, cadmium and lead are instrumental in upgrading the inherently poor compatibility of aluminum. Silicon has a beneficial effect on compatibility in addition to a moderate strengthening effect. Although not well understood theoretically, this compatibility-improving mechanism is of considerable practical value and is utilized effectively in the high-lead and tin-free alloys.

Mechanical Properties and Alloy Tempers. Conventional mechanical properties, somewhat like microstructural features, are of more value in understanding fabrication behavior of aluminum-base bearing alloys than in predicting their bearing performance. With the exception of solid aluminum alloy bearings, in which there is no steel back and where press-fit retention depends entirely on the strength of the aluminum alloy, mechanical properties of finished bearings are rarely specified—and then usually for control purposes only. Consideration of some of these properties (Table 7) does, however, contribute to an understanding of these alloys as a family of related engineering materials, and of their relationship to the better-known structural aluminum alloys. The wrought alloys as a group are low in hardness and strength compared with conventional aluminum structural alloys. With one exception (No. 14, Table 6), no use is made of heat treatment or cold working for increasing mechanical strength.

Cast aluminum-base bearing alloys are low in hardness and strength compared with conventional cast aluminum alloys, but are heat treated and cold worked to increase their yield-strength levels above as-cast values.

The majority of current commercial applications of aluminum-base bearing alloys involve steel-backed bimetal or trimetal bearings. To determine the most cost-effective aluminum material for any specific application, consideration should be given to the economic advantages of bimetal versus trimetal systems. The higher cost of the high-tin and high-lead alloys usually is offset by eliminating the cost of the lead alloy overlay plate. If the higher load capacity of a trimetal material is required, it then becomes important to select an aluminum liner alloy that provides adequate but not excessive strength, so that conformability and embeddability are not sacrificed unnecessarily. The tin-free alloy group (alloys 10 to 14 in Table 6) offers a wide range of strength properties, and the most economical choice usually is found in this group.

SILVER ALLOYS

Use of silver in bearings is largely confined to unalloyed silver (AMS 4815) electroplated on steel shells, which then are machined to very close di-

Table 6. Designations and nominal compositions of aluminum-base bearing alloys

No.	Designations SAE	AA	Other	Al	Si	Cu	Ni	Mg	Sn	Cd	Other	Form	Typical applications
High-tin aluminum alloy													
1	783	8081	...	79	...	1	...	...	20	...	...	Wrought strip, O temper, bonded to steel back	Bimetal surface layer
High-lead aluminum alloys													
2	...	...	F-66	85	4	1	...	...	1.5	...	8.5 Pb	Powder rolled and sintered strip, O temper, bonded to steel back	Bimetal surface layer
3	...	...	AL-6	88	4	0.5	...	0.5	1	...	6 Pb	Wrought strip, O temper, bonded to steel back	Bimetal surface layer
Low-tin aluminum alloys													
4	770	850.0	...	91.5	0.7	1	1	...	6.5	...	...	Cast tubes, T101 temper(a)	Solid aluminum bearings; aluminum bearing backs
5	...	A850.0	...	89.5	2.5	1	0.5	...	6.5	...	...	Cast tubes, T101 temper(a)	Solid aluminum bearings; aluminum bearing backs
6	...	B850.0	...	89.5	...	2	1.2	0.8	6.5	...	...	Cast tubes, T5 temper(b)	Solid aluminum bearings; aluminum bearing backs
7	...	...	MB-7	89	0.7	1	1.7	1	7	...	...	Cast tubes, T5 temper(b)	Solid aluminum bearings; aluminum bearing backs
8	780	828.0	...	90.5	1.5	1	0.5	...	6.5	...	...	Wrought strip and plate, H12(c) temper	Solid aluminum bearings; aluminum bearing backs
9	...	...	A300	91	1	2	...	...	6	...	...	Wrought strip, O temper, bonded to steel back	Bimetal surface layer; trimetal intermediate layer
												Wrought strip, O temper, bonded to steel back	Trimetal intermediate layer
Tin-free aluminum alloys													
10	781	4002	...	95	4	0.1	...	0.1	...	1	...	Wrought strip, O temper, bonded to steel back	Bimetal surface layer; trimetal intermediate layer
11	782	...	...	95	...	1	1	...	...	3	...	Wrought strip, O temper, bonded to steel back	Bimetal surface layer; trimetal intermediate layer
12	...	...	A250	...	...	1	1	...	...	3	1.5 Mn	Wrought strip, O temper, bonded to steel back	Trimetal intermediate layer
13	...	...	AS78	88	11	1	...	...	...	...	...	Wrought strip, O temper, bonded to steel back	Trimetal intermediate layer
14	...	4002	F-154	95	4	0.1	...	0.1	...	1	...	Wrought strip, T6 temper(d), bonded to steel back	Trimetal intermediate layer

(a) Artificially aged and cold pressed. (b) Artificially aged. (c) Strain hardened, approximately 25% cold reduction. (d) Solution treated and artificially aged.

Table 7. Approximate mechanical properties of aluminum-base bearing alloys

Classification	Tensile strength MPa	ksi	Yield strength MPa	ksi	Hardness, HB(a)
High-tin aluminum strip	114	16.5	41	6	30
High-lead aluminum strip	117	17	62	9	32
Low-tin aluminum strip	117 to 138	17 to 20	48 to 124	7 to 18	32 to 40
Tin-free aluminum strip	124 to 207	18 to 30	62 to 138	9 to 20	38 to 48
Low-tin aluminum castings	159 to 234	23 to 34	117 to 172	17 to 25	54 to 74

(a) 500-kg load, 10-mm ball.

mensional tolerances and finally precision plated to size with a thin overlay of soft metal. The overlay may be of a lead-tin-copper or lead-tin alloy. In some aircraft applications, the overlay consists of a plated layer of lead with a final layer of indium. Such bearings are then heat treated to diffuse the indium into the lead.

OTHER METALLIC BEARING MATERIALS

Gray Cast Irons. Cast irons are standard materials for certain applications involving friction and wear, such as brake drums, piston rings, cylinder liners and gears. Cast irons perform well in such applications, and thus should be given consideration as bearing materials.

Cemented Carbides. Hard materials such as cemented tungsten carbide have been used successfully for various specialized bearing and seal applications. With proper design and materials selection, performance of sleeve-type antifriction bearings, mechanical rubbing-face seals, and seals employing packings can be improved by making them from carbide.

NONMETALLIC BEARING MATERIALS

Today, nonmetallic bearing materials are widely used. They have many inherent advantages over metals, including better corrosion resistance, lighter weight, better resistance to mechanical shock, and the ability to function with very marginal lubrication or with no lubricant at all. The major disadvantages of most nonmetallics are their high coefficients of thermal expansion and low thermal conductivity characteristics.

A wide variety of plastic composites is now being used very successfully in bearing applications. Addition of fiber reinforcements and fillers such as solid lubricants and metal powders to the resin matrix can significantly improve the physical, thermal and tribological properties of these plastics. This is illustrated in Table 8, where a few of the more promising materials are listed. The PV values (stress × velocity) are for dry operation. Even with marginal lubricants, such as water, some of these compounds have truly outstanding load-carrying capacities. It should be noted, however, that certain plastics wear at a

higher rate when a lubricant is present. When the effectiveness of a lubricant depends on formation of a transferred film, its use may inhibit material transfer, resulting in higher wear.

The following paragraphs present more detailed discussions of some nonmetallic materials typically used for bearings.

Nylon. The low melting point of nylon limits its use to temperatures below about 150 °C (300 °F). To obtain dimensional stability, nylon should be stress relieved at a temperature at least 28 °C (50 °F) above the maximum temperature expected in service. This is usually accomplished by heat treating the nylon in oil or some other suitable liquid. Graphite, molybdenum disulfide and other fillers are added to nylon to improve its bearing properties. The static coefficient of friction for nylon against nylon is more than twice the kinetic friction. The friction values for steel against nylon are lower than for nylon against nylon.

Teflon, a polytetrafluoroethylene resin, is a thermoplastic material. It is used as a bearing material mainly for two reasons: (a) chemical inertness; and (b) at low speeds, an extremely low coefficient of friction with sliding metals (about 0.05). Use of Teflon as a bearing material is limited, however, because of its low thermal conductivity, high thermal expansion, thermal instability and poor resistance to cold flow. In designing Teflon bearings, consideration must be given to the fact that Teflon has a transition point at 21 °C (70 °F), which results in a linear increase of 4 mm/m (0.004 in./in.). When Teflon is heated to about 340 °C (650 °F) or higher, it gives off toxic fumes, and therefore it must be

Source: Metals Handbook Desk Edition, 1985, 20.40-20.48

Table 8. Typical PV values for plastics sliding unlubricated on steel at a velocity of 100 ft/min

Type of plastic	Operating temperature limit(a) °C	Operating temperature limit(a) °F	Typical PV(b) values
Teflon (unfilled)	230 to 260	450 to 500	1800
Nylon (unfilled)	120 to 150	250 to 300	2000 to 4000
Acetal (unfilled)	82	180	3000 to 4500
Polysulfone (unfilled)	150 to 175	300 to 350	4000 to 5000
Teflon + 30% bronze powder	230 to 260	450 to 500	28 000
Acetal + 30% glass fiber + 15% Teflon fiber	82	180	12 000
Polyester + 30% glass fiber + 15% Teflon fiber	135	275	30 000
Formulated polyphenylene sulfide (proprietary)	175	350	35 000 to 45 000
Formulated polyamide-imide (proprietary)	205	400	12 000 to 12 900
Polyimide + 15% graphite powder	260	500	>300 000

(a) Continuous use. (b) PV, stress in psi × sliding velocity in ft/min.

cooled during fabrication operations such as machining. It can be used at service temperatures (260 °C; 500 °F) higher than those for nylon, and it is not hygroscopic.

Teflon is used in bearings in several ways: (a) as a film applied by water dispersion and then cured; (b) as an impregnant in a metal matrix; and (c) as a woven layer, supported by a woven layer of glass bonded to a metal surface. Teflon applied by water dispersion generally is used as a means of preventing fretting corrosion where there is intermittent oscillation. Woven Teflon is recommended for oscillating or low-speed use, although it has been used successfully at loads as high as 400 MPa (60 ksi). However, a rule of thumb for application of this material is to work to a maximum PV value of 30 000 (load in pounds per square inch multiplied by velocity in feet per minute. Laboratory tests on metal-filled Teflon (60% Teflon; 40% metal) proved that bearings with bronze as the metal filler were greatly superior in wear resistance to those with either lead or aluminum. At low speeds, the dynamic coefficient of friction of unmodified Teflon is lower than that of the filled types; this is not true at high speeds. However, because of the increased strength provided by the fillers, it may be desirable to accept a slightly higher coefficient of friction to gain strength.

Carbon-graphite is used extensively in bearing and brush applications. Its excellent performance as a brush material confirms its desirable bearing qualities. Its service temperature, usually limited to about 370 °C (700 °F), has recently been increased by processing techniques. Carbon-graphite has good wear resistance at temperatures too high for conventional lubricants. Carbon-graphite can also operate as a bearing in water, gasoline and other nondestructive solvents. It has a low coefficient of expansion, and its thermal conductivity is between those of copper and cast iron. Although it possesses reasonable strength, its edges are likely to chip or crack during machining or installation. It is not usually considered for applications involving high impact loads. It is used in packing rings, seals, instrument bearings, and sleeve and thrust bearings.

Wood. Lignum vitae, one of the hardest woods known, has been used for centuries as a lining for various underwater bearings in ships, where metal corrodes severely. It is inexpensive and readily obtainable. Oil-impregnated wood is also used in some bearing applications.

A composition material that has a base of either paper, fabric or asbestos may be substituted for wood in applications such as ship-rudder bearings, liners for rolling mills, inking-roll bearings, bushings and pump sleeves.

Rubber often is used in bearings for devices that operate in water. Rubber can absorb shock and has fairly good resistance to abrasion and other types of wear. Rubber bearings usually are backed by a metal shell that provides additional strength. If the rubber bearing is properly designed, much of the solid contaminating material in the water can be washed out through longitudinal passages fabricated in the bearing surface.

Zinc Alloys Compete With Bronze In Bearings and Bushings

By Kenneth J. Altorfer

Fig. 1—Comparative wear of bronze bearing (left) and ZA-12 bearing after more than 4000 h of operation in an earth-moving machine. Bearings are 6 in. (150 mm) long.

About 40 years ago, alloys of zinc and aluminum were developed in Germany as substitutes for bronze at a time when its copper component was in extremely short supply. Some of the alloys, containing 10 to 30% aluminum, rivaled bronze as bearing materials and became known as white bronze.

After the shortage, a few European companies continued the development of high-aluminum zinc alloys with excellent success. In Europe, zinc alloys have been applied for decades to bearings and many other parts customarily made of bronze in North America.

Now the U.S. is catching up.

The U.S. and Canadian zinc industries have developed three high performance, high-aluminum zinc casting alloys (see Datasheet in *Metal Progress*, August 1982, p 53). Of these, ZA-12 and ZA-27 are finding increasing use as replacements for bronze in lower speed, medium temperature bearing and bushing applications.

The alloy designations, ZA-12 and ZA-27, correspond to the approximate percentages of aluminum they contain (in general, the higher the aluminum content, the better the properties). ZA-12 (ASTM B699-80) has this composition: 10.5-11.5 Al, 0.5-1.25 Cu, 0.15-0.030 Mg, 0.075 Fe max, 0.004 Pb max, 0.003 Cd max, 0.002 Sn max, bal Zn. ZA-27 (ASTM B669-82) has this composition: 25-28 Al, 2.0-2.5 Cu, 0.01-0.02 Mg, 0.10 Fe max, 0.004 Pb max, 0.003 Cd max, 0.002 Sn max, bal Zn. ZA-12 does not require heat treating under any circumstances. ZA-27 can be heat treated to improve ductility, but there are losses in other key properties, such as tensile strength. Heat treatment is confined to specific applications.

The term "high performance" used with the "ZA" alloys helps distinguish them from traditional zinc die casting alloys which do not offer equivalent engineering and physical

Table I — Properties of ZA-12, ZA-27, and SAE 660 Bronze Relating to Bearing Performance

Property	ZA-12 Sand Cast	ZA-27 Sand Cast	ZA-27 Continuously Cast[1]	SAE 660 Sand Cast	SAE 660 Continuously Cast[2]
Ultimate tensile strength, 10^3 psi (MPa)	40-45 (275-315)	58-64 (400-440)	60-64 (415-440)	30-38 (205-260)	35 (240)
Yield strength, 2% offset, 10^3 psi (MPa)	30-31 (205-215)	53-54 (365-370)	55-57 (380-395)	17-21[3] (115-145)	20[3] (140)
Elongation, %	1-3	3-6	8-11	12-20	10
Hardness, HB 10/500/30	92-97	110-120	115-130	60-70	80
Impact energy, ft-lb (J)[4]	ND	36 ± 5 (50 ± 7)	54[5] (73)	6[6] (8)	ND
Maximum operating temperature, F (C)	200 (95)	250-300 (120-150)	250-300 (120-150)	300 (150)	ND

1. Preliminary data based on specimen machined from 2 in. (50mm) in diameter solid bar (as cast).
2. ASTM B505-79 minimum mechanical requirements.
3. At 0.5% extension under load, minimum.
4. Unnotched Charpy, 0.394 by 0.394 in. (10 by 10 mm).
5. Typical.
6. Izod impact energy in ft-lb (J).
Note: ND = not determined.

properties. Companies marketing the alloys usually add a brand name to identify their products; for example, Kidd Creek Mines uses Kidd ZA-12 and Kidd ZA-27.

How These Alloys Compete With Bronzes

Lower speed bearing and bushing applications now in bronze represent a particularly promising market for the zinc alloys because of their excellent bearing properties (see Table).

For example, cast ZA-12 has a hardness range of 92 to 96 HB versus 60 to 70 HB for sand cast SAE 660 bronze.

Continuously cast preforms improve material utilization in machining. The continuous casting of Kidd ZA-27 improves tensile strength, yield strength, and hardness properties of the alloy (see Table).

The zinc alloys also cost less than bronze. This factor in combination with lower density means material cost for a given part can be reduced by about 50% and finished cost by 20 to 40%.

In housing applications it is sometimes possible to eliminate bearing and bushing inserts required for castings made of aluminum, plastic, cast iron, or conventional zinc die casting alloys. Casting walls function as a cast-in bearing material.

Performance is Verified In Laboratory Testing

To provide specific direction for bushing and bearing applications, the International Lead Zinc Research Organization (ILZRO) initiated comparison tests of ZA-12 and ZA-27 against SAE 660 bronze at an independent research laboratory.

Data are for low speed, high load applications — currently dominated by cast bronze bearings. The ZA-27 alloy exceeded the load capacity of SAE 660 bronze over the entire speed range tested and showed at least comparable wear resistance. Data indicated a lower coefficient of friction for ZA-27 at high load ranges. The ZA-12 alloy also exceeded the bronze under more limited conditions. Short term wear data favored the zinc alloys. Long term laboratory data will be gathered next.

The zinc-aluminum alloys will not compete with bronze at the highest operating temperatures, but ILZRO findings suggest the investigation of low speed applications under 250 F (120 C).

Comparative Performance In Series of Field Tests

As a primary zinc, copper, and silver producer operating its own mines, Kidd Creek has been able to test zinc alloy bearings in its own facilities as well as those of other companies.

Initial tests were at the Kidd Creek mine in Timmins, Ont., and at a Texasgulf phosphate mine near Aurora, N.C. Some tests are continuing.

Kidd Creek operates air-powered drills (more than 35 Joy VCR-150's) to prepare the ore body face for blasting. Each drill has three planetary gears with — historically — bronze bushings. They were replaced about two or three times a year in each drill during normal rebuilding; about 600 bushings were used yearly.

For the test, Kidd ZA-12 alloy bushings were substituted. Factors other than bearing wear also dictate the need for rebuilding, so it was not statistically possible to prove better performance for the zinc alloys. But it was noted that on occasion bronze bushings were destroyed, while the zinc bushings were not. Kidd Creek decided to switch to ZA-12 for its operations, and Joy now specifies ZA-12 alloy bushings for its VCR-150 drill.

At its phosphate mine in North Carolina, Texasgulf operates a Bucyrus-Erie 2550 dragline, which weighs 4300 tons (3900 Mg); its boom is 300 ft (90 m) long; it swings a 72 yard (55 m³) bucket; and its 5 ft (1.5 m) in diameter boom-deflection sheaves carry 4 in. (100 mm) cable.

The bearing test was made on the lower pair of sheaves, mounted on the dragline's housing. One sheave was equipped with zinc alloy bearings, the other with SAE 660 bronze. Bearings were 4 in. (100 mm) long, with an 8 in. (205 mm) ID and an 8 7/8 in. (225 mm) OD. The equipment was operated 10 401 h over a period of two years.

Measurements showed that the bronze bearings wore about 35% more than the ZA-12 alloy; shafts running in the bronze wore about 38% more than the ones in ZA-12 bushings. Zinc bearings have been installed in both sheaves, and the maintenance group plans to make the switch on all of its draglines.

The Kidd Creek mine in Timmins operates 17 Wagner ST8 (8 yard [6 m³] load-haul-

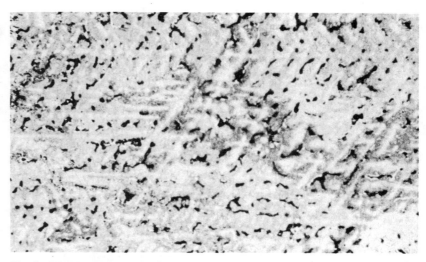

Fig. 2—Microstructure of continuous cast ZA-27. Features are a fine, heavily cored dendritic network of primary, aluminum-rich alpha phase (gray and almost black areas) uniformly dispersed in the zinc-rich eutectic (almost white areas). Note the laminar flow pattern caused by continuous casting. Etchant: solution of chromium trioxide, 50 g; sodium sulfate, 4 g; water, 1 L. 250X.

dump) earth-moving machines. In the oscillating axle of two units, ZA-12 bushings were installed side-by-side with SAE 660 bronze types. Combined load on the bushings is over 34 000 lb (150 kN) with the machine empty.

The two bushings ran for 20 months, logging 4093 h of service. At this time there was enough play in the oscillating axle to warrant a routine overhaul; the axle was disassembled and the bushings removed. The bronze bushing showed noticeable galling or pitting (Fig. 1). In contrast, the zinc alloy bushing interior surface was polished smooth and maintained complete integrity. Measurements showed that the bronze had worn 0.008 in. (0.20 mm). The zinc alloy had worn 0.007 in. (0.18 mm). At the time of installation, the bronze bushing cost $95; the ZA-12, $44.75.

Design Parameters For Zinc Alloy Bearings

Zinc alloy bearings are not a universal substitute for bronze. However, they are acceptable in virtually all oscillating applications and many rotating applications. Adoption of the same design for a bronze

bearing is frequently possible. If a design is new, parameters may be obtained from Kidd Creek Mines or other suppliers.

Two important considerations are surface speed and operating temperature. Speeds should be kept below 1400 ft/min (7.1 m/s). Higher speeds invite sharp temperature rises, particularly when lubrication becomes marginal. Conservative maximum allowable running temperatures are about 200 F (95 C) for ZA-12 and about 250 F (120 C) for ZA-27.

Originally the zinc alloys were cast using sand and permanent metal mold foundry processes. Centrifugal and continuous casting processes are also used today. Both reduce the cost of bushing and bearing production. Figure 2 shows the micro-structure of continuous cast ZA-27.

A wide range of sizes has been continuous cast from ZA-12 and ZA-27. For bearing, bushing, and other wear applications, continuous cast bar stock can enable distributors and users to meet their machining require-ments more economically. ⬩

SECTION II
Gas Turbines

An Alloy For Stationary Gas Turbines

Because stationary gas turbines usually operate on a lower grade of fuel than aircraft turbines, hot corrosion resistance is a critical consideration for stationary turbines, according to Special Metals Corp., an Allegheny Ludlum Industries Co., New Hartford, New York, U.S.A. Consequently, Special Metals developed Udimet 720, a wrought alloy combining hot workability, strength, hot corrosion resistance, excellent microstructure stability, and resistance to high temperature impact degradation during prolonged high temperature exposure.

Special Metals is a producer of vacuum melted superalloys for high temperature and high strength applications, and electroslag remelted alloys manufactured in billet, bar, remelt ingot and powder metallurgy form. Besides Udimet 720, the company has developed a number of alloys including Udimet 500, Udimet 700 and Udimet 520. Its products are marketed worldwide principally to forging and casting industries for conversion to parts for stationary and aircraft gas turbine engines, and nuclear power generation equipment.

Udimet 720 is produced by vacuum induction melting to very restrictive chemical composition limits. Ingots are consumable arc remelted for optimum control of homogeneity and microstructure. Udimet 720 exhibits typical nickel base alloy metallurgy. It is solid solution strengthened with tungsten and molybdenum and precipitation hardened with titanium and aluminum. Carbon, boron and zirconium are balanced to optimize grain boundary precipitation and properties. Like Udimet 700 and Udimet 710, the maximum useful service temperature for extended periods of time is believed to be 982°C (1800°F).

Phase stability in the alloy is controlled by matrix compositions within specified Nv limits.

Udimet 720 was specifically designed to provide excellent hot impact resistance in the as-heat-treated condition and after long periods of service at elevated temperatures. The Charpy V-Notch strength of Udimet 720 in the as-heat-

treated condition is approximately 1.5-1.8 kg/m (11-13 lb. ft.) at 899°C. After 1000 hours of exposure at 899°C, the hot impact strength will remain the same or actually improve. Typical values are in the range of 1.7-2.1 Kg/m.

The improved impact resistance of Udimet 720 after exposure is due to the

absence of deleterious carbide precipitation in the grain boundary and a generalized matrix softening due to normal overaging.

Commonly available mill forms of Udimet 720 are bar stock — 12.7 to 101.6 mm diameter, and billet. Billet sizes are available on request. ★

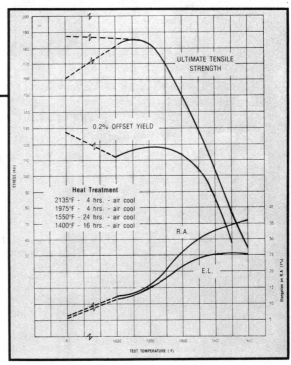

Udimet 720 wrought tensile properties. Higher room temperature ultimate tensile strengths have been achieved using heat treatments that provide improved room temperature ductility.

Alloy	Form	Charpy V-Notch[2] Impact Strength After Aging at 1600°F (871°C) for 10,000 Hrs. and Tested at 1650°F (899°C)
Udimet 720	Wrought	14 Ft.-Lbs. (1.94 Kg-m)[3]
Udimet 520	Wrought	12 Ft.-Lbs. (1.66 Kg-m)
Udimet 710	Wrought	2 Ft.-Lbs. (0.28 Kg-m)
Udimet 500	Cast	8 Ft.-Lbs. (1.10 Kg-m)
IN-738	Cast	3 Ft.-Lbs. (0.41 Kg-m)

[1] Test data supplied by Special Metals Corp. and Westinghouse Electric Corp.
[2] 0.394 in. × 0.255 in. (1.0 cm. × 0.65 cm.) Sample Cross Section.
[3] Based on Accelerated Aging at Higher Temperature.

Aged Charpy V-Notch comparison.

ALLOY	FORM	100-HOUR RUPTURE STRENGTH						1000-HOUR RUPTURE STRENGTH					
		1400°F (760°C)		1600°F (871°C)		1800°F (982°C)		1400°F (760°C)		1600°F (871°C)		1800°F (982°C)	
		KSI	(MN/m²)	KSI	(MN/m²)	KSI	(MN/m²)	KSI	(MN/m²)	KSI	(MN/m²)	KSI	(MN/m²)
Udimet 720	Wrought	84	(580)	45	(310)	18	(124)	67	(462)	32	(221)	10	(69)
Udimet 700	Wrought	78	(538)	43	(297)	16	(110)	62	(428)	29	(200)	8	(55)
Udimet 520	Wrought	69	(476)	32	(221)	9	(62)	52	(359)	19	(131)	–	(—)
IN-738	Cast	86	(593)	46	(317)	19	(131)	69	(476)	32	(221)	12	(83)
Alloy 713C	Cast	82	(566)	43	(297)	21	(145)	65	(448)	30	(207)	13	(90)
Udimet 500	Cast	65	(448)	33	(228)	13	(90)	50	(345)	24	(166)	–	(—)

Stress rupture comparison of Udimet 720 with other alloys.

Reprinted with permission from Diesel and Gas Turbine Worldwide, January/February 1982, 42, © 1982 Diesel and Gas Turbine Publications

Applications of High-Temperature Powder Metal Aluminum Alloys to Small Gas Turbines

P.P. Millan, Jr.

SUMMARY

Advanced powder metal (P/M) aluminum alloys currently are being developed using the concepts of rapid solidification and mechanical alloying technologies that offer the potential for extending the use of aluminum for high-temperature applications up to 650°F. The Garrett Turbine Engine Company high-temperature aluminum alloy program has identified compressor impellers for small gas turbines as the nearest term potential for applications of these advanced alloy systems. Substitution of aluminum impellers for titanium for steel offers the benefits of reduced component cost, weight savings, and rotating group inertia.

This paper discusses the Garrett program and describes the engine systems that have been identified as potential applications for high-temperature P/M aluminum impellers. The design criteria for these compressor impellers have been defined and serve as development goals for this program. A review of candidate alloy systems is made and current material evaluation results are presented.

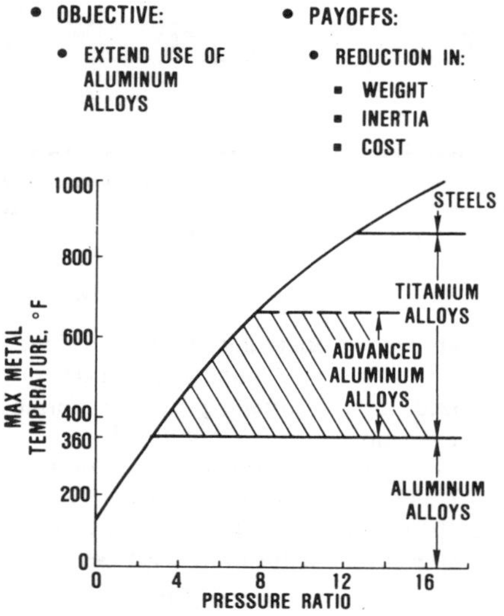

Figure 1. Aluminum alloy impeller improvement program.

INTRODUCTION

A continuing need exists for improved compressor materials that provide higher temperature strength capabilities in advanced gas turbine engines. The advent of rapid solidification technology (RST) spawned a number of novel powder production processes, which led to the development of new and unique alloys with superior properties and capabilities than otherwise would be achievable by ingot metallurgy. The extremely high cooling rates achieved in these processes result in powder particles exhibiting microcrystalline structures containing high-volume fractions of fine dispersions of stable intermetallic phases. These stable secondary phases essentially provide the elevated-temperature strength of P/M alloys. This new breed of alloys offers the potential for extending the use of aluminum in impeller applications up to 650°F.

Recognizing the potential benefits of these materials, in 1978 Garrett Turbine Engine Company initiated the selection and applications development of promising aluminum alloys that primarily have evolved from a U.S. Air Force-sponsored alloy development program. Auxiliary Power Unit (APU) compressor impellers have been identified as the nearest term applications for these alloys. As shown in Figure 1, advanced aluminum alloys potentially can replace titanium alloy impellers operating in the 360-650°F temperature range. Significant benefits of this material substitution program include:

- savings in material and fabrication costs,
- reduced component weight resulting in lower rotating group inertia requirements, and
- reduced consumption of critical titanium alloys.

It is realized that titanium alloys inherently are far superior to any aluminum alloy, advanced P/M alloys included; realistically, therefore, a one-on-one property tradeoff is not anticipated. Rather the goal is to economically replace titanium where its use at these operating temperature ranges is structurally an overdesign, but where current commercial high-strength aluminum alloys are decidedly marginal. Advanced P/M aluminum alloys potentially meet these high-temperature strength requirements.

COMPONENT APPLICATIONS

The applications development approach to this program involves identifying specific components with defined design requirements and then tailor-

Editor's Note: This paper is taken from *High-Strength Powder-Metallurgy Aluminum Alloys*, edited by Michael J. Koczak and Gregory J. Hildeman, The Metallurgical Society of AIME, Warrendale, Pennsylvania. Copyright 1982.

Reprinted with permission from Journal of Metals, March 1983, 76-81, a publication of The Metallurgical Society, Warrendale, Pennsylvania

Table I: Potential Applications for Advanced P/M Aluminum Impellers

Engine Model	User	Application
AGT101	DOE/NASA	Automotive engine
GTC36-200	U.S. Navy	F-18 APU
GTC36-50	U.S. Air Force	A-10 APU
GTCP36-55	U.S. Army	AAH attack helicopter APU
GTCP85-180	U.S. Air Force	M32-A60 APU
GT601	U.S. Army/Mack	Vehicular engine/truck engine
GTP201	U.S. Air Force/Navy/Army	Tri-service APU
GTCP165	U.S. Air Force	B-1B APU

ing a promising alloy system to meet these criteria. Table I lists some of the military and commercial engine systems currently in production or under development that represent the potential applications of P/M aluminum for compressor impellers. Current activity focuses on two component applications to demonstrate the viability of advanced P/M aluminum alloys for high-temperature service.

Under a Garrett-funded R&D program, the GTC36-200 APU was selected as the demonstration engine for centrifugal compressor impeller application (Figure 2). This APU is currently used in the U.S. Navy F-18 aircraft to provide an on-board main engine starting capability as well as accessory ground operation and ground environmental control system operation. Forged A2219-T6 aluminum alloy was initially selected for this component; however, material evaluation indicated that stress rupture and fatigue

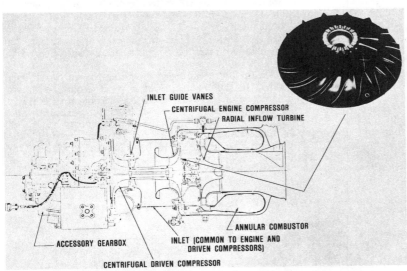

INLET GUIDE VANES
CENTRIFUGAL ENGINE COMPRESSOR
RADIAL INFLOW TURBINE
ACCESSORY GEARBOX
ANNULAR COMBUSTOR
INLET (COMMON TO ENGINE AND DRIVEN COMPRESSORS)
CENTRIFUGAL DRIVEN COMPRESSOR

Figure 2. Model GTC36-200 auxiliary power unit.

Figure 3. AGT101 Automotive Gas Turbine Engine.

Table II: APU Aluminum Impeller Minimum Property Goals

	Temperature, °F	Current Designs			Advanced Designs		
		Ultimate Strength, ksi	0.2% Yield Strength, ksi	Elongation, %	Ultimate Strength, ksi	0.2% Yield Strength, ksi	Elongation, %
Tensile Properties	75	62	45	3.0	65	50	4.0
	250	52	40	4.0	55	45	4.5
	425	40	30	4.5	45	35	5.0
	500	—	—	—	40	30	5.0

	Temperature, °F	Stress, ksi	Temperature, °F	Stress, ksi
Stress rupture 2000 hours with 2.0% minimum creep ductility	300	32	370	40
	425	20	500	25
Low-cycle fatigue (σ max) R = 0 strain control, 10^4 cycles	300	50	375	60
High-cycle fatigue (σ max) R = 0 load control Tension/tension, 10^8 cycles	250	25	250	25
	425	15	500	15

Source: Journal of Metals, March 1983, 76-81

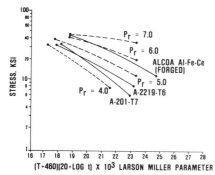

Figure 4. Stress rupture/compressor pressure ratio requirement (typical properties).

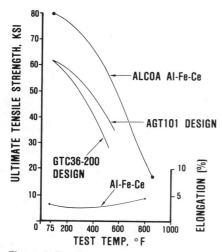

Figure 5. Tensile property of powder metal aluminum.

properties of this aluminum alloy were marginal. Consequently, a decision was made to produce the impeller from wrought Ti-6Al-4V alloy, which was structurally viable but more expensive. Successful completion of this program will facilitate substitution of P/M aluminum for the titanium alloy.

The second development effort is focused on the Advanced Automotive Gas Turbine Engine, designated AGT101, being developed under a DOE/NASA contract (DEN3-167) by the Garrett/Ford Motor Company team (Figure 3). This program is oriented at providing the United States automotive industry the technology base necessary to produce, for automotive applications, gas turbine power-trains that will have reduced fuel consumption, the ability to use a variety of fuels, low emissions, and competitive cost/performance.

The AGT101 compressor is a 4.27-in.-diameter, single-stage centrifugal configuration designed to operate at a 5:1 pressure ratio using P/M aluminum.

DESIGN CRITERIA

The material strengths and temperature characteristics of single-stage centrifugal compressor impellers are directly related to the operating pressure ratio. Higher pressure ratios are achieved by increasing the impeller tip speed, which consequently increases the stress levels both in the hub and blade regions. In addition, at higher compression ratios, the air temperature correspondingly increases through adiabatic heating. This high-temperature air raises the metal thermal profile at the highly stressed exit region of the flow passage. For these reasons, the material property goals established in this program (Table II) reflect both higher strength and temperature requirements as a function of pressure ratio illustrated in Figure 1. Figure 4 additionally shows the pressure ratios of current and advanced design APUs in Larson-Miller parameters against the stress rupture capabilities of commercial high-strength alloys and a candidate P/M aluminum alloy.

Both AGT101 and GTC36-200 compressor impellers operate within the pressure ratio range that is representative of current engine designs. A maximum metal temperature of 425°F is indicated. Although this may not appear to be too severe, the stress levels at this temperature regime become the life-limiting factors that affect alloy capability. The stress rupture properties of commercially available wrought high-strength alloys, such as A2219-T6 aluminum are shown to be marginal (Figure 4) and, therefore, unacceptable for these applications.

Advanced APUs, on the other hand, operate at significantly higher pres-

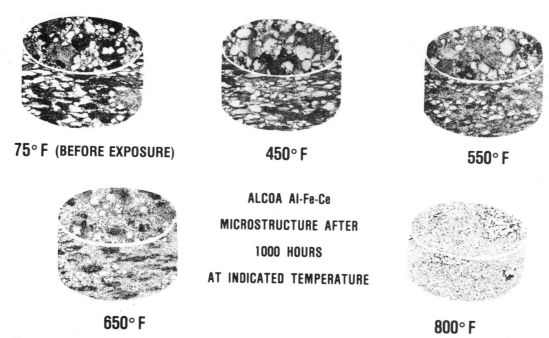

Figure 6. Alcoa Al-Fe-Ce microstructure after 1,000 hours at indicated temperature.

sure ratios and tip speeds, and consequently, higher temperature stress levels. The compressor section operating environment of these advanced APUs requires a 500°F impeller metal temperature capability. Accordingly, minimum mechanical property design limits are based on this higher temperature level at correspondingly higher stresses. At the current state of development, these new P/M alloy systems would have marginal properties. Therefore, continued alloy/process development is essential in improving the high-temperature strength capability of P/M aluminum alloys to meet the advanced impeller design criteria.

MATERIALS EVALUATION

Alcoa Laboratories, employing high-velocity gas atomization, developed rapidly solidified, high-temperature capability, P/M aluminum alloys under U.S. Air Force Contract F33615-77-C-5086. Out of this program, the ternary system Al-Fe-Ce alloy demonstrated the most attractive combination of elevated temperature tensile strength and creep resistance capability.[1] Preliminary assessments at Garrett led to selection of this alloy for applications development. This paper will present and discuss the initial results of the evaluation of Al-Fe-Ce in an attempt to establish its viability as a candidate material for this program.

Screening evaluation was initially conducted on sub-size pancake forgings (4.50 in. diameter x 1 in. high). Thermomechanical processing (TMP) of these forgings at Alcoa involved cold isostatic compaction of screened powder (−200 mesh) followed by encapsulation of the compact, evacuation, and vacuum hot pressing at a predetermined temperature. The encapsulation canister then was machined off and the preform forged in a closed die press to the final disk size.

Results of room and elevated temperature uniaxial tensile tests of specimens excised from the as-forged pancakes are graphically shown in Figure 5 in terms of ultimate tensile strength and ductility. The tests demonstrated that Al-Fe-Ce exceeds both the AGT101 and GTC36-200 APU design requirements. Material ductility, however, was marginal (2.6% average elongation at room temperature), a condition that was attributed to the thermomechanical processing response during fabrication.

Key to the elevated temperature strength of these advanced P/M aluminum alloys is the remarkably high thermal stability of the microstructures after long-term exposures occurring under thermal soak-back conditions. In a gas turbine engine environment a steep thermal gradient exists between the turbine and compressor section. Following a power shutdown the ensuing conduction of thermal energy from the hot turbine section results in a temperature rise of the centrifugal impellers, referred to as thermal soak-back. This relatively high soak-back temperature exposure cycle, although occurring at minimum stress conditions, tends to have a cumulative effect; therefore, the alloy microstructure must remain stable to preclude mechanical property degradation.

To assess this crucial thermal stability characteristic of Al-Fe-Ce alloy, specimens were exposed for 1,000 hours in an air furnace at temperatures

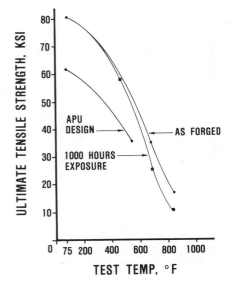

Figure 7. Effect of 1,000-hour exposure on Alcoa Al-Fe-Ce.

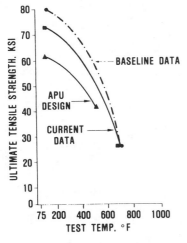
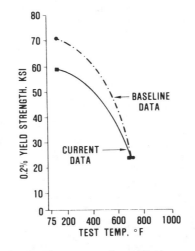
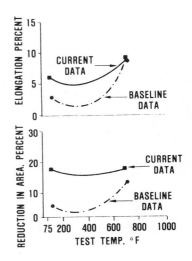

Figure 8. Tensile properties of Alcoa Al-Fe-Ce (1,000-hour overaged condition).

Source: Journal of Metals, March 1983, 76-81

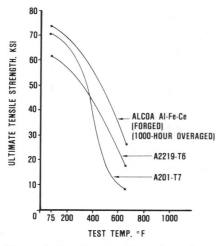

Figure 9. Tensile properties of aluminum alloys.

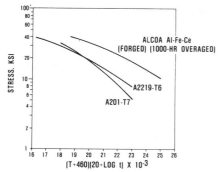

Figure 10. Stress rupture properties of aluminum alloys.

of 450, 550, 650 and 800°F. The resulting Al-Fe-Ce microstructure, as shown in Figure 6, indicates 550°F as the threshold coarsening temperature for this alloy at the current state of development. Significant microstructural degradation or phase coarsening is evident at 650°F. In-depth analysis of the microstructure either in the as-forged condition or after the thermal exposure was not covered in this study.

To verify the effect of long-term thermal exposure on mechanical properties, tensile bars were machined from the 1,000-hour exposed specimens used in the microstructural study and then tested at the same soak temperature level (i.e., 450, 550, 650, and 800°F). Results showed no significant property degradation up to the temperature range of 550°F (Figure 7); however, degradation was more pronounced at 650 and 800°F. These correlate well with the microstructural response as previously observed. Superimposing the APU design curve on Figure 7 shows that, under thermal soak-back conditions, Al-Fe-Ce has more than adequate residual strength to meet the design requirements.

Favorable results of the preliminary evaluation led to a decision to scale up the evaluation program. Component-sized forgings representative of the AGT101 compressor impeller blank dimensions (4.5 in. diameter x 3 in. high) were procured from Alcoa. Characterization of larger size forgings was envisioned to generate a second set of baseline data to assess the effect of thermomechanical processing on the increased mass. A second modification was to vary the TMP parameters with the objective of improving the alloy ductility over those obtained from the pancake forgings. It was anticipated that a partial loss of tensile strength would result, but still should be well within the program goal.

Results, shown in Figure 8 indicate that the tensile strength of these next-iteration forgings still exceeded the program goals while the ductility improvement objective was successfully met. With an average loss of 7 ksi in ultimate strength at room temperature, a two-fold improvement in ductility was achieved (increased from 2.6 to 5.9% at room temperature). Properties at elevated temperatures, particularly beyond 550°F, appear to be less sensitive to TMP variables, as evidenced by the convergence of property levels at 650°F.

Stress rupture life of the alloy likewise demonstrated promising capabilities. All tests were conducted at the same temperature level as the 1,000-hour specimen soak temperature. Based on the design criteria requiring 2,000 h/425°F/20 ksi, accelerated tests were run at 450°F/25 ksi; these tests were terminated after 500 hours with no specimen failure occurring. Statistical projection of the data generated at various stress levels (40, 35, 30, and 25 ksi) indicated a stress rupture life of 4,000 h/25 ksi; or twice the minimum requirement. At 650°F, an average life of 235 h/10 ksi was obtained and all tests at stress levels of 7.5 and 5 ksi were terminated after 500 hours without specimen failure.

Comparisons of P/M Al-Fe-Ce alloy against two of the high-strength commercial aluminum alloys (A2219-T6 and A201-T7) are shown in Figures 9 and 10, in terms of tensile and stress rupture strengths, respectively. These figures show that the P/M aluminum is superior both to A2219 and A201 at all temperature levels up to 650°F. Note that the Al-Fe-Ce data was based on 1,000-hour overexposed specimens while the other two alloys are at optimum temper conditions.

The high-cycle fatigue (HCF) test program was designed to establish the

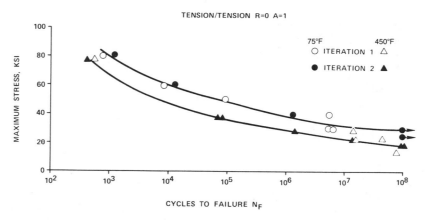

Figure 11. HCF Alcoa Al-Fe-Ce at 75°F and 450°F.

maximum stress that will yield a minimum fatigue life of 10^8 cycles. The tests were conducted in a tension/tension mode with an A-ratio of 1 (R = 0) using a sinusoidal waveform at 60 Hz frequency. Figure 11 shows the HCF test results at room temperature and 450°F. Al-Fe-Ce met the minimum impeller property goal for HCF life of 10^8 cycles at 425°F/15 ksi by demonstrating a run-out stress capability of 20 ksi/450°F. Adequate capability also is demonstrated at room temperature at a stress level of 30 ksi.

To establish a comprehensive assessment of material properties, baseline data on low-cycle fatigue capability, crack-growth resistance, and corrosion/erosion resistance are being generated for Al-Fe-Ce.

ACKNOWLEDGMENT

The author wishes to acknowledge the permission granted by the National Aeronautics and Space Administration (NASA) and the Department of Energy (DOE) to publish property data generated under the Garrett/Ford AGT101 Contract DEN3-167.

Reference

1. R.R. Sawtell, W.L. Otto, Jr., and D.J. Lege, "Elevated Temperature Aluminum Alloy Development," USAF Materials Laboratory, AFSC, Contract F33615-77-C-5086.

ABOUT THE AUTHOR

Ponciano P. Millan, Jr., Supervisor of Advanced Materials, Materials Engineering, Garrett Turbine Engine Company, Phoenix, Arizona.

Mr. Millan holds a BS degree in mining and metallurgical engineering from the Mapua Institute of Technology, Philippines and completed his graduate studies in Metallurgy in West Germany.

J. P. Van Buijtenen

Supervising Development Engineer,
Thomassen, Holland B. V.

W. M. Farrell

Chief Engineer,
Fern Engineering,
Bourne, Mass.

Mechanical Design of a High-Efficiency 7.5-MW (10,000-hp) Gas Turbine

This paper describes the mechanical design features on TF-10 gas turbine. The TF-10 gas turbine features an intercooled centrifugal compressor and an annular combustor. A regenerator is added as a standard item. The ISO rating is 7.5 MW (10,000 hp) with a rotor inlet temperature (RIT) of 1116°C (2042°F) and a thermal efficiency of approximately 44 percent. Description of the combustion, rotors, stators, bearings, and shafting is presented.

Introduction

In 1976, Thomassen Holland and Fern Engineering began the development of a high efficiency 7.5 MW gas turbine. The objective was to design a gas turbine with the highest possible thermal efficiency using proven state of the art mechanical design practice in order to achieve maximal reliability with minimal development.

Many studies were made to determine the cycle parameters such as pressure ratio and RIT, along with the basic turbomachinery flow path, number of stages, etc., which would optimize thermal efficiency. These were made in conjunction with mechanical design analysis to determine cooling air required, leakages, and bearing losses.

The final cycle selected was an intercooled, regenerative gas turbine with rotor inlet temperature of 1116°C and overall pressure ratio of 9:1.

The cycle optimization study was presented at the 1980 ASME conference [1].

The unit is a two-shaft design with an annular combustor, single stage axial high pressure turbine driving a double ended LP centrifugal compressor and a single hp centrifugal compressor, followed by a two stage load turbine.

Basic Machine Arrangement

The basic mechanical design approach was to combine the advantages of both the heavy duty industrial type design and aircraft derivatives. Features taken from aircraft design include:

- full annular combustion liner
- modular construction
- nickel base alloy turbine wheel material
- wheel space bucket cooling air boarding
- multiple fuel nozzles
- nonsplit combustor casing and exhaust frame for structural integrity

Similarly, features taken from heavy duty industrial gas turbine design include:

- hydrodynamic bearings
- low alloy steels as casing materials
- low cost construction
- simple maintenance, no special tools
- fewer stages of turbo machinery
- split turbine shell for hot gas path inspections and service
- rugged, long life design

The basic aerodynamic approach was to make full use of the parameters that make the regenerative intercooled cycle [RIC] one of the highest in thermal efficiency.

The 9:1 pressure ratio was selected by optimizing on a firing temperature [RIT], which in turn was optimized by including the effect of mechanical design in terms of cooling and sealing. For example, without realistic evaluation of the required hot gas path cooling and leakage air, the indicated optimum pressure ratio would be near 11:1. The number of stages was also selected by considering the effect of bucket cooling air, wheel space cooling air, shroud, nozzle, and stage seal leakage. The exhaust diffuser is optimized also in mechanical requirements such as strut cooling air introduction and gas path overlaps.

One of the most important aspects of a gas turbine machine arrangement is the construction and location of the rotors and bearings. In the early stage of the TF-10 design, both two shaft and single shaft versions were considered. Also, combined and separate compressor and turbine casings and simply supported and overhung wheels were studied. Twenty different shaft and bearing arrangements were considered as depicted in Fig. 2.

The single shaft versions were eliminated and the final choice was No. 17, i.e., six bearings with over hung turbine rotors and separate turbine and compressor casings. The advantages of this choice are:

1 Each of the three pieces of turbomachinery has its own rotor with two bearings of known reactions and is vibrationally isolated from the other rotors via flexible couplings.

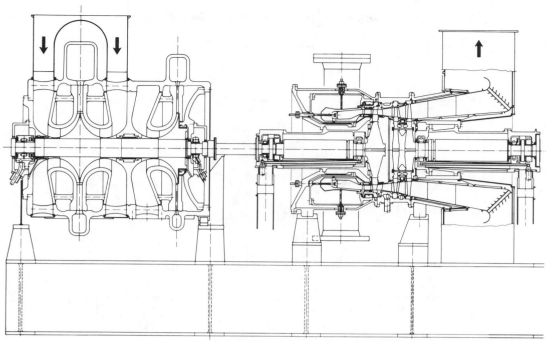

Fig. 1 TF-10 cross section

2 Problems of alignment are minimized and no shaft stresses are introduced through misalignment.

3 Better control is possible of critical speeds, rotor response, loss of bucket capability, and subsynchronous whirl.

4 Each section of turbomachinery can be designed and optimized in a modular fashion independent of each other.

5 Changes in design and application (such as a low Btu gas combustor or a different load turbine) can be accomplished without major redesign.

6 Maintenance and/or replacement of each section is independent of the other.

The arrangement then consists of a compressor with LP and hp centrifugal impellers straddle mounted on a two bearing shaft in its own independently supported casing. The compressor is connected to the turbine with a flexible coupling.

The turbine consists of two modules, hp and L.P. Six flanges break these into five major assemblies: hp casing, hp rotor cartridge, LP casing, exhaust frame, and LP rotor cartridge.

The turbine casing is also independently centerline supported and gibbed. The exhaust has an isolated hood that is sealed and supported from the exhaust frame.

Air is piped from the LP compressor outlet to the intercooler, from the intercooler to the hp compressor inlet and from the hp compressor outlet to the regenerator. Pipes return heated air from the regenerator to four inlet ports in the combustor casing.

Compressor Design

Rotor. Both impellers of the two stage centrifugal compressor are located on one shaft supported by bearings at each end. The shaft is driven from the hp end, while at the LP end, a coupling drives the accessory gear. The dimensions, shape and speed of the impellers are set by aerodynamic optimization, taking into account optimum running conditions of the HP turbine stage. So the double entry LP stage configuration was established as being necessary to approach optimum specific speed for all three turbomachinery components running at the same physical speed [2].

Wheel stress calculations were performed by finite element methods to assure mechanical and aerodynamic optimization. The results of these stress calculations indicated that a number of materials were possible, such as high alloy steels, nickel base alloys, and Titanium alloys. Other considerations, such as corrosion and erosion resistance, manufacturing and weight (critical speed), were taken into account in making the final selection: German grade 1.4405A 16/5 chromium nickel stainless steel.

An early analysis of the LP wheel showed an advantage of the LP double flow concept: the symmetric shape gives a more uniform stress distribution at the bore. Giving the backside of the hp hub a shape either inwards or outwards resulted in an improvement in the hp bore stress. (Figs. 4(a–c)). Integration of a balancing piston with hp impeller hub gave a shape which is very close to the shape of the LP hub. The result of the analysis showed a considerable decrease in bore stress. However, an unacceptable stress peak occurred in the corner between balance piston and hub (Fig. 4(d)). Different designs were investigated to solve this problem. The best solution was the shape of Fig. 4(c) [3, 4].

Aerodynamic design of the inlet channels of the radial impellers gave a lower limit to the amount of axial spacing between the impellers and/or other components. A number of different shaft configurations were considered and analyzed for critical speed with the influences of bearing span, overhang weight, and location of masses between the bearings (see Fig. 5).

Figure 5(a) was eliminated in an early stage. Apart from not fulfilling the critical speed demands. In this case, it is not possible to take full advantage of the overhung impeller concept because both shaft ends bear a coupling to another machine part.

A major remaining consideration was the thrust bearing location. Putting the thrust collar between the radial bearings increased the bearing span, while putting it outside increased the overhung weight.

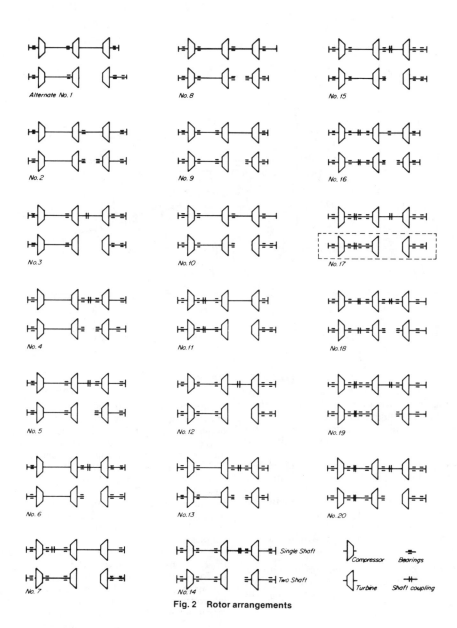

Fig. 2   Rotor arrangements

None of these locations gave a satisfactory fulfillment of critical speed demands (Figs. 5(b) and (c)).

The solution was found by separating the active and inactive thrust bearing and locating them at either side of the radial bearing (see Figs. 5(e) and (f)).

As thrust faces the shaft material itself is used, being the remaining surface between radial bearing diameter and overall shaft diameter (see Fig. 6.) Because this surface is too small to carry the full thrust as calculated for hp impeller, the latter is equipped with a thrust balancing device (see Fig. 7).

The diameter of this balancing piston is chosen in a way that under all operating conditions there will be a force present in a constant direction.

The balancing piston will be made integral with the hp impeller, which is favorable for the maximum bore stress in the wheel (see Fig. 4). Comparing Figs. 5(c) and (d) proved that the location of the hp wheel mass as close as possible to the bearing was favorable.

The use of the balancing piston gave the opportunity to level off the hp wheel exhaust pressure to atmospheric pressure in different stages, thus minimizing the amount of air leaking to atmosphere.

A scheme for the hp sealing is also shown in Fig. 7. The backface of the hp impeller is subjected to the static pressure of the wheel outlet at the outer part, while the inner part is subjected to the pressue maintained next to the bearing housing, which pressure is approximately 1.3 bar. The seal of the balance piston is split into two portions: The first seal seals from hp wheel discharge pressure to hp inlet pressure and the second seals the hp stage inlet pressure from the pressure next to the bearing housing.

Casing

The compressor module of the TF-10 has been designed fulfilling the following demands:

- horizontally split casing
- all piping connections in bottom half
- inlet hood to be mounted in any clock position

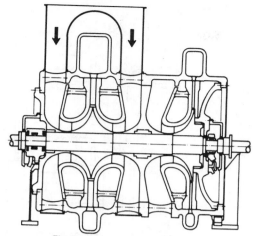

Fig. 3 Compressor cross section

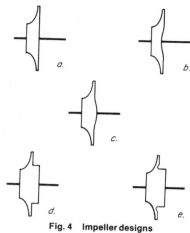

Fig. 4 Impeller designs

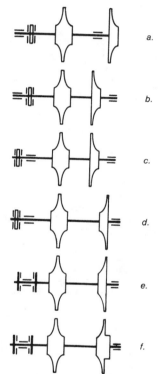

Fig. 5 Thrust bearing locations

- no bolting in flow path
- bearing inspection possible without lifting upper half

The main element of the casing is a cylinder (see Fig. 3). To allow the inlet air to flow in, openings are cast in circumferentially. The lower half has openings for the hp inlet.

Outside the cylinder, the scrolls are located for collecting the discharge air of both stages.

The piping connections of the scrolls are in the bottom half. The bearing housings are located in the end walls of the cylinder. Within the cylinder, there are three diaphragms:

- shrouds and diffusor of the LP stage
- shroud and diffusor of the hp stage
- interstage diaphragm

Pressure loading on structural parts is avoided as much as possible. The LP end-wall is only subjected to the load equal to the pressure drop across the inlet. The interstage is loaded by the pressure difference between hp and LP inlet. The hp end-wall is only subjected to hp inlet pressure at the outside. This is another advantage of the split seal on the balancing piston. Moreover, the hp diaphragm is subjected to the same pressure at both sides. The hp discharge pressure is contained between both walls of the diffusor. The only part that has to withstand full compressor discharge pressure is the hp scroll. Thrust load is taken by the LP endwall, which is shaped to do that with minimal deflection.

The LP part and hp part of the casing were analyzed by finite element methods. The pressure load and thrust load can be taken easily by the proposed design. To decrease temperature stresses, material has been removed at places where no pressure or mechanical load is of high value. Therefore, LP inlet struts were made smaller, although their number increased. Wall thickness of the main cylinder was adapted to flatten out stress peaks.

Combustion Design

The combustor liner (shown in Fig. 9) is one of the components in the TF-10 that is contributing to industrial gas turbine "state of the art" design. It is an annular design, employing aero-combustor technology. The combustor inlet temperature, due to regeneration, is high for industrial combustors, yet due to the TF-10's low pressure ratio, the air mass flow is relatively low. This, along with a relatively high firing temperature 1116°C (2042°F) RIT, provided a challenging heat transfer design objective.

To accomplish this high temperature combustor liner design and still maintain simplicity, mechanical integrity, and high degree of reliability, the liner will be made out of Hastalloy-x and will use 560°C (1040°F) inlet air to cool the liner walls.

Air is supplied to the combustor via four radial regenerator feed pipes, where it then impinges on an annular flow distributor and then turns 180 deg to approach the combustor in an axial direction.

The liner walls will be cooled using a combination of splash-impingement and film cooling. In order to keep the cooling film effectiveness high, the cooling air is introduced at 16 different axial locations (eight inner and eight outer).

This construction offers a short stiff liner that has a high

Source: Journal of Engineering for Power, July 1982, 707-711

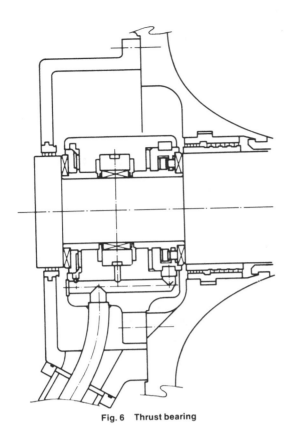

Fig. 6 Thrust bearing

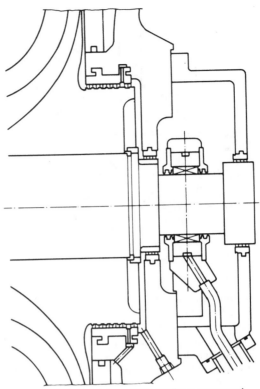

Fig. 7 Balance piston with hp sealing arrangement

plate stiffness, producing high resistance to vibration problems and creep buckling.

The annular flow distributor has also been kept short and rigid and is supported in such a manner that its thermal growth is independent from both the combustor liner and combustor casing, hence avoiding high thermal stresses.

The fuel injection system consists of 24 axial injector tubes fed from one annular manifold. The tubes are mounted for easy removal and maintenance. The ignition system consists of two radially located retractible igniters (180 deg apart). Flame detection will be provided by two radially located (again 180 deg apart) uv detectors. The detectors are mounted outside of the hp casing. Tubes, located on the flow distributor, provide sight into the primary combustion zone via a primary air injection hole.

The combustor liner, combustor casing, flow distributor, and all peripherals have been designed to allow adaption (by resizing) to a wide variety of fuels. By virtue of the TF-10 modular concept, the turbine can easily adapt even a very low Btu gas, via change in only the turbine and combustor liner, casing, and flow distributor.

Turbine Gas Path Design

The turbine stator vanes (nozzles) are cast and machined cobalt base material because of its high melting temperature for resistance to hot streaks and for its repairability (welding). The A0 nozzle is made of FSX 414 for its superior oxidation and corrosion resistance and the A1 and A2 nozzles are made from X40 for the high strength. In this section, the hp turbine is referred to as A0 stage and the power turbine is referred to as A1 and A2 stage.

The A0 nozzle and A0 shroud are hook mounted on a pin supported complete ring of Greek Ascoloy (12 percent Cr) material. This ring of low temperature expansion coefficient

is used to better maintain seal and bucket tip clearances. The A0 nozzle is integrally cast in a two-vane segment; 12 segments to an assembly with seal strip and groove type sealing between segments at the inner and outer side walls. A Hastalloy X T-shaped seal is used to seal regenerator air discharge pressure at the inner wall.

A0 nozzle is cooled utilizing a combination of impingement and film cooling techniques. The outer sidewall and the airfoil body are impingement cooled with Inconel 60 plate and insert. The inner sidewall and trailing edge of the airfoil is film cooled via drilled holes. The trailing edge uses two rows of pressure side bleed holes and one row of slots.

The A1 nozzle is made of 12 two-vane segments that are hook mounted to the forward Hastalloy X support segments and to 321 S.S. A1 stator shrouds. It is also sealed by strip-and-groove seals at the sidewalls. The A1 nozzle supports a diaphragm disc of 302 stainless steel. A body impingement insert and one row of pressure side drilled cooling holes is used to cool the A1 nozzle.

The uncooled A2 nozzle is made of 12 three-vane segments that are hook mounted on the A1 stator shroud segments and on the 304 S.S. A2 stator shroud segments. It supports the segments of 304 S.S. A1/A2 interstage seal.

The turbine rotor blades (buckets) are all of the long shank integral cover type design with one horizontal platform pin and one radial side cover pin for damping and sealing. The buckets are mounted in their wheels via a high efficiency, low stress, three tang dovetail using a loose fit. The buckets are locked axially, using a single segment locking wire hoop of Inconel X750 laced through the hookover groove that alternates between wheel and bucket dovetail ends. All three stages are cast and machined from IN 738 LC material and carry two wheel seal wings on both sides for better wheel space cooling control. The A0 bucket is cooled using ten ECM drilled holes which run from the bottom of the dovetail to a

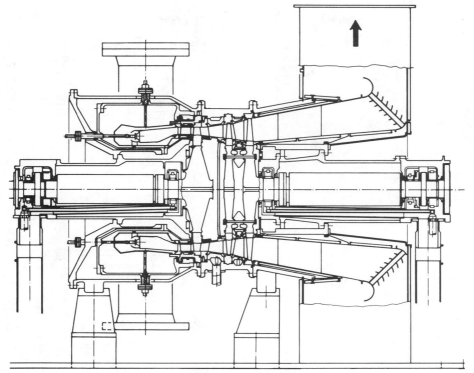

Fig. 8   Turbine cross section

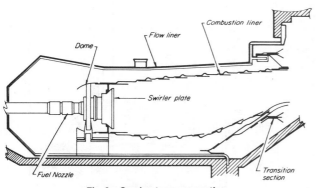

Fig. 9   Combustor cross section

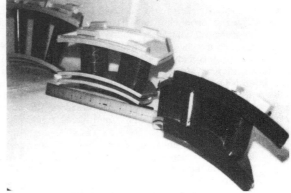

Fig. 10   A0 · A-1 · A2 nozzle

hollow cavity in the bucket tip. The cooling air is supplied from compressor discharge air and accelerated to wheel speed into the A0 forward wheel space. From there it boards and flows through holes in the wheel, one per dovetail, to a cavity at the bottom of the dovetail. There are 60 A0 buckets and 72 buckets each for the A1 and A2 stage. The A0 bucket has a free tip, whereas the A1 and A2 buckets have a Z-shaped interlocking shroud with a single rotor seal and two stator seals.

The buckets have been designed for standard and conservative static and dynamic stress levels with metal temperatures supporting long industrial parts life.

The design of the A2 bucket has been taken to the full capacity of present state of the art. It is believed to be the highest aerodynamic loading of any gas turbine in terms of annulus area carrying capacity. The parameter for this is annulus area speed squared.

Each stage of the buckets, especially the A2, will be submitted to a series of vibration tests, including a full production rotor wheel-box test for determination of the acceptability of their vibration characteristics. This will be further verified in actual prototype gas turbine operation.

The exhaust diffuser is an important part of the gas path design and its aerodynamic performance is critical to the overall gas turbine performance. The resulting diffuser as shown in Fig. 8 is an optimization of aerodynamic performance (with a recovery factor of approximately 0.6) and the mechanical design in terms of length, diameter, weight, and cost. It is fabricated of 321 S.S. with insulation jackets to help exhaust frame cooling and includes turning vanes and airfoil shaped enclosures of the exhaust frame struts.

Turbine Rotor Design

An example of the turbine rotor construction is shown with the example of the high pressure rotor in Fig. 12.

The problem of attaching a highly stressed large growth

Source: Journal of Engineering for Power, July 1982, 707-711

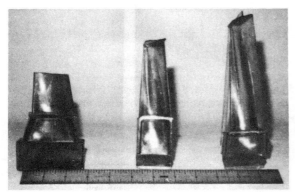

Fig. 11   A0 - A-1 - A2 bucket

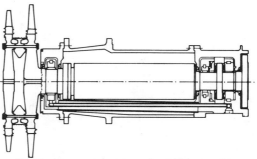

Fig. 13   Low pressure rotor and cartridge assembly

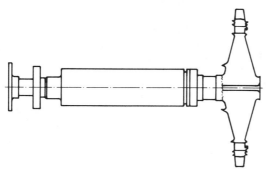

Fig. 12   High pressure rotor

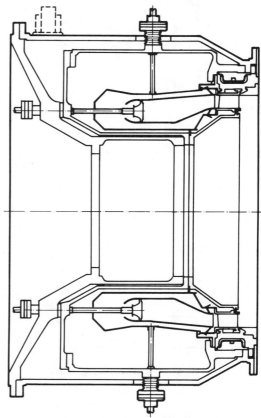

Fig. 14   High pressure casing assembly

component such as a turbine wheel to a cold, low stress shaft is solved for these rotors through the use of an integral stub. This allows the turbine wheel to be made of an excellent Nickel based material, IN718. Moreover, the wheels can be made more or less symmetrically shaped with a small bore providing excellent stress distribution. It is then bolted to a steel AMS 6304 material shaft. Therefore, the bolting flange is in an area of minimum strength requirement. The hp rotor is then a simple construction of only three components: buckets, wheel, and shaft, thereby minimizing stacking and assembly tolerances.

Both rotors are designed to have their free-free bending critical at over 150 percent speed. The two basic oil film modes, rock and bounce exhibit nearly straight-line mode shapes with little or no shaft bending and split the speed range so as to have highly damped and minimum vibration response.

All four journal bearings are highly stable, five pad, low L/D, tilt pad designs. These bearings have been designed with good industrial conservative practice in terms of clearance and oil temperature rise. The thrust bearings are also of the tilt pad design and have been sized for a specific loading considerably less than the thrust bearing capacity.

The couplings are designed to be of the dry flexible type marine flange connection for use between the rotors and the load and gas turbine compressor. These couplings are basically the only connections between the turbine section or module and the load compressor and gas turbine compressor module and allow for thermal growth mismatches exceeding those expected.

Turbine Stator Design

The turbine stator consists of five easily separated assemblies: hp rotor cartridge, LP rotor cartridge, hp casing, LP casing and exhaust frame.

Each cartridge assembly consists of two low alloy steel bearing housings, two tilting pad journal bearings, a tilting pad thrust bearing, an inactive thrust bearing, a nodular cast iron supporting cylindrical cartridge, oil feed and drain lines and seals. Tilting pad journal bearings were selected for their dynamic stability and vibration damping characteristics. The oil sump operates at a 0.5 psi vacuum to prevent oil leaks. Air ducts are cast into the hp cartridge to supply cooling air to the A0 buckets through the forward wheel space.

The hp casing assembly consists of a two piece outer cast iron support structure and a two piece inner 2¼ ´CR-Mo pressure vessel. The vessel contains the combustor assembly and the outer structure supports the hp cartridge and A0 nozzle assemblies. The casings are not split.

The LP casing assembly consists of a split nodular cast iron

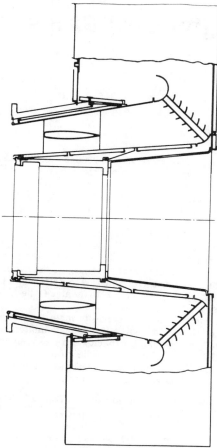

Fig. 15 Exhaust frame assembly

Maintenance

Considerable thought has been given to make the TF-10 as easy to maintain as possible. Following are some of the features which have been incorporated.

Modules. The machine has been designed around basic modules: compressor, hp turbine and LP turbine. Each can be easily removed or replaced. All modules are interchangeable without machining or dowling.

Bearings. All bearings, except the two turbine inboard journals, are accessible from outside the machine.

Compressor Rotor. Compressor split casing allows access to compressor rotor.

Turbine Blades. All blades can be inspected from outside the machine through boroscope access ports. All blades can be replaced through boroscope access ports. All blades can be replaced by removing the split LP casing.

Turbine Nozzles. A0 nozzles can be inspected through fuel nozzle tube mounting pads. A1 and A2 nozzles can be inspected through boroscope ports in the LP casing. A1 and A2 nozzles can be replaced by removing LP casing halves.

Combustor. Fuel nozzles, igniters, and flame monitors can be easily inspected and replaced from outside the engine. Combustor may be inspected through the 24 fuel nozzle mounting pads.

Conclusions

The TF-10 is designed using conservative proven design concepts. The relatively low pressure ratio and the free standing modular compressor and turbine designs provide greater flexibility for the designer and flexible maintenance options for the user.

Acknowledgments

Both authors wish to thank their colleagues at Thomassen, Holland B.V., and Fern Engineering who have been of great help in preparing this paper.

References

1 Hendriks, R., and Levine, P., "Cycle Optimization of a 10,000 SHP High Efficiency Gas Turbine System," ASME Paper 80-GT-157.

2 Balje, O. E., "A Study of Design Criteria and Matching of Turbomachines," ASME JOURNAL OF ENGINEERING FOR POWER, 1962.

3 Peterson, R. E., *Stress Concentration Factors*, Wiley, New York.

4 Thum, A., and Bautz, W., "Der Entlastungsubergang-Günstige Ausbildung des Uberganges an abgesetzen Wellen u. dgl.," *Forschung im Ingenieurwesen*, Vol. 6, 1934, p. 269.

casing, nozzle support blocks, A1 and A2 blade shroud blocks, A1 and A2 nozzles, and interstage diaphragms.

The exhaust assembly consists of a low alloy fabricated steel frame, a CRES 321 diffuser, turning vanes, and an exhaust hood. The frame has six struts and supports the LP cartridge. It is completely protected from the exhaust gas by the diffuser and is cooled by compressor leakage air. It also supports the exhaust collector hood which may be bolted to the frame so to exhaust in any direction.

Thermal History of Aircraft Engine Seal Rings of Ni-Cr-Mo Alloy

By A. M. EDWARDS
High Technology Materials Div.
Cabot Corp.
Kokomo, Ind.

At airline maintenance centers throughout the world, craftsmen overhaul modern aircraft with the care and dedication that the Wright Brothers exhibited in building their flying machine. Today's tools and materials are different, though, as are the conditions inside the engines.

The higher, more efficient operating temperatures of modern jet engines demand better heat-resistant alloys for critical areas. A good example of this a nickel-base, high-temperature alloy used for rings that experience high temperatures during flight.

The rings are fabricated by McClain International, College Park, Ga. They serve as the first-stage outer air seal for Pratt & Whitney JT8D series turbine engines which are used on the 727, 737, and DC-9 aircraft. The rings are made of nickel-chromium-molybdenum alloy, HASTELLOY® alloy S, developed and produced by the High Technology Materials Division of Cabot Corp., Kokomo, Ind.

Patented in 1973, the alloy was originally developed for rings in the JT8 series engines. This application involves severely cyclical heating conditions where components must be capable of retaining their strength, ductility and metallurgical integrity during long-time exposure. The alloy has excellent thermal stability, low thermal expansion and out· standing oxidation resistance to 2000°F (1090°C). In addition, it has good high-temperature strength and thermal fatigue resistance. Strength and ductility are retained even after aging at temperatures of 800 to 1600°F (425 to 870°C) for periods longer than 16,000 hr (1.8 years).

Primary Melting and Refining

The production of alloy S begins with melting of a 30,000 lb (13,608 kg) charge in a three-phase, electric arc furnace (see Fig. 1). The make-up of the charge is determined by computer selection of the available raw materials and is adjusted, as needed, during the melt to achieve the desired final chemistry of the alloy (see Table I).

The electric arc furnace serves to melt down and refine the raw materials. Additional refining of the molten alloy is then carried out in a 30,000 lb-capacity argon oxygen decarburization (AOD) vessel, Fig. 2, in order to remove excess carbon and "tramp" elements while retaining more of valuable metallic elements such as chromium.

Table I Chemical Composition of HASTELLOY alloy S

ELEMENT	PERCENT	ELEMENT	PERCENT
Nickel	Balance	Silicon	0.20 - 0.75
Cobalt	2.0*	Manganese	0.3 - 1.0
Chromium	14.5 - 17.0	Carbon	0.02*
Molybdenum	14.0 - 16.5	Aluminum	0.10 - 0.50
Tungsten	1.0*	Boron	0.015*
Iron	3.0*	Lanthanum	0.01 - 0.10

* Maximum

Fig. 1 Tapping of electric arc furnace in the melt shop.

Fig. 2 View of argon oxygen decarburization vessel. AOD provides a practical way to hold the extra-low carbon levels required for production of the high temperature Ni-Cr-Mo alloy.

Cabot was one of the first specialty metals producers to use the AOD process in which argon and oxygen gases are "blown" through the molten metal. The argon minimizes the reduction of the chromium and other reactive elements while the oxygen combines with the carbon to form CO which is vented through a hood at the top of the vessel.

The molten metal from the AOD vessel is tapped into an electrode mold where it is allowed to solidify. The cast 10,000 lb (4,536 kg) electrode is then removed from the mold and prepared for remelting.

Secondary Refining

Unlike ordinary metals, which normally go through a single melting process, alloy S and other high-performance alloys require secondary melting or refining. Secondary melting facilities at Cabot, among the largest in the specialty metals industry, include seven computer-controlled, electroslag remelt (ESR) furnaces.

The electroslag remelting process further refines the primary melted electrodes. In this procedure, Fig. 3, the electrode is melted dropwise through a bath of molten slag contained in a water-cooled copper crucible. This slag is kept molten by the passage of an electric current from the electrode to the copper crucible. Molten alloy droplets fall through the slag, which provides a refining action on the

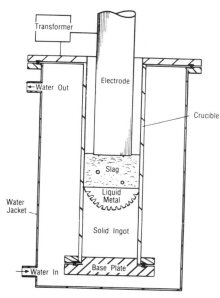

Fig. 3 Electroslag remelting of consummable electrode further refines the metal for production of forged bars used in rolling the aircraft engine seal rings.

metal, and protects the remelted material from air. In essence, a simultaneous process of melting and refining of the electrode is occurring. The molten alloy droplets collect in the bottom of the crucible and solidify into a refined ingot.

Hot Working

Hot working of the ESR ingot starts on a 2,000-ton hydraulic forge press, Fig. 4. The working temperature range for forging alloy S in ingot form is 2100°F (1150°C) to 1600°F (870°C). Three to eight reheats in soaking furnaces, Fig. 5, are usually required to maintain the workability of the ingot during forging. The final forged product is a 4-in. (10.16 cm) round-cornered-square which receives further processing on bar mills.

The 4-in. (10.16 cm) squares are reduced to 2-3/4 in. (6.99 cm) gothics on a two-stand 24-in. (60.96 cm) bar mill. The reduction is achieved while maintaining a working temperature between 1600 and 2100°F (870°C and 1150°C). Further processing is carried out by putting the bars through

Fig. 5 Gas-fired reheat furnaces in forging area.

a series of passes on a four-stand 10-in. bar mill to achieve the desired shape and dimensions.

After rolling, the bars are given a final anneal at 1950°F (1065°C) to achieve optimum mechanical properties. They are pickled in a nitric-hydrofluoric acid bath, straightened, and inspected ultrasonically for internal defects before packaging and shipment from Cabot's Kokomo plant.

Fig. 4 Ingot breakdown on a 2000-ton forging press.

Fig. 6 Flash butt welding of rolled seal ring of Ni-Cr-Mo alloy.

Fig. 7 Spot welding of Ni-Cr-Mo alloy seal ring, prior to nickel brazing in hydrogen atmosphere.

Ring Fabrication

At the McClain plant, the 83-in. (210.82 cm) rolled-shaped bars (0.62 in. or 1.57 cm thick and 2.2-in. or 5.59 cm wide) are cut to specific length, depending on which one of three seals is being fabricated, and then rolled into rings approximately 27 in. (68.58 cm) in diameter. The rings are flash-butt-welded, Fig. 6, and ground to remove excess weld material.

Thirty-six rings at a time are annealed at 1950°F (1065°C) for 30 minutes. After annealing, the rings are expanded, by about 2% of their diameter, to check the weld.

The rings are machined and then lined on the inside surface with a honeycomb ring. The honeycomb material, either HASTELLOY alloy S or X, is first spot-welded into place, Fig. 7, and then nickel-brazed (hydrogen atmosphere). Finally, the aft side of the ring is hardfaced with tungsten carbide using a plasma spray surfacing technique. The rings are inspected, packaged, and shipped to airline maintenance centers and turbine engine overhaul facilities nationwide.

McClain International, a manufacturer of FAA-approved parts for jet engine hot sections, has been fabricating first-stage engine air seal rings with HASTELLOY alloy S since 1976. Before that, they used HASTELLOY alloy C-276 which has similar thermal expansion characteristics to alloy S, but it does not have comparable thermal stability and oxidation resistance nor does it have the same degree of crack resistance in an aircraft engine environment.

CERAMICS FOR GAS TURBINE ENGINES

John L. Mason
The Garrett Corporation
and
Arthur F. McLean
Ford Motor Company

INTRODUCTION

Since the late 1940s, engineers have been intrigued with the possibility of using ceramics to improve the performance of gas turbine engines. Early attempts to utilize ceramics for this purpose were not successful, because of the inadequate thermal shock resistance of the ceramic materials available at that time and also because of inadequate design tools and technology for brittle materials. During the 1960s, new ceramic materials were developed, mainly silicon carbide and silicon nitride, that exhibited substantially improved thermal shock resistance.

This improved material availability, plus progress that had occurred in methods of design and of material analysis (scanning electron microscope, transmission electron microscope, etc.) led to an intensified interest in research and development of ceramics, directed toward gas turbine engines and other demanding, high-performance engineering applications.

This paper describes recent progress in applying ceramics to gas turbine engines. Potential benefits of ceramics to the gas turbine are described. Prior programs leading to the current Advanced Gas Turbine (AGT) are reviewed briefly. The AGT101 is described. An interim application of ceramics, to turbochargers, is considered briefly as a related subtopic.

BENEFITS FROM CERAMICS IN TURBINES

Successful application of ceramics to gas turbine engines would permit major steps forward in gas turbine engine performance and fuel economy. In addition, such use would also contribute to conservation of scarce and strategic materials. The elemental ingredients of structural ceramics (mainly silicon, carbon, and nitrogen) are non-strategic and in fact among the world's most abundant elements; they are less costly and more widely available than most of the metallic elements used in present-day gas turbines (1).

There are a number of gas-turbine applications: aircraft, missile, and ship propulsion and auxiliary power; electric power generation; gas compression for pipelines; miscellaneous emergency power applications; and vehicle propulsion. All of these could benefit from the successful use of ceramics. To date, experimental ceramic parts, made in different research programs, have been built for an aircraft turboprop engine, an electric power generating gas turbine, and several vehicular engines. In general, these programs were in the feasibility-demonstrator, or proof-of-concept category.

Of the gas turbine applications listed, aircraft propulsion is clearly the one most dominated by the gas turbine. Aircraft propulsion is expected to be one of the most challenging and difficult applications for ceramics.

As shown by Figure 1, ceramic materials maintain usable strength up to higher temperatures than do metals. Successfully applied, ceramics will therefore permit the gas turbine to operate essentially uncooled at turbine inlet temperatures well above those achievable with today's best high-temperature metallic turbine materials. Turbine inlet temperatures in the range of $1371^\circ C$ ($2500^\circ F$) are achievable with uncooled ceramics, compared to about $1038^\circ C$ ($1900^\circ F$) with uncooled metal hot-end components. With advanced aerodynamic components, the result can be an increase in the thermal efficiency of a vehicular gas turbine from about 40 percent in a metal engine to 46 percent or better in a ceramic engine (2), as shown in Figure 2. At the same time, per Figure 3, the airflow through the engine and therefore the size of the engine can be substantially reduced below the airflow and size for a comparable metal engine (2).

The vehicular gas turbine requires ceramics in order to realize competitive or better levels of fuel economy relative to other types of advanced vehicular engines such as the adiabatic diesel. Also, the vehicular turbine must have exhaust heat recovery, by means of a regenerator or recuperator, in order to obtain good fuel economy at part power. A passenger car engine in particular runs most of the time at low power levels. For even a large automobile, the road load is only about 10 hp at 30 mph, yet maximum power has to be many times greater to provide good hill climbing and acceleration.

Projections were made (3) of the thermal efficiencies obtainable with various types of present and candidate automobile engines, and, as shown in Figure 4, the advanced ceramic gas turbine appeared most attractive.

In the utility power field, Westinghouse (4) reported that a large combined-cycle combustion steam power plant employing a high-temperature ceramic turbine could operate at 50 percent thermal efficiency using fuel derived from coal -- a significant increase over today's 42 percent.

The small, auxiliary turbine engine without exhaust heat recovery (the so-called simple cycle) is an existing, proven type of power plant used either for emergency-standby or continuous power for a wide variety of applications: generator sets, pump and compressor drives, air-supply units, industrial power plants, very small portable power plants, cogeneration applications, and marine and hydrofoil auxiliary engines. Use of ceramics in this simple-cycle engine, with the permitted increase in turbine inlet temperature from $1000^\circ C$ ($1832^\circ F$), all metal, to $1371^\circ C$ ($2500^\circ F$), all-ceramic, increases the engine thermal efficiency from 28 percent to 33 percent, as shown in Figure 5 (2). Furthermore, as shown in Figure 6 (2), the power output of the ceramic turbine engine is almost 80 percent larger than that of its metal counterpart.

EXPERIMENTAL PROGRAMS

A pioneering ceramic activity of the 1970s was the Ford-DARPA program (5), in which the objective of 200 hr durability was achieved on a multiplicity of ceramic components on a complete stationary ceramic flowpath. The 200 hr durability test consisted of 175 hr at $1055^\circ C$ ($1930^\circ F$) and 25 hr at $1371^\circ C$ ($2500^\circ F$).

The stationary components comprised the following:

    1. A reaction-bonded Si_3N_4 (RBSN) nose cone.

    2. A reaction-bonded SiC (RBSC) combustor.

    3. Two stators. Both RBSN and RBSC were demonstrated.

    4. Two RBSN tip shrouds.

An all-ceramic, dual-density rotor was also hot spin-tested for 200 hr at a rim temperature of $1000^\circ C$ ($1832^\circ F$), blade temperatures of $1200^\circ C$ ($2192^\circ F$), and at speeds up to 50,000 rpm. The dual-density rotor consisted of RBSN blades with a hot-pressed Si_3N_4 (HPSN) hub.

In addition, a complete ceramic flowpath including a dual-density Si_3N_4 rotor was successfully tested in a turbine engine for 37 hr, including 2 hr at a turbine inlet temperature of $1371^\circ C$ ($2500^\circ F$) and at 50,000 rpm (6). Subsequently, several dual-density Si_3N_4 rotors were tested for 25 hr with actual blade temperatures of $1316^\circ C$ ($2400^\circ F$).

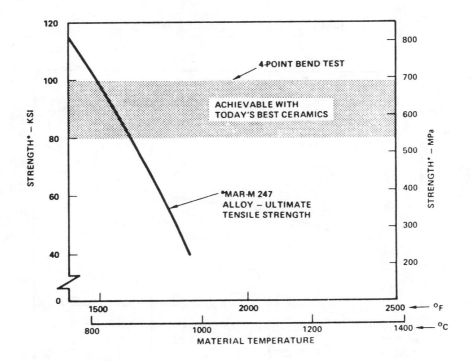

Figure 1. Strength vs temperature, metals and ceramics

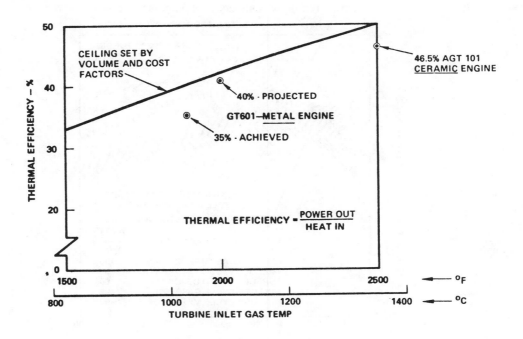

Figure 2. Best thermal efficiency -- vehicular gas turbines

Source: Materials and Society, February 1982, 201-217

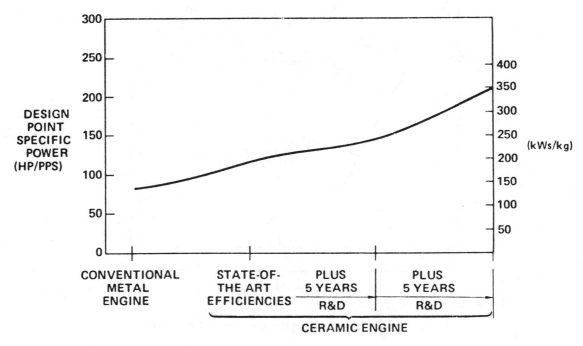

SPA 7897-1

Figure 3. Engine airflow vs temperature

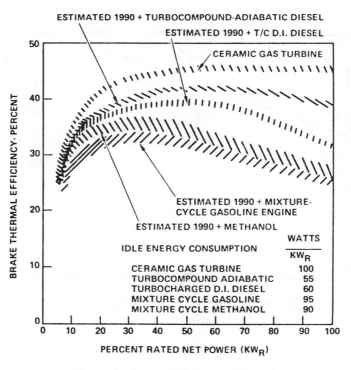

Figure 4. Advanced engine efficiencies

96

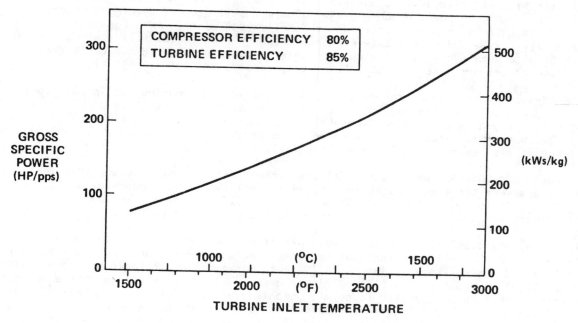

SPA 7897-4

Figure 5. Simple cycle efficiency vs temperature

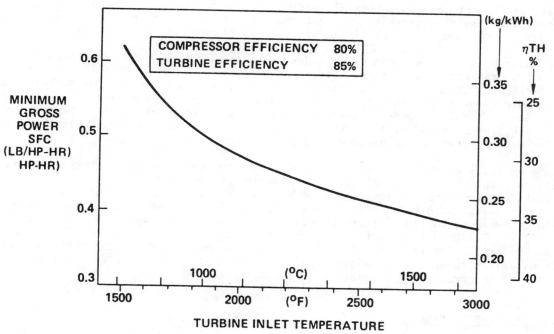

PA 7897-3

Figure 6. Simple cycle specific power vs temperature

Source: Materials and Society, February 1982, 201-217

Another gas turbine ceramic material under development was lithium aluminum silicate (LAS). In particular, a derivative of this material, aluminum silicate, was developed by Corning and applied to ceramic regenerator cores. The aluminum silicate was developed to resolve the corrosion problem that had been encountered with earlier LAS cores. In a Ford-DOE-NASA supported program (7), five aluminum silicate regenerators were each successfully tested for more than 10,000 hr on a duty cycle with regenerator inlet temperatures up to 800°C (1472°F). Also, four cores were tested for more than 5000 hr at temperatures to 1000°C (1832°F).

In a Ford-supported program, the concept of an uncooled LAS housing was introduced as a means of containing and ducting internal hot gases within a turbine engine while providing a stable platform for the regenerator sealing faces (8). The housing was a large 23 kg (50 lb) LAS component. Because of its size, it represented a pioneering fabrication effort. One such housing was successfully tested for 1000 hr without failure.

In another Ford-DOE-NASA program to evaluate ceramic stators under severe transient operation, RBSN stators were built and tested for 30,000 thermal cycles to temperatures of 1200°C (2192°F), and 9000 cycles to temperatures of 1371°C (2500°F) (9). A stator was supplied to Ford by Garrett's AiResearch Casting Company (ACC) as a part of this program (10).

An in-house interdisciplinary program was initiated at Garrett in 1973 to evaluate the potential of ceramics for short-life engine applications. Included were studies of design, fabrication, properties and inspection. In late 1975, Garrett was awarded a ceramic engine demonstrator program by the DARPA and the Navy (11). This program involved the redesign of the hot section of the Garrett T76 turboprop engine to permit uncooled operation at average turbine inlet temperatures of 1204°C (2200°F). Figure 7 shows a cross section of the ceramic engine, designated TSE331C-1. The redesigned engine contained 102 separate ceramic components. It was successfully operated in two-hour cycles to full design conditions. A 30 percent increase in power and a 7 percent reduction in fuel consumption were demonstrated, compared to the baseline metallic engine.

However, longer-time cyclic operation was not achieved because of the sensitivity of the design to contact stress fractures at static structure interfaces (12). A DOE-NASA-sponsored test program was carried out to impose biaxial loading (Figure 8) on various ceramic materials, and then to measure the resulting strength degradation. The stresses resulting from the biaxial loading were modeled and found to be friction-dependent, as illustrated in Figure 8.

Under a subsequent AFWAL program (13), redesign of the TSE331C-1 was undertaken to alleviate this problem, and the ceramic static structure components were successfully operated under severe cyclic conditions with no contact stress problem. In a parallel DARPA-funded effort (14), ceramic rotor blades inserted into a superalloy disc were successfully tested in a modified T76 engine for 15 hr under severe cycles to an average turbine inlet temperature of 1204°C (2200°F). Figure 9 shows the test cycle that was used. Figure 10 shows calculated stress contours for a first-stage rotor blade during a starting transient.

These engine tests established that it was feasible to use ceramics in short-life gas turbine engines, but that much additional development would be required to achieve reliability and durability in long-term applications. One problem was the potential susceptibility of the ceramic materials to oxidation over a long time period. In 1978, Garrett undertook a program for DOE and NASA to evaluate the longer-term exposure of commercial SiC and Si_3N_4 to combustion gases at 1204 to 1371°C (2200 to 2500°F) (15). Cyclic testing was carried out for 3500 hr in the burner rig shown schematically in Figure 11. This testing demonstrated the capability of the generic ceramic materials to survive in a gas-turbine environment. The positive results of this testing were a factor supporting the initiation of the AGT program.

ADVANCED GAS TURBINE PROGRAM

In late 1979, based on prior accomplishments and on the potential of ceramics, DOE, through NASA Lewis Research Center, initiated an Advanced Gas Turbine (AGT) program. An R and D contract (from the AGT101) was awarded to a team consisting of the Garrett Corporation and the Ford Motor Company (16). A parallel contract (for the AGT100) was awarded to the Allison and Pontiac divisions of General Motors (17). Each of these programs involved the design, development, and

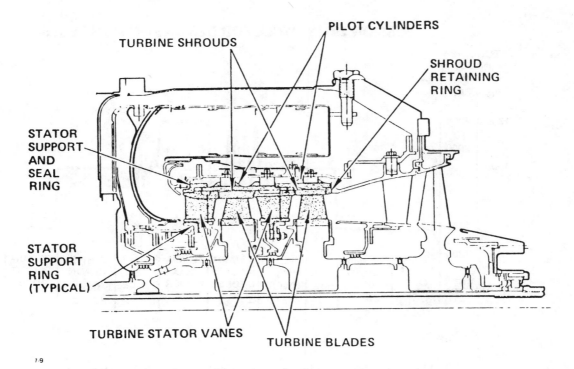

TURBINE SHROUDS

PILOT CYLINDERS

SHROUD
RETAINING
RING

STATOR
SUPPORT
AND
SEAL
RING

STATOR
SUPPORT
RING
(TYPICAL)

TURBINE STATOR VANES

TURBINE BLADES

7-9

Figure 7. TSE331C-1 ceramic turbine section

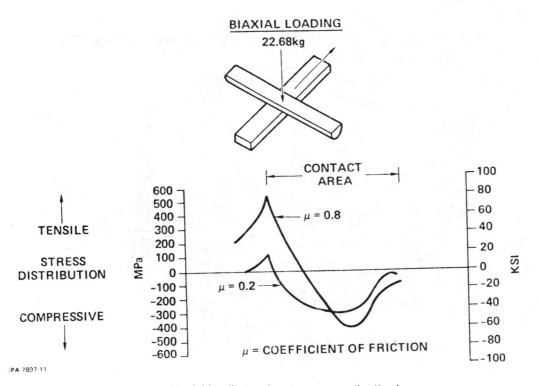

BIAXIAL LOADING
22.68kg

CONTACT
AREA

TENSILE

STRESS
DISTRIBUTION

COMPRESSIVE

μ = 0.8

μ = 0.2

μ = COEFFICIENT OF FRICTION

PA 7897-11

Figure 8. Biaxial loading and contact stress distribution

MISSION DUTY CYCLE FOR DEMONSTRATION TESTS

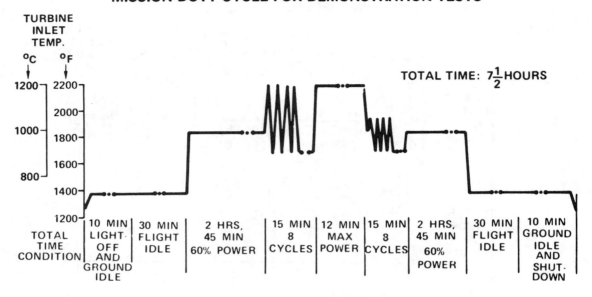

SPA 7897-12

Figure 9. Test cycle for 15-hour engine run (ceramic first-stage rotor).

WORST CASE CONDITION 125 SECONDS AFTER ENGINE START

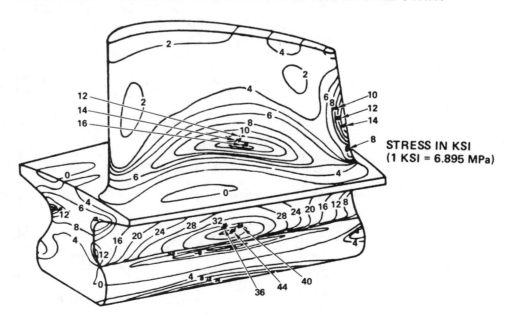

Figure 10. First-stage rotor blade maximum principal stress distribution

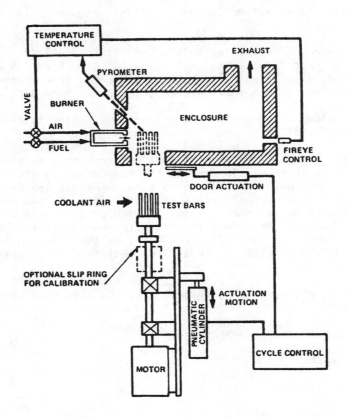

Figure 11. Schematic of ceramic test bar
durability test facility (3500-hour test)

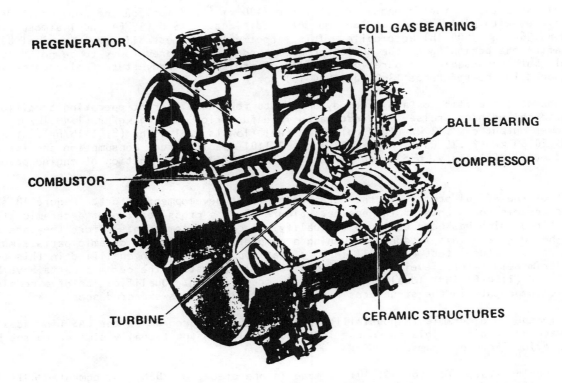

Figure 12. The Garrett-Ford AGT101

Source: Materials and Society, February 1982, 201-217

testing of an advanced ceramic automotive gas turbine. Figure 12 is a cutaway drawing of the AGT101.

Figure 13 shows the program organization of the Garrett/Ford AGT 101. Garrett is the prime contractor, through its Garrett Turbine Engine Company (GTEC). Ford, as major subcontractor, is responsible for vehicle integration design aspects, for the regenerator subsystem, and for a number of ceramic components. Other ceramic contractors include ACC, Carborundum, and Corning.

Figure 14 shows major elements of the program schedule. Engine test-stand demonstration is scheduled for the end of 1985.

The AGT101 is a 75 kW (100 hp) rated single-shaft gas turbine. In the fully ceramic version, all hot-end parts will be ceramic. To develop the non-ceramic subsystems, an interim metal version of the AGT101 has been built and run for a total of about 150 hr. Figure 15 illustrates the concept of introduction of ceramic parts into the engine in two phases. Three cross sections are shown. The ceramic parts are shaded in each cross section. The cross-section on the left, labeled 871°C (1600°F), represents the all-metal engine that is being run now. The cross-section on the right, labeled 1371°C (2500°F), is the all-ceramic engine. The center cross-section, labeled 1149°C (2100°F), is an intermediate configuration with all parts ceramic except for the turbine rotor.

On the previously shown cutaway drawing (Figure 12), the flow path components of the AGT101 are identified as follows:

1. Single-stage centrifugal compressor.

2. Rotary regenerator.

3. Single-can combustor.

4. Single-stage radial turbine.

5. Various ducts and baffles connecting the above components.

The AGT101 has a target fuel economy of 5.5 liters/100 km (42.8 mpg) on the combined federal driving cycle, with diesel fuel, at ISO ambient conditions -- 15°C (59°F) and 1 atmosphere pressure--in a 1360 kg (3000 lb) automobile. This target value was established in early 1980, shortly after the beginning of the AGT101 program, and is considered today to be not only achievable but surpassable. Figure 16 is a plot of projected steady-state fuel economy versus vehicle speed in the reference 1360 kg (3000 lb) car.

Engine schematic diagrams contain thermodynamic data for two different operating conditions, idle and so-called high cruise. The estimated idle fuel flow is 0.68 kg/hr (1.49 lb/hr), at an engine power output of 1.13 kW (1.52 hp). The fuel flow is 3.91 kg/hr (8.61 lb/hr), at a power output of 20.69 kW (27.75 hp). Figure 17 shows AGT101 specific fuel consumption and its counterpart, thermal efficiency, projected for the ceramic AGT101 as a function of engine power output.

The heart of the AGT101 program is the ceramic component development effort. Figure 18 shows the major ceramic components of the engine. Figure 19 is a cross section of a ceramic structures test rig which has been designed to qualify all ceramic components before they are run in the test-bed engine. This is done by running a complete set of static ceramic parts simultaneously in the structures test rig. All ceramic static parts have been qualified in this test rig to an interim temperature level of 871°C (1600°F). Also, most of the ceramic parts have been run to 1149°C (2100°F) individually in a thermal screening rig. Qualification of a complete set of ceramic components in the structures rig at 1093°C (2000°F) is expected soon.

Included in the Ford ceramic responsibilities on the AGT 101 program is the LAS flow separator housing made by Corning. This component is structurally and functionally similar to the housing used in previous Ford programs.

The Ford turbine stator for the AGT 101 is made in one piece, of RBSN. A segmented AGT101 stator was made of RBSN by ACC. The one-piece stator approach is favored over the segmented

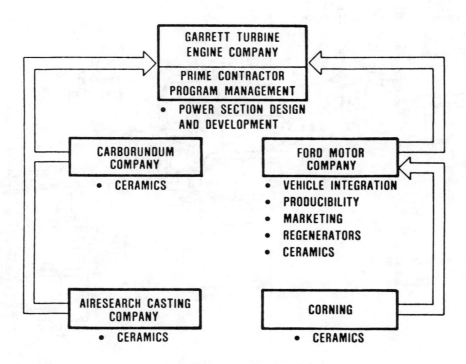

PA 7897-17

Figure 13. AGT101 program organization

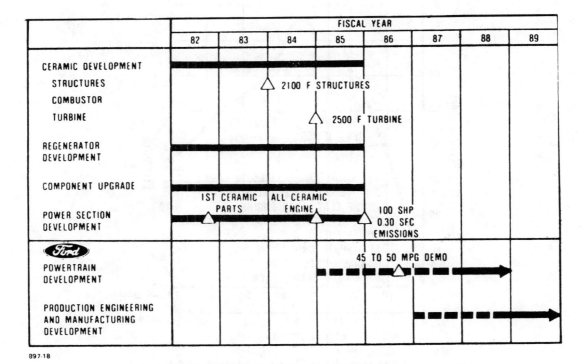

897-18

Figure 14. AGT101 program schedule

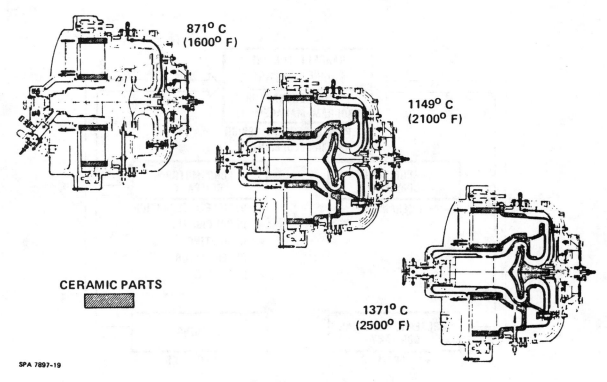

871° C
(1600° F)

1149° C
(2100° F)

1371° C
(2500° F)

CERAMIC PARTS

SPA 7897-19

Figure 15. AGT101 engine evolution

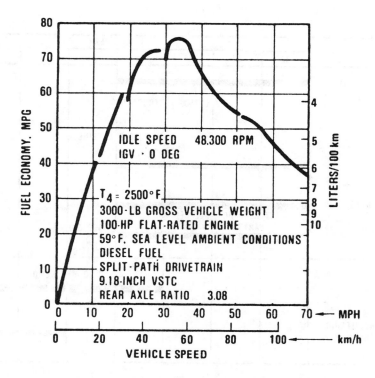

IDLE SPEED   48.300 RPM
IGV · 0 DEG

T_4 = 2500°F
3000-LB GROSS VEHICLE WEIGHT
100-HP FLAT-RATED ENGINE
59°F. SEA LEVEL AMBIENT CONDITIONS
DIESEL FUEL
SPLIT-PATH DRIVETRAIN
9.18-INCH VSTC
REAR AXLE RATIO   3.08

Figure 16. AGT101 steady-state fuel economy vs vehicle speed

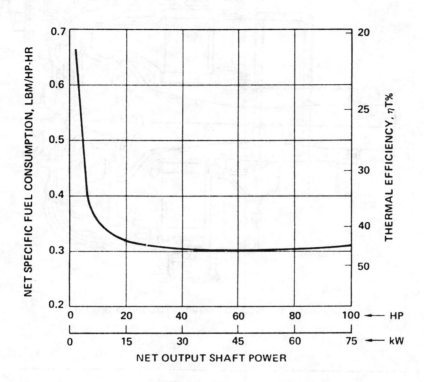

Figure 17. AGT101 sfc and thermal efficiency

Flow Separator Housing
Turbine Shroud
Turbine Rotor
Inner Diffuser Housing
Outer Diffuser Housing
Combustor Liner
Stator Vane Segments
Turbine Transition Liner
Combustor Baffle
Turbine Baskshroud
Bolts
Regenerator Shield

Fig. 18. AGT 101 Ceramic Parts

Source: Materials and Society, February 1982, 201-217

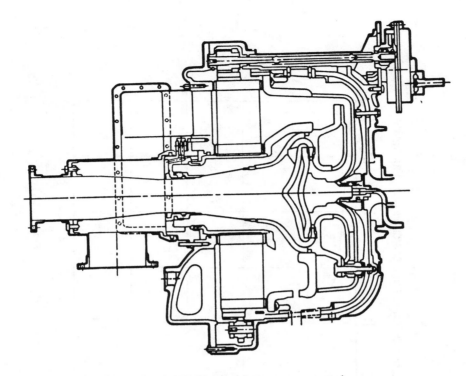

Figure 19. AGT101 ceramic structures test rig

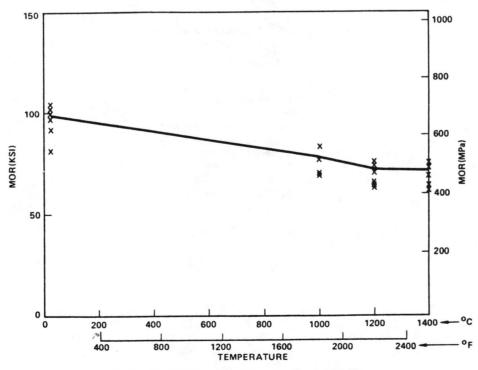

Figure 20. MOR vs temperature: Ford SRBSN

approach from a cost standpoint, but is subject to higher operating stresses than are the segmented stator vanes. It is planned that the Ford one-piece stator will be made from sintered Si_3N_4 (SSN), which is stronger than RBSN.

The radial turbine rotor is the most difficult component to make for the AGT101 program. Both Ford and ACC are working on this component. Ford is developing a series of sintered reaction-bonded Si_3N_4 (SRBSN) rotors. The ACC material is SSN. Both materials are showing promise in terms of strength (modulus of rupture, MOR) and Weibull modulus. Figure 20 shows Ford MOR data versus temperature for test bars of RM3, which is one of the Four SRBSN materials. Figure 21 shows comparable data for ACC's SSN. Both materials show test-bar strengths in excess of program requirements.

Early in the AGT101 program, dummy, bladeless rotors were made in order to evaluate the fabrication process on a spinnable part comparable in size to the actual rotor. Both Ford and ACC made successful dummy rotors. Two such rotors were made by Ford; one in the green, as-cast condition, and the other after sintering. The shrinkage from the last process is about 10 percent. Ford spin-tested one of their dummy rotors to 134,000 rpm without failure. Figure 22 shows stress contours calculated for this test condition. A large fraction of the volume of the part is seen to be exposed to principal tensile stresses of over 345 MPa (50,000 psi), with a peak stress of 402 MPa (58,300 psi). This is considered a significant milestone.

The present challenge is to replicate these test-bar and dummy-rotor properties in an entire bladed rotor, and then to fabricate a number of consistently good rotors. Work aimed at these objectives is underway. Problems being addressed include cracking during the drying of the cast preform, and incomplete blade fill during the casting of the preform. A rotor at Ford with known blade imperfections was spin-tested to a speed above the design value of 100,000 rpm. Three single-blade fractures occurred, at speeds of 78,850, 95,900, and 106,760 rpm, respectively. Several ACC rotors have been spin-tested, one to 115,000 rpm. Several rotors have passed a series of hot turbine rig tests. Additional test rotors are now being processed.

CERAMIC TURBOCHARGER DEVELOPMENT

Ceramic turbocharger development and evaluation are now underway for passenger car turbochargers with the following objectives:

1. Improving turbocharger transient performance.

2. Reducing turbocharger installed cost.

In a ceramic turbocharger, the high-temperature rotating component (the turbine) is ceramic, replacing the presently used nickel-base alloy. Such replacement, made across the board on all passenger-car turbochargers made by Garrett, would save about 200 tons of nickel annually.

The ceramic turbine for a typical passenger-car turbocharger is considerably smaller than the ceramic rotor for an AGT-type turbine. An experimental ceramic rotor for the Garrett T2 turbocharger is smaller compared to the previously shown AGT101 ceramic rotor. The turbocharger rotor also typically has lower design tip speeds, and therefore lower design stresses, than does the AGT-type gas-turbine rotor.

For these and other reasons, ceramic-turbocharger development is a short-range proposition compared to ceramic gas-turbine development. Multi-unit ceramic-turbocharger testing in vehicles is a present reality. This kind of testing, and indeed the evaluation of ceramic turbochargers in general, is expected to yield information that should accelerate considerably the application of ceramics to gas turbines.

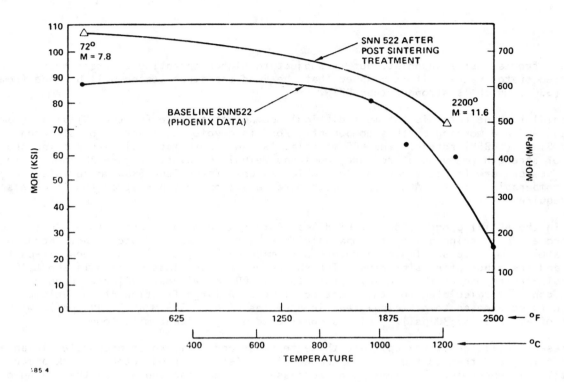

Figure 21. MOR vs temperature: ACC SSN

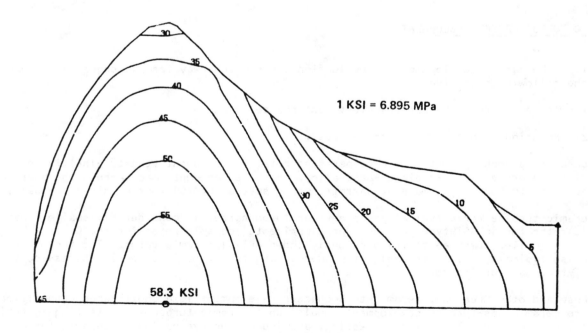

1 KSI = 6.895 MPa

58.3 KSI

Figure 22. Stresses calculated for Ford dummy rotor at 134,000 rpm

CONCLUSIONS AND RECOMMENDATIONS

The AGT program is a prime technical challenge. It has been called a long-term, high-risk program. The payoff of a successful program would be commensurately high. Considerable U.S. government and private money has been and is being spent on the AGT program and on other structural ceramic programs, including those listed in this paper. Concurrently, there are major ceramic R and D and applications efforts elsewhere, especially in Japan. We urgently need to understand Japanese versus U.S. incentives for and ways of implementing new high technology in general and ceramic technology in particular; this needs to be done quickly and then followed up quickly with appropriate action. Failure to apply what we now know about ceramics, and equally, failure to continue to pursue research toward more reliable and more advanced ceramics, could have a significantly adverse future effect on the U.S. economy, on our material and energy resources, and even on our national defense.

ACKNOWLEDGEMENT

The assistance of D. W. Richerson and J. R. Kidwell in the preparation of this paper is gratefully acknowledged.

REFERENCES

1. R. N. Katz, "Recent Developments in High-Performance Ceramics," AMMRC Report MS 77-4, March 1977.
2. A. F. McLean and D. A. Davis, "The Ceramic Gas Turbine--A Candidate Power Plant for the Middle and Long-Term Future," SAE Paper 760239, Automotive Engineering Congress and Exposition, Detroit, February 1976.
3. R. C. Ronzi and C. J. Rahnke, "Worldwide Automotive Powertrain Directions," SAE Paper 811377, International Pacific Conference on Automotive Engineering, Honolulu, November 1981.
4. R. J. Bratton and D. G. Miller, "Brittle Materials Design, High-Temperature Gas Turbine," AMMRC CTR 76-32, Final Report, December 1976.
5. E. C. Van Reuth, "The Advanced Research Projects Agency's Ceramic Turbine Program," p. 1, Ceramics for High-Performance Applications, J. J. Burke, A. E. Gorum, and R. N. Katz, Editors, Brook Hill Publishing Co., 1974.
6. A. F. McLean and E. A. Fisher, "Brittle Materials Design, High-Temperature Gas Turbine," AMMRC TR 81-14, Final Report, March 1981.
7. C. A. Fucinari, C. J. Rahnke, V. D. N. Rao, and J. K. Vallance, "Ceramic Regenerator System Development Program," NASA CR-a65a39, Final Report, October 1980.
8. A. F. McLean, "Ceramic Turbine Housings," ASME Paper 82-GT-293, International Gas Turbine Conference, London, April 1982.
9. R. Arnon and W. Treia, "Evaluation of Ceramics for Stator Applications--Gas Turbine Engines," NASA CR-168140 Final Report, March 1983.
10. Gersh, H. M., "NASA Gas Turbine Stator Vane Ring," International Gas Turbine Conference, Houston, 1981.
11. D. W. Richerson and K. M. Johanson, "Ceramic Gas Turbine Engine Demonstration Program," final report, Contract No. N00024-76-C-5352, May 1982.
12. D. W. Richerson, L. J. Lindberg, W. D. Carruthers, and J. Dahn, "Contact Stress Effects on Si_3N_4 and SiC Interfaces," Ceramic Engineering and Science Proceedings, July-August, 1981.
13. J. W. Wimmer, D. W. Richerson, P. Ardans, and L. J. Lindberg, "Ceramic Components for Turbine Engines," Final Report, AFWAL-TR-83-4012, March 1983.
14. D. W. Richerson and J. M. Wimmer, "Ceramic Component Development for Limited-Life Propulsion Engines," AIAA Paper 82-1050, 18th Joint Propulsion Conference, Cleveland, June 1982.
15. W. D. Carruthers, D. W. Richerson, and K. W. Benn, "3500-Hour Durability Testing of Commercial Ceramic Materials," Interim Report, NASA CR-159785.
16. R. A. Rackley and D. M. Kreiner, "Advanced Gas Turbine Technology Development (Systems and Components)" Proceedings of the 20th Automotive Technology Development Contractors' Coordination Meeting, Dearborn, October 1982.
17. H. E. Helms and R. A. Johnson, "Advanced Gas Turbine Technology Development: AGT 100 Systems and Components," ibid.

Evaluation of sulfidation corrosion inhibitors by small burner rig testing

M. J. Zetlmeisl, D. F. Laurence, and K. J. McCarthy

Corrosion inhibitors to prevent metal attack in sulfidation environments were evaluated using a burner rig followed by metallographic examination of corrosion specimens. Of the nine different additive treatments studied under a set of standard sulfidation conditions, Cr-Si (Cr/Na = 2.25, Si/Na = 1) and Cr (Cr/Na = 4.5 and 2.25) were the most successful, with the Cr-Si showing slight superiority over Cr alone. Some of the other materials such as Mg retarded the attack but did not prevent advanced stages of the corrosion as did Cr-Si and Cr. Materials such as Ti and Ce, which are beneficial alloying additions against hot corrosion by sulfidation, were unsuccessful inhibitors as fuel additives.

TABLE 1 — Test conditions for all experiments

Fuel	Diesel (5.5 mL/min)
Contaminants (ppm)	125 Na, 15 Mg, 4.8 Ca, 4.1 K, 225 Cl, 1% S
Air:fuel ratio	50:1
Temperature	874 ± 6 C
Duration	Baseline—240 hours; with additive—1000 hours or shorter if the additive is obviously failing
Alloys tested	B-1900, Udimet 700, Udimet 500, Inconel 738, X-40
Frequency of cycling to room temperature	At least once every 100 hours

Introduction

THE MOST COMMON FORM OF HOT corrosion attack of metal gas turbine components is that caused by molten alkali sulfate (Na_2SO_4 + K_2SO_4) condensing on the metal parts at temperatures in the 850 to 900 C range.[1] This kind of corrosion is primarily a localized attack which causes subsurface loss of alloy integrity, unlike the general surface attack caused by molten V_2O_5.[2,3] Alkali sulfate hot corrosion is more common than the V_2O_5 type because alkali metal contamination can come from either the fuel or the combustion air, whereas V_2O_5 can occur only in certain types of crude and residual fuel oils.[4]

Alkali sulfate hot corrosion caused by fuel-borne alkali metal can be prevented adequately by desalting the fuel because the majority of the contaminants are present as water soluble salts.[2,5] Airborne alkali is more difficult to prohibit, especially in coastal and marine environments, but inlet air treatment by filtration can go a long way to prevent this source of alkali.[6] In addition, turbine manufacturers are constantly upgrading the metallurgy, seeking ways to extend the service life of industrial turbines which are almost inevitably exposed to alkali sulfates.[7]

Despite the existence of these engineering solutions, gas turbine manufacturers and users have always been interested in fuel additives which would produce ash components, rendering the alkali sulfates noncorrosive. The purpose of this work is: (1) to investigate in small burner rig tests possible additive materials, singly and in combination, with a view to evaluating their relative effectiveness; (2) to establish optimum dosage levels for materials that are effective; and (3) to determine whether any combination of materials has any unique properties.

Experimental

The burner rig, test specimens, and measurement techniques were essentially the same as some which have been in use for years; a detailed description has been given elsewhere.[8]

The object of the test is to burn a fuel with a specified level of contaminants and additive at a given temperature for a period of time and to measure the depth and type of penetration into the test specimens as a result of exposure to this environment. Test conditions common to all of these experiments are given in Table 1. The chemical composition of the alloys is given in Table 2.

Oil soluble forms of contaminants and additives were used: Na as Na-Petronate, Mg as KI-16, Cr as KI-55, Al as KI-91, Si as KI-39, Ti as Tyzor TOT, Ca as Ca-naphthenate, K as K-nonyl phenylate made by reacting KOH with nonyl phenol, Cl as ethylene dichloride, and S as di-tert-butyl disulfide.

Test specimens were machined into one in. (2.54 cm) discs, 60 mils (0.15 cm) thick, from bar stock of the following materials: U-700, U-500, X-40, and In-738. B-1900 was cut from tensile bars into 1/4 in. (0.64 cm) discs, 100 mils (0.254 cm) thick. Specimen thickness was measured with a micrometer before the test opposite a punch mark for reference so that the same part of the specimen could be measured after the test. Diameter was also measured. The specimens were cleaned in pickling solution at 60 C, rinsed in xylene, acetone, then water, cleaned again in pickling solution at 60 C, rinsed again in water, and air dried.

Reprinted with permission from Materials Performance, June 1984, 41-44, © 1984 National Association of Corrosion Engineers

Source: Materials Performance, June 1984, 41-44

TABLE 2 — Chemical composition[1] of alloys used in rig tests

Alloy	Ni	Cr	Co	Mo	W	Ta	Cb	Al	Ti	Fe	Mn	Si	C	B	Zr
In-738	61	16.0	8.5	1.7	2.6	1.7	0.9	3.4	3.4	—	—	—	0.17	0.010	0.10
Udimet 500[2]	52	18.0	19.0	4.2	—	—	—	3.0	3.0	—	—	—	0.07	0.007	0.05
Udimet 700	53.4	15.0	18.5	5.2	—	—	—	4.3	3.5	—	—	—	0.08	0.030	—
B-1900	64	8.0	10.0	6.0	—	4.0	—	6.0	1.0	—	—	—	0.10	0.015	0.10
X-40	10.5	25.5	54.0	—	7.5	—	—	—	—	—	0.75	0.75	0.50	—	—
													0.25		

[1]Compositions given in wt%.
[2]U-500 values given for "cast" version. Slightly different values for "wrought."

TABLE 3 — Summary of alloy penetration data from burner rig experiments

Test	Treatment	Time (h)	Alloy penetration (mils/1000 h) B-1900	U-700	U-500	In-378	X-40
1	None	100	111	—	—	—	—
		170	97	129	29	—	—
		240	83	—	20	35	Slight
2	Cr/Na = 4.5	477	67	2.9	—	—	—
		750	60	1.1	2.6	—	—
		1000	45	1.2	2.1	0.9	0.8
3	Cr/Na = 2.25	500	4.4	1.0	—	—	—
		750	40	1.0	0.5	—	—
		1000	45	0.6	1.0	0.12	1.8
4	Cr/Na = 2.25, Si/Na = 1	500	1.7	1.6	—	—	—
		750	26	1.2	1.9	0.5	—
		1000	24	1.3	0.2	0.4	2.6
5	Ti/Na = 3	230	112	—	—	—	—
		490	92	64	4.1	8.2	—
		750	64	—	4.8	13.3	14
6	Al/Na = 2.25	204	68	79	—	—	—
		350	49	41	49	34	—
		400	47	79	75	75	9.4
7	Mg/Na = 2.25	300	111	3.9	—	—	—
		413	105	1.8	2.7	3.6	—
		530	103	56	0.6	4.0	3.1
8	Si/Na = 2.25	340	53	—	—	—	—
		437	46	69	15	5.0	—
		530	38	65	10	6.0	4.5
9	Mg/Na = 2.25, Si/Na = 1	361	52	54	—	—	—
		425	44	52	22	21	—
		600	47	52	36	21	16
10	Ce/Na = 2.25	399	52	61	—	—	—
		520	—	—	31	32	—
		615	43	61	35	24	14

After exposure in the small burner test, the specimens were cross sectioned through a diameter containing the punch mark, mounted in thermosetting resin, and polished for metallographic examination. The surface of the cross section was examined under a microscope, and two photomicrographs were taken for each of the specimens, one at 62.5X and one at 250 or 500X. From the low magnification photomicrograph, the minimum thickness of sound material remaining was measured as described elsewhere.[8] From the high magnification, a qualitative evaluation of the type of corrosion was made. Great caution was taken to present the specimen perpendicular to the polishing plane because deviations can cause significant error in measuring penetration.

Results and discussion

There are ten burner rig experiments discussed in this report: one experiment to establish baseline conditions without treatment and nine experiments with different treatments. A summary of the depth of penetration (mils/1000 h) data from these tests is given in Table 3. In cases where the test did not last 1000 hours, extrapolated values are given.

Test 1, baseline, no additive

The results from the baseline test as presented in Table 3 are close to expected. One slight deviation is that B-1900 was expected to do somewhat worse than U-700, whereas the opposite was true. The smaller size of the B-1900 specimen might have caused it to be shielded from the gas stream by the larger specimens, thus accounting for the reversal.

The B-1900 and U-700 were virtually destroyed in 240 hours. There is a thick, outer porous oxide layer, deep depletion of gamma prime, and a large area of chromium sulfide in the depletion zone.

U-500 and In-738 showed similar rates of penetration, as indicated in Table 3. Qualitatively, these specimens had the same features of porous oxide, gamma prime depletion, and large areas of chromium sulfide found in the B-1900 and U-700.

As expected, the X-40 specimen was hardly penetrated after 240 hours. According to published data, a typical penetration rate for this alloy under similar conditions would be about 2.1 mils/1000 h.[8]

Rig test 2, Cr/Na = 4.5

The use of Cr as a sulfidation inhibitor dates back to the early 1950s.[9] More recent investigations have studied the effectiveness of Cr in burner rig tests.[10,11] An examination of the data from this experiment (Table 3) reflects the dramatic protection provided by Cr. Even at 1000 hours, U-700, U-500, and In-738 showed at most only slight internal oxidation and gamma prime depletion; there were no internal sulfides in evidence. B-1900, on the other hand, was attacked, but less than in the baseline test. This result is not surprising in view of the notoriously poor corrosion resistance of B-1900. The photomicrographs from this experiment are similar to those from rig test 3.

Rig test 3, Cr/Na = 2.25

Both the general desirability of minimizing the total amount of ash and the expense of the chromium additive give ample incentive for reducing the treating rate from Cr/Na = 4.5 to 2.5.

Comparing the penetration data for this experiment with those at Cr/Na = 4.5 in Table 3, the protection is at least as good at the lower dosage. The similarity in rates of penetration is corroborated by the appearance of the specimen microstructures.

An attempt to lower the Cr treatment rate to Cr/Na = 1 failed: After 600 hours, all of the specimens had signs of severe sulfidation attack. For the sake of brevity, these results are not discussed in detail.

Rig test 4, Cr/Na = 2.25, Si/Na = 1

This combination of Cr-Si represents the most successful sulfidation inhibitor tried in this investigation. Penetration rates for most of the alloys are about as good or better than the other Cr tests. The appearance of the photomicrographs is more significant, however, than the quantitative data. B-1900 was similar to the other B-1900 specimens, but the attack is less serious. U-700 had no internal sulfidation and far less internal oxidation than the corresponding specimen in test 2 or 3. There is some depletion of gamma prime. U-500 had virtually no internal oxidation and only shallow gamma prime depletion. The specimen is significantly superior to the U-500 speci-

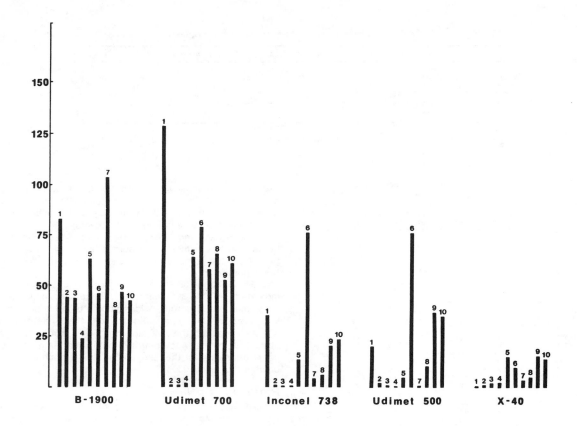

FIGURE 1 — Summary of burner rig data for different additive treatments and alloys under sulfidation conditions: (1) no additive; (2) Cr/Na = 4.5; (3) Cr/Na = 2.25; (4) Cr/Na = 2.25, Si/Na = 1; (5) Ti/Na = 3; (6) Al/Na = 2.25; (7) Mg/Na = 2.25; (8) Si/Na = 2.25; (9) Mg/Na = 2.25, Si/Na = 1; and (10) Ce/Na = 2.25.

mens from the other tests. In-738 is similar to U-500, and X-40 is virtually untouched.

The results of this experiment are perhaps the most significant of all the burner rig tests; blocking the early stage of internal oxide formation and slowing down depletion of Cr and gamma prime will greatly retard the formation of internal sulfides.

Rig test 5, T/Na = 3

The rationale for the use of Ti is that optimization of the Ti:Al ratio in alloys improves resistance to internal sulfidation.[12] Unfortunately, as the data from the test presented in Table 3 show, the promise of Ti as an additive was not realized. There were rather significant levels of attack after 750 hours. Nevertheless, this material did offer some protection as the specimens were not nearly as badly penetrated as the baseline samples.

Rig test 6, Al/Na = 2.25

The results of this experiment represent a spectacular failure. During the course of this test, the accumulated deposits became cobalt blue, and the depth of the color was in direct proportion to the amount of cobalt in the alloys (Table 2).

X-ray fluorescence analysis of the deposits confirmed the presence of Co. Notice that U-700, U-500, and In-738 had an abrupt increase in rate of penetration between 350 and 400 hours. At 300 to 350 hours, corrosion was severe, but there was a substantial amount of material still intact. In the next 50 to 100 hours, the samples actually fractured as if they had reached some critical threshold and then abruptly fell apart.

Apparently, the Al leached Co out of the alloys to a critically low level at which they failed.

From these results, one could hardly justify using Al at these temperatures and in these matrices, even if the Al were only a minor component of the additive. This reaction between Co and Al was reproduced in a crucible test on the same alloys. Al was mixed with Na_2SO_4 as Al_2O_3 such that Al/Na = 2.25. The blue color appeared at the alloy/deposit interface after several hundred hours at 900 C.

Rig test 7, Mg/Na = 2.25

The data from this experiment show that, as usual, B-1900 was not protected. U-700 did well for about 400 hours and then fell apart in about the next 100 hours. Apparently, there was an incubation period or the occurrence of some sort of critical threshold as in the Al test. The deposits were not blue, as they were with the Al specimens. U-500, In-738, and X-40 were only mildly attacked in comparison to baseline. The reason for the partial, yet significant, success of Mg against this type of attack is probably its basicity, i.e., its ability to tie up SO_3 and render the deposits less acidic. X-ray diffraction analysis of deposits, however, showed MgO as the stable phase rather than $MgSO_4$, just as would be predicted from available thermodynamic data[13] and SO_3 pressures in the range of 10^{-8}-10^{-5} atm.

Rig test 8, Si/Na = 2.25

As reflected in the data presented in Table 3, there was some protection from Si over baseline, but not as much as from Mg.

Rig test 9, Mg/Na = 2.25, Si/Na = 1

The specimens were attacked more severely in this test than in the tests with Mg alone or Si alone, as reflected in Table 3.

Rig test 10, Ce/Na = 2.25

It was reported recently that superficial application of CeO_2 powder was found to retard metal attack in sulfidation environments by decreasing scale growth rate, increasing scale adhesion, and reducing oxide grain size in much the same manner as alloying additions of Ce.[14] The oxidation rates were measured on a continuously weighing microbalance. With these results in mind, a rig test was undertaken with Ce/Na = 2.25. As reflected by the data from this experiment in Table 3, Ce did not prove to be a good inhibitor when added to the fuel. It would be interesting in the future to see if the results would have been different if the test specimens had been first coated superficially with CeO_2 and then put through the burner rig test.

Overview and summary

To get an overall view of the data generated from these burner tests, a bar graph was made of depth of penetration in mils/1000 hours for the different alloys under the different conditions of the experiments (Figure 1). The following observations can be made:

1. Additives containing Cr were by far the most effective.
2. Cr-Si had a marginal edge over the other Cr combinations, especially for In-738 and U-500. (The appearance of the photomicrographs is even more convincing evidence for the superiority of Cr-Si.)
3. Mg, Ti, and Si gave a reasonable amount of protection for the three better alloys, In-738, U-500, and X-40.
4. No additive protected B-1900 effectively.
5. Ranked in order of increasing corrosion resistance, the alloys fall in this sequence: B-1900 (least resistant), U-700, In-738, U-500, and X-40 (most resistant). This ranking agrees with the experience of others.
6. Ranked in order of increasing corrosion protection, the additive treatments fall in this sequence: Al or no additive (least protective), either Mg-Si, or Ce, Ti or Si, Mg, and Cr, with Cr/Na = 4.5 or 2.25 giving approximately equal results. Cr-Si has a slight edge over the other Cr treatments.

References

1. P. A. Bergman, High Temperature Metallic Corrosion of Sulfur and Its Compounds, Z. A. Foroulis, Ed., The Electrochemical Society, Princeton, New Jersey, p. 224 (1970).
2. A. M. Beltran and D. A. Shores, The Superalloys, C. T. Sims and W. C. Hagel, Eds., John Wiley & Sons, New York, New York, p. 317 (1972).
3. J. Stringer, Properties of High Temperature Alloys, Z. A. Foroulis, Ed., The Electrochemical Society, Princeton, New Jersey, p. 513 (1976).
4. P. I. Fontaine and E. G. Richards, Hot Corrosion Problems Associated with Gas Turbines, ASTM, Philadelphia, Pennsylvania, p. 246 (1967).
5. S. S. Dreymann and J. W. Hickey, Heavy Fuel Treatment Systems, GER-2484B (1978).
6. R. B. Tatge, et al, Gas Turbine Inlet Air Treatment, GER-2490D (1980).
7. Z. A. Foroulis and F. S. Pettit, Properties of High Temperature Alloys, The Electrochemical Society, Princeton, New Jersey, 1976.
8. C. T. Sims, High Temperature Gas Turbine Engine Component Materials Testing Program, Task 1, "Fireside 1," Final Report, prepared for the US Department of Energy under Contract No. EX-76-C-01-1765, July 1978.
9. B. O. Buckland, Industrial and Engineering Chemistry, Vol. 46, p. 2163 (1954).
10. C. E. Lowell and D. L. Deadmore, Corrosion Science, Vol. 18, p. 747 (1978).
11. N. S. Bornstein and M. A. DeCrescente, US Patent No. 3,5189,491 (1971).
12. J. Billingham, J. Lauridsen, R. E. Lawn, and M. A. P. Dewey, Deposition and Corrosion in Gas Turbines, A. B. Hart and A. J. B. Cutler, Eds., Wiley, New York, New York, p. 229 (1973).
13. I. Barin and O. Knacke, Thermochemical Properties of Inorganic Substances, Springer Verlag, Berlin (1973).
14. G. M. Ecer, R. B. Singh, and G. H. Meier, Oxid. Met., Vol. 18, p. 55 (1982).

SECTION III
Gears

STATUS OF UNDERSTANDING FOR GEAR MATERIALS

Dennis P. Townsend
National Aeronautics and Space Administration
Lewis Research Center
Cleveland, Ohio 44135

SUMMARY

Today's gear designer has a large selection of possible gear materials to choose from. The choice of which material to use should be based on the requirements of the application and will include the operating conditions of load, speed, and temperature in addition to reliability, weight, noise limitation, accuracy, and cost. The plastic materials are generally low in cost with low strength capabilities and are suitable for many light-duty applications. Die-cast alloy and sintered powder-metal gears are also fairly inexpensive and will operate at higher loads and temperatures than plastic gears. The three types of cast iron offer a medium-strength gear at a cost that varies with the accuracy of machining requirements. Gears can be manufactured from several aluminum alloys for light weight and medium cost and may be anodized for improved load capacity. The copper alloys, bronze and brass, are more costly but have good sliding and wear properties that are useful for worm gear applications. The hot-forged powder-metal gears have the advantage of medium cost with good accuracy and high strength. Several low- to medium-alloy steels are available for gear design and most can be heat treated for added strength. The medium-alloy gear materials offer high strength when case hardened and will satisfy most high-load medium-temperature applications. For more severe load, speed, and temperature requirements the advanced high-temperature alloys must be used. These include EX-53, CBS 600, Vasco X-2, Super Nitralloy (5Ni-2Al), and forged AISI M-50. As the requirements become more stringent, the cost will also increase. It is necessary that the gear designer have a working knowledge of the various gear materials in order to match the most economical material with the design requirements.

INTRODUCTION

A wide variety of gear materials is available today for the gear designer. Depending on the application the designer may choose from materials such as wood (fig. 1), plastics, aluminum, magnesium, titanium, bronze, brass, gray cast iron, nodular and malleable iron, and a whole variety of low-, medium-, and high-alloy steels (fig. 2). In addition there are many different ways of modifying or processing the materials to improve their properties or to reduce the cost of manufacture. These include reinforcements for plastics, coating and processing for aluminum and titanium, and hardening and forging for many of the iron-based (or ferrous) gear materials.

In many applications the main reason for selecting a specific gear material is economic (the material strength and gear accuracy being secondary requirements). Applications such as home appliances, automobiles, recreational vehicles, instruments, and toys are but a few of the areas where high-production, low-cost, lightto medium-duty gears are used.

When selecting a gear material for an application, the gear designer will first determine the actual requirements for the gears being considered. The design requirements for a gear in a given application will depend on such things as accuracy, load, speed, material, and noise limitations. The more stringent these requirements are, the more costly the gear will become. For instance, a gear requiring high accuracy because of speed or noise limitations may require several processing operations in its manufacture. Each operation will add to the cost. Machined gears, which are the most accurate, can be made from materials with good strength characteristics. However, these gears are very expensive. The cost is further increased if hardening and grinding are required as in applications where noise limitation is a critical requirement. Machined gears should be a last choice for a high-production gear application because of cost.

Some of the considerations in the choice of a material include allowable bending and Hertz stress, wear resistance, impact strength, water and corrosion resistance, manufacturing cost, size, weight, reliability, and lubrication requirements. In aircraft applications, such as helicopter, V/STOL, and turboprops, the dominant factors are reliability and weight. Cost is of secondary importance. Off-the-road vehicles, tanks, and some actuators may be required to operate at extremely high loads with a corresponding reduction in life. These loads may produce bending stresses in excess of 150 000 psi and Hertz stresses in excess of 400 000 psi for portions of the duty cycle. This may be acceptable because of the relatively short life of the vehicle and a deemphasis on reliability. (As a contrast, aircraft gearing typically operates at maximum bending stresses of 65 000 psi and maximum Hertz stresses of 180 000 psi). Considerable research has been done on advanced aircraft gear materials at NASA Lewis Research Center in recent years (refs. 1 to 6). Some of these data are presented here.

GEAR MATERIALS

Plastics

There has always been a need for a lightweight, low-cost gear material for light-duty applications. In the past, gears were made from wood or phenolic-resin-impregnated cloth. However, in recent years with the development of many new polymers, many gears are made of various "plastic" materials. Table I lists plastic materials used for molded gears. The most common molded plastic gears are the acetate and nylon resins. These materials are limited in strength, temperature resistance, and accuracy. The nylon and acetate resins have a room-temperature yield strength of approximately 10 000 psi. This is reduced to approximately 4000 psi at their upper temperature limit of 250° F. Nylon resin is subject to considerable moisture absorption, which reduces its strength and causes considerable expansion. Larger gears are made with a steel hub that has a plastic tire for better dimensional control. Plastic gears can operate for long periods in adverse environments, such as dirt, where other materials would tend to wear excessively. They can also operate without lubrication or can be lubricated by the processed material as in the food industry. The cost of plastic gears can be as low as a few cents per gear for a simple gear on a high-volume production basis. This is probably the most economical gear available. Often a plastic gear is run in combination with a metal gear to give better dimensional control, low wear, and quiet operation.

Polyimide is a more expensive plastic material than nylon or acetate resin, but it has an operating temperature limit of approximately 600° F. This makes the polyimides suitable for many adverse applications that might otherwise require metal gears. Polyimides can be used very effectively in combination with a metal gear without lubrication because of polyimide's good sliding properties (refs. 7 to 9). However, polyimide gears are more expensive than other plastic gears because they are difficult to mold and the material is more expensive.

Nonferrous Metals

Several grades of wrought and cast aluminum alloys are used for gearing. The wrought alloys have higher strength and good machinability. The most common wrought alloy used in gearing is 2024-T4. The wrought alloy 7075-T6 is stronger than 2024-T4 but is also more expensive.

Aluminum does not have good sliding and wear properties. However, it can be anodized with a thin, hard surface layer that will give it better operating characteristics. The coating is thin and brittle and may crack under excessive load. Anodizing gives aluminum good corrosion protection in salt water applications.

Magnesium is not considered a good gear material because of its low elastic modulus and other poor mechanical properties.

Titanium has excellent mechanical properties, approaching those of some heat-treated steels with a density nearly half that of steel. However, because of its very poor sliding properties, producing high friction and wear, it is not generally used as a gear material. Several attempts have been made to find a wear-resistant coating, such as chromium plating, iron coating (refs. 10 to 13), and nitriding for titanium, with no real success. Titanium would be an excellent gear material if a satisfactory coating or alloy could be developed to provide improved sliding and wear properties.

Several alloys of zinc, aluminum, magnesium, brass, and bronze are used as die-cast materials. Many prior die-cast applications now use less expensive plastic gears. The die-cast materials have higher strength properties, are not affected by water, and will operate at higher temperatures than the plastics. As a result, they still have a place in moderate-cost, high-volume applications. Most die-cast gears are made from lower cost zinc or aluminum alloys. Copper alloys can also be used at a somewhat higher cost. The main advantage of die casting is that the requirement for machining is either completely eliminated or drastically reduced. The high fixed cost of the dies makes low production runs uneconomical. Some of the die-cast alloys used for gearing are listed in references 14 and 15.

Copper Alloys

Several copper alloys are used in gearing. Most are the bronze alloys containing varying amounts of tin, zinc, manganese, aluminum, phosphorous, silicon, lead, nickel, and iron. The brass alloys contain primarily copper and zinc with small amounts of aluminum, manganese, tin, and iron. The copper alloys are most often used in combination with an iron or steel gear to give good wear and load capacity especially in worm gear applications where there is a high sliding component. In these cases the steel worm drives a bronze gear. Bronze gears are also used where corrosion and water are a problem.

Several copper alloys are listed in table II. The bronze alloys are either aluminum bronze, manganese bronze, silicon bronze, or phosphorous bronze. These bronze alloys have yield strengths ranging from 20 000 to 60 000 psi and all have good machinability.

Ferrous Alloys

Cast iron is used extensively in gearing because of its low cost, good machinability, and moderate mechanical properties. Many gear applications can use gears made from cast iron because of its good sliding and wear properties, which are in part a result of the free graphite and porosity. There are three basic cast irons distinguished by the structure of graphite in the matrix of ferrite. These are (1) gray cast iron, where the graphite is in flake form; (2) malleable cast iron, where the graphite consists of uniformly dispersed fine, free-carbon particles or nodules; and (3) ductile iron, where the graphite is in the form of tiny balls or spherulites. The malleable iron and ductile iron have more shock and impact resistance. The cast irons can be heat treated to give improved mechanical properties. The bending strength of cast iron ranges from 5000 to 25 000 psi (ref. 16), and the surface fatigue strength ranges from 50 000 to 115 000 psi (ref. 17). In many worm gear drives a cast iron gear can be used to replace the bronze gear at a lower cost because of the sliding properties of the cast iron.

Sintered Powder Metals

Sintering of powder metals is one of the more common methods used in high-volume, low-cost production of medium-strength gears with fair dimensional tolerance (ref. 18). In this method a fine metal powder of iron or other material is placed in a high-pressure die and formed into the desired shape and density under very high pressure. The green part has no strength as it comes from the press. It is then sintered in a furnace under a controlled atmosphere to fuse the powder together for increased strength and toughness. Usually, an additive (such as copper in iron) is used in the powder for added strength. The sintering temperature is then set at the melting temperature of the copper to fuse the iron powder together for a stronger bond than would be obtained with the iron powder alone. The parts must be properly sintered to give the desired strength.

There are several materials available for sintered powder-metal gears that give a wide range of properties. Table III lists properties of some of the more commonly used gear materials although other materials are available. The cost for volume production of sintered powder metal gears is an order of magnitude lower than that for machined gears.

A process that has been more recently developed is the hot-forming powder-metals process (refs. 19 and 20). In this process a powder-metal preform is made and sintered. The sintered powder-metal preformed part is heated to forging temperature and finished forged. The hot-formed parts have strengths and mechanical properties approaching the ultimate mechanical properties of the wrought materials. A wide choice of materials is available for the hot-forming powder-metals process. Since this is a fairly new process, there will be undoubtedly be improvements in the materials made from this process and reductions in the cost. Several promising high-temperature, cobalt-base alloy materials are being developed.

Because there are additional processes involved, hot-formed powder-metal parts are more expensive than those formed by the sintered powder-metal process. However, either process is more economical than machining or conventional forging while producing the desired mechanical properties. This makes the hot-forming powder-metals process attractive for high-production parts where high strength is needed, such as in the automotive industry.

Accuracy of the powder-metal and hot-formed processes is generally in the AGMA class 8 range. Better accuracy can be obtained in special cases and where die wear is limited, which would tend to increase the cost. Figure 3 shows the relative cost of some of the materials or processes for high-volume, low-cost gearing.

Hardened Steels

A large variety of iron or steel alloys are used for gearing. The choice of which material to use is based on a combination of operating conditions such as load, speed, lubrication system, and temperature plus the cost of producing the gears. When operating conditions are moderate, such as medium loads with ambient temperatures, a low-alloy steel can be used without the added cost of heat treatment and additional processing. The low-alloy material in the non-heat-treated condition can be used for bending stresses in the 20 000-psi range and surface durability Hertz stresses of approximately 85 000 psi. As the operating conditions increase, it becomes necessary to harden the gear teeth for improved strength and to case harden the gear tooth surface by case carburizing or case nitriding for longer pitting fatigue life, better scoring resistance, and better wear resistance. An improved lubrication system may also be required to remove the heat generated by the meshing of gear teeth. There are several medium-alloy steels that can be hardened to give good load-carrying capacity with bending stresses of 50 000 to 60 000 psi and contact stresses of 160 000 to 180 000 psi. The higher alloy steels are much stronger and must be used in heavy-duty applications. AISI 9300, AISI 8600, Nitralloy N, and Super Nitralloy are good materials for these applications and can operate with bending stresses of 70 000 psi and maximum contact (Hertz) stresses of 200 000 psi. The high-alloy steels should be case carburized for AISI 8600 and 9300 or case nitrided for Nitralloy for a very hard wear-resistant surface. Gears that are case carburized will usually require grinding after the hardening operation because of distortion occurring during the heat-treating process. The nitrided materials offer the advantage of much less distortion during nitriding and therefore can be used in the as-nitrided condition without additional finishing. This is very helpful for large gears with small cross sections where distortion can be a problem. Since case depth for nitriding is limited to approximately 0.020 in., case crushing can occur if the load is too high. Some of the steel alloys used in the gearing industry are listed in table IV.

Gear surface fatigue strength and bending strength can be improved by shot peening (refs. 5 and 21). Figure 4 from reference 5 is a plot of the surface fatigue life of standard ground AISI 9310 gears and standard ground and shot-peened AISI 9310 gears. The 10-percent surface fatigue life of the shot-peened gears was 1.6 times that of the standard ground gears.

The low- and medium-alloy steels have a limited operating temperature above which they begin to lose their hardness and strength, usually around 300° F. Above this temperature the material is tempered, and early bending failures, surface pitting failures, or scoring will occur. To avoid this condition, a material is needed that has a higher tempering temperature and that maintains its hardness at high temperatures. The generally accepted minimum hardness required at operating temperature is Rockwell C58. In recent years several materials have been developed that maintain a Rockwell C58 hardness at temper- atures from 450° to 600° F (ref. 3). Some of these materials have been or will be tested for surface fatigue life in the NASA gear test facility and the results compared with those for the standard aircraft gear material AISI 9310. Several materials have shown promise of improved life at normal operating temperature. The hot hardness data indicate that they will also provide good fatigue life at higher operating temperatures.

AISI M-50 has been used successfully for several years as a rolling-element bearing material for temperatures to 600° F (refs. 22 to 26). It has also been used for lightly loaded accessory gears for aircraft applications at high temperatures. However, the standard AISI M-50 material is generally considered too brittle for more heavily loaded gears.

AISI M-50 is considerably better as a gear material when forged with integral teeth. Figure 5(a) is a cross section through a forged AISI M-50 gear showing the excellent grain flow and good tooth shape that improves the bending strength. The grain flow from the forging process improves the bending strength and impact resistance of the AISI M-50 considerably (ref. 27).

The M-50 material can also be ausforged with gear teeth to give good bending strength and better pitting life (refs. 2 and 28). Figure 5(b) is a cross section through an ausforged gear tooth. The tooth is poorly formed because of the lower ausforging temperature (around 1400° F), and some of the good grain flow had to be cut away during finishing of the gear. Further, the ausforging temperature is so low that forging gear teeth is very difficult and expensive. As a result, ausforging for gears has considerably limited application (ref. 2).

Figure 6 is a Weibul plot comparing the forged AISI M-50 and ausforged AISI M-50 gear surface fatigue test data with the standard AISI 9310 steel data. The Weibul plot shows the percentage of specimens failed as a function of the system (two gears) life in stress cycles. These test data were taken with a superrefined naphthenic mineral oil lubricant, a speed of 10 000 rpm, a temperature of 170° F, a maximum Hertz stress of 248 000 psi with an 8-in. diametral pitch, and a 3.5-in.-pitch-diameter gear. The results show that the forged and ausforged gears can give lives approximately 5 times those of the standard AISI 9310 gears (ref. 2).

Nitralloy N is a low-alloy nitriding steel that has been used for several years as a gear material. It can be used for applications requiring temperatures of 400° to 450° F. A modified Nitralloy N called Super Nitralloy or 5Ni-2Al Nitralloy was used in the United States Supersonic Aircraft Program for gears. It can be used for gear applications requiring temperatures to 500° F. A Weibul plot of the surface fatigue data for the Super Nitralloy gears is shown in figure 7.

Two materials that were developed for case-carburized tapered roller bearings but also show promise as high-temperature gear materials are CBS 1000M and CBS 600 (refs. 29 and 30). These materials are low- to medium-alloy steels that can be carburized and hardened to give a hard case of Rockwell C60 with a core of Rockwell C38. Weibul plots of the surface fatigue test results for CBS 600 and AISI 9310 are shown in figure 7. The CBS 600 has a medium fracture toughness that could cause fracture failures after a surface fatigue spall has occurred.

Two other materials that have recently been developed as advanced gear materials are EX-53 and EX-14. Reference 31 reports that the fracture toughness of EX-53 is excellent at room temperature and improves considerably as temperature increases. These materials are presently being evaluated for surface fatigue in the NASA gear test facility. The EX-53 surface fatigue results show a 10-percent life that is twice that of the AISI 9310.

Vasco X-2 is a high-temperature gear material (ref. 32) that is currently being used in advanced CH 47 helicopter transmissions. This material has an operating temperature limit of 600° F and has been shown to have good gear load-carrying capacity when properly heat treated. The material has a high chromium content (4.9 percent) that oxidizes on the surface and can cause soft spots when the material is carburized and hardened. A special process has been developed that eliminates these soft spots when the process is closely followed (ref. 33). Several groups of Vasco X-2 with different heat treatments were surface fatigue tested in the NASA gear test facility. All groups except the group with the special processing gave poor results (ref. 4). Figure 9 is a Weibul plot of these data showing the variation of surface fatigue life with different heat treatment processes. Vasco X-2 has a lower fracture toughness than AISI 9310 and is subject to tooth fracture after a fatigue spall.

REFERENCES

1. Townsend, D. P.; and Zaretsky, E. V.: A Life Study of AISI M-50 and Super Nitralloy Spur Gears With and Without Tip Relief. J. Lubr. Technol., vol. 96, no. 4, Oct. 1974, pp. 583-590.
2. Townsend, D. P.; Bamberger, E. N.; and Zaretsky, E. V.: A Life Study of Ausforged, Standard Forged, and Standard Machined AISI M-50 Spur Gears. J. Lubr. Technol., vol. 98, no. 3, July 1976, pp. 418-425.
3. Anderson, N. E.; and Zaretsky, E. V.: Short-Term, Hot-Hardness Characteristics of Five Case-Hardened Steels. NASA TN D-8031, 1975.
4. Townsend, D. P.; and Zaretsky, E. V.: Endurance and Failure Characteristics of Modified Vasco X-2, CBS 600 and AISI 9310 Spur Gears. J. Mech. Des., vol. 103, no. 2, Apr. 1981, pp. 506-515.
5. Townsend, D. P.; and Zaretsky, E. V.: Effect of Shot Peening on Surface Fatigue Life of Carbonized and Hardened AISI 9310 Spur Gears. NASA TP-2047, 1982.
6. Townsend, D. P.; Coy, J. J.; and Zaretsky, E. V.: Experimental and Analytical Load-Life Relation for AISI 9310 Steel Spur Gears. J. Mech. Des., vol. 100, no. 1, Jan. 1978, pp. 54-60.
7. Fusaro, R. L.: Effects of Atmosphere and Temperature on Wear, Friction and Transfer of Polyimide Films. ASLE Trans., vol. 21, no. 2, Apr. 1978, pp. 125-133.
8. Fusaro, R. L.: Tribological Properties at 25° C of Seven Polyimide Films Bonded to 440C High-Temperature Stainless Steel. NASA TP-1944, 1982.
9. Fusaro, R. L.: Polyimides. Tribological Properties and Their Use as Lubricants. Presented at the 1st Technical Conference on Polymides (Ellenville, N.Y.), Nov. 10-12 1982.
10. Johansen, K. M.: Investigation of the Feasibility of Fabricating Bimetallic Coextruded Gears. AFAPL-TR-73-112, AiResearch Mfg. Co., Dec. 1973. (AD-776795.)
11. Hirsch, R. A.: Lightweight Gearbox Development for Propeller Gearbox System Applications Potential Coatings for Titanium Alloy Gears. AFAPL-RT-72-90, General Motors Corp., Dec. 1972. (AD-753417.)
12. Delgrosso, E. J., et al.: Lightweight Gearbox Development for Propeller Gearbox Systems Applications. AFAPL-TR-71-41-PH-1, Hamilton Standard, Aug. 1971. (AD-729839.)

13. Manty, B. A.; and Liss, H. R.: Wear Resistant Coatings for Titanium Alloys. FR-8400, Pratt & Whitney Aircraft, Mar. 1977. (AD-A042443).
14. Dudley, D. W.: Gear Handbook, The Design, Manufacture and Applications of Gears. McGraw Hill Book Co., 1962.
15. Michalec, G. W.: Precision Gearing Theory and Practice. John Wiley & Sons Inc., 1966.
16. AGMA Standard for Rating the Strength of Spur Gear Teeth. AGMA 220.02, American Gear Manufacturers Association, Aug. 1966.
17. AGMA Standard for Surface Durability (Pitting) of Spur Gear Teeth. AGMA 210.02, American Gear Manufacturers Association, Jan. 1965.
18. Smith, W. E.: Ferrous-Based Powder Metallurgy Gears. Gear Manufacture and Performance, P. J. Guichelaar, B. S. Levy, and N. M. Parikh, eds., American Society for Metals, 1974, pp. 257-269.
19. Antes, H. W.: P/M Hot Formed Gears. Gear Manufacture and Performance, P. J. Guichelaar, B. S. Levy, and N. M. Parikh, eds., American Society for Metals, 1974, pp. 271-292.
20. Ferguson, B. L.; and Ostberg, D. T.: Forging of Powder Metallurgy Gears. TARADCOM-TR-12517, TRW-ER-8037-F, TRW, Inc., May 1980. (AD-A095556.)
21. Straub, J. C.: Shot Peening in Gear Design. AGMA Paper 109.13, June 1964.
22. Hingley, C. G.; Southerling, H. E.; and Sibley, L. B.: Supersonic Transport Lubrication System Investigation. (AL65T038 SAR-1, SKF Industries, Inc.; NASA Contract NAS3-6267.) NASA CR-54311, May 1965.
23. Bamberger, E. N.; Zaretsky, E. V.; and Anderson, W. J.: Fatigue Life of 120-mm-Bore Ball Bearings at 600° F with Fluorocarbon, Polyphenyl Ether and Synthetic Paraffinic Base Lubricants. NASA TN D-4850, 1968.
24. Zaretsky, E. V.; Anderson, W. J.; and Bamberger, E. N.: Rolling-Element Bearing Life from 400° to 600° F. NASA TN D-5002, 1969.
25. Bamberger, E. N.; Zaretsky, E. V.; and Anderson, W. J.: Effect of Three Advanced Lubricants on High-Temperature Bearing Life. J. Lubr. Technol., vol. 92, no. 1, Jan. 1970, pp. 23-33.
26. Bamberger, E. N.; and Zaretsky, E. V.: Fatigue Lives at 600° F of 120-Millimeter-Bore Ball Bearings of AISI M-50, AISI M-1 and WB-49 Steels. NASA TN D-6156, 1971.
27. Bamberger, E. N.: The Development and Production of Thermo-Mechanically Forged Tool Steel Spur Gears. (R73AEG284, General Electric Co.; NASA Contract NAS3-15338). NASA CR-121227, July 1973.
28. Bamberger, E. N.: The Effect of Ausforming on the Rolling Contact Fatigue Life of a Typical Bearing Steel. J. Lubr. Technol., vol. 89, no. 1, Jan. 1967, pp. 63-75.
29. Jatczak, C. F.: Specialty Carburizing Steels for Elevated Temperature Service. Met. Prog., vol. 113, no. 4, Apr. 1978, pp. 70-78.
30. Townsend, D. P.; Parker, R. J.; and Zaretsky, E. V.: Evaluation of CBS 600 Carburized Steel as a Gear Material. NASA TP-1390, 1979.
31. Culler, R. A.; Goodman, E. C.; Hendrickson, R. R.; and Leslie, W. C.: Elevated Temperature Fracture Toughness and Fatigue Testing of Steels for Geothermal Applications. TERRATEK Report No. Tr 81-97, Terra Tek, Inc., Oct. 1981.
32. Roberts, G. A.; and Hamaker, J. C.: Strong, Low Carbon Hardenable Alloy Steels. U.S. Patent No. 3,036,912, May 29, 1962.
33. Cunningham, R. J.; and Lieberman, W. N. J.: Process for Carburizing High Alloy Steels. U.S. Patent No. 3,885,995, May 27, 1975.

TABLE I. – PROPERTIES OF PLASTIC GEAR MATERIALS

Property	ASTM	Acetate	Nylon	Polyimide
Yield strength, psi	D 638	10 000	11 800	10 500
Shear strength, psi	D 732	9510	9600	11 900
Impact strength (Izod)	D 256	1.4	0.9	0.9
Elongation at yield, percent	D 638	15	5	6.5
Modulus of elasticity, psi	D 790	410 000	410 000	460 000
Coefficient of linear thermal expansion, in./in.$^\circ$F	D 696	4.5×10^{-5}	4.5×10^{-5}	2.8×10^{-5}
Water absorption (24 hr), percent	D 570	0.25	1.5	0.32
Specific gravity	D 792	1.425	1.14	1.43
Temperature limit, $^\circ$F	-----	250	250	600

TABLE II. – PROPERTIES OF COPPER ALLOY GEAR MATERIALS

Material	Modulus of elasticity, psi	Yield strength, psi	Ultimate strength, psi
Yellow brass	15×10^6	50 000	60 000
Naval brass	15	45 000	70 000
Phosphor bronze	15	40 000	75 000
Aluminum bronze	19	50 000	100 000
Manganese bronze	16	45 000	80 000
Silicon bronze	15	30 000	60 000
Nickel-tin bronze	15	25 000	50 000

TABLE III. – PROPERTIES OF SINTERED POWDER–METAL ALLOY GEAR MATERIALS

Composition	Ultimate tensile strength, psi	Apparent hardness, Rockwell	Comment
1 to 5Cu-0.6C-94Fe	60 000	B60	Good impact strength
7Cu-93Fe	32 000	B35	Good lubricant impregnation
15Cu-0.6C-84Fe	85 000	B80	Good fatigue strength
0.15C-98Fe	52 000	A60	Good impact strength
0.5C-96Fe	50 000	B75	Good impact strength
2.5Mo-0.3C-1.7N-95Fe	130 000	C35	High strength, good wear
4Ni-1Cu-0.25C-94Fe	120 000	C40	Carburized and hardened
5Sn-95Cu	20 000	H52	Bronze alloy
10Sn-87Cu-0.4P	30 000	H75	Phosphorus-bronze alloy
1.5Be-0.25Co-98Cu	80 000	B85	Beryllium alloy

TABLE IV. – PROPERTIES OF STEEL ALLOY GEAR MATERIALS

Material	Tensile strength, psi	Yield strength, psi	Elongation in 2 in., percent
Cast iron	45 000	------	---
Ductile iron	80 000	60 000	3
1020	80 000	70 000	20
1040	100 000	60 000	27
1066	120 000	90 000	19
4146	140 000	128 000	18
4340	135 000	120 000	16
8620	170 000	140 000	14
8645	210 000	180 000	13
9310	185 000	160 000	15
440C	110 000	65 000	14
416	160 000	140 000	19
304	110 000	75 000	35
Nitralloy 135M	135 000	100 000	16
Super Nitralloy	210 000	190 000	15
CBS 600	222 000	180 000	15
CBS 1000M	212 000	174 000	16
Vasco X-2	248 000	225 000	6.8
EX-53	171 000	141 000	16
EX-14	169 000	115 000	19

C-83-1244

Figure 1. - Early wooden gears (courtesy Smithsonian Institution).

C-77-117

Figure 2. - High-alloy, precision-manufactured aircraft spiral-bevel gear set.

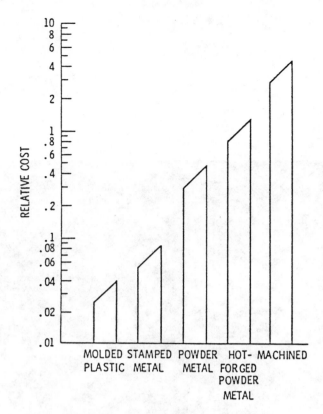

Figure 3. - Relative costs of gear materials.

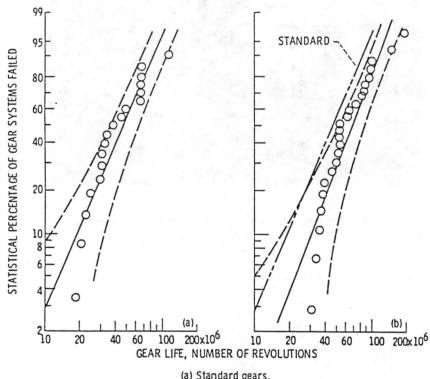

(a) Standard gears.
(b) Shot-peened gears.

Figure 4. - Surface (pitting) fatigue lives of standard ground and shot-peened carburized and hardened CVM AISI 9310 steel spur gear systems. Speed, 10 000 rpm; lubricant, synthetic paraffinic oil; gear temperature, 170° F; maximum Hertz stress, 248 000 psi.

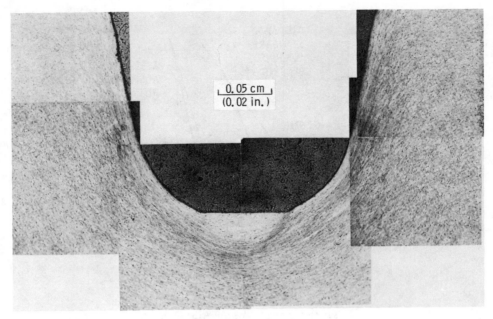

0.05 cm
(0.02 in.)

(a) Standard forged gear

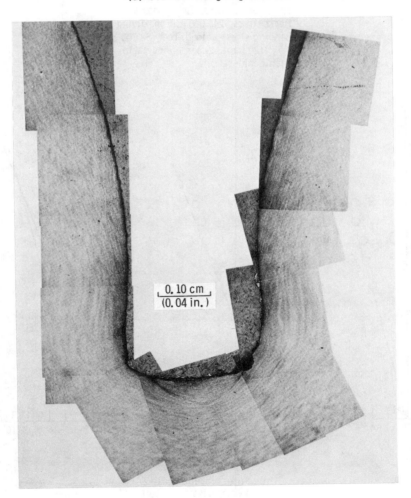

0.10 cm
(0.04 in.)

(b) Ausforged gear

Figure 5. - Photomicrographs of macro grain flow patterns. Etchant, 3 percent nital.

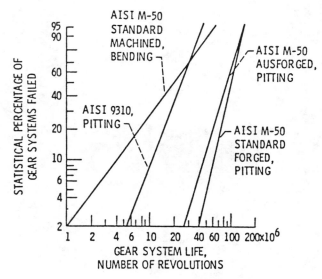

Figure 6. - Fatigue lives of spur gear
systems made of VIM-VAR AISI M-50
and VAR AISI 9310. Speed, 10 000
rpm; temperature, 70° F; maximum
Hertz stress, 248 000 psi; maximum
bending stress at tooth root, 38 700
psi; lubricant, superrefined naphthenic mineral oil.

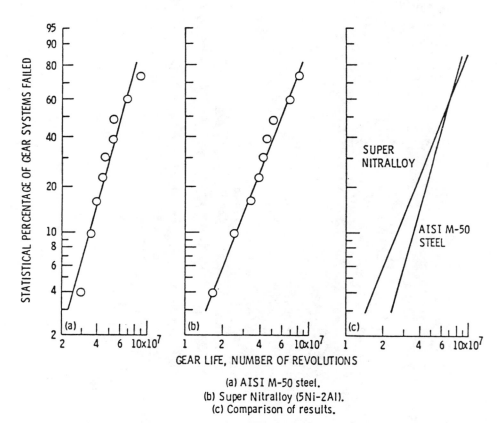

(a) AISI M-50 steel.
(b) Super Nitralloy (5Ni-2Al).
(c) Comparison of results.

Figure 7. - Surface (pitting) fatigue lives of spur gear systems made of CVM AISI M-50
steel and CVM Super Nitralloy (5Ni-2Al). Speed, 10 000 rpm; temperature, 170° F;
maximum Hertz stress, 275 000 psi; lubricant, superrefined naphthenic mineral oil.

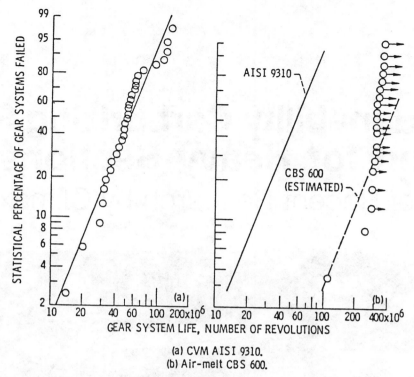

(a) CVM AISI 9310.
(b) Air-melt CBS 600.

Figure 8. - Surface (pitting) fatigue lives of spur gear systems made of
CVM AISI 9310 and air-melt CBS 600. Speed, 10 000 rpm; temperature, 170° F; maximum Hertz stress, 248 000 psi; lubricant,
synthetic paraffinic oil with 5 percent EP additive.

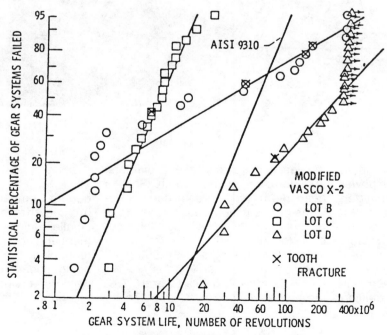

Figure 9. - Surface (pitting) fatigue lives of CVM modified Vasco
X-2 spur gears heat treated to different specifications. Pitch
diameter, 3.5 in. ; speed, 10 000 rpm; maximum Hertz stress,
248 000 psi; lubricant, synthetic paraffinic oil.

High Hardenability Carburizing Steels Developed for Heavy Sections

A Highlight of Recent Research by Climax

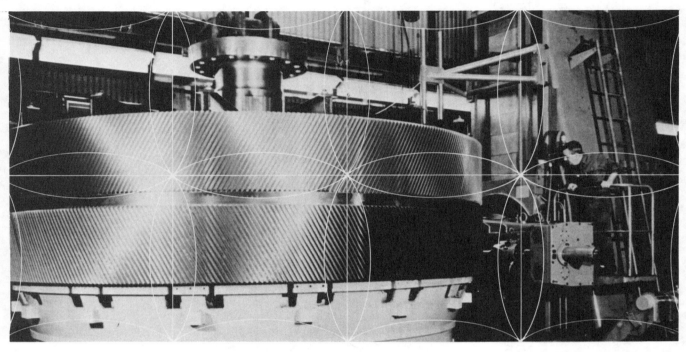

Machining of a large "bull" gear for marine power train.

Courtesy Transamerica DeLaval, Inc.

Pictured is a large "bull" gear similar to those used in marine vessel power trains. Traditionally made of normalized and tempered medium carbon steels, these gears are large in diameter and heavy in section to transmit high loads. If the rim and gear teeth are carburized, quenched, and lightly tempered to achieve high hardness, the teeth will transmit higher loads, thus permitting the design of gears smaller in diameter and narrower in face width. Smaller gears permit more streamlined hull designs, and smaller tooth size leads to quieter gears, which is important especially for submarines.

The carburizing steel EX 55, nominally containing 0.17% C, 0.87% Mn, 0.25% Si, 1.8% Ni, 0.55% Cr and 0.75% Mo and developed by Climax Molybdenum Company, a division of AMAX Inc., is the leading candidate for smaller, quieter "bull" gears. Development and adoption of carburizing steel technology for such applications is being sponsored by the U.S. Navy, which is working closely with Climax on this project. Data for EX 55 steel, and other EX steels developed by Climax, are contained in a recently issued bulletin.[1]

Other Recent Research Achievements in Carburizing Steels

The design and production of better steels for carburizing will be greatly facilitated by laboratory studies that have explained the role of metallurgical factors on the fracture and fatigue properties of carburized steels.[2-4] A fully martensitic structure at the carburized surface is essential, and most readily achieved in chromium-molybdenum steels.[5] Yet chromium-free steels will meet design requirements, if chromium supplies should become threatened or inadequate.[6] Heavy gears require high-hardenability steels, but such steels must be readily annealed for better machinability; proper alloying can aid annealability.[7] Gears and bearings that operate above ambient temperatures, e.g. in helicopter drivetrains, require good hot hardness and temper resistance, as well as good fatigue resistance; steels for such applications typically contain substantial molybdenum additions.

Research resulted in development of a Ni-Cr-V-1%Mo grade that will retain hardness and toughness during prolonged service at elevated temperature, and at the same time the room-temperature properties are even better than those of EX 55.

For more information on molybdenum carburizing steels, consult the following references. Reference 1 is available upon request. Others are in the open literature, or will be published soon. For specific questions on molybdenum carburizing steels, contact Dr. Thomas G. Oakwood, Development Manager, or Dr. Daniel E. Diesburg, Research Supervisor, Climax Molybdenum Company, P.O. Box 1568, Ann Arbor, Michigan 48106; telephone 313 761 2300.

References to Research on Carburizing Steels

1. EX Steels, a look at four carburizing grades, Bulletin M-584, Climax Molybdenum Company, Ann Arbor, Michigan 48106.

2. C. Kim and D. E. Diesburg, "Fracture of Case-Hardened Steel in Bending," Engineering Fracture Mechanics, Vol. 18 (1), 1983, pp. 69-82.

3. C. Kim, D. E. Diesburg, and G. T. Eldis, "Effect of Residual Stress on Fatigue Fracture of Case-Hardened Steels," published in ASTM STP 776 (1981), Residual Stress Effects in Fatigue, pp. 224-234.

4. T. B. Cameron, D. E. Diesburg, and C. Kim, "Fatigue and Overload Fracture of Carburized Steels," Journal of Metals, Vol. 35 (7), July 1983, pp. 37-41.

5. G. T. Eldis and Y. E. Smith, "Effect of Composition on Distance to First Bainite in Carburized Steels," Journal of Heat Treating, V. 2 (2), p. 62-72.

6. D. E. Diesburg, G. T. Eldis, and H. N. Lander, "Chromium-Free Steels for Carburizing," presented at Trends in Critical Materials Requirements for Steels of the Future—Conservation and Substitution Technology for Chromium, a public workshop held at Vanderbilt University, Nashville, Tennessee, October 4-7, 1982 (to be published in proceedings).

7. J. M. Tartaglia, T. Wada, D. E. Diesburg, and G. T. Eldis, "Influence of Alloying Elements on the Annealability of Carburizing Steels," Journal of Metals, Vol. 34 (5), May 1982, pp. 30-35.

A Carburizing Gear Steel For Elevated Temperatures

By Walter E. Burd

Photo of AH-1T+ courtesy Bell Helicopter Textron Inc. Bell has studied the fatigue life of Alloy 53 to gage its potential for use in helicopter transmission gears.

Few applications demonstrate more dramatically the need for reliable gearing materials than a helicopter transmission. Few applications offer a more hostile environment to gearing materials than a helicopter transmission that has suffered a total loss of lubricant. Responding to a need for improved gearing materials capable of performing over a wide temperature range, Carpenter Technology Corp. has developed a new carburizing alloy called Pyrowear alloy 53.

The alloy has good temper resistance and high case hot hardness — 55 HRC at 600 F (315 C) — while maintaining high core impact strength and fracture toughness. Case temper resistance is significantly greater than that of other gearing alloys such as AISI 9310, 3310, or 8620. The new alloy has the necessary properties to function as a gear in an elevated temperature environment, giving it a distinct advantage over the more common carburizing materials.

The alloy is expected to have widespread application in the aerospace industry and wherever designs call for more compact transmission systems than are now possible with conventional gear materials. Bearing characteristics are also excellent. Properties of the steel have been confirmed in a number of standard laboratory tests.

Development Programs Started in the Seventies

Advanced aircraft and helicopter requirements continually demand improved transmission gear materials with increased load carrying capacity, longer life, and the ability to retain their strength at elevated temperatures. Ability of a material to retain its hardness and fatigue resistance at elevated temperatures is an important design criterion for high performance gears and rolling element bearings such as those found in helicopter transmissions. Low hardness can result in permanent surface deformation and distress during operation, which lead to premature failure.

Carpenter embarked on a program in the 1970s to develop an improved material for helicopter transmission gears. Transmissions normally operate at about 250 F (120 C). Loss of lubricant elevates transmission operating temperature to the 500 F (260 C) range. Common gear steels undergo a rapid drop in mechanical properties at elevated temperatures, so a gearing material that would sufficiently extend transmission operating life even with total loss of lubricant was sought.

Reprinted from *Metal Progress*, May 1985, 33-35, © 1985 American Society for Metals

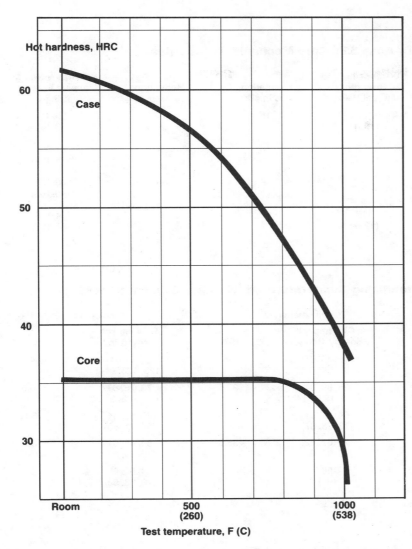

Fig. 1 — *Hardness retention at elevated temperatures is a key property of Carpenter's new carburizing steel. Potential uses include helicopter transmission and other aerospace gearing applications and rock bits for geothermal well drilling. (Note: case hardness values were converted from Rockwell A data.)*

core impact strength properties, while the lesser chromium content (1%) helps to improve carburization of the case.

The alloy is designed for greater case temper resistance than conventional types. It attains the core mechanical properties shown in Table I following heat treatment.

Room temperature Charpy V-notch impact strengths for various austenitizing temperatures are compared for air cooling and oil quenching in Table II.

Figure 1 shows case and core hot hardness values from room temperature to 1000 F (540 C).

Improved properties are a result of Alloy 53's chemical composition. For example, compared with 9310, a popular gearing material, the new alloy's higher copper content (2% vs 0.20%) provides greater impact strength. Its higher molybdenum content (3.25% vs 0.10%) enhances elevated temperature properties. And the vanadium in Alloy 53 (9310 has none) helps maintain a fine grain structure.

Two other important comparisons of the two gearing materials:

1. At room temperature, the true fracture toughness of the new steel is 90 ksi$\sqrt{}$in. (100 MPa$\sqrt{}$m), while that of 9310 is about 70 ksi$\sqrt{}$in. (75 MPa$\sqrt{}$m).

2. While the new alloy requires slightly longer carburizing time than 9310, total case depth achievable is the same for both.

New Alloy Evaluated By Government/Industry Labs

Standard property tests by Carpenter as well as several independent evaluations demonstrate that Alloy 53 matches or exceeds predicted performance and meets all design criteria for room and elevated temperature applications.

The National Aeronautics & Space Administration (NASA) conducted endurance life tests on gears made from the alloy, comparing the results with similar data for 9310 and a group of

Following experimentation with analysis and melting methods, Carpenter's R&D efforts resulted in the new alloy. Composition of Pyrowear 53: 0.10 C, 0.35 Mn, 1Si, 1 Cr, 2 Ni, 3.25 Mo, 2 Cu, 0.10V. It is designed to function as a helicopter transmission gear following total lubricant loss, and offers advantages in other applications even at room temperature.

Engineering Properties Of the New Alloy

At 600 F (260 C) the material exhibits a case hardness of 55 HRC and a core hardness of 35 HRC. Optimum impact properties are obtained by double vacuum melting (vacuum induction melting plus vacuum arc remelting) and heat treating with oil quenching. Deformation and warpage during heat treatment can be minimized by air hardening, but with some loss in impact strength.

The alloy must be carburized to produce a rich case of 0.80 to 0.90% surface carbon prior to hardening.

The relatively high percentage of copper (2%) in the analysis contributes significantly to high

Table I – Alloy 53's Core Mechanical Properites

Test Temperature, F (C)	Yield Strength, 10^3 psi (MPa)	Tensile Strength, 10^3 psi (MPa)	Elongation, % in 4D	Reduction in Area, %	Charpy V-Notch Impact Energy, ft-lb (J)
−65 (−55)	160 (1103)	193 (1331)	18	63	39-41 (53-56)
Room temperature	140 (965)	170 (1172)	16	66.5	87-95 (118-129)
212 (100)	—	—	—	—	103-120 (140-163)
350 (175)	130 (896)	175 (1207)	12	46	114-116 (155-157)

Table II – How Austenitizing Temperature Affects Alloy 53's Toughness[1]

Austenitizing Temperature, F (C)[2]	Oil Quenched		Air Cooled	
	CVN Impact Energy, ft-lb (J)	HRC	CVN Impact Energy, ft-lb (J)	HRC
1650 (900)	106,97,93 (144, 132, 126)	34	91,91,84 (123, 123, 114)	34.5
1700 (925)	96, 98, 84 (130, 133, 114)	37.5	73, 75, 67 (99, 102, 91)	36.5
1750 (955)	54, 54, 64 (73, 73, 87)	36.5	40, 52, 44 (54, 71, 60)	36
1800 (980)	71, 67, 62 (96, 91, 84)	38.5	56, 44, 51 (76, 60, 69)	38
1850 (1010)	68 (92)	39	54, 61, 65 (73, 83, 88)	39

1. Charpy V-notch (CVN) impact energy at room temperature.

2. 25 min at temperature.

competitive alloys (the study by Dennis Townsend of NASA's Lewis Research Center in Cleveland has not been published). Wiebel plots indicate that the Carpenter alloy has superior properties. Similar results were obtained in rolling element fatigue tests on a standard rig machine.

Bell Helicopter (under US AVRADCOM TR-83-D-31A and 31B) also studied the fatigue life of the alloy, using a standard R.R. Moore rotating beam fatigue tester. These evaluations included carburized as well as electron beam welded specimens along with the core properties. It was reported that Alloy 53 had over 25% longer fatigue life than 9310.

Rock Bits — The elevated temperature fracture toughness of the alloy and that of several other steels was evaluated for possible use in geothermal well drilling applications. This study was done by Terra Tek Inc., Salt Lake City, under a U.S. Dept. of Energy contract (DOE.EY-76-C-02-0016 and TR-80-87).

Conventional drill bit steels exhibit increased wear and reduced toughness when run at elevated temperatures in geothermal wells. For these reasons, they are operated at lower speeds and lighter loads, resulting in lower penetration rates than those obtained in conventional rock drilling.

The new alloy was tested, along with several other steels, for cones and lugs in conventional roller cone rock bits. Short rod fracture toughness measurements were made on each of the tested steels between room temperature and 750 F (400 C). Fatigue crack resistance was determined at 570 F (300 C) for high temperature steels and at room temperature for conventional steels. Scanning electron microscopy analyses of the fractured short rod specimens were correlated with observed crack behavior from the test records. Test results indicate that the properties of Alloy 53 are consistent with good performance in this application.

Fracture Toughness and Fatigue Properties of VASCO X-2 Steel Gear Materials

A. A. ANCTIL and M. AZRIN

Fracture toughness and fatigue properties were obtained for two types of Vasco X-2 tool steel: the standard 0.24 wt pct carbon level and a modified version in which the carbon was lowered to 0.15 wt pct. The heat treatment consisted of austenitizing between 900 °C and 1120 °C followed by an oil quench and a two-hour double temper at 316 °C. The Charpy impact K_Q Fracture toughness (precracked Charpy), and endurance limit were higher for the 0.24 carbon Vasco X-2. Only the modified Vasco X-2, with lower carbon level, had the hardness level of approximately 40 HRC that is specified for the core of high performance carburized aircraft gears. However, when toughness or fatigue is the main concern, then the 0.24 carbon Vasco X-2 should be the recommended material.

INTRODUCTION

The need for improved materials for high performance gears is a major concern of aircraft designers. A requirement exists for greater helicopter transmission component reliability during abnormally high operating temperature conditions. This condition is experienced during loss of transmission lubricant. Materials having improved tempering resistance would be highly desirable in this case. There are, however, other important considerations in gear material selection.

Gear failure can occur by a number of failure modes depending on the operating conditions and/or material factors. Wear at the mating tooth surfaces occurs at relatively low speeds and high contact loads. Pitting, a form of surface fatigue, occurs at intermediate speeds and loads. At sufficiently high loads the beam bending type loading on the gear teeth can produce bending fatigue. At high surface speeds scoring or scuffing can occur due to a breakdown of the lubricant film. Some of these failure mechanisms are not simply a result of material limitations. For example, failure due to wear or scoring can be minimized by the proper choice and maintenance of lubricants. However, other solutions to prevent or at least delay the onset of failure are material dependent. High gear tooth surface loads require a hard surface. Fatigue life is improved by imposing a residual compressive stress at the surface region. The usual method of obtaining a combination of high hardness and compressive residual stress is by carburizing a moderate hardness material.

Screening or evaluating carburized gear materials involves a combination of metallurgical examination and some form of gear testing using either complete gears or gear sections. Gear testing, although a reliable method of evaluation, is extremely costly and time consuming. Its advantages are that these tests simulate actual operating conditions and permit variations in these operating conditions. Gear tests are not suitable where the concern is for alloy development or the development of new heat treatments. Here the large number of evaluations required preclude the costs and delays that are required in fabricating, machining, heat treating, final machining, inspecting, testing, and evaluating actual gears.

Vasco X-2 is a secondary hardening tool and die steel selected for use as helicopter transmission gears. This

A. A. ANCTIL and M. AZRIN are with the Metals Research Division, Army Materials and Mechanics Research Center, Watertown, Massachusetts 02172.

material, a low carbon version of H12 tool steel, is a consumable vacuum-arc remelted tool steel produced by Teledyne-Vasco Corp. Interest in Vasco X-2 was a result of its high hot hardness and scoring resistance. After early studies on Vasco X-2, it became apparent that a lower carbon level would be necessary to obtain a conventional core hardness level of Rockwell C 40. Lowering the standard Vasco X-2 carbon level from 0.24 to 0.15 decreases the Rockwell C hardness, HRC, about ten points and changes the temperature range over which the equilibrium phases ferrite and austenite are stable. Lower carbon (lower hardness) material generally has higher toughness. An exception to this hardness-toughness relationship will be observed in the data to be reported here. The microstructure for various heat treatments has been reported by Cunningham[1] and Fopiano and Kula.[2] Both of these efforts have been important in determining the optimum heat treatment, in terms of mechanical properties, for carburized helicopter transmission gear application. The emphasis of this study is on mechanical property characterization, particularly fracture toughness and rotating beam fatigue.

Materials and Procedures

The chemical compositions of Vasco X-2 and the low carbon modification of this alloy are given in Table I. The alloy modification was accomplished primarily by decreasing the carbon content to 0.15 wt pct. The plate material was received in the annealed condition, 30.5 cm square, 0.95 cm and 1.59 cm thick. Standard size Charpy V-notch and compact tension specimens, used to determine the impact energy, static/dynamic fracture toughness, and crack propagation data, were machined such that the fracture plane was through the thickness and perpendicular to the plate rolling direction. The R. R. Moore rotating beam fatigue specimen axis was parallel to the plate rolling direction. Heat treatment consisted of austenitizing for one hour at a range of temperatures (900 °C to 1120 °C), oil quenching, cooling in liquid nitrogen (−196 °C) for 20 minutes followed by a double temper (two hours + two hours) at 316 °C. Impact and fatigue specimens in blank form were sealed in quartz tubing, while oversized compact tension specimens were ceramcoated for surface protection during austenitization. These specimens were subsequently finished machined.

Impact and static/dynamic fracture toughness specimens were tested at −54 °C, 24 °C, and 232 °C in accordance with accepted test methods. The −54 °C test temperature represents the minimum helicopter transmission operating temperature while 232 °C is well above the normal operating temperature. Fracture toughness data, K_Q, were obtained from precracked Charpy specimens using a Manlabs, Inc. Physmet Slow-Bend tester. Dynamic fracture toughness data, K_{Id}, were obtained using precracked Charpy specimens and a computerized impact testing system. The hammer impact velocities were 1.52 mps at −54 °C, 3.3 mps at 24 °C, and 5.18 mps at 232 °C.

Crack propagation data were obtained from 1.52 cm thick compact tension specimens. The specimens were precracked 0.127 cm. The specimens were subjected to sinusoidal loading with a constant mean load at three cycles per second with an axial-fatigue, hydraulic, closed-loop testing machine. Stress ratios of 0.07 and 0.10 were used. Crack length measurements were made every 0.0254 cm with a traveling microscope (30 times magnification) and stroboscopic illumination.

Crack growth rates da/dn were determined graphically from the "a" vs N (number of cycles) curves. The stress intensity factor was calculated from $K = \sigma_g \sqrt{a}\, \gamma$ where σ_g is the gross section stress and "a" is the crack length. The term γ relates the crack length and geometric variables to correct for the finite width of the specimen.[3] The quantity ΔK is the stress intensity factor difference between maximum and minimum load at a particular crack length. The crack propagation rate for the accumulation of fatigue damage was related to ΔK by the empirical power law relationship as proposed by Paris:[4] $da/dn = C \cdot (\Delta K)^m$ where C in this equation is the straight line intercept at $\Delta K = 1.0$ MPa$\sqrt{m}$, and m is the growth rate exponent.

Results and Discussion

Table II shows the hardness for the two carbon levels. The 0.15C alloy generally meets the core hardness requirement of approximately 40 HRC. Figure 1 shows how impact energy curves vary with carbon level and hardening temperature. A relatively smooth curve has been drawn through the limited data to indicate a smooth ductile-brittle transition region. The absence of a sharp transition is consistent with reported observations on a similar alloy.[1] Although only a limited number of duplicate tests was performed, it should be noted that relatively little scatter in the data occurred.

Table I. Chemical Composition of Vasco X-2 Alloys (wt pct)

Alloy Designation	C	Si	Mn	S	P	W	Cr	V	Mo	Co	Ni	Cu
0.15C	0.15	0.96	0.23	0.009	0.013	1.28	4.92	0.42	1.34	0.03	0.07	0.08
0.24C	0.24	0.88	0.32	0.007	0.010	1.38	4.92	0.45	1.43	—	—	—

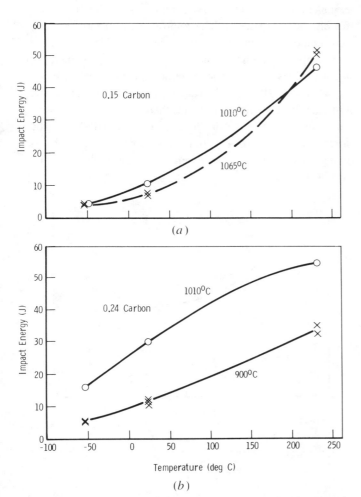

Fig. 1 — Charpy impact energy *vs* test temperature. (a) 0.15 carbon level; 1010 °C and 1065 °C austenitizing temperature, (b) 0.24 carbon level; 900 °C and 1010 °C austenitizing temperature.

The precracked Charpy fracture toughness, K_Q, results of Figures 2a and 2b, have the same relative ranking, with respect to test temperature, as those in Figure 1, although now a plateau appears at the higher test temperatures. Valid fracture toughness data using compact tension specimens were determined only at room temperature, for three austenitizing treatments (Figure 2c). This toughness is higher at the 955 °C hardening temperature. The dynamic fracture toughness K_{Id} (Figure 3) shows a difference between the 1010 °C and 1065 °C hardening temperatures for the 0.15C

alloy. The 0.24C level has a significantly higher K_{Id} and K_Q for the 1010 °C hardening temperature.

Figure 4 shows curves of rotating beam fatigue stress *vs* the number of cycles. The maximum endurance limit of approximately 830 MPa occurred at a hardening temperature of 1010 °C. The endurance limit is slightly lower for the 1065 °C condition. The endurance limit was not obtained at the highest hardening temperature considered, 1120 °C. The 0.24C has a slightly higher endurance limit at 1010 °C (Figure 5) than observed for the 0.15C alloy.

The experimentally determined crack growth curves for the 0.24C alloy, showing the crack length *vs* the number

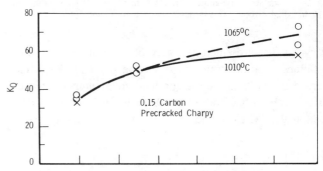

Fig. 2a — Fracture toughness, K_Q, *vs* test temperature. Precracked Charpy specimens of 0.15 carbon level austenitized at 1010 °C and 1065 °C.

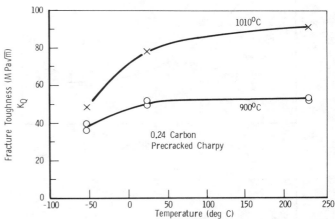

Fig. 2b — Fracture toughness, K_Q, *vs* test temperature. Precracked Charpy specimens of 0.24 carbon level austenitized at 900 °C and 1010 °C.

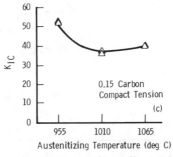

Fig. 2c — Fracture toughness, K_{IC}, *vs* austenitizing temperature. Compact tension specimens of 0.15 carbon level tested at room temperature.

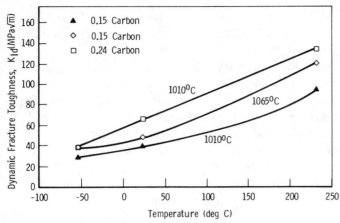

Fig. 3—Plane-strain dynamic fracture toughness *vs* test temperature for 0.15 and 0.24 carbon levels austenitized at 1010 °C and 1065 °C.

of cycles, are presented in Figure 6. The number of cycles to failure, at a maximum gross section stress of 17.6 MPa, were essentially the same for hardening temperatures of 1010 °C, 1065 °C, and 1120 °C. However, material hardened at 955 °C exhibited twice the fatigue crack propaga-

tion life. Lowering the gross section stress from 17.6 MPa to 13.0 MPa for steel austenitized at 1065 °C increased the number of cycles to failure considerably, from 35,000 cycles to 87,500 cycles.

The crack growth rate was measured at various points on each crack growth curve and plotted *vs* the change in stress intensity factor about the crack tip on log-log coordinates (Figure 7). The data for each specimen (except 1120 °C) follow the empirical power law relationship with the exponent ranging from 2.5 to 2.9. The values of the exponent correspond to those reported in the literature[5] for high strength steels. The improvement in fatigue crack propagation rate for the specimen having a 955 °C hardening temperature (A) is indicated by its displacement to higher values of ΔK. The specimens having similar crack growth curves (B, D, and E) are grouped together. The crack growth rate curve for specimen E, hardened at 1120 °C, was not linear in the range of ΔK 27.5 to 33.0 MPa$\sqrt{m}$. The lower maximum gross section stress for the 1065 °C hardening temperature was an extension of curve D (1065 °C) to lower values of ΔK.

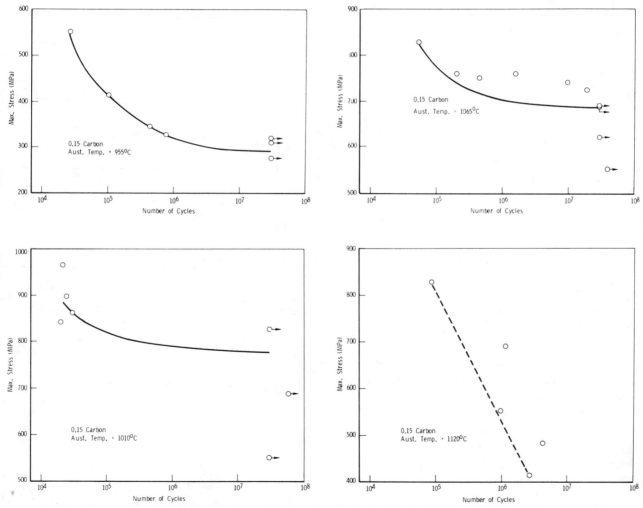

Fig. 4—S-N curves for R. R. Moore specimens of 0.15 carbon level austenitized at 955 °C, 1010 °C, 1065 °C, and 1120 °C.

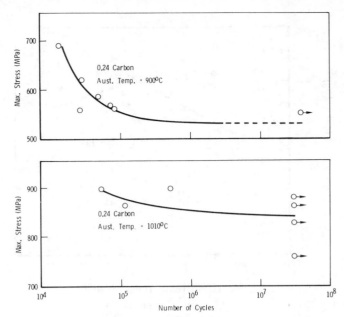

Fig. 5 — S-N curves for R. R. Moore specimens of 0.24 carbon level austenitized at 900 °C and 1010 °C.

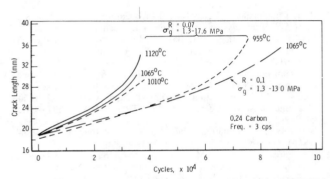

Fig. 6 — Crack growth curves for 0.24 carbon level austenitized at 955 °C, 1010 °C, 1065 °C, and 1120 °C.

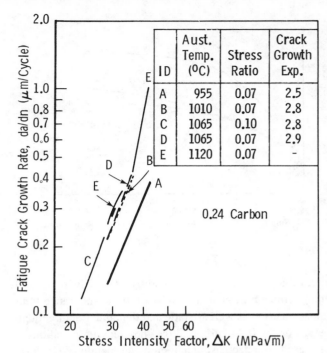

ID	Aust. Temp. (°C)	Stress Ratio	Crack Growth Exp.
A	955	0.07	2.5
B	1010	0.07	2.8
C	1065	0.10	2.8
D	1065	0.07	2.9
E	1120	0.07	-

Fig. 7 — Fatigue crack growth rate *vs* stress intensity factor.

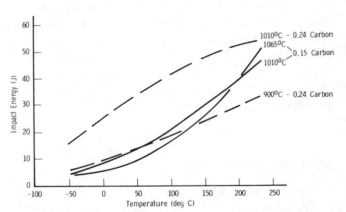

Fig. 8 — Summary of Charpy impact energy *vs* test temperature.

Figures 8 to 10 summarize the detailed presentations of Figures 1 to 6. Although the 0.24C alloy at 1010 °C in Figures 8 to 10 has superior properties, the hardness level is considered too high for core material of carburized gears.[2,6] In terms of toughness there is no advantage in reducing the carbon level to lower the hardness since the higher carbon Vasco X-2 generally results in improved toughness. These results are consistent with toughness data reported elsewhere.[1,2]

The Vasco X-2 steels have a duplex microstructure of alloyed ferrite dispersed in a tempered martensite matrix.[1,2] There is also some retained austenite. The change in carbon from 0.24C to the 0.15C level produces changes in hardness, retained austenite, and the duplex microstructure. The position of the secondary hardening peak of both alloys fall in the 510 °C to 540 °C tempering temperature range. Although the exact location of the peak may shift slightly with hardening temperature, tempering has no effect on the properties reported here since the 316 °C temper used was always well below the peak region. The percentage of primary ferrite in the duplex microstructure is a function of the austenitizing temperature and is higher for the lower carbon alloy which therefore requires a higher austenitizing treatment to keep the ferrite to a minimum level.[2] Cooling rate was not a factor in this study since all specimens were relatively small and oil quenched from the hardening temperature. Ferrite volume decreases only slightly with time at austenitizing temperature.[1] The amount of retained austenite in both alloys has been minimized by the subzero cooling and double temper, and therefore is not considered a problem.

SUMMARY

Presented in this paper is a collection of impact, fracture toughness, and fatigue data, obtained at a range of hardening temperatures followed by a 316 °C double temper, that can be used in the design of high performance carburized

Source: Journal of Heat Treating, December 1982, 331-336

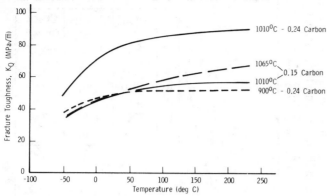

Fig. 9 — Summary of fracture toughness, K_Q, vs test temperature.

aircraft gears. The 0.24 carbon alloy, with a 1010 °C hardening temperature, has the highest fatigue endurance limit, impact energy, as well as static and dynamic fracture toughness. For the 0.24 carbon alloy, the crack propagation life increased with decreasing hardening temperature. However, only the modified X-2 had a hardness level normally considered appropriate for the core material of high performance carburized aircraft gears. It should be emphasized that in terms of toughness there is no advantage in going to a lower carbon level since the higher carbon material generally has a higher toughness. In addition, at the lower carbon level a higher austenitizing temperature is required to reduce the volume of primary ferrite.

ACKNOWLEDGMENTS

The authors thank Drs. Gregory B. Olson and Eric B. Kula for helpful discussions during the preparation of this paper.

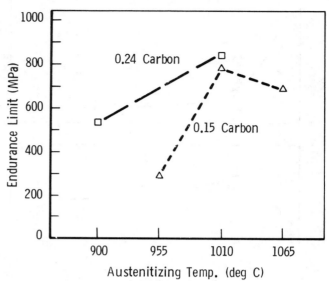

Fig. 10 — Endurance limit vs austenitizing temperature.

REFERENCES

1. R. J. Cunningham: "Vasco X-2, 0.15% Carbon (BMS 7-223) Steel HLH/ATC Transmission Gear Material Evaluation, Test Results and Final Report", Number D301-10036-2, July 8, 1974. Prepared by The Boeing Co., Vertol Division, for U.S. Government under Contract DAAJ01 71–C–0840 (P6A).
2. P. J. Fopiano and E. B. Kula: "Heat Treatment, Structure, and Properties of Standard and Modified Vasco X-2 Carburizing Grade Gear Steels", ASME Publication 77–DET–151, or Army Materials and Mechanics Research Center, AMMRC TR 77–8, May 1977 (ADA 044949).
3. B. Gross: "K Calibration, Plane Strain Crack Toughness Testing", ASTM STP 410, Am. Soc. Testing Mats., 1966.
4. P. C. Paris: "The Fracture Mechanics Approach to Fatigue", Fatigue, an Interdisciplinary Approach, Syracuse University Press, New York, 1964.
5. A. A. Anctil and E. B. Kula: "Fatigue Crack Propagation in 4340 Steel as Affected by Tempering Temperature", ASTM STP 462, Am. Soc. Testing Mats., 1970.
6. D. P. Townsend and E. V. Zaretsky: "Comparison of Modified Vasco X-2 and AISI 9310 Gear Steels", Report NASA-TP-1731, Lewis Research Center, Cleveland, OH, November 1980 (N81–14322).

SECTION IV
Generating-Plant Components

Modified 9Cr-1 Mo Steel for Power Plants

Small additions of niobium and vanadium to the standard 9Cr-1Mo elevated temperature steel result in a stronger and more ductile material. The new modified 9Cr-1Mo steel also exhibits improved long term creep properties, lower thermal expansion, higher thermal conductivity and better resistance to stress corrosion cracking.

Elevated temperature steels must retain their strength, toughness and corrosion resistance through years of service in boilers, steam-driven power generation equipment and reaction vessels. The ferritic chromium-molybdenum steels, such as 2¼Cr-1Mo, have traditionally been chosen for this kind of service at temperatures up to 565 C (1050 F). However, to maintain comparable or the same strength at higher temperatures, an austenitic stainless steel such as AISI Type 304 had to be used. The variations of estimated design allowable stress intensity with temperature for the modified and standard 9Cr-1Mo steels, 2¼Cr-1Mo steel and AISI Type 304 stainless steel are shown in the accompanying graph. The austenitics can do the job, but they are more expensive and somewhat less efficient in terms of heat transfer. But an even more significant disadvantage is that the austenitic stainless steels are difficult to join to the ferritic steels. The differences in thermal expansion between the two types of material result in the cracking of joints that are often thermally cycled during service. The modified 9Cr-1Mo steel was developed to meet the elevated temperature strength of AISI Type 304 but without the problems associated with the use of austenitic stainless steels.

The new steel, developed by Oak Ridge National Laboratory and Combustion Engineering with assistance from Climax, is now in the process of being qualified for the ASME Boiler and Pressure Vessel Code. The accompanying table compares the composition of the new steel and the standard grade as specified in ASTM A213-79b, Grade T9. With Code approval, the modified 9Cr-1Mo steel is expected to find applications beyond the originally targeted steam tubing for fossil fuel fired boilers. Other proposed applications include breeder reactor systems, pressure vessels for coal liquefaction and gasification, oil refining hydrotreating equipment, solar power towers, geothermal energy systems, and in the very long term, fusion reactors. Current predictions are that this new steel will account for 250,000 pounds of molybdenum per year in normal maintenance replacement programs for power plant steam tubing and an additional 90,000 pounds per year for new fossil fuel fired power plant construction.

Allowable design stresses for 9Cr-1Mo and 2¼ Cr-1Mq steels.

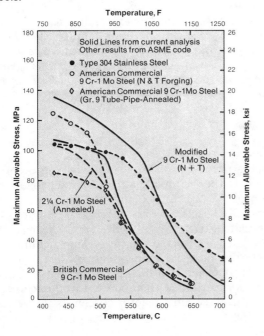

Compositions of standard and modified 9Cr-1Mo elevated temperature steels

	Modified	Standard[a]
Carbon	0.08-0.12 %	0.15 % max
Chromium	8.00-9.00	8.00-10.00
Molybdenum	0.85-1.05	0.90-1.10
Silicon	0.20-0.50	0.25-1.00
Manganese	0.30-0.50	0.30-0.60
Niobium	0.06-0.10	—
Vanadium	0.18-0.25	—
Sulfur	0.010 max	0.030 max
Phosphorus	0.020 max	0.030 max
Nitrogen	0.03-0.07	—

(a) ASTM A213-79b, Grade T9

STEAM GENERATOR MATERIALS PERFORMANCE IN HIGH TEMPERATURE GAS-COOLED REACTORS

J. E. CHAFEY and D. I. ROBERTS
General Atomic Company, P.O. Box 81608
San Diego, California 92138

High temperature gas-cooled reactor (HTGR) systems feature a graphite-moderated, uranium-thorium, all-ceramic core and utilize high pressure helium as the primary coolant. The steam generators in these systems are exposed to gas-side temperatures approaching 760°C (1400°F) and produce superheated steam at 538°C (1000°F) and 16.5 MPa (2400 psi). Thus, the design and development of steam generators for these systems require consideration of time-dependent materials behavior, corrosion, fretting, wear, and other related phenomena of concern in all steam generators.

The prototype Peach Bottom Unit No. 1 40-MW(electric) HTGR was operated by the Philadelphia Electric Company for a total of 1349 equivalent full power days during a 7-yr period. Upon planned decommissioning of that plant, the forced-recirculation U-tube steam generators and other components were subjected to destructive properties tests and metallurgical examinations. These tests and examinations showed the steam generators to be in very satisfactory condition. .

The 330-MW(electric) Fort St. Vrain HTGR, owned and operated by Public Service Company of Colorado, and now in the final stages of startup, has achieved 70% power and generated more than 1.5 × 10^6 MWh of electricity. The steam generators in this reactor are once-through units of helical configuration, and their design and development required considering a number of new materials factors including creep fatigue. Also, because of the once-through design, water chemistry control needed special consideration.

Current designs of larger HTGRs also feature steam generators of helical tube once-through design. Materials issues that are important in these designs include detailed consideration of time-dependent behavior of both base metals and welds, as required by current American Society of Mechanical Engineers Code rules, evaluation of bimetallic weld behavior, evaluation of the properties of very large tubesheet forgings, consideration of the gaseous corrosion effects of the primary coolant, and other related factors.

INTRODUCTION

Figure 1 illustrates the principal features of high temperature gas-cooled reactors (HTGRs). The reactor core consists of graphite blocks, which act as both moderator and structure and contain the fuel that generates fission heat. The fuel used may be either the carbides or oxides of uranium and thorium. The system is cooled by high pressure helium which, in a typical steam-cycle HTGR, enters the core at temperatures of ~320°C (~650°F) and is heated to temperatures of ~760°C (1400°F) by the core. The helium is then circulated to heat exchangers, where the heat is transferred and can be converted into energy for useful work. The cool helium is then recirculated back to the core, and the circuit continues. Because of the relatively high helium temperatures, the heat exchangers can be steam generators capable of generating high pressure, high temperature steam compatible with modern turbomachinery. Indeed, the all-ceramic reactor core gives the system the potential to go to very high core outlet temperatures

Reprinted with permission from Nuclear Technology, October 1981, 37-49, © 1981 American Nuclear Society, Inc.

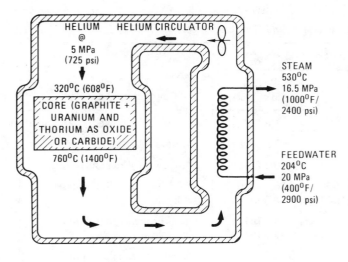

Fig. 1. Simplified schematic diagram of steam generating HTGR.

construction or development, as shown in Table I. Of the reactor systems shown in Table I, only four feature steam generators that produce modern high pressure, high temperature steam (see Table II). These are the Peach Bottom, FSV, THTR, and large HTGR systems. The Peach Bottom reactor was operated for seven years and has been shut down, decommissioned, and the steam generators examined. The FSV reactor has entered commercial operation, and substantial operating experience has been accumulated. The THTR, in the Federal Republic of Germany, is still under construction and, therefore, has yet to accumulate operating experience with its steam generators. Similarly, the large HTGRs are currently only in the design stage. Accordingly, this paper focuses on the units that feature these modern steam generators and pays most attention to the Peach Bottom and FSV reactors, although other systems are referred to when appropriate.

[approaching 1000°C (1832°F)], and currently, significant effort is being devoted to developing the reactor for high temperature process heat applications. However, this paper focuses on the steam generating (electricity producing) version of this reactor system.

The HTGR has been under development for more than 20 years. During that time, a number of units have been developed, and several are still under

PEACH BOTTOM UNIT NO. 1

The Peach Bottom Unit No. 1 (PB-1) prototype HTGR (Fig. 2) was operated by Philadelphia Electric Company for a total of 1349 equivalent full power days over seven years between January 1967 and October 1974, when it was shut down for planned decommissioning. Table III gives relevant operating statistics, and Table IV gives the principal features of the operating environment.

TABLE I

Principal Steam-Generating HTGRs

Name	Size [MW(thermal)]	Sponsor/Owner	Location	Type	Status
Dragon	20	Organization for Economic Cooperation and Development/Nuclear Energy Agency	Winfrith, U.K.	Experimental unit	Operated 1966 to 1976, now decommissioned
Arbeitsgemeinschaft Versuch Reaktor (AVR)	46	Kernforschungsanlage	Jülich, Federal Republic of Germany (FRG)	Experimental unit	In operation since 1967
Peach Bottom	115	U.S. Atomic Energy Commission (AEC) [U.S. Department of Energy (DOE)]/Philadelphia Electric Company	Pennsylvania	Small prototype	Operated 1967 to 1974, now decommissioned
Fort St. Vrain (FSV)	842	AEC (DOE)/Public Service Co. of Colorado	Colorado	Demonstration reactor	In operation since December 1976
Thorium Hoch Temperatur Reaktor (THTR)	762	Vereinigte Elecktriziatswerke Westfalen	Schmeehausen, FRG	Demonstration reactor	Scheduled for startup 1982
HTGR-Steam Cycle (SC)	2000-3000			Power reactor	In design

TABLE II

Principal Features of HTGR Steam Generators and Heat Exchangers

| Plant | Maximum Primary Coolant Temperature [°C (°F)] | Steam/Water Conditions | | Steam Generator Configuration and Type |
		Temperature [°C (°F)]	Pressure [MPa (psi)]	
Dragon	830 (1526)	203 (397)	1.6 (232)	(Not used for power generation)
AVR	950 (1742)	505 (941)	7.5 (1088)	Involute tube
Peach Bottom	700 (1292)	536 (997)	10.2 (1480)	U-tube forced-recirculation
FSV	775 (1427)	Main steam 538 (1000) Reheat steam 538 (1000)	16.5 (2400) 4.9 (711)	Once-through helical bundles with reheat
THTR	750 (1382)	Main steam 530 (986) Reheat steam 530 (986)	18.0 (611) 5.6 (812)	Once-through helical bundles with reheat
HTGR-SC	690 (1274)	540 (1004)	16.5 (2393)	Once-through helical bundles

Fig. 2.   Prototype 40-MW(electric) HTGR at Peach Bottom, Pennsylvania.

Figure 3 shows the general layout of the Peach Bottom primary coolant system. The two steam generators, which were housed in separate pressure vessels, were forced-recirculator boilers with pendant U-tube economizer, evaporator, and superheater sections (see Fig. 4). Figure 5 is a photograph of a tube bundle, which shows the original Type 304H stainless steel superheater bundle. However, prior to reactor

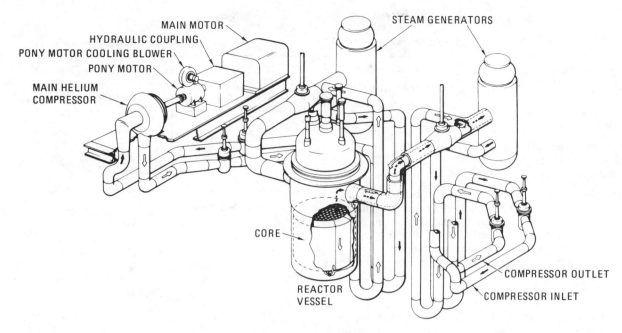

Fig. 3.   Isometric view of primary coolant systems.

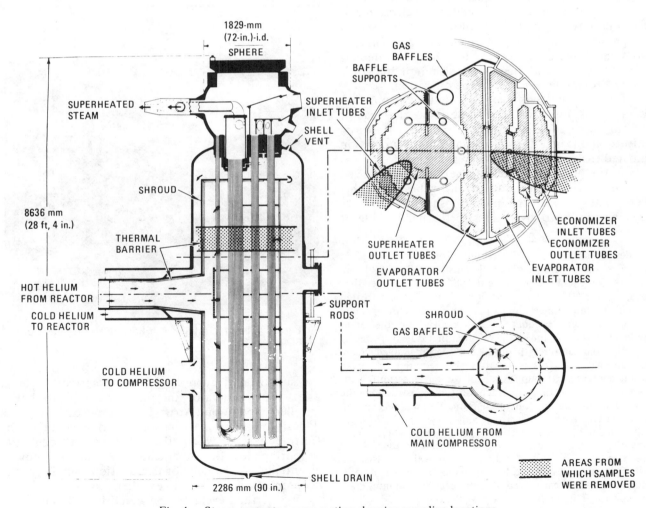

Fig. 4.   Steam generator cross section showing sampling locations.

TABLE III

Peach Bottom Operating Statistics

Operation period	January 1967 to October 1974
Total power generated	1.38×10^6 MW(electric)-h
Average gross thermal efficiency	37.2%
Total time of operation with coolant at high temperature	~35 000 h
Steam generator leakage experience	Two small leaks from the superheater tubesheets that were present at startup persisted at the same level throughout life.

TABLE IV

Peach Bottom Environments

Primary Coolant: Helium at 2.2 MPa (319 psi)	
Impurity	Partial Pressure (Pa)
H_2	20
H_2O	1
CO	1
CO_2	<0.06
CH_4	2

Secondary Coolant:
Feedwater at 12 MPa (1741 psi)

pH	9.4
Hydrazine	50 to 200 ppb
Dissolved O_2	0 (none detected)

Steam Drum Conditions

SiO_2 ion	100 to 500 ppb
Chloride ion	100 to 600 ppb
pH	8.8 to 9.2

commissioning, the superheater section encountered stress corrosion cracking; as a result, the superheater section was retubed with Alloy 800.[a] Table V indicates the final steam generator materials selection and generating environment. The Alloy 800 tubes served throughout the operating life without any failures. Two small leaks occurred due to fabrication defects at the tubesheet, but these were so minor that they remained tolerable throughout the operation.

After decommissioning, a destructive examination program was undertaken to determine the reactor condition. The DOE and the Electric Power Research Institute sponsored this program. This examination evaluated the condition of one steam generator.

The top head of the steam generator was removed to examine the steam side of the tubes and tubesheets. The side of the pressure shell was cut, as shown in Fig. 6, to give access to the helium side of

---

[a]The designation of Alloy 800 has undergone many changes in the years subsequent to the building of the Peach Bottom reactor. For example, in modern U.S. codes and specifications, "Alloy 800" refers to a form of material heat treated at relatively low temperatures to produce a fairly fine-grained high-yield-strength product. "Alloy 800H," on the other hand, refers to a version of the material that is heat treated at very high temperatures to produce a coarser grained creep-resistant material. This material is also required to have a carbon content in the upper half of the range permitted for Alloy 800. At the time of Peach Bottom construction, the importance of heat treatment to properties was recognized, while the significance of carbon content was not. Thus, for high temperature service, under conditions where creep strength was important, a material designated either "Alloy 800, Grade, 2" or "Alloy 800-solution annealed" was specified. This material required high temperature heat treatment, but did not have carbon control or specific minimum grain size control. The latter material was that employed in Peach Bottom.

Fig. 5. Manufacturing Peach Bottom steam generator.

the superheater, economizer, and evaporator tubing. The steam generator internals appeared to be in excellent condition. Apart from some carbonaceous deposit on the tubes and some evidence of minor fretting damage where the tubes pass through the support plates, the superheater tube bundle showed essentially no evidence of degradation. Similarly, the economizer section was in excellent condition.

To further evaluate the condition of the steam generator construction materials, metallographic and

TABLE V

Peach Bottom Steam Generator Materials and Operating Conditions

Heat Transfer Tubing			
	Material	Mean Wall Temperature [°C (°F)]	Environments
Superheater tubing	Alloy 800 to B163-64 and code case 1325	415 to 557 (779 to 1035)	Helium on o.d., steam on i.d.
Evaporator tubing	C-Si steel to SA-192	329 to 343 (624 to 649)	Helium on o.d., steam/water on i.d.
Economizer tubing	C-steel to SA-179	232 to 315 (450 to 559)	Helium on o.d., water on i.d.

Other Components		
	Material	Temperature [°C (°F)]
Shrouds	1 Cr–$\frac{1}{2}$ Mo to SA-387 Gr B	358 (676)
Thermal insulation	Type 304 stainless steel foil	400 to 700 (752 to 1292)
Superheater tubesheet	Low alloy steel clad with INC 82	540 (1004)
Evaporator and economizer tubesheets	Carbon steel	230 to 340 (446 to 644)

Fig. 6. Superheater tubes visible through hole cut through steam generator shell and shroud after service.

mechanical test specimens were removed and subjected to detailed metallurgical evaluation. In general, this examination confirmed that all construction materials were in excellent condition. As Fig. 7 indicates, no evidence of significant interaction was given between the Alloy 800 superheater tubing and the external helium environment, although these helium-exposed surfaces showed carbon deposits (believed to have resulted from minor oil ingress into the reactor during operation). The internal, steam-exposed surfaces of the superheater tubing were covered by thin, protective, entirely satisfactory oxide films. Examination of these oxide films showed that the apparent rate of oxide growth was entirely consistent with the observations made by others[1] (see Fig. 8).

Similarly, the economizer and evaporator tubing showed no evidence of adverse interaction with the primary coolant environment, and they showed the presence of protective oxide films of the expected thickness on the steam and water sides (see Figs. 9 and 10).

Both low-carbon steel tubes shown in Fig. 10 exhibit slightly different microstructures at the helium side and at the steam side. Also, the Nital etchant delineated the grain boundaries more prominently on tube No. 50 than on tube No. 78. These

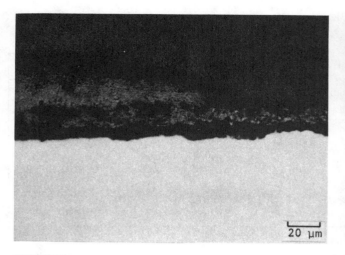

UNETCHED

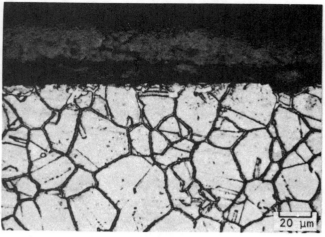

ELECTROLYTIC OXALIC ACID ETCH

HELIUM SIDE

UNETCHED

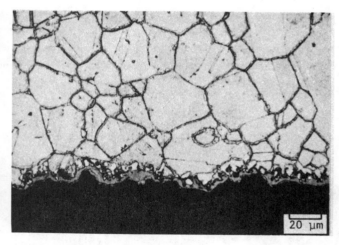

ELECTROLYTIC OXALIC ACID ETCH

STEAM SIDE

Fig. 7.   Surface conditions on superheater outlet tube No. 20 (Alloy 800).

differences are chance occurrences and are judged to be because the evaporator tubes had sustained minor, but somewhat variable, decarburization during tube manufacture.

Evaluation of the mechanical properties of the exposed ferritic steels showed that the service had had relatively little effect. In the case of the Alloy 800 superheater tubing, substantial age-hardening was noted to have occurred because of service exposure, resulting in a considerable strengthening (see Figs. 11 and 12) and some loss of ductility (see Fig. 13). However, despite this thermal aging effect, the residual ductility of the material remained at entirely satisfactory values.

The overall conclusion from the post-decommissioning examination of the PB-1 reactor was that all construction materials, including those used in the steam generator, appeared to have performed entirely satisfactorily and, from a materials viewpoint, were capable of many additional years continued exposure.

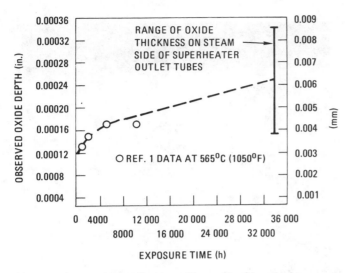

Fig. 8. Comparison of steam-side oxide film thickness on superheater outlet tubing with published data.

THE FSV REACTOR

The 330-MW(electric) FSV HTGR (Fig. 14), now operating on the Public Service Company of Colorado grid, has features that differ significantly from the PB-1 unit. Figure 15 shows the flow diagram for this reactor.

As indicated, the reactor utilizes a PCRV (Fig. 16). The PCRV concept was chosen for its important inherent safety advantages. However, because this vessel type was chosen, the steam generators needed to be as compact as possible. To achieve this in the FSV plant, the steam generators chosen were once-through units comprising helically wound tubing.

The reactor contains 12 heat exchanger modules. Figure 17 is a simplified schematic drawing and Fig. 18 a photograph of one such module. Table VI indicates the construction materials and relevant operating environment for these units. As shown,

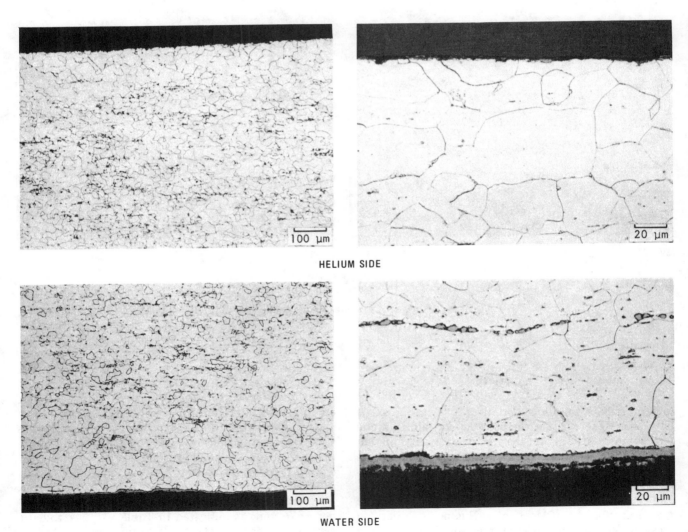

Fig. 9. Surface features on typical economizer outlet tubing (carbon steel) (Nital etch).

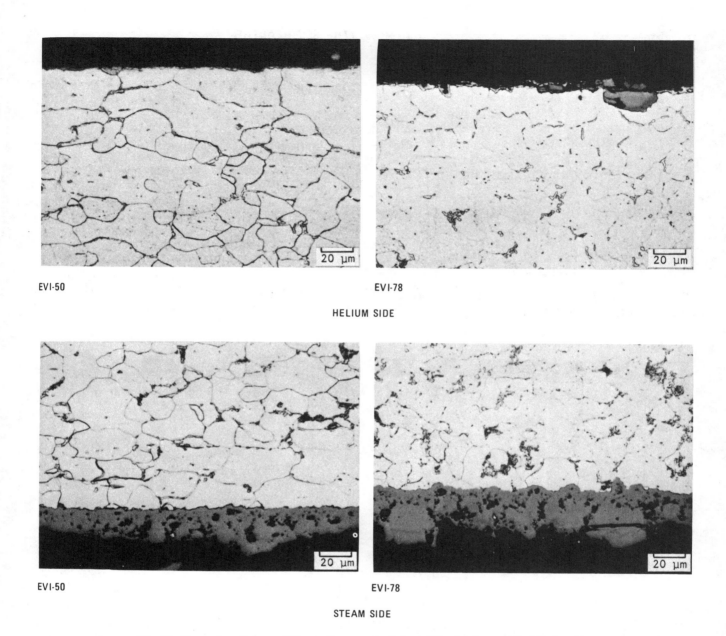

EVI-50   EVI-78

HELIUM SIDE

EVI-50   EVI-78

STEAM SIDE

Fig. 10. Typical surface conditions of evaporator inlet tubing (carbon steel) (Nital etch).

Alloy 800 is used in the hottest sections of the superheater and reheater. The intermediate temperature superheater is $2\frac{1}{4}$ Cr−1 Mo, and the remainder of each module is low alloy steel.

Table VI shows that the pressure-containing metal temperatures in this reactor are significantly higher than they were in the PB-1 units. As a result, in the design of the FSV units, considerable attention was directed toward elevated temperature design considerations. These steam generators were built before American Society of Mechanical Engineers Boiler and Pressure Vessel Code rules for elevated temperature construction had approached the level of complexity

that now exists in Code Case N-47. Nevertheless, the design of the FSV units considered fatigue, long-term, time-dependent properties, and to some extent, creep/fatigue interaction. In addition, because these steam generators are subject to high cross-tube gas flows, special attention was directed toward flow-induced vibration and related fretting and wear damage. After much testing, wear protection was provided by protective sleeve devices.

This reactor started power generation in December 1976 and, as indicated in Table VII, has, to date, generated very substantial amounts of electricity. Table VIII shows the reactor operating environments.

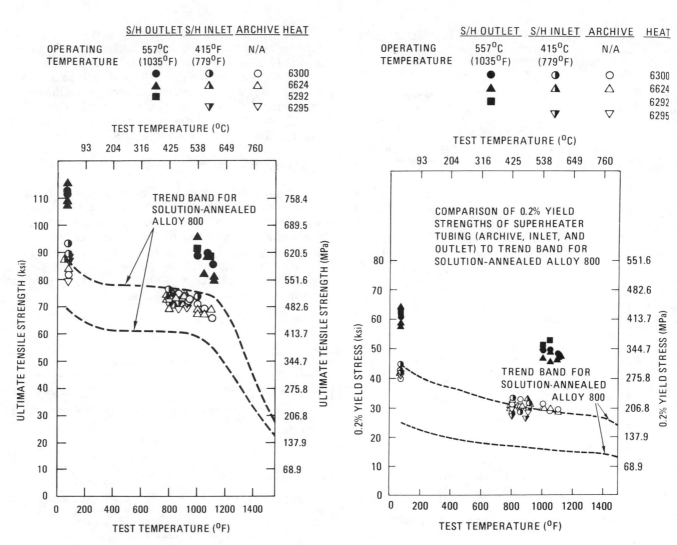

Fig. 11. Comparison of tensile strength data from superheater tubing with trend band for solution-annealed Alloy 800.

Fig. 12. Comparison of 0.2% yield strengths of superheater tubing with trend band for solution-annealed Alloy 800.

TABLE VI

FSV Steam Generator Materials and Operating Conditions

Components	Material	Maximum Operating Temperature [°C (°F)]	Environments
Superheater II and reheater	Alloy 800—solution annealed	730 (1346) (tubes) 760 (1400) (supports)	Helium and steam
Superheater II and evaporator	$2\frac{1}{4}$ Cr–1 Mo	540 (1004) (tubes) 620 (1148) (supports)	Helium and steam/water
Economizer	$\frac{1}{2}$ Cr–$\frac{1}{2}$ Mo	400 (752)	Helium and water
Feedwater leads and ringheader	$\frac{1}{2}$ Cr–$\frac{1}{2}$ Mo/carbon steel	260 (500)	Helium and water
Steam leads and ringheader	Alloy 800—mill annealed	540 (1004)	Helium and steam
Wear protection coatings	Chromium carbide	750 (1382)	Helium

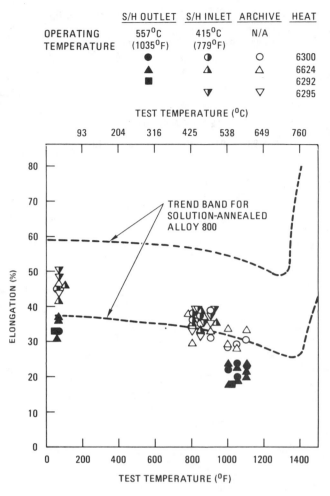

	S/H OUTLET	S/H INLET	ARCHIVE	HEAT
OPERATING TEMPERATURE	557°C (1035°F)	415°C (779°F)	N/A	
	●	◑	○	6300
	▲	△	△	6624
	■			6292
		▽	▽	6295

Fig. 13. Comparison of tensile elongation of superheater tubing with trend band for solution-annealed Alloy 800.

The steam generators have shown very satisfactory operation. One tube leak was experienced in November 1977. The exact location of the tube leak was difficult to discern; however, the leak was located in the Alloy 800 superheater bundle section. Indications at the time suggested that this was simply a random failure, and subsequent operation without evidence of further leakages tends to confirm this view.

FUTURE LARGE HTGRs

A number of designs have evolved in recent years for large steam generating HTGRs. Due to the very satisfactory prior experience, the steam generators in these units are expected to have materials and configurations generally similar to the FSV reactor. For instance, the units will be helically wound, once-through units with Alloy 800H and $2\frac{1}{4}$ Cr−1 Mo as the principal construction materials.

TABLE VII

FSV Operating Statistics

Initial power operation	December 1976
Total power generated to date	$>1.2 \times 10^6$ MW(electric)-h
Gross thermal efficiency (at 70% power)	37%
Expected gross thermal efficiency at 100% power	39%
Steam generator leakage experience	One leak from a superheater tube occurred in November 1977

Fig. 14. Fort St. Vrain power plant.

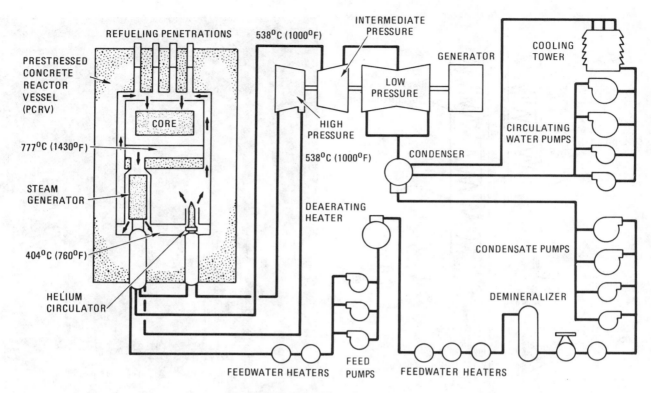

Fig. 15. Fort St. Vrain flow diagram.

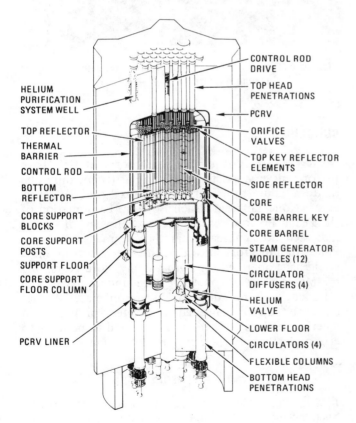

Fig. 16. Fort St. Vrain prestressed concrete pressure vessel.

TABLE VIII

FSV Environments

Primary Coolant: Helium at 5.0 MPa (725 psi)	
Impurity	Partial Pressure (Pa)
H_2	40 to 120
H_2O	1 to 10
CH_4	4 to 20
CO	4 to 20
CO_2	2 to 40
Secondary Coolant: Feedwater at 20 MPa (2901 psi)	
Steam Generator Inlet Characteristics	
pH	9.1 to 9.3 (NH_3)
Hydrazine	10 to 20 ppb
Conductivity	0.1 μmho/cm (max)
Dissolved O_2	5 ppb (at deaerater outlet)

However, the design of these future steam generators will be governed by the increasingly complex and sophisticated design rules of Code Cases N-47, N-48, etc. These requirements impose the need for very complex and complete finite element analyses

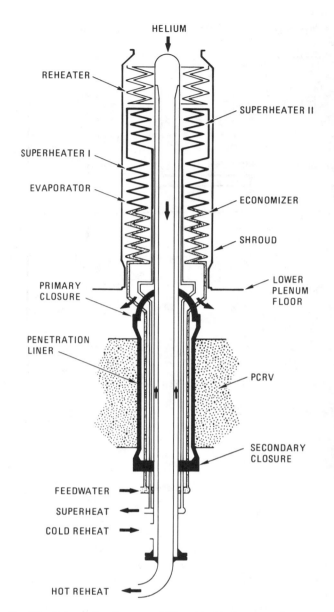

Fig. 17. Schematic diagram of Fort St. Vrain steam generator module.

HELIUM

REHEATER

SUPERHEATER II

SUPERHEATER I

EVAPORATOR

ECONOMIZER

SHROUD

PRIMARY CLOSURE

LOWER PLENUM FLOOR

PENETRATION LINER

PCRV

SECONDARY CLOSURE

FEEDWATER

SUPERHEAT

COLD REHEAT

HOT REHEAT

Fig. 18. Steam generator module and reheater standpipe during manufacture.

including consideration and analysis of creep/fatigue interaction phenomena, creep, ratchetting, and other failure modes. Moreover, the effects of specific environments on properties must be considered, as must the effects of fabrication factors, such as welding and bending, on materials behavior. Such analyses will require that existing data bases for construction materials be considerably extended.

Also, economic factors will almost certainly dictate that future HTGR steam generators have larger modules than were used in FSV. This will necessitate evaluating the properties of very large forgings, developing wear protection systems appropriate for the new configurations, and related factors.

CONCLUSIONS

Overall, HTGR steam generator materials have performed very satisfactorily. In particular, only one random-failure tube leak has occurred on FSV generators. None failed at Peach Bottom. Examination of the Peach Bottom reactor after seven years of service indicated no significant materials degradation of any kind. All indicators show that the Peach Bottom units could have operated for a great many more years.

Materials selection for future units appears well founded, and the additional work on materials that currently appears needed is related primarily to meeting modern code requirements and responding to economically induced configuration changes.

REFERENCE

1. W. L. PEARL et al., "General Corrosion of Materials for Nuclear Superheat Applications," GEAP-4760, General Electric Company (1965).

Requirements for Structural Ceramics in Advanced Power Plants

W. T. Bakker, G. G. Trantina, and D. Kotchick

During the last decade, considerable research and development has been carried out to demonstrate the use of structural ceramics, especially silicon carbide and silicon nitride, in high temperature gas turbines. A smaller but significant effort has been conducted to demonstrate the feasibility of ceramic heat exchangers. Design methods have been developed to take into account the brittle nature of ceramics. Thus, the viability of ceramic components in small turbines and subscale heat exchanger modules has been demonstrated for short test times. Use of ceramics in large-scale advanced power plants will require considerable additional effort to ensure long-term durability and high reliability.

INTRODUCTION

The central theme of EPRI's research policy is to ensure the supply of economically competitive electricity in the near and long term, using systems which meet environmental requirements and are less dependent on foreign energy resources.[1] Broad criteria for the development of new power generation equipment are therefore:

1. High systems efficiency
2. Low pollution potential
3. High reliability
4. Low maintenance cost
5. Use of domestic energy resources

Materials programs, including the development of structural ceramics must support one or more of the above-mentioned criteria, preferably for a specific application. The use of ceramics in electric power gen-eration systems is, therefore, of interest for the following reasons:

1. *Higher Efficiency.* The maximum potential use temperature of presently available ceramics ranges from 1200 to 1400 °C for most potential applications in power generation, that of most metals, including super-alloys, is about 850 to 1000 °C. Thus, the use of ceramics should result in higher overall efficiencies, especially in the gas turbine area.

2. *High Corrosion and Erosion Resistance.* Intrinsic characteristics, such as hardness and chemical compat-ibility, indicate that ceramics are more resistant to corrosion and erosion in low purity fuels. Since future power generation fuels will be increasingly derived from coal, their impurity content may be expected to in-crease.

3. *Availability and Cost.* Raw materials for structural ceramics, mainly Si, C, and N_2, are readily available domestically and intrinsically inexpensive. Critical in-gredients of high temperature alloys, especially Ni, Co, and Cr, are imported. Their availability and cost may be subject to interruption and escalation.

W. T. BAKKER, G. G. TRANTINA, and D. KOTCHICK are with the Electric Power Research Institute, General Electric Company, and AiResearch Co. of California.

Reprinted from Journal of Materials for Energy Systems, September 1980, 41-50, © 1980 American Society for Metals

Based on the above criteria, the following potential applications for structural ceramics have been tentatively identified:

1. *Gas Turbines*. For gas turbines, the possiblity exists to cool metallic components to relatively low temperatures, at which both hot strength and corrosion resistance are adequate. Thus, the efficiency of gas turbines made with metal parts can be increased by cooling critical high temperature components. This allowed an increase in turbine inlet temperature to about 1200 °C at present. Development work is in progress to increase turbine inlet temperatures further to the 1400 to 1600 °C range, using either water or air cooling. The use of cooling will decrease the overall systems efficiency. The difference with systems using uncooled ceramics is generally projected to be about 2 to 3 pct. Such an efficiency increase is significant and can result in considerable fuel savings over the life of a power plant. However, the incentive to develop an all-ceramic turbine for power generation is not as great as in the aircraft and automotive fields where ceramics have additional advantages because of their lower weight/power ratios. The introduction of ceramics in large power turbines will, therefore, most likely be relatively slow and start with low stress applications such as combustor linings, nozzles, and shrouds.

2. *Heat Exchangers*. In the heat exchanger area, the need for ceramics is more clear cut. Metallic heat exchangers are limited to 850 to 900 °C in clean environments and to considerably lower temperatures in corrosive gas streams. Using presently available ceramics such as siliconized silicon carbide, maximum materials temperatures of 1200 °C can be maintained easily, while semi-commercial materials such as alpha silicon carbide promise continuous use temperatures to at least 1400 °C. Thus, the efficiency of indirect fired gas turbines can be boosted by about 5 pct in absolute terms. This will result in a fuel saving of about 17 pct. An especially promising application is the use of ceramic heat exchangers in solar thermal central receivers. Here the capital cost per kWh is very high. Thus, the use of a high efficiency Brayton cycle engine is the preferred method to convert sunshine to electricity. With metallic heat exchangers the turbine inlet temperature is limited to 820 °C and the turbine efficiency is about 27 pct. With a ceramic heat exchanger using commercially available ceramic tubes, the turbine inlet temperature can be increased to 1030 °C and the turbine efficiency increased to about 35 pct.[2] Since corrosion is nonexistent and the systems pressure is only 0.7 to 1.0 MPA (100 to 150 psi), the long-term durability of the receiver appears promising.

Other potential applications for ceramic heat exchangers are in relatively small 5 to 15 MW dispersed thermal-mechanical power plants, especially where unconventional fuels such as wood waste or other by-products or waste materials are used as fuels.

3. *First Wall and Blanket Components for Fusion Reactors*. The main attraction of ceramics, mainly silicon carbide, here is that the half life of radioactive isotopes produced during irradiation is very short (Fig. 1).[3] Thus, repair and replacement of first wall and blanket components is much easier and waste materials do not require special storage facilities. Only through the use of low activation first wall materials can the fusion reactor fulfill its promise as a truly "clean" source of primary energy. The use of silicon carbide instead of aluminim or vanadium alloys has the additional advantage that high coolant temperatures can be used. The possibility exists, for instance, to use a closed cycle gas turbine with helium as the working fluid and a turbine inlet temperature of about 1000 °C using presently available ceramic materials

Requirements for all the applications of ceramics mentioned above are naturally diverse. However, there are common features which will be discussed in this paper. Firstly, ceramic components, both large and small, must be sufficiently strong to withstand the mechanical and thermal stresses during service. Secondly, the ceramics must be able to withstand degradation by corrosion, erosion, and irradiation over long periods of time without excessive material and strength losses.

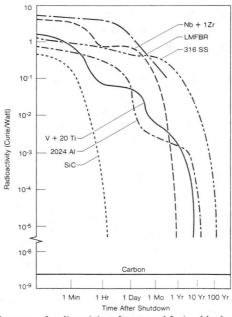

Fig. 1—Decrease of radioactivity of proposed fusion blanket and first wall materials.

STRENGTH REQUIREMENTS

Obviously, a ceramic component should not fail under the stresses imposed during service if service conditions are within design limits. During the last 10 years our understanding of the mechanical behavior of brittle solids has increased markedly. Nevertheless, occasionally, ceramic components still fail unexpectedly at stress levels far below the nominal strength level of the material. Large undetected flaws or unusual stress concentrations or poor design and fabrication are generally blamed for such failures. However, it is not yet generally recognized that the allowable maximum design stress, especially in large components, is only a small fraction of the nominal strength of the material, as determined from modulus of rupture data, despite excellent recent papers on this subject.[4,5,6] Thus, larger "safety" factors are required than for metal structures. To estimate the order of magnitude of such "safety" factors, the design methodology generally used for ceramic structures will be briefly reviewed.

In general, the strength of ceramics depend on the stress required to propagate a critical pre-existing flaw rapidly throughout the ceramic specimen. The strength is thus controlled by the flaw size distribution. Unfortunately, the flaw size distribution is not known and cannot be easily determined by nondestructive testing as the size of critical flaws is generally below the detection limit of available NDE equipment. Therefore, one must devise a scheme to statistically simulate the strength variation in ceramics caused by variations in flaw size distributions from specimen to specimen. By now, the Weibull[7] distribution is widely used to characterize the strength variation. Use of the Weibull strength distribution parameters derived from a set of *representative* test specimens, allows calculation of the failure probability of components and structures of different size and stress distribution. Thus, the probability of failure for a given component is:[4]

$$P = 1 - e^{-kV} \left(\frac{(\sigma \max)^m}{\sigma o} \right) \qquad [1]$$

in which

P	= Probability of failure
k	= Load factor indicating the uniformity of the stress distribution, *i.e.*, $k = \int (\sigma/\sigma \max)^m dV/V$
V	= Volume of component or specimen
$\sigma \max$	= Maximum stress in component or specimen
σo	= Normalizing constant, representing 63.2 pct cumulative failures for a given set of failures of specimens with KV = 1
m	= Weibull modulus = a measure of the scatter

of the data and, thus, of the flaw size distribution. A small m indicates a large scatter.

A rearrangement and simplification of the basic Eq. [1] is used to devise the Weibull modulus m, from the strength distribution of a set of test specimens:

$$\ln \ln \frac{1}{1 - P} = m \ln\sigma + \text{Constant} (V, K, \sigma o) \qquad [2]$$

Here σ is the fracture strength of the individual specimens. The failure probability P is calculated for each specimen as $n/N + 1$, where n is the ordering number and N is the total number of specimens. The Weibull modulus is obtained as the least square estimate of the slope of the plot of strength against failure probability. V, K, and σo are constants for a given set of identical specimens.

The accuracy of the Weibull modulus thus calculated, is dependent on the number of specimens tested.[5] Unfortunately, the confidence limits for m are broader than those for the characteristic strength calculated from the same data as shown in Fig. 2. This means that a large sample population is needed for reliable determination of the Weibull modulus. A sample size of 30 is generally accepted as the minimum. Unfortunately, for practical reasons, most Weibull moduli reported are based on only 10 to 15 samples per data point.

Equation [1] indicates that the fracture probability of a material will vary with sample size and stress distribution. Generally, the probability of failure will in-

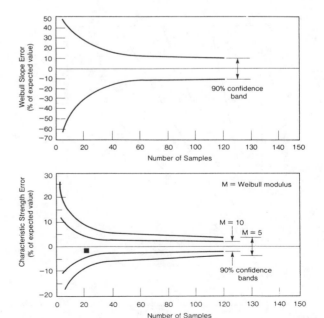

Fig. 2—Confidence limits of Weibull Modulus and characteristic strength as a function of sample size.

Source: Journal of Materials for Energy Systems, September 1980, 41-50

crease with increasing sample size and increasing load factor. However, the increase in failure probability is also dependent on the Weibull modulus. For materials with Weibull moduli in the 7 to 10 range, the decrease in allowable fracture strength with increasing component size is very dramatic indeed. Trantina[4] has developed the following simplified relationship:

$$\frac{\sigma_f}{\sigma max} = \left(\frac{KS \cdot VS}{K_B \cdot V_B}\right)^{1/m} \qquad [3]$$

σ_f is the mean fracture strength of a set of standard bend specimens with a volume V_B and a load factor of K_B and σmax is the expected maximum fracture stress in a component with a volume V_s and a load factor K_s.

Figure 3 shows a plot of this equation for various Weibull moduli. Thus, the expected fracture strength of a medium size component (for instance, a cylinder about 6 in. in diameter and 10 in. long) is only about 20 to 40 pct of that of a small MOR sample for $m = 5$-7. However, for design purposes, a much lower failure probability is required. Further manipulation of the statistics leads to Fig. 4, giving the probability of failure as a function of relative fracture strength for 'small' and "large" turbine components. For large components, the allowable design stress is only about 10 pct of the nominal strength, *i.e.*, the allowable design stress for a component made from a ceramic with a nominal strength of 35,000 psi and a Weibull modulus of 7 is only 3500 psi, unless the component is proof tested at higher stress levels to decrease the probability of failure of the remaining components.

The significant effect of the Weibull modulus on the

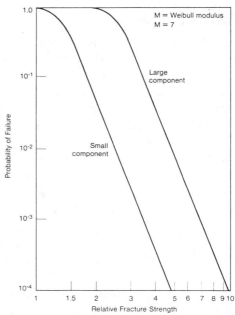

Fig. 4—Probability of failure as a function of relative fracture strength ($\sigma f / \sigma$ max).

allowable design stress is also clearly illustrated in Fig. 5, showing the strength required for a turbine rotor disk as a function of Weibull modulus.[5]

From the above, it is clear that an excellent theoretical basis exists for the design of ceramic components. As mentioned earlier, extensive test programs are needed to verify the design methodology, as a large number of samples are needed for each data point. Figure 6 shows the experimental confirmation of the dependence of fracture strength on the volume and load factors of various laboratory size components.[8]

Results of an extensive test program on siliconized silicon carbide are given in a recent EPRI report on the development of a ceramic heat exchanger.[9] Figure 7 summarizes the pertinent results. There are large vari-

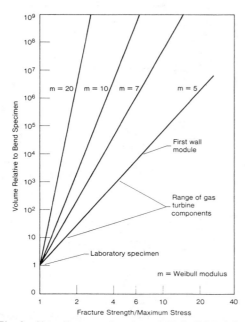

Fig. 3—Size effect as a function of Weibull Modulus.

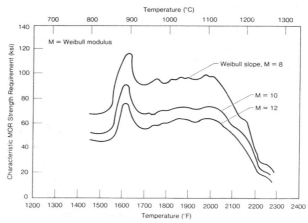

Fig. 5—Nominal strength requirements for first stage turbine rotor disk at 2500 °F TIT, 100 pct speed and 90 pct reliability.

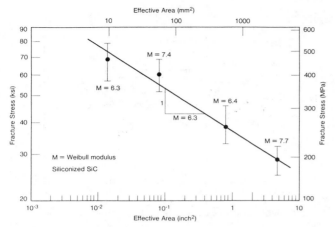

Fig. 6—Average Fracture Stress *vs.* effective area for small and large M.O.R. bars, spin bars and spin disks.

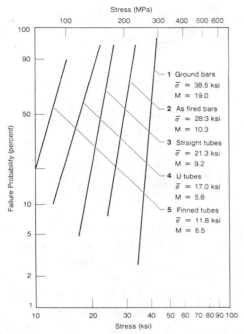

Fig. 7—Probability of failure of test specimen and heat exchanger components made from siliconized silicon carbide.

ations not only in fracture strengths, but also in Weibull modulus. As is well known, surface grinding (*i.e.*, elimination of surface flaws) can improve both fracture strength and Weibull modulus, but grinding is impractical except for very critical, relatively small parts (curve *1vs2*). The decrease in strength between as fired MOR bars and straight heat exchanger tubes is entirely due to the size effect as the Weibull modulus is similar. However, the U tubes and finned tubes apparently have both lower strength and lower Weibull modulus, indicating a different flaw size distribution. To use the nominal strength data for such components, an experimentally determined correction factor is needed to calculate the allowable fracture stress, as shown in

Fig. 8. A 25 pct reduction of the original MOR test data is needed to bring the experimental data in line with the predicted data. This indicates that verification testing is needed to determine variations in strength and Weibull modulus based on component size and complexity.

It is concluded that the allowable design strength of ceramic components varies greatly with Weibull modulus and component size. Without special techniques such as proof testing, "safety factors" as high as 5 to 10 may be required to derive allowable design stresses from nominal strength data from MOR tests. Analytical procedures based on the Weibull probabilistic technique should be used to determine the maximum allowable stress for each component. For unusually large or complex components, the calculated stresses must be confirmed experimentally. Since the confidence limits on the Weibull modulus are relatively large, the nominal strength and Weibull modulus of each material should be based on a sample population of at least 30 specimens.

LONG-TERM DURABILITY AT ELEVATED TEMPERATURES

So far, we have only discussed the general design requirements for ceramics in utility applications independent of temperature, service conditions and length of service. In this section, we will discuss the limitations imposed on the use of structural ceramics by the requirement of long-term service at elevated temperatures, under potentially corrosive conditions.

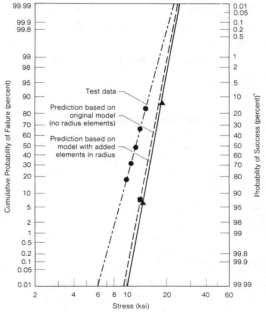

Fig. 8—Comparison between test data and predicted strength for finned tubes.

Maximum Service Temperature

Figure 9[10] shows that structural ceramics, such as silicon carbide, potentially have a much higher maximum service temperature than metals. Pure silicon carbide and nitride have a service limit potential well over 1600 °C. However, most commercially available materials contain additives, designed to aid sintering and densification. Unfortunately, they generally also lower the service limit.

Impurities, mostly oxides, such as FeO, and CaO, have the same effect. An extreme example is siliconized silicon carbide. This material contains about 10 pct free silicon. Since the melting point of silicon is about 1400 °C, a sharp drop in strength is found at 1400 °C. Hot pressed silicon nitride, containing MgO as a densification aid and up to 4 pct additional impurities, suffers an increasingly rapid loss in strength at temperatures above 1000 °C. Sintered silicon carbide and reaction bonded silicon nitride, which contain a smaller amount of harmful impurities, retain their short-term strength at least up to 1600 °C.

The effect of impurities and sintering aids on the high temperature strength of ceramics manifests itself also in the occurrence of slow or 'subcritical' crack growth, which is usually caused by grain boundary sliding at elevated temperatures, when the impurities at the grain boundaries become viscous. For many materials slow crack growth can be characterized as follows:[9]

$$V = A K^n \qquad [4]$$

Here V is the crack propagation velocity, K is the stress intensity factor and A and n are constants depending on environment and temperature. The crack growth exponent, n, can be determined most conveniently from constant strain rate tests using:

$$\frac{\sigma_1}{\sigma_2} = \left(\frac{e_1}{e_2}\right)^{n+1} \qquad [5]$$

Here, σ_1 is the fracture stress at stress rate e_1, and σ_2 the fracture stress at e_2. The crack growth exponent thus determined can then be used to predict time to failure for different stress levels at the same temperature and environmental conditions:

$$\left(\frac{\sigma_1}{\sigma_2}\right)^n = \frac{t_2}{t_1} \qquad [6]$$

Here t_1 is the service life at the applied stress σ_1, and t_2 is the service life at an applied stress of σ_2.

The decrease in fracture stress, with time is illustrated in Fig. 10, for materials with different slow crack growth exponents.[10] The effect of the reduction in

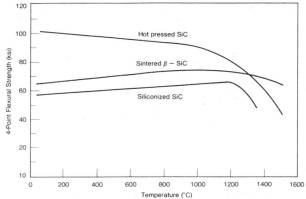

Fig. 9—Short-term flexural strength of structural ceramics.

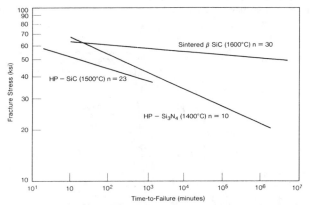

Fig. 10—Decrease in fracture stress due to slow crack growth.

fracture stress with time on the probability of failure, calculated by Trantina[4] for a service life of 10,000 h and a Weibull modulus 7 is shown in Fig. 11.

It is clear that even a relatively high crack growth exponent will cause some reduction in allowable stress because of the long service life required in most utility applications. Ceramics should preferably not be used at temperatures where their slow crack growth exponent is less than 30 unless the Weibull modulus is unusually high. This limits the maximum use temperature of siliconized silicon carbide to about 1200 °C, that of sintered silicon carbide to about 1400 °C. These guidelines apply only in clean oxidizing environments. Corrosive environments may accelerate crack growth, as will be discussed later.

There is frequently a temperature region where some ceramics get stronger. This strengthening is generally contributed to blunting of surface cracks by oxidation of the ceramic. This strengthening is most commonly indicated by an increase in mean strength as shown in Fig. 12 for siliconized silicon carbide.[11] Sometimes, an increase in Weibull modulus occurs also as is shown in Fig. 13, where cyclic fatigue tests at 1060 °C resulted in a significant increase in Weibull modulus, without a loss

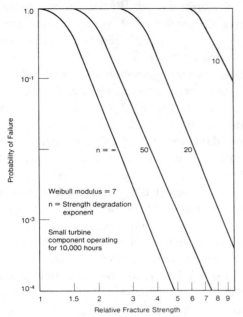

Fig. 11—Probability of failure of small turbine component as a function of crack growth exponent.

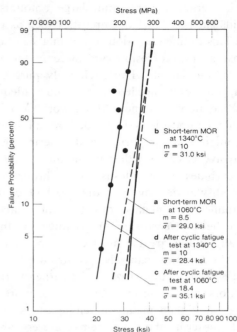

Fig. 13—Results of cyclic fatigue tests of siliconized silicon carbide at 1060 and 1340 °C.

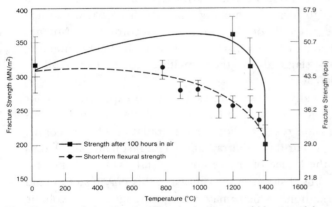

Fig. 12—Strength increase in siliconized silicon carbide caused by oxidation at high temperatures.

Environmental Effects

Both carbides and nitrides are thermodynamically unstable in oxidizing atmospheres. However, in highly oxidizing environments, a protective SiO_2 layer is formed on the surface, which controls the actual oxidation rate. Figure 14[12] gives a compilation of oxidation rate constants obtained by several investigations on silicon carbide and silicon nitride under a wide variety of conditions. The data generally indicate that the oxidation rates of the ceramics are significantly lower than those of high temperature alloys. However, high temperature data above 1100 °C are generally not available. There are many reports that localized pitting corrosion may occur during oxidation above 1000 °C,

in strength.[9] At 1340 °C, where slow crack growth occurs, a marked decrease in strength is evident. Recent data by Trantina indicate an increase in strength of about 20 pct at 1200 °C, when samples are oxidized under stress for 300 h.

It may be concluded that presently available ceramics cannot be used for long periods (10,000 h) at very high temperatures because of the occurrence of extensive slow crack growth. On the other hand, strengthening of ceramics may occur at temperatures just below the onset of extensive slow crack growth, but well above the maximum use temperature of most metals. This temperature regime appears presently the most suitable area for the first application of ceramics.

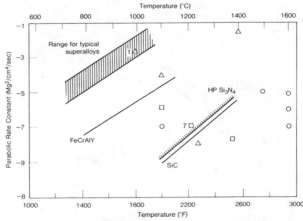

Fig. 14—Oxidation behavior of ceramics.

Source: Journal of Materials for Energy Systems, September 1980, 41-50

especially in ceramics containing large amounts of sintering aids such as MgO or impurities such as CaO and FeO.[13] This pitting corrosion results in a significant strength loss as the corrosion pits increase the flaw population in the sample. Thus, relatively pure compositions appear to be needed to guarantee adequate long term oxidation resistance. Much work on promising ceramic compositions is needed as all present materials contain some impurities and sintering aids.

The corrosive effect of other contaminants on the stability of carbides and nitrides is mainly through their ability to modify or destroy the protective SiO_2 coating on the ceramic. For metals, this type of attack is commonly called hot corrosion. In combustion atmospheres, alkali sulfates are the most common corrosive species, but vanadates can also contribute to hot corrosion. Figure 15[12] summarizes some earlier work by various authors, indicating that ceramics are more hot corrosion resistant than metals at temperatures below 1100 °C. Long-term data at temperatures above 1100 °C are again exceedingly rare. A more recent paper by Brooks and Meadowcroft[15] indicates that porous reaction bonded silicon nitride and siliconized silicon carbide are seriously corroded ($>$50 mill/1000 h) at temperatures as low as 900 °C. Dense HP silicon nitride was less seriously corroded. The 1000 h test data reported by Ward, et al.[16] indicate no corrosion of siliconized silicon carbide for 1000 h at 1200 °C, but considerable thinning of alpha silicon carbide. Experimental conditions were: gas velocity 27 m/s, fly ash 0.688 g/h, sea salt 0.0025 g/ft³ gas, V_2O_5 0.0019 g/ft³ gas). It is not clear if the wear of the alpha silicon carbide was due to erosion or corrosion. According to Tressler[17] both reaction sintered and hot pressed silicon nitride suffer high strength losses after exposure to a NaCl - Na_2SO_4 eutectic mixture for 144 h at 1000 °C, indicating that hot corrosion accelerates the formation of corrosion pits by oxidation. Again, it appears that the presence of sintering aids and/or impurities control corrosion resistance in presently available ceramics.

Ceramics are generally considered considerably more erosion resistant than metals at elevated temperatures. For temperatures below 1000 °C, this has been well demonstrated by Barkalow, et al.[18] in Fig. 16. Corrosion/erosion data above 1000 °C are again lacking. Insidious interactions between erosion and corrosion cannot be ruled out without experimental evidence to the contrary, as is indicated by the unexplained wear of alpha silicon carbide in a burner rig test at only 27 m/s.[16]

Most erodents, such as coal slag, will be, at least partially, liquid at temperatures of interest (1100-1400 °C). Thus the interaction between molten oxides and silicon carbide or nitride should also be considered. In a previously mentioned burner rig test, little or no interaction was noted between coal slags and silicon carbide at 1200 °C.[16] However, at higher temperatures, the iron oxide in the slag is reduced by silicon carbide. The resulting Fe reacts with free silicon in siliconized SiC to form low melting eutectics. This could degrade the strength of the ceramic.[19] According to Muan,[20] silicon carbide will rapidly dissolve in FeO containing coal slags above their melting point, i.e., 1250 to 1400 °C.

Finally, a few remarks about the expected behavior of silicon carbide in first wall and blanket applications infusion reactors. Potential degradation of the ceramic can occur through irradiation with fast neutrons. Atoms in the lattice of the ceramic crystals will be displaced, causing dislocations. Eventually, this may result in void formation, which may affect strength and volume stability. Thermal conductivity is also reduced. All the affected properties are important design parameters. In

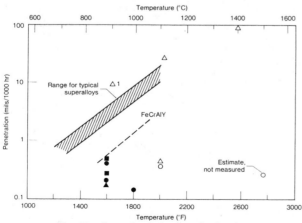

Fig. 15—Corrosion behavior of ceramics.

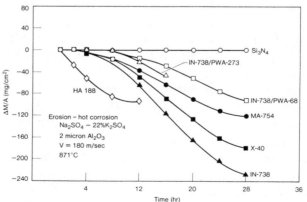

Fig. 16—Corrosion-erosion of hot pressed silicon nitride and super alloys at 871 °C.

addition, transmutation products such as He and H will form, because of the high energy of the neutrons (14 MeV).

The damage of neutron irradiation caused solely by atomic displacements has been extensively studied in gas cooled fission reactors, which use CVD silicon carbide to coat fuel pellets.[21] Generally, the strength of CVD silicon carbide is not affected by irradiation. After a small initial expansion, the material appears to be volume stable up to about 1050 °C. At higher temperatures, continuous expansion appears to occur (Fig. 17). The thermal conductivity of the material is reduced to about 50 pct of the original value. The effect of helium and hydrogen, generated during irradiation has not been determined at this point.

A program was started by EPRI to study irradiation damage in more conventional silicon cabide products, such as siliconized silicon carbide.[22] Initial results indicate a 25 pct decrease in strength and also a decrease in Weibull modulus after irradiation. Annealing at 1200 °C reduces the decrease in strength to about 10 pct (Fig. 18). The total neutron fluence used so far was quite low. Longer exposure times are needed to provide useful design information.

CONCLUSIONS

A. Ceramic design technology now allows a realistic determination of allowable design stresses for most high temperature applications. Depending on the size of the component and the Weibull modulus of the material, "safety factors" as high as 5 to 10 may be needed to provide a probability of failure less than 10^{-4}. Highly stressed or very large parts must, therefore, be proof tested to allow higher design stresses.

B. The service limit of most ceramics for long-term utility applications is generally determined by the onset of relatively rapid subcritical crack growth. Thus, the practical service limit of siliconized silicon carbide is about 1200 °C while that of alpha silicon carbide is about 1400 °C for long term(>5 years) utility applications.

C. Strengthening of silicon carbide and silicon nitride ceramics may occur through oxidation and crack blunting when held at temperatures below the onset of extensive slow crack growth, especially while under stress.

D. In general, silicon carbide and silicon nitride ceramics are much more resistant to oxidation and hot corrosion than metal alloys. However, long-term data at temperatures above 1000 °C are still rare. In some ceramics, especially those containing high levels of impurities or sintering aids, oxidation and hot corrosion cause corrosion pits, which markedly reduce the strength of the material. More well-controlled, long-term exposure data, especially under stress, are needed.

E. Pure forms of silicon carbide, such as CVD silicon carbide, are quite resistant to irradiation. However, previous data on CVD silicon carbide cannot be used for design purposes on other, less pure and less dense forms of silicon carbide. Thus, additional long term radiation data are needed to determine design properties of irradiated materials for use in fusion reactors.

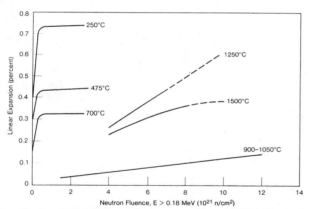

Fig. 17—Irradiation-induced linear expansion of CVD silicon carbide.

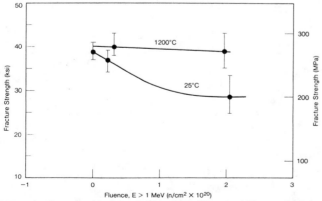

Fig. 18—Fracture strength of siliconized silicon carbide as a function of neutron fluence.

REFERENCES

1. EPRI Report No. PS-1141-SR, Overview and Strategy 1980–1984. Research and Development Program Plan.
2. Black & Veatch Consulting Engineers, *Solar Thermal Conversion to Electricity, Utilizing a Central Receiver, Open Cycle Gas Turbine Design*, EPRI Report ER-387-SY, March 1977.
3. McDonnell Douglas Astronautics Co., Conference Proceedings: Low Activiation Materials Assessment for Fusion Reactors, EPRI Report No. ER-328-SR, March 1977.
4. G. G. Trantina, H. G. de Lorenzi: *Design Methodology for Ceramic Structures*, Paper No. 77 GT-40, Proc. Gas Turbine

Conference ASME, Philadelphia, PA, March 1977.

5. R. A. Jeryan: "Use of Statistics in Ceramic Design and Evaluation," *Ceramics for High Performance Applications—II*, J. J. Burke, E. N. Lenoe, R. N. Katz, ed., Brook Hill Publishing Co., Chestnut Hill, MA, 1977.

6. M. Coombs, D. M. Kotchick: *Materials Design Requirements for Ceramics in Turbines and Heat Exchangers*, Proc. Fourth Annual Conference on Materials for Coal Conversion and Utilization, Oct. 9-11, 1979. DOE Report No. Conf. 791014.

7. W. Weibull: "A Statistical Distribution Function of Wide Applicability," *J. Appl. Mech.*, vol. 18, Sept. 1951, pp. 293−297.

8. G. G. Trantina, C. A. Johnson: "Spin Testing of Ceramic Materials," *Frac. Mech. Ceram.*, vol. 3, 1978, pp. 177−188.

9. AiResearch Co. of California, High Temperature Ceramic Heat Exchanger, EPRI Report FP-1127.

10. S. A. Borz, D. C. Larsen: *Properties of Structural Ceramics*, Proc. 4th Annual Conference on Materials for Coal Conversion and Utilization, Oct. 9-11, 1979, Gaithersburg, MD.

11. Norton Co., NC430, Densified Silicon Carbide, Sales Literature 1979.

12. C. T. Sims, J. E. Palko: *Surface Stability of Ceramics Applied to Energy Conversion Systems*, Proc. Workshop on Ceramics for Advanced Heat Engines, Jan. 24-26, 1977, Orlando, FL, ERDA Report No. Conf. 770110, p. 187.

13. S. W. Freimann, J. J. Mecholski, W. J. McDonough, and R. W. Rice: "Effect of Oxidation on the Room Temperature Strength of Hot Pressed Si_3N_4 - MgO and Si_3N_4 - ZrO_2," *Ceramics for High Performace Applications II*, Brook Hill Publishing Co., Chestnut Hill, MA, 1978, p. 1069.

14. D, W. Richels: *Frac. Mech. Ceram.*, vol. 3, 1978, pp. 177–88.

15. S. Brooks, D. B. Meadowcroft: *Corrosion of Hot Pressed and Reaction Bonded Si_3N_4 and Self Bonded SiC in Oil Fired Environments*, Royal Navy, U. S. Navy Gas Turbine Materials Conference, Bath, England, September 1976.

16 M. E. Ward, N. G. Solomon, A. G. Metcalfe: *Develpment of a Ceramic Tube Heat Exchanger With Relaxing Joint*, DOE Reports FE-2556-17, June 15, 1978, and -26, July 15, 1979.

17. R. E. Tressler, M. D. Meiser, T. Yonushonis, "Molten Salt Corrosion of Silicon Nitride," *J. Amer. Ceram. Soc.*, 1976, vol. 59, p. 441.

18. R. H. Barkalow, F. S. Pettit: *Corrosion-Erosion of Materials in Coal Combustion Gas Turbines,* Proc. Conference on Corrosion/Erosion of Coal Conversion Systems Materials NACE, 1440 South Creek Drive, Houston, Texas, 1979, p. 139.

19. R. A. Perkins, H. W. Lavendel: *Reaction of Silicon Carbide with Fused Coal Ash*, EPRI Report No. AF294, November 1976.

20. A. Muan, W. Toker: *Reactions Between Slags and SiC*, Paper 20-R-79F, Am. Cer. Soc., Refr. Div. Fall Meeting, Bedford Springs, PA, 1979.

21. L. H. Rovner, G. R. Hopkins: "Ceramic Materials for Fusion," *Nucl. Technol.*, vol. 29, June 1976, p. 274.

22. R. A. Matheny, J. C. Corelli, C. G. Trantina: "Radiation Damage in Silicon Carbide and Graphite for Fusion Firstwall Applications," *J. Nucl. Mater.*, 1979, p. 9.

Heat Treatment, Aging Effects, and Microstructure of 12 Pct Cr Steels

J. W. SCHINKEL, P. L. F. RADEMAKERS, B. R. DRENTH,
and C. P. SCHEEPENS

The 12 pct Cr steels are attractive materials for advanced steam generators. In support of the DEBENE project for the development of a sodium-cooled fast reactor, a materials program is in progress to show the applicability of these steels. The program comprises mechanical, corrosion, and weldability tests. Additionally, extensive microstructural investigations are carried out, including optical microscopic examinations, scanning electronmicroscopy, and transmission electronmicroscopy on thin foils and extraction replicas are used. In this paper, some of the results, mainly obtained from a number of heats of 12Cr1MoV steel (DIN X20CrMoV121) have been described. Information has been given concerning general features of the microstructure, in relation to the chemical composition and the heat treatment; microstructural changes due to thermal aging; and influence of melting method, product form, and size of the steel on structure and properties.

INTRODUCTION

For advanced steam generator design for fast breeder reactors, high-alloy martensitic steels with a high chromium content appear to be interesting constructional materials. Rather extensive research programs are necessary, however, to show the feasibility of these steels for nuclear components and to obtain optimum properties.

The main parts of a steam generator are:

(1) The *shell,* which will be fabricated from rolled plate (25- to 40-mm thickness)

(2) The *heat transfer tubing,* consisting of thinwalled extruded or pilgered tubes (wall thickness, 2 to 3 mm)

(3) For specific designs, the *tube sheets,* which will be fabricated from forged discs (thickness, 350 mm or more)

J. W. SCHINKEL, P. L. F. RADEMAKERS, B. R. DRENTH, and C. P. SCHEEPENS are all with Metaalinstituut TNO, Apeldoorn, The Netherlands. Originally presented at The International Conference on Production, Fabrication, Properties, and Application of Ferritic Steels for High-Temperature Applications, Warren, PA, October 6–8, 1981.

In our program, attention is paid to various products in relation to these parts. The effect of differences in hardening or tempering treatment and the influence of thermal aging have been established on the mechanical properties and the stress corrosion performance. In addition, preheat and post-weld heat-treatment procedures for manual metal arc and tungsten inert gas welding are optimized. Also, the applicability of high-temperature brazing is studied.

Most work is performed on steel X20CrMoV121, a 12 pct Cr steel with 0.2 pct C, 1 pct Mo, and 0.3 pct V. Extensive experience has been acquired with this steel in conventional power plants in Europe, and it has been chosen as an interesting reference material in the design of post-SNR 300 components.

Some results of our investigations on mechanical properties and stress corrosion tests have been given in References 1 and 2. The mechanical testing program, especially, was accompanied by extensive microstructural studies. This paper deals with some of the results of the microstructural studies, and relations with mechanical properties are discussed. Information is given on the examined 12 pct Cr steels concerning:

(a) General features of the microstructure (effect of chemical composition; austenitizing, hardening, and tempering)

(b) Microstructural changes during thermal aging

(c) Effect of melting method of the steel on the test data

(d) Effect of product form and size on the test results

Details of the heats of X20CrMoV121 are given in Table I.

EFFECT OF ALLOYING ELEMENTS AND HEAT TREATMENT ON MICROSTRUCTURE

Applicability of Phase Diagrams

From the concentration sections of the Fe-Cr-C system in Figure 1,[3] it can be concluded that the austenite phase field is most expanded in the temperature range 1000° to 1200 °C. This means that the highest chromium content can be dissolved in the austenitic lattice in this temperature range. For higher chromium contents than about 16 pct and for higher austenitizing temperatures, delta-ferrite is stable in the microstructure. At lower temperatures, the contraction of the austenite phase field is determined by the solubility limit of carbon, and carbide formation can occur, so that austenitizing temperatures up to about 1100 °C can be necessary to obtain a homogeneous austenite.

The elevated temperature strength of the hardenable 12 pct Cr steels is determined by molybdenum, vanadium, and, eventually, tungsten and niobium. These elements are forming special carbides and carbonitrides during the tempering treatment and creep.[4,5]

Because the mentioned carbide-forming elements contract the austenite phase field, the content of these elements must be adjusted to each other. The austenite-forming elements nickel, carbon, and nitrogen are used to affect the microstructure in such a way that it is almost free from delta-ferrite (stable ferrite directly coagulated from melt at high temperature). Only steels that are fully austenitic at the solution temperature will harden fully on cooling. Any delta-ferrite present at the solution temperature remains in the structure, reduces the attainable strength, and may have other unfavorable effects. A maximum amount of 5 pct delta-ferrite generally is thought to be acceptable in finished products.[5,6]

The influence of the alloying elements on the austenite phase boundary at 1050 °C has been described by Bungardt *et al.*[7]

For martensitic chromium steels with a Cr content up to 15 pct and additions of Mo, V, W, Ni, and so forth, the following formula can be derived from these data for the chromium equivalent:

$$Cr_{equ} = Cr + Si - 0.3Mn - 2.3Ni + 2Mo + 4V + 1.5W \quad \text{(element contents in wt pct)}$$

Calculations carried out with this formula showed that the air-melted heats 12V/1, 12V/2, 12W/1, and 12W/2 with a

Table I. Heats of 12Cr1MoV Steel, Type X20CrMo(W)V121

TNO Code No.	Products	Melting Practice[a]	Chemical Composition, Pct									
			C	Si	Mn	P	S	Cr	Ni	Mo	V	Others
12V/1	Pipe @ 114 × 20.5 mm	AM	0.19	0.48	0.49	0.024	0.006	11.1	0.50	1.0	0.32	0.023N
12V/2	Bar @ 25 mm	AM	0.22	0.44	0.52	0.019	0.005	11.2	0.51	0.93	0.28	
12V/3	Bar @ 20 mm	ESR	0.20	0.11	0.51	0.023	0.005	12.4	0.54	1.2	0.31	
12V/10	Plate 30 mm	ESR	0.17	0.38	0.54	0.018	0.004	11.32	0.58	1.00	0.31	0.032N
12V/11	Pipe @ 20 × 2 mm	AM	0.20	0.47	0.56	0.008	0.019	11.45	0.50	0.95	0.32	
12V/12	Pipe @ 20 × 2 mm	ESR	0.20	0.39	0.56	0.008	0.005	11.45	0.50	0.95	0.32	
12V/13	Pipe @ 20 × 2 mm	VAR	0.20	0.47	0.50	0.008	0.019	11.45	0.50	0.95	0.32	
12V/14	Hot rolled square 100 × 100 mm	ESR	0.18	0.41	0.56	0.007	0.017	11.7	0.63	1.02	0.30	
12V/15	Forged disc 690 × 140 mm	VAR	0.21	0.31	0.40	0.023	0.007	12.14	0.38	0.97	0.33	
12V/16	Forged bar 129 × 213 mm	VAR	0.21	0.31	0.40	0.023	0.007	12.14	0.38	0.97	0.33	
12V/17	Bloom @ 125 mm	AM	0.20	0.48	0.56	0.007	0.018	11.4	0.50	0.94	0.32	
12V/18	Bloom @ 125 mm	ESR	0.20	0.36	0.56	0.009	0.004	11.4	0.49	0.92	0.31	
12V/19	Bloom @ 125 mm	VAR	0.20	0.49	0.49	0.004	0.018	11.3	0.49	0.95	0.32	
12V/20	Strip 15 mm[b]	ESR	0.174	0.2	0.6	0.016	0.005	11.0	0.9	0.9	0.22	0.038N
12V/24	Strip 15 mm[b]	ESR	0.21	0.2	0.5	0.018	0.005	11.3	0.7	1.0	0.26	0.033N
12W/1	Pipe @ 54 × 11 mm	AM	0.20					10.9	0.52	1.0	0.36	0.51W
12W/2	Bar @ 25 mm	AM	0.21	0.37	0.50	0.016	0.005	11.8	0.48	0.99	0.28	0.51W

[a]AM = air melted, ESR = electroslag remelted, VAR = vacuum-arc remelted.
[b]Experimental heats.

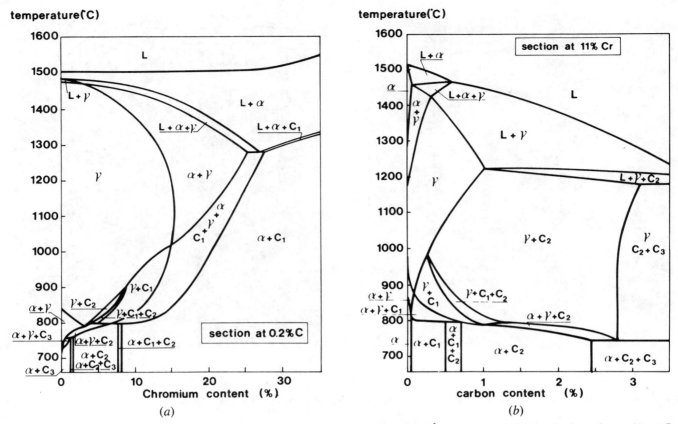

Fig. 1.—Quasi-binary phase diagrams, Fe-Cr-C system (courtesy of Bungardt, Kunze, Horn[3]). (a) Section at 0.2 pct C, (b) section at 11 pct Cr. L = Liquid, α = ferrite, λ = austenite, $C_1 = M_{23}C_6$, $C_2 = M_7C_3$, $C_3 = M_3C$.

carbon content of about 0.2 pct lie at a sufficient distance from the $\gamma/\alpha + \gamma$-boundary in the phase diagrams (see, for example, Figure 2). Nevertheless, minor amounts of delta-ferrite were observed in these heats. On the other hand, an ESR heat (mat. 12V/3) with a Cr_{equ} of about 15 did not contain any primary delta-ferrite, and segregation is not observed. These observations confirm that the micro-segregation in air-melted products is more pronounced than in ESR heats.

Electron microprobe investigations on an air-melted heat with 0.2 pct C and a Cr_{equ} of 14.2 for the bulk composition (heat 12W/1) showed that fluctuations of more than $1Cr_{equ}$ were not measured, with the exception of ferrite containing areas in which a maximum Cr_{equ} of 16 was found and other areas with a Cr_{equ} of 12.5. These observations also have been evaluated in Figure 2, together with the Cr_{equ}s of some other melts. Figure 2 is a section of the Fe-Cr-C phase diagram at 1050 °C. Instead of the Cr-content, we used the Cr_{equ} in this figure, which is partly based on data from Bungardt *et al.*[3]

It can be concluded from the Cr_{equ} calculations and the electron microprobe investigations that areas with inter-dendritic segregation, with a $Cr_{equ} < 15$, can be fully aus-tenitized, while segregated areas with a $Cr_{equ} > 15$ can form some delta-ferrite at 1050 °C. Microstructural areas with deviating chemical composition can be seen in Figure 3.

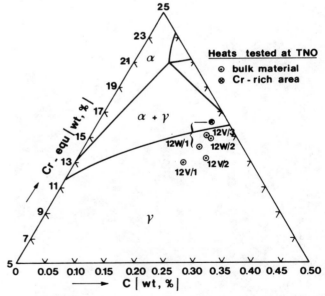

Fig. 2—Calculated Cr_{equ} values plotted in Fe-Cr-C ternary phase diagram for 1050 °C. Diagram based on data from Reference 3.

Figure 3(a) shows a hardened 12 pct Cr steel with a segre-gated area. The delta-ferrite particles are surrounded by a white etching area, which disappears when the steel is tem-pered, Figure 3(b).

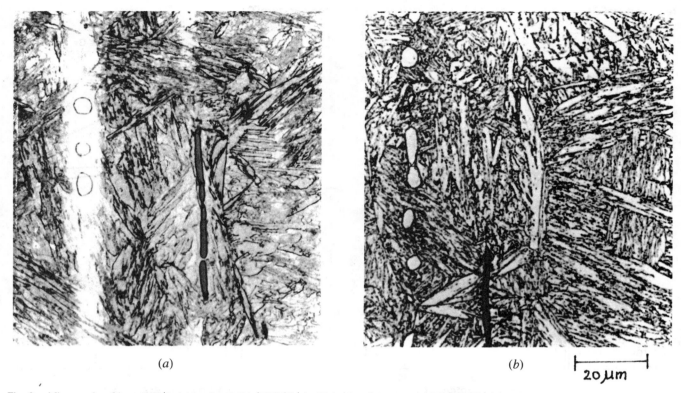

(a)   (b)   ⊢——20 μm——⊣

Fig. 3—Micrographs of heat 12W/1, (a) hardened ½ h/1050 °C/air; (b) hardened + tempered 2 h/750 °C/air.

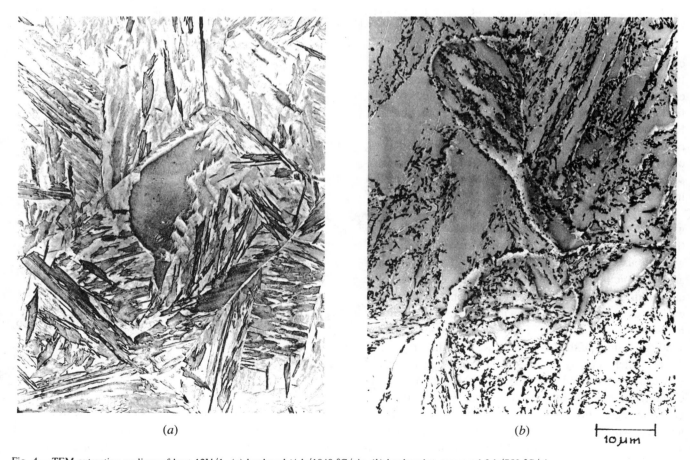

(a)   (b)   ⊢——10 μm——⊣

Fig. 4—TEM extraction replicas of heat 12V/1, (a) hardened ½ h/1040 °C/air, (b) hardened + tempered 2 h/750 °C/air.

172

Hardening

Normally, the austenitizing temperature is selected to give complete solution of carbides without excessive grain growth or formation of delta-ferrite. Figure 4(a) gives a microstructure of austenitized and air-cooled (hardened) 12 pct Cr steel. It can be seen that carbides have been completely dissolved and no carbides formed during the cooling treatment. Most of the austenite has transformed in a lath-shaped martensite.

In Figure 4, we also see some ferrite in the neighborhood of former austenite boundaries. The color of the ferrite in the electron micrograph is dark gray.

After sufficient holding time, the hardening of 12CrMoV and 12CrMoWV steels with 0.2 pct C is carried out from 1020° to 1070 °C. Within this range, increasing the austenitizing temperature results in coarser grains and higher as-hardened hardness values. In most cases, ASTM grain size numbers 4 to 5 are measured. The higher austenitizing temperatures produce better hot strength but, normally, poorer low-temperature toughness. Therefore, the proper temperature will depend on the mechanical requirements.

Transformation diagrams for continuous cooling are useful to estimate the type of cooling needed for complete transformation to martensite. The diagram in Figure 5 is adopted from Reference 8. In the cooling transformation diagram in Figure 5 for a heat of X20CrMoV121 steel, no bainite appears. It can be concluded from Figure 5 that, for a wide range of cooling rates, the austenite is transformed into martensite with a hardness of 520 to 600 HV. This holds for air hardening of bars up to 300 mm, plates up to 150 mm, and pipes with a wall thickness up to 60 mm. Small parts, therefore, can be air cooled from the austenitizing temperature, although larger ones preferably will have to be cooled in oil or saltbath. Consequences with respect to the impact properties are discussed in a later section of this paper.

Tempering

Our experiments concerning the effect of the tempering condition on the hardness, impact toughness, and tensile properties of X20CrMoV121 steel are described previously.[1] One of the conclusions is that a tempering treatment of two to four hours at 760° to 780 °C results in optimum properties for most heats.

Figure 4(b) shows the microstructure of a tempered material which reveals the following features: Carbides are present on the martensite lath boundaries, the lathes themselves are free from precipitates. Furthermore, we see some heterogeneous carbide distribution in the ferrite-containing area [compare with Figure 4(a)]. Selected area diffraction revealed that, in the tempered martensite, only $M_{23}C_6$ carbides are precipitated. Between the precipitation-free ferrite and carbide-rich area, no grain or phase boundary was present. A pronounced carbide formation along the grain and phase boundaries also could not be established. Special hardening and tempering treatments (low hardening temperature or long tempering) were tried out recently in Germany and Switzerland[9,10,11] to improve the low-temperature toughness without destroying the creep properties.

CHANGES DUE TO THERMAL AGING

Mechanical Tests

The influence of thermal aging is extensively studied on heats 12V/1 and 12V/10 (Table I), by exposing the material in a furnace for up to 30,000 hours. Before the exposure, the material was hardened one-half hour at 1050 °C and tempered two hours at 760 °C. In this condition, material 12V/1 consisted of tempered martensite with less than 5 pct ferrite, grain size ASTM 4 to 5, and hardness 235 HV. The aging conditions of heat 12V/1 and some mechanical test results are given in Table II. From the table, it can be seen that the impact values show a substantial decrease, whereas the tensile elongation values only show minor effects. The LCF tests did not reveal a distinct effect. The hardness values, which are not in the table, did not show any effect, either. The effects of long-term aging treatments in the temperature range 500° to 650 °C on hardness and impact toughness of X20CrMoV121 steel recently have been examined by other authors.[10,12,13] They, too, found minor effects on hardness and a decrease of the impact value of about 50 pct, as well.

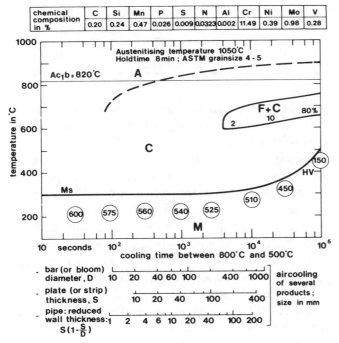

chemical composition in %	C	Si	Mn	P	S	N	Al	Cr	Ni	Mo	V
	0.20	0.24	0.47	0.026	0.009	0.0323	0.002	11.49	0.39	0.98	0.28

Fig. 5—Cooling transformation diagram for a X20CrMoV121 steel,[8] combined with equivalent product sizes.

Table II. Aging Conditions and Mechanical Test Results of 12V/1

Material Condition	ISO-V Impact (J)					Tensile Test		LCF
	−40 °C	0 °C	20 °C	40 °C	80 °C	20 °C (A5, Pct)	450 °C (A5, Pct)	550 °C (Nf at $\Delta \varepsilon_t = 1$ Pct t)
Initial	35	68	106	128	130	23	17	1235
+ 1000 h/450°	—	—	114	128	136	26	18	1810
+ 3000 h/450°	—	—	107	128	132	24	19	1510
+10,000 h/450°	—	—	126	134	132	22	17	1500
+30,000 h/450°	33	—	108	130	137	22	18	—
+ 1000 h/500°	—	—	114	128	135	25	18	1910
+ 3000 h/500°	—	—	93	123	128	25	18	1920
+10,000 h/500°	—	—	57	88	107	22	18	1550
+30,000 h/500°	17	—	39	57	92	21	17	
+ 1000 h/550°	—	—	66	86	110	24	18	1525
+ 3000 h/550°	—	—	43	72	105	24	19	1930
+10,000 h/550°	—	—	38	57	94	23	19	1410
+30,000 h/550°	16	—	36	57	94	22	17	
+ 1000 h/600°	23	40	64	100	120	23	16	—
+ 3000 h/600°	21	37	56	88	113	22	17	—
+10,000 h/600°	—	—	—	—	—	23	16	1825

Table III. Aging Conditions and Various Toughness Criteria for Heat 12V/10

Material Condition	ISO-V Impact, 20 °C (J)	COD Test, 20 °C (δ_{ci} mm)	Pellini Test, NDT °C	Tensile Test, 20 °C (A5, Pct)	450 °C (A5, Pct)
Initial	83	0.16	−30	23	16
+ 3000 h/500°	58	0.22	—	22	17
+10,000 h/500°	41	0.17	—	21	16
+ 1000 h/550°	47	0.13	—	21	15
+ 3000 h/550°	33	0.21	−15	21	15
+10,000 h/550°	31	0.17	—	20	15
+10,000 h/600°	42	0.24	—	22	16

In addition, for heat 12V/10, the influence of thermal aging on various toughness criteria is determined. Results are given in Table III.

From the table, it can be seen that aging only has an unambiguous influence on the impact ductility. The other tests hardly show an influence (tensile, LCF, COD tests) or indicate only a minor effect (pellini tests). The loading rate is thought to be responsible for these differences.

Metallographic Examinations

The metallographic examinations are carried out on specimens resulting from the 20 °C impact tests on heat 12V/1. Special attention is given to the longest aging times and to the various aging times at 550 °C. The following aspects are studied: fractography (macroscopic and SEM examinations) and microstructure, in particular precipitation processes and dislocation structures (microscopy, SEM, TEM on thin foils).

Macroscopic examinations of the fracture surfaces are carried out with binoculars. From the light and glittering parts of the fractographs, the percentage of brittle fracture is estimated. Average values, resulting from three examinations, are given in Figure 6. SEM examinations in the brittle fractured area revealed the percentage intergranular fracture, which concerns fracture along former austenite grain boundaries. Values calculated from the SEM screen are also incorporated in Figure 6.

Aging at 450 °C has no influence on the fracture appearance. After aging at 500 °C, the percentage brittle fracture is increased, whereas hardly any intergranular fracture is observed. After exposure at 550 °C, both the percentage

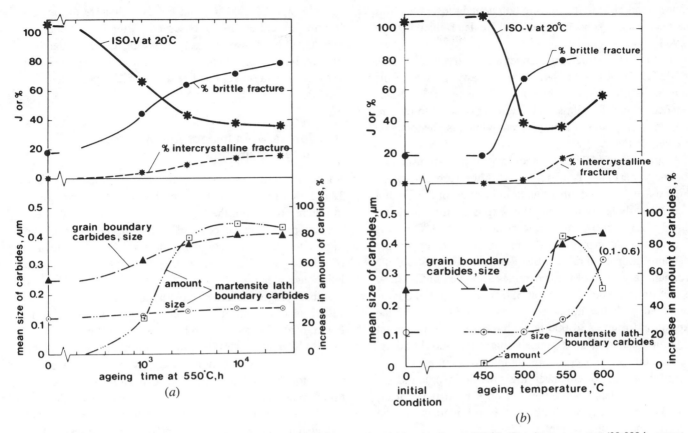

Fig. 6—Impact toughness and microstructural parameters for heat 12V/1 in relation to (a) aging time at 550 °C, (b) aging temperature (30,000 h, except 10,000 h at 600 °C).

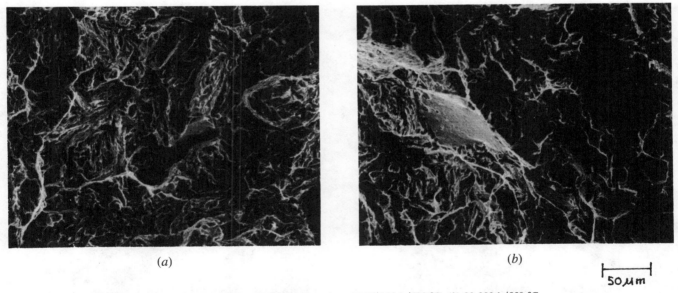

Fig. 7—SEM fractographs of an impact specimen of heat 12V/1 after aging, (a) 30,000 h/500 °C, (b) 30,000 h/550 °C.

brittle fracture and the amount of intergranular fracture has increased, whereas hardly any intergranular fracture is observed. After exposure at 550 °C, however, both the percentage brittle fracture and the amount of intergranular fracture has increased, although for 30,000 hours the impact

value is comparable with the 500 °C value. Typical SEM micrographs of fracture surfaces are given in Figure 7.

The precipitation processes due to aging are studied on etched cross sections (SEM) and on thin foils (TEM). The size and relative amount of precipitates on former austenite

Source: Journal of Heat Treating, June 1984, 237-248

grain boundaries, on martensite lath boundaries, and within martensite laths (subcell boundaries) are determined for heat 12V/1.

The first microstructural changes are observed after aging at 500 °C. Under these conditions, the average size of the carbides is not changed, but the martensite laths are surrounded by carbide films with a thickness of 20 to 60 nm. A TEM micrograph of a thin foil showing this phenomenon is given in Figure 8(a).

After aging at 550 °C, a minor growth of martensite lath boundary carbides is observed without film formation [Figure 8(b)]. On the other hand, an increase of the amount of these carbides with increasing exposure time is determined [Figure 6(a)]. Furthermore, aging at 550 °C results in a substantial growth of the precipitates on the former austenite grain boundaries. SEM pictures to illustrate the pronounced growth of these carbides are given in Figure 9. In this figure, material aged at 550 °C for 10,000 hours is compared with the initial condition. Continuation of the exposition for up to 30,000 hours has not caused important additional changes, so it appears that, after 10,000 hours, the precipitation processes are completed (Figure 6).

Aging at 600 °C enhances the growth of precipitates. On martensite lath boundaries and on subcell boundaries, coagulation occurs, resulting in a lower increase in the relative amount of precipitates [Figure 6(b)].

In the initial condition, dislocations are ordered in irregular subgrains within the martensite laths (Figure 8). After aging at 450 °C, 500 °C, and 550 °C, changes in the dislocation structures are not observed. After 10,000 hours at 600 °C, only minor changes consisting of less-entangled dislocations are observed. These observations are consistent with the already-mentioned absence of hardness changes.

Relation between Impact Toughness and Structural Parameters

Impact toughness and microstructural parameters of heat 12V/1 as a function of the aging time at 550 °C are given in Figure 6(a). From the figure and the microstructural examinations, it can be concluded that the decrease of the impact toughness is caused by the growth and nucleation of carbide precipitates on grain boundaries and martensite lath boundaries.

Impact toughness and microstructural parameters of heat 12V/1 as a function of the aging temperature are shown in Figure 6(b). The aging time has been 30,000 hours or, for the 600 °C treatment, 10,000 hours. From the figure and the metallographic examinations, it can be concluded that the impact toughness decrease at 500 °C must be due to carbide film formation along martensite lath boundaries, resulting in pure transgranular fracture. At higher temperatures, growth and coagulation of carbides, not only at martensite lath boundaries, but also at former austenite grain boundaries, results in a partial intergranular fracture without additional

(a)

(b)

0.5 μm

Fig. 8—TEM thin foils of heat 12V/1 after aging, (a) 30,000 h/500 °C, (b) 30,000 h/550 °C.

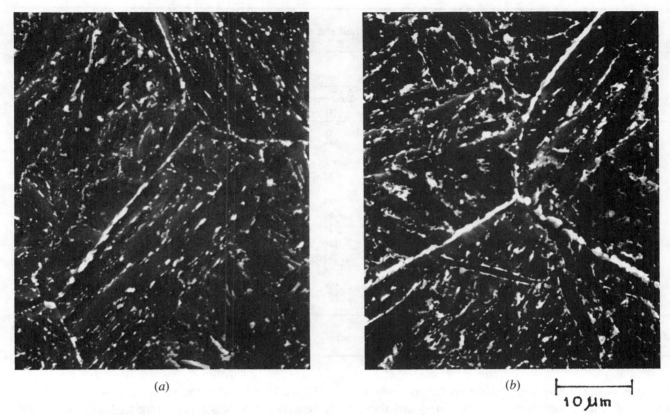

(a)

(b)

Fig. 9 — SEM micrographs of heat 12V/1 (etched), (a) initial condition, (b) aged 10,000 h/550 °C.

decrease of the impact toughness. Other authors describe similar film formation and impact toughness decrease in a 12 pct Cr steel with 0.15 pct C, which, after hardening, was directly tempered at 500 °C.[14]

EFFECT OF MELTING METHOD AND PRODUCT SIZE

Melting Method

A series of heats has been examined with respect to differences in microstructure, inclusion content, and sulfur content. Three melting processes are taken into consideration: air melting (AM), vacuum-arc remelting (VAR), and electroslag remelting (ESR).

In Table IV, ISO-V upper-shelf energy values for several products are given. Figure 10 shows the ratio of the upper-shelf energies in the transverse and longitudinal direction, with regard to the melting method and the sulfur content. The influence of the sulfur content on the transverse impact toughness for X20CrMoV121 steel also has been described by Kalwa.[13] For 0.005 to 0.010 pct sulfur, he mentioned toughness values which were twice the value for sulfur contents of 0.015 pct and more.

The influence of sulfur content and inclusions is demonstrated by comparing ESR heat 12V/18 with AM heat

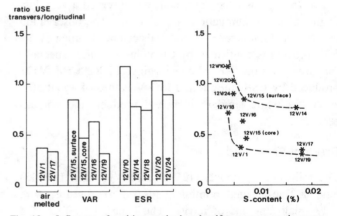

Fig. 10 — Influence of melting method and sulfur content on the upper-shelf energy ratio transverse/longitudinal direction of X20CrMoV121.

12V/17 or VAR heat 12V/19. The former contains 0.004 pct sulfur, and practically no inclusions are found; the latter both contain 0.018 pct sulfur, and many elongated inclusions are observed. Electroslag remelting obviously can lead to a lower sulfur content than vacuum-arc remelting or air melting. However, special melting techniques, e.g., argon-oxygen decarburization (AOD), can lead to sulfur contents of 0.010 and lower, as well.

The influence of the sulfur content on the USE ratio is not fully consistent (see, for example, heats 12V/1 and

Table IV. Effect of Melting Method and Sulfur Content on Ductility

Steel	Melting Method[a]	Sulfur Pct	Upper Shelf Energy (USE) ISO-V at 80 °C		Transv./Long.
			Transverse (J)	Longitudinal (J)	
12V/1	AM	0.006	48	130	0.37
12V/17	AM	0.018	37	102–114	0.34
12V/15 Surface	VAR	0.007	82	97–101	0.83
12V/15 Core	VAR	0.007	43	88–97	0.46
12V/16	VAR	0.007	70	107–119	0.62
12V/19	VAR	0.018	36	110–126	0.31
12V/10	ESR	0.004	140	120	1.17
12V/14	ESR	0.017	102–106	132–142	0.76
12V/18	ESR	0.004	66	80–101	0.73
12V/20	ESR	0.005	108	106	1.02
12V/24	ESR	0.005	107	122	0.88

[a]AM = air melted, ESR = electroslag remelted, VAR = vacuum arc remelted.

12V/14). It should be emphasized that other factors, such as the total amount of inclusions (e.g., silicates) and their shape, the melting method (in particular, the homogeneity), and the degree of deformation play a role as well.

As already has been described, we observed a very low degree of microsegregation in ESR heats. The observation that pure and homogeneous materials can be obtained by the ESR process is confirmed by Choudhury et al., especially for large forgings.[15] For large forgings of X20CrMoV121 produced by ESR, they found ISO-V values of about 50 J at 20 °C, while similar air-melted products did not exceed 25 J.

Product Form and Size

The thinwalled products (12V/11, 12V/12, and 12V/13) have a fine-grained homogeneous microstructure, with only small amounts of ferrite. The beneficial effect of a considerable deformation followed by a proper heat treatment can be seen by comparing Figures 11(a) and (b), which show the microstructure of a semifinished product (bloom 12V/19) and its final product (tube, 12V/13).

Microstructural examinations of a rolled plate of an ESR heat (12V/10) revealed zones which locally contain relatively coarse carbides. Nevertheless, the ductility values in the longitudinal and transverse direction showed up well (>80 J at 20 °C).

The thickwalled products which are examined sometimes show relatively low ductility values in the as-delivered condition. On a microscale, areas with substantial amounts of delta-ferrite are found frequently.

Figure 12 illustrates the microstructure in the core of a vacuum-arc remelted large forging (12V/15). Besides delta-ferrite, grains with a dendritic structure can be distinguished. These inhomogeneities, as a consequence of segregation and a low hot-working ratio in the core of large products, are partially responsible for a poor low-temperature toughness. An additional reason is the low cooling rate of large products, which can result in pronounced carbide formation on grain boundaries in the temperature range 750° to 600 °C (Figure 5).

In Figure 13, the influence of the cooling rate on the impact toughness is illustrated (data from References 1 and 12). Compared with specimens from simulated cooling tests, the real products reveal an additional decrease of the toughness in the range 10^3 to 10^4 seconds. Correlation of these findings with Figure 5 leads to the conclusion that for diameters >100 mm or plate thicknesses >50 mm, the effect of the hot-working ratio on ductility values is more pronounced than for smaller products.

CONCLUSIONS

(1) Variations in transformation behavior (especially the ferrite-forming tendencies) of X20CrMoV121 steel have been explained by using microprobe analyses and Cr-equivalent calculations in relation to phase diagrams.

(2) It is found that local variations in Cr equivalent can occur. Regions with a Cr equivalent of 15 and below can be fully austenitized, and, in regions with Cr equivalents above 15, the formation of ferrite can be expected.

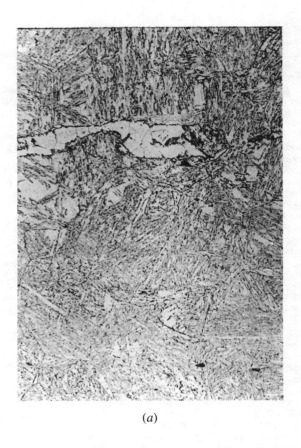

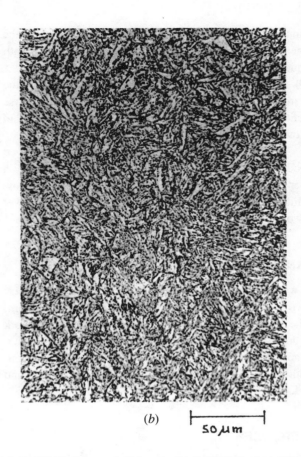

(a)   (b) ⊢——50μm——⊣

Fig. 11—Micrographs of VAR X20CrMoV121 steel, (a) semifinished product 12V/19 (bloom φ 125 mm), (b) finished product 12V/13 (tube φ 20 × 2 mm).

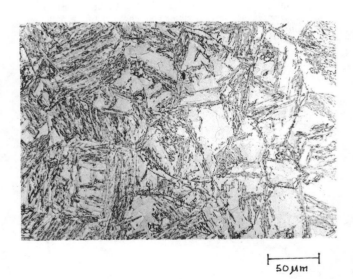

⊢——50μm——⊣

Fig. 12—Microstructure of the core of a large forging (heat 12V/15).

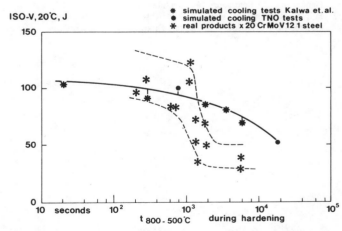

Fig. 13—Impact toughness of hardened + tempered X20CrMoV121 as a function of the cooling time during hardening.

(3) It has been confirmed that microsegregation in air-melted products is more pronounced than in ESR heats. When delta-ferrite formation has to be suppressed in air-melted heats, the Cr equivalent must be limited more than in ESR heats.

(4) Aging treatments up to 30,000 hours in the temperature range 500° to 600 °C lower the impact toughness, but have no unfavorable effect on the results of tensile tests, COD tests, and low-cycle fatigue tests.

Minimum impact toughness is attained after about 10,000 hours in the temperature range 500° to 550 °C. At 500 °C, this is a consequence of an almost continuous film of carbides on the martensite lath boundaries; at 550 °C, pronounced carbide precipitation and coagulation on the former austenite grain boundaries and martensite lath boundaries occur.

(5) In comparison with other methods, ESR leads to the best improvement of the impact ductility in the transverse direction, as a result of the more homogeneous structure and the low inclusion content which can be achieved.

(6) Thinwalled tubes and rolled plates generally showed a good ductility after hot and cold working followed by a proper heat treatment. Thick sections showed relatively low ductility values, due to a low cooling rate and a lower hot-working ratio.

ACKNOWLEDGMENT

This work is performed under contract of the Dutch Ministry of Economic Affairs, in cooperation with B. V. Neratoom, for the development of LMFBR steam generators. The work is coordinated by the Project Group Nuclear Energy TNO. The authors would like to acknowledge gratefully the contributions of Mrs. T. Krisch, Mr. J. C. van Wortel, and Mr. J. van der Veer.

REFERENCES

1. J. Vrijen, J. K. van Westenbrugge, L. van der Wiel, P. L. F. Rademakers, C. P. Scheepens, and J. W. Schinkel: "Materials Selection and Optimization for Post SNR-300 Steam Generators," International Conference on Materials Performance in Nuclear Steam Generators, St. Petersburg, FL, October 5–9, 1980.
2. P. L. F. Rademakers *et al.*: "SNR Research Activities on Steam Generator Corrosion and Water Chemistry," *Water Chemistry of Nuclear Power Stations*, ADERP Seminar, Seillac, France, March 17–21, 1980.
3. K. Bungardt, E. Kunze, and E. Horn: "Untersuchungen über den Aufbau des Systems Eisen-Chrom-Kohlenstoff," *Arch. Eisenhüttenwes.*, 1958, vol. 29, pp. 193–203, 261.
4. R. Petri, E. Schnabel, and P. Schwaab: "Zum Legierungseinfluss auf die Umwandlungs-und Ausscheidungsvorgänge bei der Abkühlung warmfester Röhrenstähle nach dem Austenitisieren, Part II," *Arch. Eisenhüttenwes.*, January 1981, vol. 52, pp. 27–32.
5. W. Wessling: "Wärmebehandlung und mechanische Eigenschaften der hochwarmfesten Vergütungsstähle mit 12% Chrom," *Sie and Wir*, 1976, no. 17, pp. 4–12.
6. E. Krainer and H. Scheidl: "Einfluss des Deltaferritgehaltes auf die mechanischen Eigenschaften der 12% Chromstähle sowie Dämpfungsuntersuchungen an Turbinenwerkstoffen," *Radex-Rundsch.*, 1979, no. 3, pp. 246–68.
7. K. Bungardt, E. Kunze, and E. Horn: "Einfluss verschiedener Legierungselemente auf die Grösse des γ-Raumes im System Eisen-Chrom-Kohlenstoff; Part I, System Fe-Cr-C-Mo," *Arch. Eisenhüttenwes.*, 1967, vol. 38, pp. 309–20.
8. G. Kalwa and E. Schnabel: "Wärmebehandlung und Eigenschaften dickwandiger Bauteile aus warmfesten Röhrenstählen," *VGB Kraftwerkstech.*, August 1978, vol. 58, pp. 604–13.
9. B. Walser *et al.*: "Der Einfluss von kurzzeitigen Uebertemperaturen auf das Zeitstandverhalten von Stahl X22CrMoV121," *Arch. Eisenhüttenwes.*, June 1979, vol. 50, pp. 249–53.
10. V. J. Granacher *et al.*: "Langzeitverhalten warmfester Stähle für den Kraftwerksbau," *VGB Werkstofftagung*, 1980.
11. H. Brandis *et al.*: "Technologie der Wärmebehandlung warmund hochwarmfester Stähle," *Thyssen Edelst. Tech. Ber.*, 1981, vol. 7, pp. 28–40.
12. V. K. Kussmaul and D. Blind: "Schweiss-und Bruchmechanikversuche an Proben und Bauteilen aus Werkstoff X20CrMoV121 bei Wanddicken bis 270 mm," *VGB Kraftwerkstech.*, April 1980, vol. 60, pp. 305–25.
13. G. Kalwa: "Entwicklungstendenzen der warmfesten Stähle für nahtlose Rohre im Kraftwerksbau," *VGB Werkstofftagung*, 1980.
14. V. G. Prabhu Gaukar *et al.*: "Role of Carbon in Embrittlement Phenomena of Tempered Martensitic 12Cr-0.15% C Steel," *Met. Sci.*, July 1980, pp. 241–52.
15. A. Choudhury *et al.*: "Verwendung des Elektro-Schlacke-Umschmelzverfahrens zur Herstellung von schweren Turbinenrotoren aus Stahl mit 12% Cr," *Stahl Eisen*, September 1977, vol. 97, pp. 857–66.

The Sasol III coal liquefaction plant, in South Africa, will produce 44,000 barrels of oil, kerosene, and other fuels daily.

Will better alloys meet coal conversion needs?

Before plants can convert coal to oil and gas commercially, engineers must upgrade construction materials

by CARL R. WEYMUELLER, *senior editor*

FOUR WAYS TO CONVERT COAL TO OIL AND GAS

The simplest of the four coal conversion processes is **pyrolysis.** Reactors heat the coal to burn out tar, generate coal gas, and leave coke and ashes. While the coal heats, processors may add hydrogen to improve the generated gas.

In **direct hydrogenation,** or hydroliquefaction, coal reacts with hydrogen, producing liquid fuel.

Extraction calls for coal to be dissolved in a solvent, which is then separated to remove the liquid portion, leaving solid impurities. Separate reaction of the liquid with hydrogen produces the synfuel, usually a liquid.

Indirect liquefaction (Fischer-Tropsch synthesis) burns coal with oxygen and steam in a gasifier to produce carbon monoxide and hydrogen. Subsequently catalyzed, these compounds form the fuels, liquids or gases. This process is the basis for the commercial Sasol plants operating in South Africa.

While the world appears to have plenty of oil and natural gas for the foreseeable future, farsighted thinkers prepare for the day when these convenient fuels begin to run out. Because this country has immense reserves of coal—it's a one-country OPEC in that fuel, some say—processes that convert coal to gas, oil, kerosene, and other hydrocarbon-based fuels catch the attention of U.S. developers.

Last October, 200 research and development scientists and engineers met at the National Bureau of Standards, Gaithersburg, Md., to review progress in coal conversion processes. NBS, the U.S. Department of Energy, the Electric Power Research Institute, and the Gas Research Institute sponsored this sixth yearly *Materials for Coal Conversion and Utilization* conference. Representatives of energy producers, government labs, and research groups reported encouraging results in materials and design developments, and agreed that much more work remains to be done before we see large coal conversion plants operating in this country.

Coal conversion is easy

. . . at least in theory. All conversion processes add hydrogen, from steam decomposition, to carbon-rich coal. At high temperatures and pressures, sometimes assisted by catalysts, hydrogen and carbon combine to form the light hydrocarbons that make up oil, gas, kerosene, and other fuels.

Coal contains 70 to 93 percent carbon, more than enough to make synthetic fuels in large amounts. Unfortunately, coals also contain appreciable quantities of sulfur, chlorine, sand, and clay. Sulfur and chlorine combine with elements in structural alloys that go into the process equipment, lowering mechanical properties. Sand and clay particles, as they speed through reactor vessels and pipes at high velocities, eat up the best materials money can buy.

The tougher the better

Coal conversion calls for the most rugged alloys and refractories available. Hard work starts at crushers, which grind coal into fine particles that feed to pretreatment. Equipment puts through great quantities of coal, but temperatures and pressures remain low. Wear, by abrasion and erosion, is the major failure mode. Construction alloys require high strength to resist heavy loads and impacts.

Pretreatment equipment heats, dries, and stores crushed coal for conversion. Temperatures run to 800 F, drying and oxidizing the coal enough so that it will not cake in storage hoppers. Equipment suffers oxidation, sulfidation, wear from abrasion and erosion, and corrosion.

Alloys must resist heat

Converters operate at high temperatures (1,800 F and above) and pressures, to 1,500 lb/in.^{2} Parts fail due to excessive sulfidation, oxidation, and, at lower temperatures, carburization. Many components, furthermore, contact molten slag and ash particles, which corrode and erode even the hardiest high-temperature alloys.

Post treatment includes multiple-stage quenching to cool the synfuel, and cleaning to remove impurities, solid and gaseous. Entrance temperatures being high, at least 1,800 F in most reactors, excessive sulfidation and oxidation do their worst. In gas scrubbers, water sprays take out ash particles and corrosive gases: H_2S, NH_3, SO_2, and $HCNO$. Equipment that handles molten slag and hot ash must be constructed of alloys with good high-temperature strength and resistance to corrosion and thermal shock.

A bumpy road to production

Pilot plants turn coal into synthetic oil gas today, but scaling their operations up into full-size conversion plants calls for more durable materials. The search for such alloys and refractories, and ways to fabricate them, goes forward on several fronts, supported by the government and industry.

HOW COAL GASIFIERS FAIL

Plant subsystem	Construction materials	Failure modes
Coal handling	Carbon and low-alloy steels, austenitic and ferritic stainless steels	Abrasive and erosive wear, general corrosive attack, mechanical overload.
Coal pretreatment	Austenitic and ferritic stainless steels, carbon and low-alloy steels, Incoloy 800, hardfacing alloys	Sulfidation and oxidation, abrasive and erosive wear.
Gasifier	Austenitic stainless steels, Incoloy 800, RA330, carbon and low-alloy steels (typically refractory-lined)	Sulfidation and oxidation, slag attack, erosive wear, creep and creep fatigue, aqueous corrosion in water and steam lines
Gas-stream quench, cleanup, and solids waste disposal	Austenitic and ferritic stainless steels, Incoloy 800, carbon and low-alloy steels (often refractory-lined)	Aqueous corrosion, erosive wear, sulfidation and oxidation, thermal shock.

Source: D. R. Diercks, Argonne National Laboratory

Big-name research participants include Battelle Columbus Laboratories, Argonne National Laboratory, Exxon Research and Engineering Co., National Bureau of Standards, IIT Research Institute, Oak Ridge National Laboratory, General Electric Co., Westinghouse Electric Corp., U.S. Bureau of Mines, Babcock & Wilcox, and several schools: University of California, Iowa State University, and the Colorado School of Mines, among others. Guided by sponsors such as the U.S. Department of Energy, Electric Power Research Institute, and Gas Research Institute, engineers of these organizations and others search for alloys and refractories that will withstand the demands of coal conversion.

Pilot plant work

. . . weeds out the weak materials. The EPRI funds two large-scale (250 tons of coal daily) coal liquefaction plants, now undergoing test runs. These and other coal conversion pilot plants tell engineers what alloys and refractories will do the job. So far, tests confirm what many predicted—

alloys, even highly alloyed stainlesses and superalloys, take a beating in coal conversion atmospheres.

Investigators at Oak Ridge National Laboratory, working under the direction of R. A. Bradley, ran stress corrosion tests of commercial alloys in Exxon Donor Solvent (EDS) and H-Coal plants. Their conclusions: Low-alloy steels corrode and crack moderately, nickel-base alloys sulfidize, and some austenitic stainlesses crack. Good performers, all stainlesses, included stabilized high-alloy austenitic, duplex ferritic-austenitic, and ferritic types.

A report by G. Sorell, Exxon Research and Engineering Co., offers more hope, at least for materials in the EDS 250-ton/day plant in Baytown, Texas. He cites good materials performance, except for some localized erosion and corrosion. Refractory linings eliminated erosion in some areas, and Type 321 stainless successfully replaced corroded carbon steel in a tower.

Busy days for failure analysts

Analyzing failures of components in coal conversion plants

keeps investigators hopping. D. R. Diercks, metallurgist, has, with others at Argonne National Laboratory, studied failures in parts from several types of gasifiers. They encompass a variety of plant designs, capacities, process conditions, and construction materials. Diercks reported some findings at the conference.

● Welds in a coal-char transfer line (RA 330 and Incoloy 800 base metal, Inconel 182 filler) failed after operating for several thousand hours at temperatures to 1,800 F. Because failure occurred due to sulfidation of the high-nickel filler, Diercks suggested replacing it with a lower-nickel filler, nearer that of the base metal.

● An expansion joint in an 8-inch-diameter gas line of Type 304 stainless developed a 270° crack around a pipe-to-T weld. Intergranular cracks and grain-boundary sulfides led Diercks to diagnose stress-corrosion failure from exposure to polythionic acid, formed when hydrocarbon condensates reacted with sulfidic scale during plant shutdown. His solution: At each shutdown, purge the line with dry nitrogen to remove condensate before it contacts the scale.

● Three flame-sprayed (with Cr_2O_3) molybdenum-tube thermocouple protectors corroded and embrittled in service to 2,250 F. Diercks found that surfaces of moly tubes had been primed with an Ni-Al-Cr coat to bond the flame-sprayed chrome oxide layer to tubes. When erosion broke through the oxide, it exposed the high-nickel substrate to hot gases, which sulfidized the nickel, helping to destroy the coating and embrittle the moly tube. Diercks suggests flamespraying the coating directly, or using a silicide primer.

Big plants needed

Once pilot plant operations and studies determine optimum processes, designs, and materials, fabricators can start to work on full-scale plants. These will be big—some planners envision reactors towering 220 feet high with diameters to 23 feet and walls to 8 inches thick. Building such monsters challenges welding to develop high-production processes.

Reporting at the conference on one such process, U.A. Schneider, Westinghouse Electric Corp., described an automatic gas tungsten arc process developed at the Nuclear Components Div., Tampa, Fla., to fabricate thick-walled pressure vessels of 2-1/4 Cr-1 Mo plate. The process calls for a narrow-gap joint: 6° included angle. Westinghouse, Schneider reported, has prepared a handbook that details procedures for welding 4- and 8-inch plate in different positions. It also lists specs for base material, shielding gas, filler wire, and nondestructive testing, and describes installation, operation, and maintenance of the equipment.

Electron beam welding offers promise for joining thick-walled pressure vessels. Working at the Alliance (Ohio) research center, Babcock & Wilcox engineers developed EB welding procedures for joining plates of 2-1/4 Cr-1 Mo, 8 inches thick from two sides. Described by Charles M. Weber of B&W, in a government report (DOE/OR 10244-10), EB welding requires a vacuum chamber, so equipment remains to be developed to weld reactor sections in the field. An alternative might be the out-of-vacuum EB process developed by Hitachi, Ltd., Tokyo. A 110-kW EB machine welds 4-inch-thick walls of pressure vessels. The beam feeds through a long focusing tube, called a zoom gun, which concentrates the beam at the joint seam.

Big business for welding

Challenges abound for the welding industry in coal conversion. Not only will huge plants have to be constructed, engineers need to devise better inspection techniques for thick welds, work out ways to clad large surface areas with alloys that withstand corrosion and wear, and develop methods for heat treating in the field. In his survey, *Coal Conversion* (Metal Construction, April 1981, p. 31-36), J. L. Robinson, The Welding Institute, states that designers of weldments will have to know much more about fatigue and fracture properties at high temperatures, properties of thick joints, and the effects of corrosion and erosion.

High reliability and safety, Robinson says, are essential to operate conversion plants economically. This means effective fracture mechanics assessments of fatigue and fracture properties, probably leading to development of fitness-for-purpose standards for evaluating defects that appear during fabrication. We may see new tests for determining properties of welds in thick plates. These should help quantify properties, such as fracture toughness, fatigue, creep, hydrogen embrittlement, and susceptibility to stress corrosion cracking.■

SECTION V
Heat Exchangers

Materials Development for HTGR Heat Exchangers

W. R. Johnson

D. I. Roberts

GA Technologies, Inc.
San Diego, Calif. 92138

High-temperature, gas-cooled reactors (HTGR's) are uranium/thorium-fueled, graphite-moderated, helium-cooled systems capable of producing high-temperature primary coolant. Several variants of this system are under active development in the United States and worldwide. In one version, the primary coolant heat is transferred to steam generators producing 538°C/16.5 MPa steam for use in electricity generation or process heat applications. The materials and design technology for steam generators in this system are well developed, relying heavily upon prior experience with fossil-fired steam generators and the steam generators of the commercial HTGR's. The major work that remains to be done is to complete qualification of the materials and to respond to evolving rules pertinent to elevated-temperature nuclear design and construction. Other versions of the HTGR generate much higher primary coolant gas temperatures (850° to 950°C) and exchange this heat, through intermediate heat exchangers (IHX's), to a secondary loop for higher temperature process heat applications. Although IHX's for these systems are typically pressure-balanced (low-stress) units, their design involves several challenges, including the potential interactions between structural materials and impurities present in the HTGR primary coolant. Considerable work is required to qualify materials for IHX applications, including detailed mechanical property characterization, determination of environmental influences on performance, provision of welding materials and procedures for producing joints of adequate strength and integrity, and provisions for wear protection. Some of the work currently under way addressing these issues is described.

Introduction

High-temperature gas-cooled reactor systems have been under development for 20 years or more, and during that time several have been constructed (see Table 1). The three early reactors were small prototype units. The FSV and THTR reactors are demonstration plants designed for commercial power generation.

Operating experience with the reactor systems has generally been very good. Although the FSV reactor experienced several problems during its early startup phase that caused some delay in the reactor schedule, performance has been satisfactory since the plant began operation in 1976. The relatively low fission product levels in the primary coolant systems in these reactors have allowed plant operation and maintenance with minimum exposure of operating personnel to radiation. During the later stages of operation of the Dragon reactor and during recent operation of the AVR, the potential of these reactor systems for generating very high primary coolant outlet temperatures (up to 950°C) has also been demonstrated.

As a result of this prior experience, interest exists in several countries of the world in further developing this reactor

system for a variety of potential applications. In the United States interest has focused upon four potential variants (Table 2). The HTGR-SC is aimed at electricity generation and is similar in concept to the FSV, Peach Bottom, and THTR reactors. The HTGR-SC/C is aimed at exploiting the HTGR in a situation where maximum economic benefit can be derived from the potential of the system for electricity generation and for the generation of steam for chemical process applications. The HTGR-GT and HTGR-PH variants are more advanced systems characterized by significantly higher primary coolant outlet temperatures. Both these systems have received considerable study in recent years and the process heat system, in particular, continues to receive considerable effort in the United States, Germany, and Japan.

Each HTGR variant features heat exchangers that must perform efficiently and reliably for the advantages of the reactor system to be realized. In the HTGR-SC and SC/C versions, the heat exchangers are steam generators that transfer heat from the high-temperature (~700°C) helium gas primary coolant to produce 538°C/16.5 MPa steam in the secondary coolant. In the HTGR-GT direct cycle, there are large recuperators that exchange heat between the hot and cold loops of the helium coolant and precoolers that extract heat from the helium gas immediately prior to compression and recirculation. In the indirect cycle version of the HTGR-

Table 1 HTGR's constructed to date

Name	Size [MW(t)]	Sponsor/Owner	Location	Type	Status
Dragon	20	Organization for Economic Cooperation and Development/Nuclear Energy Agency	Winfrith, U.K.	Experimental unit	Operated 1966 to 1976, now decommissioned
Arbeitsgemeinschaft Versuch Reaktor (AVR)	46	Kernforschungsanlage	Julich, Federal Republic of Germany (FRG)	Experimental unit	In operation since 1967
Peach Bottom	115	U.S. Atomic Energy Commission (AEC) [U.S. Department of Energy (DOE)]/Philadelphia Electric Company	Pennsylvania	Small prototype	Operated 1967 to 1974, now decommissioned
Fort St. Vrain (FSV)	842	AEC (DOE)/Public Service Company of Colorado	Colorado	Demonstration reactor	In operation since December 1976
Thorium Hoch Temperatur Reaktor (THTR)	762	Vereinigte Elecktriziatswerke Westfalen	Schmeehausen, FRG	Demonstration reactor	Scheduled for startup 1985

Table 2 HTGR systems

Reactor Type	Maximum Primary Coolant Temperature (°C)	Remarks
Steam cycle, electricity generation (HTGR-SC)	700 to 750	FSV and Peach Bottom of this type
Steam cycle/cogeneration (HTGR-SC/C)	700	
Gas turbine (HTGR-GT)	850	May be direct or indirect cycle
Process heat (HTGR-PH)	850 to 950	Indirect cycle

Table 3 Principal features of HTGR steam generators

Plant	Maximum Primary Coolant Temperature (°C)	Steam/Water Conditions Temperature (°C)	Steam/Water Conditions Pressure (MPa)	Steam Generator Configuration and Type
Dragon	830	203	1.6	Not used for power generation
AVR	950	505	7.5	Involute tube
Peach Bottom	700	536	10.2	U-tube forced recirculation
FSV	775	Main steam 538	16.5	Once-through helical bundles with reheat
		Reheat steam 538	4.9	
THTR	750	Main steam 530	18.0	Once-through helical bundles with reheat
		Reheat steam 530	5.6	
HTGR-SC	690	540	16.5	Once-through helical bundles

Table 4 Behavior considerations important in selection of materials for HTGR steam generators

1. Tensile, creep-rupture, low-cycle fatigue and creep-fatigue interaction behavior
2. Interaction of materials with environments (impure helium, steam/water, temperature) and effects on mechanical properties
3. Thermal aging and embrittlement
4. Fabricability (weldability, formability, availability of required product forms, etc.); matching or appropriate properties in welds
5. Resistance to friction wear and adhesion in HTGR helium
6. Cost

GT and the HTGR-PH systems, IHX's transfer heat from the primary to the secondary helium loops at temperatures up to 950°C.

The heat exchangers are critical components from performance, economic, and safety viewpoints and achievement of the necessary integrity requires the availability of substantial materials and fabrication technology, particularly for the most challenging of the heat exchangers - steam generators and the IHX. This paper reviews the status of the development of pertinent heat exchanger materials technology for HTGR steam generators and IHX's.

Steam Generator Materials Technology

Four of the reactor systems shown in Table 1 (Peach Bottom, FSV, THTR, and HTGR-SC) have featured steam generators that produce modern, high-pressure, high-temperature steam (Table 3). The Peach Bottom reactor was operated for 7 years and has been shut down and decommissioned, and the steam generators have been examined. The FSV reactor has entered commercial operation, and substantial operating experience has been accumulated. The THTR is under construction in the FRG and substantial fabrication-related experience has been developed.

As a result of this prior experience, considerable knowledge and technology exist in all aspects of HTGR steam generators, and this is reflected in current reactor design activities.

Several versions of the HTGR feature a multicavity prestressed concrete reactor pressure vessel (PCRV), which houses the components of the primary coolant system. Figure 1 shows a typical layout of a steam-generating version of the plant, and Fig. 2 presents a schematic flow diagram.

Since the cost of an HTGR plant is influenced by the size of the PCRV, it is desirable that the steam generator occupy a relatively small volume. As a result, compact high-performance steam generator designs are employed. In current U.S. HTGR designs, the steam generators are once-through, uphill boiling units in which a large fraction of the heat transfer section consists of densely packed, helically wound tubes with the water/steam on the inside of the tube and the primary coolant helium on the outside. A typical wound, once-through steam generator module used in the FSV reactor is shown in Fig. 3.

The key considerations governing the selection of materials for steam generators are summarized in Table 4. In addition, selection is strongly influenced by earlier HTGR experience supplemented by relevant operating experience with fossil-fired steam generators. (It should be noted that HTGR steam generators generate steam at temperatures and pressures similar to those found in most modern fossil-fired power plants.) Alloy 800H, which was used as the superheater/reheater of the Peach Bottom, FSV, and THTR plants, will be employed for the higher-temperature superheaters of the HTGR-SC/C system. The lower-temperature superheater, together with the evaporator and economizer sections, will be fabricated from 2-1/4 Cr - 1Mo in the annealed condition, based on prior HTGR and fossil plant experience.

In HTGR's, the steam generator tubing constitutes the primary coolant boundary, and as such, design, analysis, and fabrication are governed by Section III and Code Cases N-47

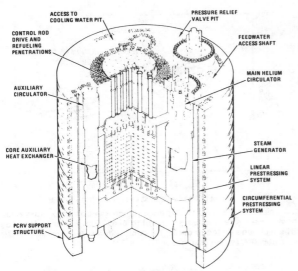

Fig. 1   HTGR-SC/C nuclear steam supply system

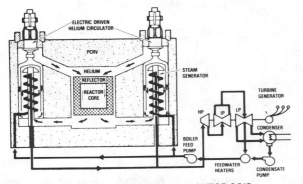

Fig. 2   Schematic flow diagram of HTGR-SC/C

Fig. 3   FSV steam generator module

Table 5   Variables under study in fatigue and creep-fatigue program on Alloy 800H

Temperature range	538° to 760°C
Test conditions	Continuous cycling
	Hold time (creep-fatigue) tests with holds in tension, compression, or both
Environmental effects	Tests in air and HTGR helium

Table 6   Expected range of impurities in primary coolant of HTGR-SC/C

Impurity	Expected Range (Pa)
H_2	9-18
H_2O	0.1-5
CO	5-10
CO_2	1-2
CH_4	0.5-1

and N-48 of the ASME Boiler and Pressure Vessel Code. Analysis in accordance with these rules requires the availability of complete data adequate to form a basis for assessment of time-dependent and time-independent properties, such as tensile, creep, rupture, fatigue, creep-fatigue, and deformation behavior throughout design life. In addition, Case N-47 requires that the designer explicitly account for any environmental degradation of materials that may occur in service. Since HTGR steam generators are designed for a 40-year life, a substantial body of data is required to permit all the required analyses to be performed. Accordingly, even though there is substantial extant data on Alloy 800H and 2-1/4 Cr - 1Mo, additional information is required to allow a complete assessment in accordance with these rules. Generation of this information is the purpose of an important part of the U.S. DOE HTGR materials programs.

With respect to the properties noted in Table 4, the basic mechanical property data for both Alloy 800H and 2-1/4 Cr - 1Mo are contained in ASME Code Case N-47. However, because of the strong influence of creep-fatigue in component design, considerable work is under way to better characterize this behavior aspect of Alloy 800H. The variables under current study are shown in Table 5.

The tests employ parallel-section (as opposed to hourglass) specimens with axial extensometers. This permits accurate testing at the small strain ranges that are relevant to the design of reactor components. As this testing progresses, it has become apparent that the observed sum of creep and fatigue damage fractions at failure for Alloy 800H test specimens can often be substantially less than unity, particularly in tests that

are performed at near design conditions, such as at lower temperatures and lower strain ranges and with larger hold times. The significance of this is under current study.

A unique feature of the HTGR is the presence of the helium primary coolant gas. Helium is, of course, chemically inert. However, there are practical limits to the purity levels that can be obtained in a circulating primary coolant gas, and interactions can occur between the impurities and the structural materials. Table 6 illustrates some of the typical impurities expected to be present in the primary coolant of an HTGR-SC/C plant during normal operation. Examination of these impurity species from a thermodynamic standpoint indicates, as shown in Fig. 4, that a wide range of material responses can be expected. For example, within the expected impurity range for the SC/C system it is possible to find environments that are decarburizing to both high- and low-chromium alloys and environments in which high-alloy materials may carburize while low-chromium ferritic steels will be decarburized. Analysis of the data generated in laboratory experiments using a range of possible simulated HTGR primary coolant gases shows that these phenomena are, in fact, observed. Figure 5 shows data obtained on the exposure of 2-1/4 Cr - 1Mo to a simulated HTGR helium

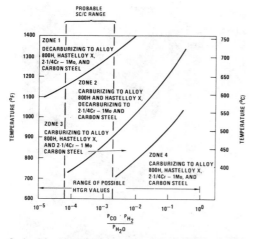

Fig. 4   Carburizing/decarburizing potential of HTGR primary coolant based on the CO + H ⇌ C + H_2O reaction (ignoring the influence of CH_4)

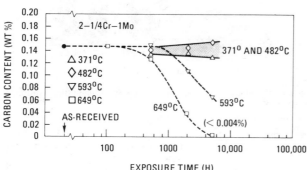

Fig. 5   Bulk carbon content of 2-1/4 Cr - 1Mo after exposure in an HTGR helium environment containing 20 Pa H_2, 10 Pa CO, 2 Pa CH_4, 5 Pa H_2O, and 0.5 Pa CO_2 for various times in comparison with that of as-received material. (Carbon content detection limit is 0.004 wt percent)

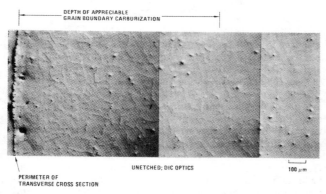

Fig. 6   Carburization in Alloy 800H creep specimen (0.53 percent ϵ_c) observed after 19 000 hrs exposure in dry, controlled-impurity helium at 800°C

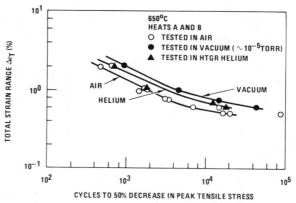

Fig. 7   Number of cycles to 50 percent decrease from maximum value of peak tensile stress as a function of total axial strain range for Alloy 800H specimens tested under strain control in air, vacuum, and HTGR helium at 650°C (strain rate = 4 × 10^{-3} s^{-1})

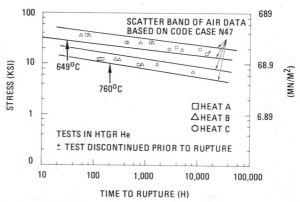

Fig. 8   Stress-rupture data of Alloy 800H (9.5-mm-dia specimens) tested in HTGR helium containing 150 Pa H_2, 45 Pa CO, 5 Pa CH_4, 5 Pa H_2O, and 0.5 Pa Co_2

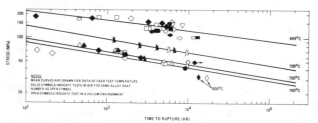

Fig. 9   Stress-rupture values for Alloy 800 and Alloy 800H in air and in controlled-impurity helium test environments. (Test materials were Alloy 800H or Alloy 800 having wt percent C ≥ 0.025 and grain size coarser than ASTM No. 5.)

environment at a variety of temperatures. As indicated, at high temperatures significant decarburization occurred. Figure 6 shows the microstructure of Alloy 800H exposed to simulated HTGR coolant gas for 19,000 hrs and illustrates carburization that occurred as a result of that exposure.

Environmental interactions such as carburization and decarburization can exert a significant influence on the properties of structural materials, and substantial testing has been performed to assess these effects. The influence of the helium environment on fatigue properties is generally benign (Fig. 7). With respect to creep-rupture properties, data now available suggest that, while there is some influence, effects

seem generally to be fairly small. This is illustrated, in the case of Alloy 800H, by Figs. 8 and 9. Figure 8 shows that data generated in a variety of helium environments generally fall within the expected scatterband of air data for this alloy. Similarly, Fig. 9 shows that data from air and helium tests on the same heats of material generally fall within the same scatterband. However, more detailed examination (Fig. 10) shows that the HTGR helium environment does have some minor degrading effect. Figures 11 and 12 illustrate similar effects for 2-1/4 Cr - 1Mo steel.

At the temperatures that prevail in HTGR steam generators, many of the high-temperature alloys of interest undergo microstructural changes that may cause embrittlement, which occasionally can be serious. Because it is necessary to show that reactor structures can absorb high-strain-rate events (such as seismic loads) after extended high-

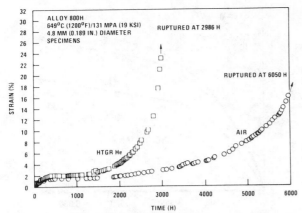

Fig. 10 Comparison of creep curves for one heat of Alloy 800H tested in helium and air environments (helium impurities 150 Pa H_2, 45 Pa CO, 5 Pa CH_4, 5 Pa H_2O, and 0.5 Pa CO_2)

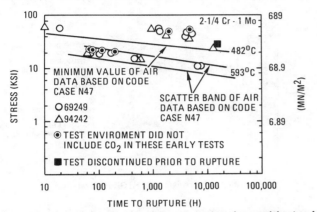

Fig. 11 Comparison of stress-rupture behavior of several heats of annealed 2-1/4 Cr - 1Mo steel in HTGR helium and the scatter band of air data for the alloy

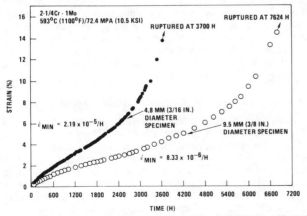

Fig. 12 Comparison of creep curves for one heat of 2-1/4 Cr - 1Mo (4.8 and 9.5-mm dia specimens) tested at 593°C and 72.4 MPa in HTGR helium (helium impurities same as Fig. 8)

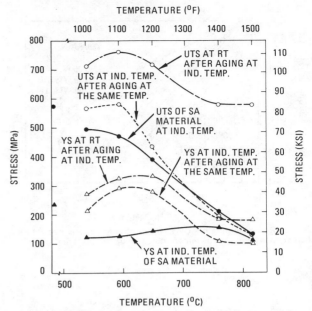

Fig. 13 Effects of 20,000-hrs thermal aging (in air) on the strength of Alloy 800H at room temperature and at the aging temperature (grain size No. 3, 0.05 wt percent C, 0.39 wt percent Al, 0.44 wt percent T)

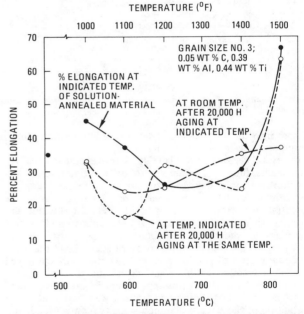

Fig. 14 Effect of thermal aging (in air) on the ductility of Alloy 800H at room temperature and at the aging temperature

temperature operation, it is important to understand the extent to which aging causes changes in properties. Figures 13 and 14 indicate changes produced in the tensile characteristics of Alloy 800H as a result of 20,000 hrs aging at various temperatures. Figure 15 illustrates a general trend of toughness of this alloy with extended aging at different temperatures. The Charpy V-notch data shown in Fig. 15 are a convenient way of illustrating the changes that occur. However, for design purposes, more quantifiable fracture toughness information (such as that generated by compact tension testing) is required, and such data are currently being generated. However, in general, it appears that this alloy retains adequate toughness after long-time aging.

The HTGR steam generator components are, of course, fabricated by welding, and it is important to establish that the properties of the resulting weld joints are compatible with those of the base metal. This is being assessed in a program of elevated-temperature testing of all-weld-metal and cross-weld specimens on Alloy 800H and 2-1/4 Cr - 1Mo. Data available to date suggest that, on the average, the rupture strengths of the weld metals used to join Alloy 800H to itself and to join 2-1/4 Cr - 1Mo may be slightly lower than those of the base metal at the longest times/highest temperatures. Accordingly, consideration is being given to the introduction of weld metal

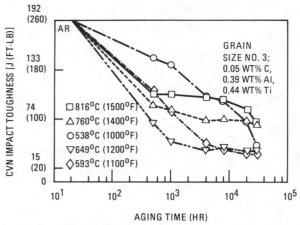

Fig. 15   Room-temperature CVN impact strength of one heat of Alloy 800H aged in air at various temperatures for up to 30,000 hrs

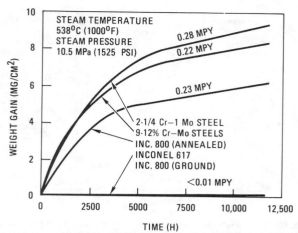

Fig. 17   Corrosion rate of candidate HTGR steam generator materials in superheated steam at 538°C

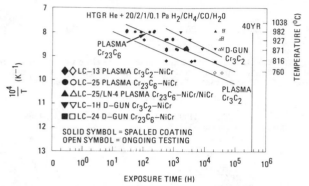

Fig. 16   Spallation chart of chromium carbide · nichrome coatings on Alloy 800H

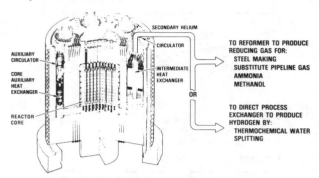

Fig. 18   Nuclear heart of an HTGR-PH system

factors into the allowable stresses used for long-time, high-temperature design with these materials.

In several regions of the HTGR steam generators, relative movements (due, for example, to differential thermal expansion) can occur between component surfaces. There are also a number of locations where vibratory forces create the potential for fretting wear. In the HTGR helium environment, such surfaces can, if left unprotected, suffer considerable damage, particularly if the surfaces are operating in the higher-temperature ranges. To decrease the potential effects of friction and wear, coatings are employed at critical locations. Chromium-carbide-based coatings are currently under study for this application. Evaluation of these coatings consists of sliding and fretting wear tests, including evaluation of the effects of sliding velocity, contact load, etc. In addition, it has been noted that after extended elevated-temperature exposure to HTGR helium, some of these coatings will undergo spallation (Fig. 16), and care must be taken to select a coating suitable for the intended service conditions.

The materials in the HTGR-SC/C heat exchangers are also exposed to water and steam environments, and it is important to establish the compatibility of the materials employed with these environments. The feed-water employed in the steam generator is, of course, of ultra-high purity, with pH and oxygen control. In general, the compatibility of the materials used in the steam generator with environments of this kind is well established. However, it was considered prudent to determine experimentally the long-time corrosion behavior of the materials used in the highest-temperature region of the steam generator in a representative steam environment. To this end, pertinent tests were performed by the Oak Ridge

National Laboratory. These tests, which extended for 28,000 hrs, showed that corrosion rates of Alloy 800 were low, predictable, and satisfactory (Fig. 17) [1]. Another important consideration in the steam generators is the influence of steam flow velocity relative to causing erosion-corrosion of the low alloy steel components. Extensive tests were performed to show that erosion-corrosion behavior was satisfactory provided that pH was maintained in a specified regime, flow velocities were maintained at moderate values, and where appropriate, low alloy steels were employed.

Intermediate Heat Exchanger Materials Technology

The layout of the nuclear heart of a typical GA-designed HTGR-PH system is shown in Fig. 18. The function of the IHX in this system is to transfer heat from the primary to the secondary helium coolant. For process applications, it is economically desirable for the secondary loop to be able to deliver heat to the user process at relatively high temperatures (significantly above 800°C). As a result, the goal for the HTGR-PH system is to operate with core outlet temperatures of at least 850°C and, ultimately, 950°C. The key component limiting the ability to achieve these temperatures is the IHX.

To function at such elevated temperatures, several important features are designed into the IHX. For example, for all normal operation, the pressure in the primary and secondary coolant loops is balanced so that the pressure differential carried across the heat transfer tubing is essentially zero. As a result, relatively thin-walled tubing can be employed. In fact, in an actual IHX design, the wall thickness of the tubing is sized by the ability to carry the short-time loads that occur due to the pressure differential resulting from the accidental depressurization of one loop. The use of thin-walled tubing allows temperature differentials across tube

Table 7 Impurity levels expected in the primary coolant of the HTGR-PH system

Impurity	Concentration (Pa)
H_2O	0.05
H_2	50
CO	5
CH_4	5
CO_2	<0.1
N_2	1.5
O_2	<<<0.1

Table 8 Carburization-resistant coatings under study

Aluminized coatings	Preformed oxide coatings
Al	In air
Al + Cr	In controlled environments
Al + Pt	
Al + Rh	
M-Cr-Al-Y coatings	Claddings
M = Fe or Ni	Fe-Cr-Al-Mo-Hf
	Fe-Cr-Al-Y
	Udimet 720
	Inco clad 671/800H

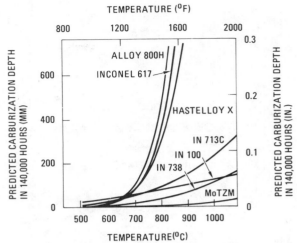

Fig. 19 Comparative carburization behavior of metals in HTGR-PH helium

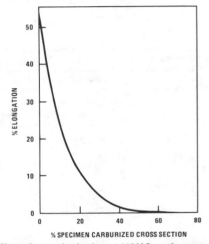

Fig. 20 Effect of precarburization at 1100°C on the room-temperature elongation of Hastelloy X

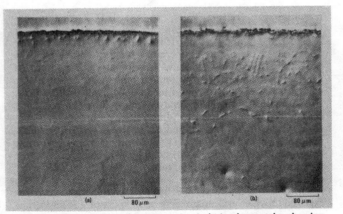

Fig. 21 Differential interference contrast photomicrographs showing Hastelloy X specimens exposed for 3000 hrs at 900°C in controlled-impurity helium containing 50 Pa H_2, 5 Pa CO, 5 Pa CH_4, and <0.05 Pa H_2O. Prior to exposure, specimen surface condition was (a) preoxidized for 10 hrs at 900°C and (b) standard machined surface.

walls to be minimized, thus minimizing creep-fatigue effects.

Another recognized feature of the IHX is that because of the very high operating temperatures, it may be appropriate to design the heat exchanger for a shorter life than the reactor as a whole.

Despite these design features, the design of a 950°C IHX remains very challenging, particularly with respect to materials selection. The types of material considerations that are important in an IHX are generally similar to those shown in Table 4 as pertinent to the steam generator. However, because of the much higher operating temperature of the IHX, many of the property changes and environmental effects considered are accelerated. In addition, new phenomena, such as evaporation of alloying elements, can become important considerations in very high-temperature IHX design.

One aspect of materials behavior over which temperature exercises a strong influence is the interaction of materials with primary coolant impurities. One reason for this is that, in an HTGR-PH system with 950°C core outlet temperature, it is expected that the combination of impurities present in the primary coolant gas will tend to be more reducing (less oxidizing) and more carburizing than in an HTGR-SC system. Typical expected HTGR-PH impurity values are shown in Table 7.

Extensive experimental evaluations of materials behavior in this environment have been carried out over the past ~10 years. The results of these programs clearly show that predicted carburization rates in an HTGR-PH environment such as that shown in Table 7 become very high for commercially available wrought alloys such as Alloy 800H, Inconel 617, and Hastelloy X at temperatures above ~700°C, as shown in Fig. 19.

Other alloys, such as the cast nickel-base superalloys and MoTZM show much lower carburization rates. However, these alloys are not suitable for the construction of large welded heat exchangers of thin-walled tubing.

The occurrence of carburization is undesirable from several standpoints. Severe carburization causes embrittlement of many wrought alloys. Figure 20 shows the effects of carburization on the room-temperature tensile ductility of Hastelloy X. Similar effects are observed for carburized Alloy 800H and Inconel 617.

Carburization has also been observed to produce dimen-

sional changes (swelling) in some alloys [2] and has the potential to affect other key properties such as creep, rupture, fatigue, and creep-fatigue.

The potentially deleterious effects of carburization on wrought alloy candidate materials for the higher-temperature

Source: Journal of Engineering for Power, October 1983, 725-734

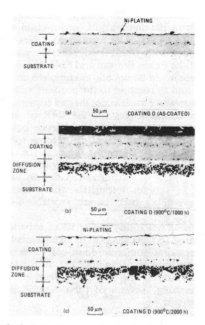

Fig. 22 Optical photomicrographs of aluminide-coated Hastelloy X showing (*a*) no pores in the as-coated condition and (*b*, *c*) extensive pores in the diffusion zone after exposure to controlled-impurity helium (50 Pa H_2, 5 Pa CO, 5 Pa CH_4, <0.1 Pa H_2O) at 900°C for 1000 and 2000 hrs, respectively (unetched)

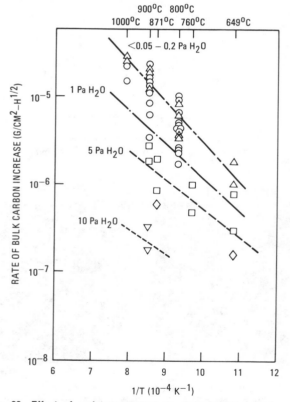

Fig. 23 Effect of moisture content on carburization behavior of Hastelloy X in helium containing 50 Pa H_2, 5 Pa CH_4, and 5 Pa CO

reactor systems have prompted a search for protective coatings. Table 8 lists the coatings and claddings being studied at GA.

Preformed oxides offer the potential for a simple means of protection against carburization, and preliminary results are very encouraging. For example, it has been shown that

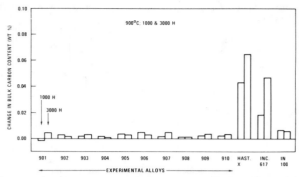

Fig. 24 Change in bulk carbon content of experimental alloys Hastelloy X, Inconel 617, and IN100 after 1000 and 3000 hrs exposure in simulated HTGR-PH helium at 900°C

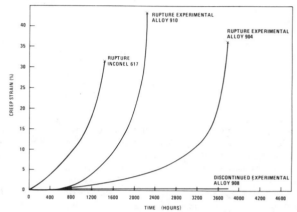

Fig. 25 Creep-rupture behavior in HTGR-PH helium of several alloys at 51.7 MPa and 900°C. (Alloy 908 discontinued after 3800 hrs)

preoxidation of Hastelloy X in air at 900°C for 10 hrs produces an oxide coating that is effective in protecting against carburization in HTGR helium. This is illustrated in Fig. 21, which shows the cross-sectional appearance of Hastelloy X exposed at 900°C for 3000 hrs with and without preoxidized coating.

Diffused coating methods, such as aluminizing, offer another potentially attractive solution, and exposures in HTGR helium show that such coatings can provide effective protection. However, a difficulty that has been observed with these coatings is the tendency during long-time exposure for formation of pores at the coating-substrate interface (Fig. 22). These pores are believed to be Kirkendall voids, and their formation and development can ultimately lead to coating separation. It is believed that this process will limit use of diffused coatings for this application unless coating compositions can be developed that minimize the interdiffusion of elements causing this effect.

Another possible approach to the control of carburization in wrought alloys is based on the observation that carburization rates in HTGR helium are very sensitive to the water content of the gas. As shown in Fig. 23, carburization rates for Hastelloy X can be reduced by almost two orders of magnitude with sufficient water present. Unfortunately, there are limits to the permissible water content in HTGR helium imposed by control of graphite oxidation. At present time, more work is needed to determine if a "window" exists within which satisfactory corrosion rates can be obtained in both metal and graphite components. It is also important to consider, in this regard, possible behavior in crevice regions or regions of semistagnant gas wherein the water content of the gas may be depleted, allowing rapid local carburization to occur.

A further approach to control of the environmental interaction issue is to develop 'tailored' alloys intrinsically suitable for this application. Work in this direction is underway and preliminary results look encouraging. A series of experimental alloys have been developed in which the surface protection against carburization results from the formation of surface films of aluminum and titanium oxides. (These oxides are stable under all credible HTGR operating conditions.) To permit the preferential formation of these oxides, the alloys have a low chromium content. Strengthening in the alloys is achieved through the use of molybdenum and tungsten together with the effects of aluminum, titanium, carbon, and zirconium. Initial results for these alloys are most encouraging. Tests in HTGR-PH helium have shown that carbon pickup in the experimental alloys is very low, compared to Hastelloy X, Inconel 617, and Alloy 800H (Fig. 24). Mechanical property tests have shown several of the experimental alloys to possess excellent creep and rupture strength (Fig. 25) with good retention of toughness and ductility. Tests on additional heats of these alloys are currently underway to more fully establish their capability and potential.

Clearly, for an IHX, there are many materials issues that will require development and qualification. These will include protective coatings to decrease the effects of friction and wear and vibration, development of welds with adequate high-temperature properties, and an improved understanding of creep-fatigue behavior at these very elevated temperatures and related effects. However, establishment of these data must await completion of the basic studies to identify viable candidate materials of construction.

Conclusions

Heat exchangers play a key role in HTGR's, and considerable materials technology is required to support their design. For the HTGR-SC steam generators, there is considerable prior experience and established knowledge. Accordingly, materials and design technology are well developed. Principal materials of construction are 2-1/4 Cr - 1Mo and Alloy 800H. The technology relies heavily upon prior experience with fossil-fired generators and the steam generators of the Peach Bottom and FSV HTGR's. The major work that remains to be done is to complete the qualification of materials and to respond to the evolving rules pertinent to elevated-temperature nuclear design and construction.

The materials challenges for the IHX of an HTGR-PH system are far greater and the experience base is far less. Accordingly, significant effort is needed to develop the required technology. Foremost among the issues to be resolved is finding a material and operating mode that will mitigate the effects of carburization due to interaction between primary coolant impurities and materials of construction. Several alternatives are available and are being explored.

One promising possibility is a series of experimental alloys that are tailored to meet the HTGR requirements and appear to offer the required combination of corrosion resistance and strength. Other approaches include use of existing alloys with coatings or modifying the primary coolant gas impurities to be less carburizing. Several of these approaches are under current study.

Acknowledgment

This work was performed in part under Department of Energy Contract DE-AT03-76ET35301.

References

1 Greiss, J. C., and Maxwell, W. A., "The Corrosion of Several Alloys in Superheated Steam at 482° and 538°C," Oak Ridge National Laboratory Report ORNL/TM-6465, Nov. 1978.

2 Inouye, H., and Rittenhouse, P. L., "Relationship Between Carburization and Zero-Applied-Stress Dilation in Alloy 800H and Hastelloy X," in *Proceedings of the IAEA Specialists Meeting on High-Temperature Metallic Materials for Application in Gas-Cooled Reactors, Vienna, Austria, May 4–6, 1981*," Austrian Research Center, Seibersdorf, Austria, p. L-1.

Materials Selection for Metallic Heat Exchangers in Advanced Coal-Fired Heaters

I. G. Wright

Battelle-Columbus Laboratories,
Columbus, Ohio 43201

A. J. Minchener

Coal Research Establishment,
Cheltenham, England

The application of advanced coal-fired heaters to heat the working fluid for a closed-cycle gas turbine provides some challenging problems for the selection of metallic heat exchanger materials. The requirements of a working fluid temperature of 1550°F (1116 K) at a pressure of 300–600 psig (2.07–4.14 MPa/m²) necessitate that the alloys used for the hottest part of the heat exchanger possess high-temperature strength in excess of that available in widely used alloys like alloy 800. The maximum-duty alloys must therefore be selected from a group of essentially nickel-base alloys for which there is scant information on long-term strength or corrosion resistance properties. The susceptibility to corrosion of a series of candidate heat exchanger alloys has been examined in a pilot plant size fluidized-bed combustor. The observed corrosion behavior confirmed that at certain locations in a fluidized-bed combustor nickel-base alloys are susceptible in varying degrees to rapid sulfidation attack, and must be protected by coating or cladding.

Introduction

Closed-cycle systems, in which a high-temperature fluid is used to drive a turbine to generate electricity, appear to offer an attractive prospect for high-efficiency, modular units which have little or no water requirement. A further advantage of such systems is that virtually all of the high-temperature compatibility problems are transferred to the heat exchanger, which could assume several different configurations and burn a range of fuels.

A particular design of high-temperature heat exchanger intended to burn coal to heat air to a temperature of 1550°F (1116 K) is the subject of a current combined combustion technology-materials-engineering study [1, 2]. The chosen maximum temperature yields a potential closed combined cycle system efficiency of 51 percent at a working fluid pressure of 600 psig (4.14 MPa/m²). More importantly, perhaps, it represents the maximum capability of practical metallic heat exchanger materials in terms of high-temperature strength. Assuming a maximum metal temperature of 1650°F (1172 K) calculated ASME Boiler Code-type strength requirements reveal a very limited choice of available alloys. In this study, alloy candidates for the top temperature duty, considering availability, fabricability, and cost are (in descending order of maximum allowable stress): Inconel 617, Inconel 618E, Nimonic 86, and Incoloy 800H. A heat exchanger construction is envisaged in which a minimum amount of one or two of these alloys is used in the outlet passes, and where the more conventional 300 series stainless steels and low-alloy steels are used as system temperature and pressure permit.

The coal-fired heat exchanger under consideration is a fluidized-bed combustor, using limestone as a sulfur sorbent. Experience in the use of metallic materials in this type of combustor is fairly sparse and is increasing at a slow rate, since only a few developmental units are in operation. Unfortunately, the materials compatibility studies linked to these units have been concerned with alloys required for steam-cycle duty [3], or for different air heater cycles which require less strength [4] than in the system of current interest. Results of direct interest, however, concern the widespread experience of rapid but sometimes unpredictable corrosion of nickel-rich alloys in the fireside environment of coal-fired, fluidized-bed combustion. Since the alloys of choice for the maximum temperature duty are unavoidably nickel-base, it is essential that their behavior in the fireside environment be understood so that decisions on strategies of location and protection can be properly made. The results described here represent one part of a materials compatibility study designed to support the materials selection decisions for this heat exchanger design.

Experimental

The alloys chosen for corrosion testing were four high-strength wrought alloys representing possible choices for very high temperature heat transfer tubing for advanced heat exchanger designs. The compositions are shown in Table 1.

Specimens 2 in. by 1.5 in. (51 mm by 38 mm) were cut from plate stock, and 0.25-in. (6.4-mm) holes drilled on 1.5-in. (38-mm) centers to allow for support in the fluidized-bed combustor. The cut edges were deburred, and an identifying code engraved on one face, but no machining was done on the major exposed surfaces, which were exposed in the mill-finished condition. The specimens were mounted on a refractory-lined door which comprises one wall of the

Reprinted with permission from Journal of Engineering for Power, July 1983, 446-451, © 1983 American Society of Mechanical Engineers

Source: Journal of Engineering for Power, July 1983, 446-451

Table 1 Alloy compositions

Alloy	Heat No.	C	P	S	Si	Mn	Ni	Fe	Co	Cr	Mo	Cu	Aℓ	Ti	W	Ce
Inconel 617	XX19A7UK	.06	–	.001	.1	.01	55.4	.34	12.12	21.53	9.03	.02	1.11	–	–	–
Inconel 618E	–		.05	–	.005	.08	.04	55.5	14.97	–	22.78	–	.01	–	.32	6.24	–
Nimonic 86	UNJ 4360	.06	–	.004	.35	.48	65.0	1.76	.53	25.4	10.4	.11	.02	.05	–	.03
Incoloy 800	HH0229A	.06	–	.002	.21	.91	30.4	46.58	–	20.7	–	.44	.29	.41	–	–

Table 2 Summary of fluidized-bed operating conditions*

		Test No. 1	Test No. 2	Test No. 3
Fuel		Wellbeck (UK) coal (0.53% Cl, 1.6% S)	Illinois No. 5 coal (0.07% Cl, 2.9% S)	Char containing 2% volatiles (produced from Illinois No. 5)
Sulfur sorbent		None	Penrith limestone	Penrith limestone
Excess air level (%)		10–20	Substoichiometric	10–20
Ca:S Mole Ratio		0	3:1	3:1
Flue Gas Analysis (Vol.%)	O_2	2.6	0.1	2.7
	CO	–	3.7	–
	CO_2	16.0	17.0	17.1
	SO_2	0.117	0.0357	0.046
	CH_4	–	0.7	–
	H_2S	–	0.0366	–

* In all runs, the fluidizing gas velocity was 3 ft/sec and the bed temperature was 1650 F.

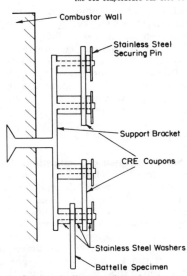

Fig. 1 Schematic diagram of specimen assembly

Combustor Wall
Stainless Steel Securing Pin
Support Bracket
CRE Coupons
Stainless Steel Washers
Battelle Specimen

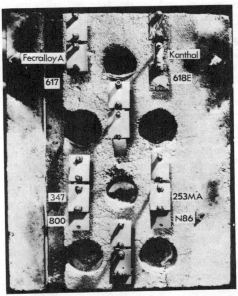

Fig. 2 Arrangement of specimens on combustor wall (after test no. 2)

fluidized-bed combustor. Figure 1 shows schematically the arrangement of the specimens, and their relationship to other test coupons, on the support bracket. A point worthy of note is that the specimens described here were all mounted with an area of overlap with test coupons from the main test program; these overlapping coupons were of alloys which have shown good resistance to corrosion in previous testing. The mounted specimens were located in the lower part of the tube bank, as can be judged from their positioning with respect to the holes in the door which accommodated the in-bed tubes in Fig. 2. The specimens of interest are labeled 617, 618E, 800, and N86; the lowest specimen being some 11 in. above the distributor plate. In test number 3, the location of specimens was slightly different, with Inconel 617 and 618E occupying the positions shown in Fig. 2 for Incoloy 800H and Nimonic 86, respectively; Nimonic 86 was in the overlapped position of the upper center mounting, while Inconel 800H was in the overlapped position of the lower center mounting, some six inches above the distributor plate.

The actual fluidized-bed combustor used was a 1-ft (0.3-m) square, atmospheric, fluidized-bed combustor at the Coal Research Establishment, near Cheltenham. The bed was run for a series of 250 hr (9×10^5 s) corrosion tests to investigate the effects of different operating conditions on candidate

alloys for in-bed heat exchangers [5], and the alloys described here were included in three of these runs. The conditions of operation during the three runs of interest are listed in Table 2.

In test number 1, a high chlorine coal was burned without a sulfur absorbent. Coals of this type have caused corrosion problems in pulverized-coal steam boilers in the United Kingdom, possibly because of the direct action of the chlorine or because high chlorine often implies a high level of alkali metal salts in the coal.

Test number 2 was run with a high sulfur coal and limestone, but under conditions of substoichiometric combustion air flow, in order to subject the corrosion coupons to the type of combustion conditions expected to be encountered near the coal feed ports, or near the walls in future commercial combustors. In this test, the fuel to air ratio was approximately twice that used in normal operation and 3 to 4 percent CO was measured in the flue gas stream.

The third test was run with a high-temperature char, produced by coking Illinois No. 5 coal at 1690°F (1194 K), and limestone. The use of this fuel was intended to minimize the emission of volatiles from the coal feed entering the fluidized-bed combustor, so removing one possible source of the localized oxygen-deficient (and possibly sulfur-rich) conditions previously measured in this bed [6].

Results and Discussion

After removal from the fluidized-bed combustor at the end of a 250 hr (9×10^5 s) exposure, the specimens were visually examined, reweighed, and then sectioned for metallographic examination, along the longitudinal axis where pretest thickness measurements were made. The thickness of remaining metal at specific locations was measured from photographs made of the cross sections.

Test No. 1: High-Chlorine Coal, No Limestone. The specimens exhibited a very thin, uniform, orange-brown deposit after this test. The extent of material loss in this test, as measured by specimen thickness change, is presented in Table 3. The losses were very small except for a few locations on Nimonic 86. Examination of the cross sections shown in Fig. 3 confirm that all four alloys formed thin protective oxide scales, with localized breakdown observed only in the overlapped regions and on the free end of some of the specimens. Over the majority of the surface, Inconel 617 formed a thin, uniform, protective external oxide scale, and a thin (10–20 μm) layer of internal oxide subscale, as shown in Fig. 3. The only sulfidation attack of this alloy occurred near the top attachment hole in the overlapped region; here sulfide penetration occurred to a depth of $\sim$ 150 μm.

Table 3 Summary of specimen thickness changes

	Thickness Loss (μm/Side)		
Alloy	Free End	Center	Overlapped End
Test No. 1 (High Cl Coal, No Sorbent)			
Inconel 617	38	36	27 (180)[A]
Inconel 618E	43	17	+4 (100)
Nimonic 86	98	79	61 (650)
Incoloy 800H	50	60	33
Test No. 2 (Substoichiometric Combustion)			
Inconel 617	>886	>880	>885 (Thin specimen)
Inconel 618E	582 (1433)	488 (956)	766 (1007)
Nimonic 86	+4	+4	885
Incoloy 800H	28	0	372
Test No. 3 (Low-Volatility Char)			
Inconel 617	46	42	3 (292)
Inconel 618E	55	44	33 (461)
Nimonic 86	3	5	>883
Incoloy 800H	46	46	53

A: numbers in brackets indicate maximum loss including internal penetration.

Inconel 618E formed a thin, uniform, perfectly protective oxide over most of the exposed surface, except at the free end, and in the overlapped zone near the attachment hole and on the back face, where sulfidation occurred to about 100 μm. The behavior of Nimonic 86 was very similar to this, except that in the overlapped region around the hole and on the back face massive sulfidation attack produced large pools of internal sulfide and grain boundary penetration extending some 600–700 μm into the alloy. At the free end, sulfides were present some 100–150 μm beneath the outer scale. Incoloy 800H, on the other hand, exhibited completely protective behavior in this environment, though the oxide scale on the front face was thickened and nonuniform compared to that on the back face.

Clearly, little adverse effect is evident of the high-chlorine (high alkali metal content) coal on corrosive deposit formation or on the formation of adherent, protective oxide scales on the freely exposed surfaces of these alloys. In locations where deposits might accumulate by the action of the fluidized-bed, however, sulfidation attack was initiated. The susceptibility of Nimonic 86, in particular, to this attack is surprising.

Attempts to measure the prevailing oxygen partial pressure in the bed, through the top center port shown in Fig. 2, using a stabilized zirconia electro-chemical probe, were thwarted by rapid loss of the outer platium electrode. Presumably, reaction occurred with the alkali chlorides or chlorine released from the coal. Previous measurements [5] under oxidizing conditions (using a sulfur sorbent) indicated a rapidly fluctuating oxygen partial pressure with a range of 10^{-13} to 4×10^{-8} atm., while the flue gas analysis indicated 3.4×10^{-2} atm. oxygen. At the lower central port (Fig. 2), previous readings ranged from 5×10^{-4} to 1×10^{-1} atm. oxygen.

Test No. 2: Substoichiometric Combustion. The surfaces of the specimens, except Inconel 617, after this run appeared light grey, and were quite smooth, with no obvious build up of deposit. During this test run, the specimen of Inconel 617 broke in half, separation occurring just below the bottom of the overlapping specimen. Visual examination of the remaining piece, and of the severed piece after recovery indicated very heavy corrosive attack. The surface of Inconel 618E exhibited gross attack with voluminous grey-appearing external scale at the top of the specimen, around the region of overlap, at the bottom of the specimen, and along the edges (see Fig. 2). The specimens of Nimonic 86 and Incoloy 800 were attacked only in very localized spots on the freely ex-

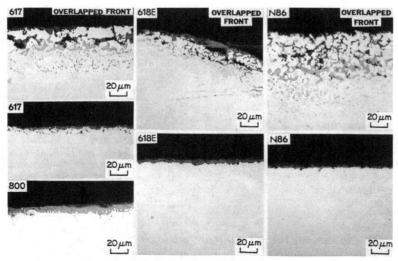

Fig. 3 Cross sections of alloys from test number 1

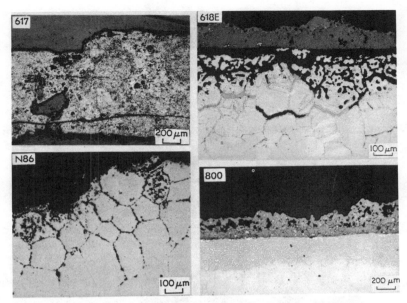

Fig. 4 Cross sections of the alloys from test number 2, showing corrosion morphologies in the overlapped areas

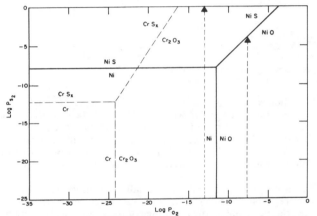

Fig. 5 Equilibrium thermodynamic diagram showing range of stability of phases in the Cr-O-S and Ni-O-S systems at 1650°F (1172 K).

posed surfaces, but were heavily corroded in the areas of overlap.

Cross sections of the alloys are presented in Fig. 4. The specimen of Inconel 617, which was only 0.15-in. (0.38-cm) thick initially (as indicated in Table 3), had been completely converted to an oxide/sulfide, with only isolated areas of metal remaining. The areas of Inconel 618E which were visibly attacked in fact exhibited extensive sulfidation. A thick (100–200 μm) external scale of mixed oxide/sulfide was formed in these areas, while sulfide penetration of alloy grain boundaries extended some 750 μm into the substrate. Large pools, apparently of chromium sulfide surrounded by nickel and iron essentially devoid of chromium, were also observed near the alloy/external scale interface, indicating the existence of intensely sulfidizing conditions in these locations.

Nimonic 86 formed a thin protective scale, with virtually no subsurface penetration, over most of its surface, as can be judged by the thickness change data in Table 3. In the overlapped end zones, however, very extensive corrosive loss had occurred. The external scale was missing, but pools of chromium-rich sulfide were observed in the surface of the remaining alloy, with further sulfide penetration extending down alloy grain boundaries a further 600 μm into the substrate.

The specimen of Incoloy 800H similarly exhibited quite protective behavior over the majority of the exposed face. Small, isolated blisters associated with accelerated oxidation and penetration of sulfur to form small sulfide pools at the near surface were also present, however, with sulfide precipitates in the alloy grain boundaries to a depth of 580 μm, indicating localized initiation of accelerated attack. In the area of overlap, massive sulfidation attack had occurred, with the formation of thick (200–400 μm), external scales, and a thick zone of sulfide subscale.

Measurements of the prevailing in-bed oxygen partial pressure at the upper central port in Fig. 2 showed a fairly steady value, in the range 10^{-13} to 5×10^{-12} atm. oxygen [7]. Major differences from the data measured under "normal" combustion conditions were the reduced frequency of the fluctuations and a great reduction in excursions to higher oxygen partial pressures. It is difficult, however, to reconcile the greatly increased severity of sulfidation attack of Inconel 617 and 618E specimens under the substoichiometric conditions with this observed change in frequency of fluctuation of the oxygen partial pressure, since the minimum achieved was similar in both tests.

As suggested by the superimposed phase stability diagrams for Cr-O-S and Ni-O-S shown in Fig. 5 [8, 9], the formation of chromium sulfide on the surface of an alloy is not possible in an oxygen-and sulfur-bearing gas mixture at 1650°F (1172 K), when the prevailing oxygen partial pressure is 10^{-12} to 10^{-13} atm. Only when the oxygen partial pressure is reduced below $\sim 10^{-16}$ atm. does this become possible at very high sulfur partial pressures; as the oxygen partial pressure is reduced further, the necessary sulfur partial pressure also falls. In practice, sufficiently low oxygen partial pressures exist beneath protective Cr_2O_3 scales, so that if sulfur can transport through these scales to create a sufficient partial pressure, chromium sulfide formation can occur in this location.

Nickel sulfide is thermodynamically less stable than chromium sulfide, and will, therefore, not form under equilibrium conditions in a Ni-Cr alloy unless the chromium activity is reduced to a very low level. Kinetically, however, nickel sulfide can form in areas of local chromium depletion in an alloy, providing a sufficient flux of sulfur to the alloy is maintained. Since nickel sulfide is molten at the temperature

Source: Journal of Engineering for Power, July 1983, 446-451

199

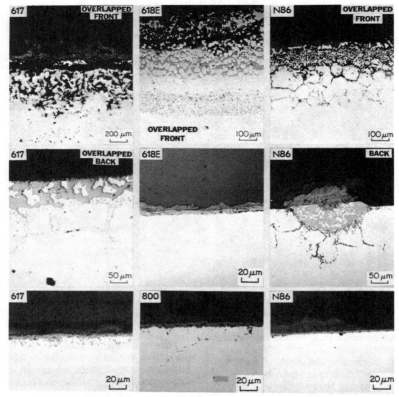

Fig. 6 Cross sections of the alloys from test number 3, showing variation of corrosion morphologies over the specimen surfaces

of interest, its presence ensures rapid degradation of the alloy. The NiO/NiS line in Fig. 5 indicates that NiS-formation is possible at oxygen partial pressures of 10^{-8} atm. and that the necessary sulfur partial pressure is reduced as the oxygen partial pressure is reduced down to $\sim 5 \times 10^{-11}$ atm., after which it remains constant. Since the mean oxygen partial pressure in test 2, and under "oxidizing" conditions, was $\sim 10^{-13}$ and $\sim 10^{-11}$ atm., respectively, the sulfur partial pressure required for nickel sulfide formation would have been expected to be similar in both cases.

Test No. 3: Combustion of Low-Volatility Char. The surface appearance of the specimens after this test was very similar to that after Test Number 1. The thickness change data, shown in Table 3, show that the three nickel-base alloys were heavily attacked in the overlapped areas, while the corrosion of Incoloy 800H in the same location was much less severe. Figure 6 illustrates the massive sulfidation/oxidation which occurred in the overlapped region on the front surface of Inconel 617. The corrosion penetration of the rear surface in the same location was less, but the extent of sulfidation was nevertheless quite massive. Almost all the freely exposed surface of this alloy exhibited protective behavior, with the formation of a thin, uniform oxide and a subscale of internal oxide; slight sulfidation occurred of the freely exposed end of the specimen.

The sulfidation attack of the front and rear surfaces of the overlapped zone of Inconel 618E was massive (see Fig. 6) with sulfide penetration to $\sim 300~\mu m$, and chromium depletion of the alloy extending a further $100~\mu m$. Slight sulfidation of the freely exposed end of this alloy was also observed. Over the rest of the free surface, perfectly protective behavior was exhibited, with the formation of a thin, uniform scale, with no subscale. The adherence of this scale appeared poor, however, since it detached in large pieces (still adhering to the bed-deposit) on handling.

Nimonic 86 was also grossly sulfidized on the front face in the overlapped region, while on the back face in the same location isolated pits of very heavy sulfidation occurred. The free end of the specimen was also slightly sulfidized. Over the rest of the free surface, a thin protective scale was evident, along with very localized pits associated with massive sulfidation.

Incoloy 800H formed an essentially protective oxide over its entire surface, except the line marking the start of the overlapped zone on the front face of the specimen. Here a few pits associated with localized sulfide formation and accelerated oxidation were formed.

The oxygen partial pressure at this location in the combustor was typical for 10–20 percent excess air combustion conditions, that is, rapidly oscillating between 10^{-13} and 10^{-8} atm.

Summary

The results of the corrosion tests described were intended to provide information on the susceptibility to corrosion attack of the four candidate heat exchanger alloys in the fireside environment of a coal-fired fluidized-bed combustor and cannot be used with any confidence to predict corrosion rates. It should also be noted that the combustion conditions used in these three tests were not normal operating conditions, but were designed to evaluate materials behavior in specific environments which might be encountered in some regimes of fluidized-bed operation.

All four alloys appeared susceptible to sulfidation related accelerated attack to some degree. In the two test runs which employed nominally oxidizing conditions of 10–20 percent excess air, the freely-exposed surfaces in general exhibited protective oxidation behavior. Sulfidation occurred in the areas where the specimens were overlapped with others, and at the upstream edge, areas where thick deposits of bed material or erosion damage, respectively, might be expected. Whether these deposits were coal ash (test no. 1) or CaO/CaSO$_4$ (test no. 3) did not materially affect the susceptibility to sulfidation.

The combustion of char was intended to eliminate the plume of combusting volatiles, with its associated extremes of oxygen deficiency, expected to emanate from coal entering the bed. No noticeable differences occurred in the local oxygen partial pressure near the exposed specimens, however, though the free surface of one alloy, Nimonic 86, showed signs of localized sulfidation in the char run, which were absent in the coal run in test number 1.

In the substoichiometric run massive sulfidation attack was initiated on the free surfaces of Inconel 617 and 618E but, surprisingly, not on Nimonic 86 nor Incoloy 800H. While the local oxygen partial pressure near the specimens fluctuated less than under normal combustion conditions, the minimum levels were similar and, from a thermodynamic viewpoint, not low enough to initiate massive sulfidation attack unless the expected protective oxide scales were damaged or removed. It is difficult to envisage significant differences in erosion conditions arising between normal and substoichiometric combustion, unless the physical nature of the bed material is greatly changed. It is possible that under substoichiometric conditions, calcium sulfide may exist as a possibly transient species in a limestone bed. If this were included in the deposits formed on the alloy surfaces, it might constitute a local source of sulfur, or a possible reactant with the otherwise protective Cr$_2$O$_3$ scale.

Of the four alloys investigated, Inconel 800H with the lowest high-temperature strength capabilities, was the least attacked, although it was extensively sulfidized in the overlapped region in the substoichiometric run. The three nickel-base alloys were very susceptible to sulfidation in the overlapped regions under all three conditions examined. Substoichiometric combustion conditions led to comprehensive sulfidation of Inconel 617, and to massive localized sulfidation of Inconel 618E. Nimonic 86 appeared to be initiating catastrophic sulfidation conditions in the substoichiometric and char combustion runs, which suggests extreme sensitivity to the local environment, hence to location in the bed.

Conclusions

Conclusions which appear justified from this investigation are:

(*i*) The high-strength nickel-base alloys, Inconel 617 and 618E, are extremely susceptible to accelerated sulfidation attack at 1650°F (1172 K) in the bed of a fluidized-bed combustor where conditions of substoichiometric combustion are encountered. These alloys must be protected by a cladding or coating of oxidation/sulfidation-resistant alloy.

(*ii*) Nimonic 86 exhibited less susceptibility to sulfidation than the Inconels, but showed signs of initiating massive sulfide-related attack locally over its surface, and could not be used with confidence in this environment without a protective cladding or coating.

(*iii*) Incoloy 800 appeared to be fairly resistant when freely exposed but was sulfidized to a slight extent on leading edges.

(*iv*) All four alloys were susceptible to rapid sulfidation-related attack in occluded or stagnant locations, which are unavoidable in real systems.

Acknowledgments

The authors would like to express their appreciation to Dr. J. Stringer of EPRI for his permission to participate in the fluidized-bed combustor tests at CRE. The alloys were evaluated as part of a DOE program, Contract DE-AC01-80-ET15020, which is addressing the problems associated with coal-fired closed-cycle gas turbine systems; Carey Kinney is the program monitor.

References

1 Stambler, I., *Gas Turbine World*, Vol. 10, No. 2, 1980, p. 20.

2 Campbell, J., Hastings, G. A., and Holt, C. F., "Coal-Fired Heaters for CCGT Cogeneration Service," ASME Paper No. 81-GT-212, presented at the 26th International Gas Turbine Conference, Houston, 1981.

3 Rogers, W. A., Page, A. J., and LaNauze, R. D., "The Corrosion Performance of Heat Exchanger Tube Alloys in Fluidized Combustion Systems," 6th European Corrosion Congress, London, 1977.

4 Moskowitz, S., "Recent Developments in PFB Technology, 5th Annual Conference on Materials for Coal Conversion and Utilization, Gaithersburg, 1980.

5 Minchener, A. J., et al., "Materials Problems in Fluidized-Bed Combustion Systems," Report No. EPRI CS 1853 1981.

6 Cooke, M. J., Cutler, A. J. B., and Raask, E., *J. Inst. Fuel*, Vol. 45, No. 373, 1972, p. 153.

7 Lloyd, D., and Minchener, A. J., "Corrosion Within Coal-Fired Fluidized-Bed Combustion Systems," 6th Annual Conference on Materials for Coal Conversion and Utilization, Gaithersburg, 1981.

8 Gulbransen, E. A., and Jansson, S. A., "General Concepts of Oxidation and Sulfidation Reactions, A Thermochemical Approach," *High-Temperature Metallic Corrosion of Sulfur and Its Compounds*, edited by Z. A. Foroulis, The Electrochem. Soc., 1970.

9 Rapp, R. A., "High-Temperature Gaseous Corrosion of Metals in Mixed Environments," *Materials Problems and Research Opportunities in Coal Combustion*, NSF/OCR Workshop, Columbus, Ohio, 1974.

WHAT'S NEW IN HEAT EXCHANGER DESIGN

Stiff competition has spurred manufacturers to offer more efficient, cost-effective units. Design concepts are surveyed for shell-and-tube, plate, extended surface, and regenerative heat exchangers

R.K. SHAH
Technical Director of Research
Harrison Radiator Division
General Motors Corp.
Lockport, N.Y.

The increasing need for energy conservation, use of energy recovery, and development of new energy sources have engendered many new advances in heat exchanger technology in the last decade. These include an extended range of operating conditions, a variety of additional materials, including plastic, and significantly advanced manufacturing techniques. The use of compact heat exchangers (usually compact plate-fin, tube-fin, and regenerative) has also become widespread [1]. Recent advances in design theory and analysis methods can be found in [2, 3].

In this article, new developments are discussed for the following major types of heat exchangers: shell-and-tube, plate, extended surface (plate-fin and tube-fin), and regenerative. Emphasis is placed primarily on developments that have led to commercial single-phase applications.

SHELL-AND-TUBE EXCHANGERS

Discussion will focus on nuclear heat exchangers, tube bundle design to eliminate flow-induced vibrations, leak-free tube-to-tubesheet joints suitable for mass production, and a variety of low-finned tubing. Since 1958 no new shell configuration has been reported by the Tubular Exchanger Manufacturers Association (TEMA) [4]. In 1978, however, TEMA did introduce a new designation for the crossflow E shell as X shell, to recognize the pure crossflow arrangement.

Nuclear Nearly all nuclear heat exchangers are vertically oriented. The heating fluid flows down and the coolant flows up, promoting natural circulation of the flow. Different tube shapes are used in reducing thermal stresses among the shell, tubes, and tubesheets in a nonremovable tube bundle. If straight tubes are used for design simplicity and reduced cost, an expansion joint or bellows is used in the central downcomer or in the shell.

Different design features are incorporated depending on the type of reactor. In pressurized water/heavy water reactors, the steam generators are either of U-tube recirculation or straight-tube, once-through type. High-temperature gas reactors usually employ helical water/steam coils. The heat exchangers in liquid metal fast breeder reactors are normally of single-pass, counterflow types, with tube configurations such as straight, bent, hockey stick, J shape, and helical [5, 6]. The internals of a liquid sodium-to-liquid

Fig. 1 A liquid sodium-to-liquid sodium breeder heat exchanger with bent tubes. The exchanger is operating in the U.S. Dept. of Energy's Fast Flux Test Facility, Richland, Wash. (Courtesy Foster Wheeler Energy Corp.)

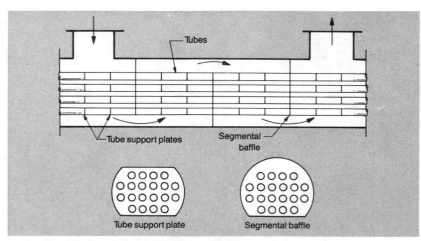

Fig. 2 A no-tubes-in-window segmental baffle exchanger.

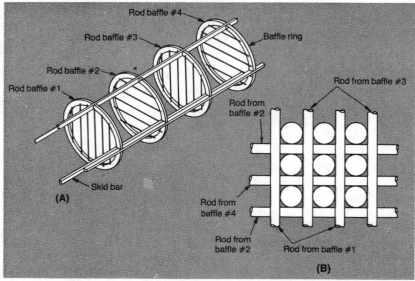

Fig. 3 RODbaffle heat exchanger: (A) Four rod baffles held by skid bars (no tubes shown), (B) Square layout of tubes with rods.

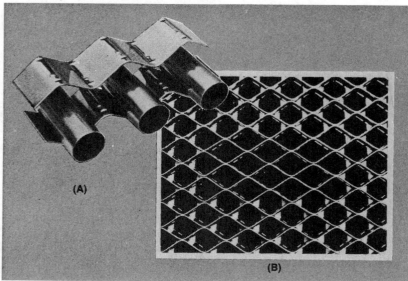

Fig. 4 NESTS® assembly baffles: (A) NESTS elements with tubes. (B) Close-up view with selectively positioned baffle tabs (no tubes are shown).

breeder heat exchanger with bent tubes are shown in Fig. 1.

In order to prevent a possible sodium-water reaction due to leaks between shellside and tubeside fluids, some straight-tube breeder-steam generators have double-walled (duplex) tubes, with inner and outer walls in good contact [7]. Similar double-tube designs have been used in solar heat exchangers and automotive heat exchangers that use exhaust gas.

Vibration Flow-induced vibration has become an important problem in shell-and-tube exchangers. Due to cost cutting measures and the quest for high thermal performance, high fluid flow rates on the shell side are sometimes accommodated in an undersized shell or an inadequate baffle arrangement. Problems occur in two forms: tube vibration, which can lead to excessive wear of the tube walls and fluid leaks between the tube and shell sides, or acoustic vibration.

In segmentally baffled exchangers, the flow-induced vibration is usually associated with the tubes in the baffle window zone. The unsupported length of these tubes is typically twice as long as tubes in the crossflow zone (the central portion of the tube bundle). As a result, they have a much lower natural frequency and a higher tendency for vibration excitation. A design that virtually eliminates flow-induced vibration calls for the removal of tubes in the window zone, and perhaps the addition of support plates (baffles with windows on both sides) between baffles to reduce the unsupported span of the remaining tubes. Such a no-tubes-in-window baffle exchanger is shown in Fig. 2.

Tube vibration has been shown to result more frequently from flow across the tubes rather than parallel to them. Therefore, a recent approach has been substitution of tube-support baffles with a grid-support structure. In the RODbaffle, developed by Phillips Petroleum Co. [8, 9], the grid is made of parallel rods mounted on support rings (Fig. 3). The rings are closely spaced, at 15 to 30 cm (6–12 in.), and the pattern is repeated every four rings. Positive support is provided to all tubes along the length of the bundle, and the shellside fluid is allowed to flow parallel to the tubes. In addition to eliminating any significant tube vibration, this extremely stiff bundle reduces pressure drop and fouling rates of the shellside fluid.

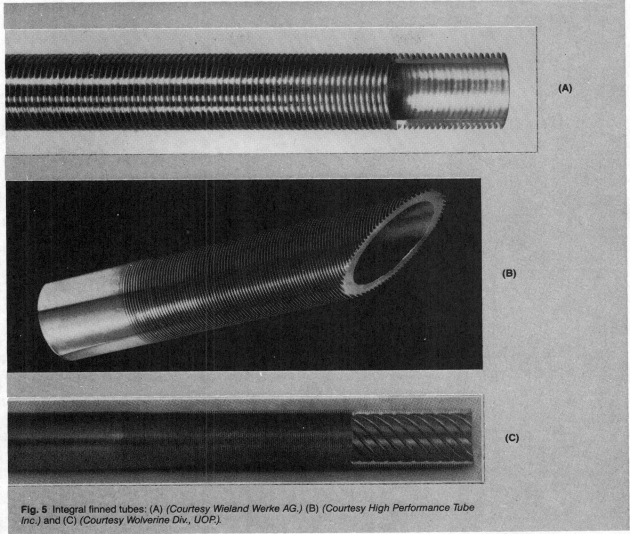

Fig. 5 Integral finned tubes: (A) *(Courtesy Wieland Werke AG.)* (B) *(Courtesy High Performance Tube Inc.)* and (C) *(Courtesy Wolverine Div., UOP.)*.

Another support arrangement that counters potential tube vibration problems is made of formed strips of steel. In Nests (Neoteric Endo-Stratiformed Tube Supports), developed by Ecolaire Heat Transfer Co. [10], preformed strips are put together and locked into a honeycomb structure (Fig. 4). Conventional baffles are replaced and all tubes are supported at each Nests. The flow is parallel to the tubes with Nests baffles. Nests can be used with industrially common tight pitch ratios (1.25, for example) in the triangular tubefield layout.

In acoustic vibration there is usually no mechanical damage to the tubes or the shell. However, the noise may be so intense (up to 150 dB) as to make the vicinity uncomfortable for workers. Acoustic vibration can be prevented in bundle design or by modification. Deresonating baffles can be designed to raise the acoustic frequency

of the shell higher than any flow-induced frequency. To correct an existing problem, one successful approach is the selective removal of a small number of tubes to inhibit the formation of acoustic standing waves [11].

Manufacture In addition to mechanical rolling, two manufacturing methods have been developed recently for a tube-to-tubesheet joint of uniform quality and suitable for mass production: hydraulic expansion and impact welding [6, 9, 12]. In hydraulic expansion, the tube end in the tubesheet is constrained by an O-ring seal on the inside of the tube, and then pressurized to about 115 percent of the tube yield stress with a fluid (usually demineralized or distilled water).

The currently available intensifier can provide water pressures up to 483 MPa (70,000 psig) [12]. One advantage of this method over mechani-

cal rolling is that there is no limit to the tube length that can be expanded at one time, and noncircularity in the tube or tube hole is accommodated.

In the impact (explosion) welding process, controlled detonation of a suitably positioned explosive charge produces a high-velocity impact on the surface to be joined. The resulting bond between the tube and tubesheet is a metallurgical bond. Impact welding is used to obtain a good bond when the operating pressure on the shell or tube side is over 14 MPa (2000 psig), or to join materials that are not weldable.

Low Finned Tubes Integral low finned tubes for shell-and-tube exchangers increase the surface area on the shell side two or three times over bare tubes. Moore [13] reports that, contrary to common belief, low finned tubes actually resist fouling, and usually remain on stream two or

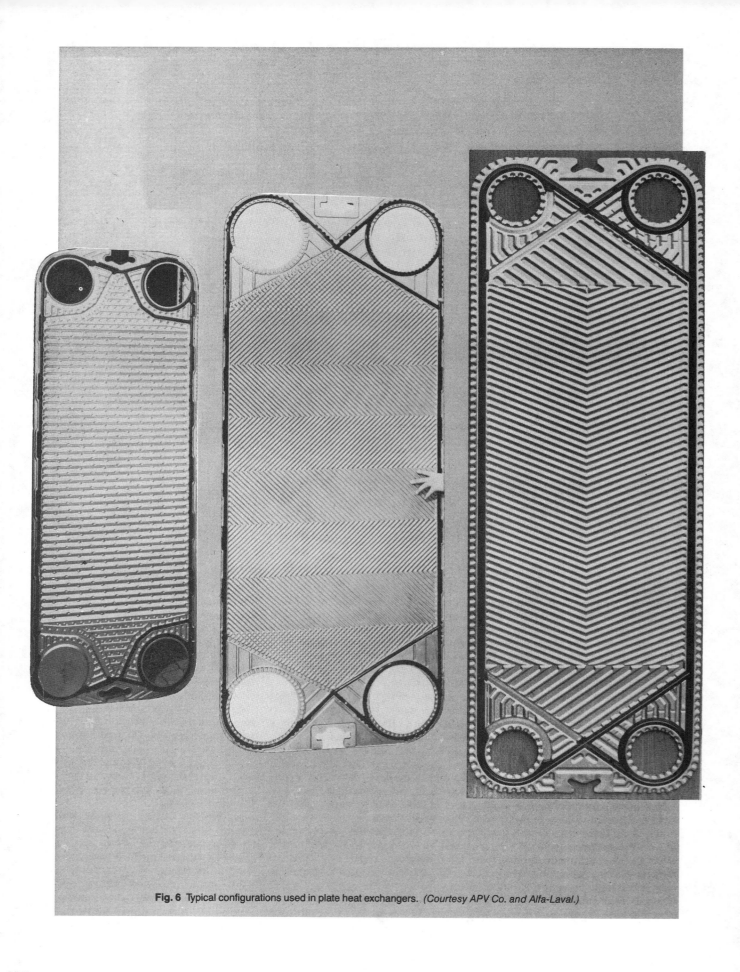

Fig. 6 Typical configurations used in plate heat exchangers. *(Courtesy APV Co. and Alfa-Laval.)*

three times longer than bare tubes in comparable conditions. He also notes that some finned tube bundles can be cleaned in 1/10 the time it takes for a bare tube bundle.

The technology of integral finned tubing has advanced considerably in the past decade. Exchangers are available in fin densities up to 1575 fins/m (40 fins/in.), fin heights from 0.76 to 1.55 mm (0.03 to 0.06 in.), and tube outside diameters from 9.5 to 25.4 mm (0.375 to 1 in.). An example is shown in Fig. 5A. Fine-Fin tubing, developed by High Performance Tube Inc., is an integral finning with hard-to-work materials (such as titanium, Inconel, Hastelloy, etc.). The fin outside diameter is the same as the tube plain-end diameter as shown in Fig. 5B.

Integral low-finned tubing with internal enhancement, such as the Turbo-Chil finned tube by Wolverine Division of UOP, is a recent arrival. The Turbo-Chil-enhanced tubing is prominent in the refrigeration industry in applications with condensation or vaporization outside the tubes and single-phase fluid inside the tubes. Other high-performance tubings are GEWA-T by Wieland-Werke AG of W. Germany, Thermoexcel-E, C by Hitachi Corp., Japan, and High Flux tubing by Union Carbide Corp. These and many other developments in enhanced boiling and condensing surfaces are discussed in [14–16].

There is considerable automation in the manufacture of shell-and-tube exchanger components, especially in welding and drilling. The size of the exchanger is increasing with the advancing technology, limited only by weight (the largest weighs about 16 t) or transportation constraints. Shell diameters of 4.2 m (13.75 ft) or tube lengths of 18.2 m (60 ft) have been seen.

A family of ferritic and austenitic stainless steel has been introduced for pit corrosion resistance. Other materials have been introduced for high pressure, high temperature, and special applications.

A new development in metal tubular exchangers (without high-pressure shell) is the use of bayonet tubes—tubes within tubes—to alleviate thermal stresses at the tubesheet during waste heat recovery at very high temperatures [17]. A bank of bayonet tubes is suspended from the air chambers into the flue gas stream. Usually the flue gas flows normally to the tubes. Because each tube is attached only at

the top of the chamber, it expands or contracts with temperature changes. The Hazen metallic recuperator, for example, manufactured by C-E Air Preheater, operates at 1260°C (2300°F) flue gas temperature.

PLATE EXCHANGERS

Plate heat exchangers (also known as plate-and-frame exchangers) have been commonly used in dairy, beverage, food, pharmaceutical, and other industries where hygiene requires frequent cleaning of the exchanger. However, increasing use is being made of this exchanger in process industries where stainless steel, titanium, and other expensive materials are necessary and the operating pressure and temperature ranges are moderate.

A plate exchanger consists of a stack of thin rectangular metal plates separated and sealed by gaskets. In an overall counterflow arrangement, the two fluids flow in alternate channels. Each plate is stamped out of sheet metal with some form of corrugation (Fig. 6)—over 60 different corrugation patterns are now in use. Patterns such as herringbone, chevron, and others have many contact points between adjacent plates to provide structural rigidity. This allows the plates to be thinner and increases maximum allowable pressures.

In order to obtain the required heat transfer using the available pressure drop on each side, plate mixing was introduced in the 1970s by Alfa-Laval. In this type of exchanger, plates of different corrugation patterns are mixed in one frame to produce the exact thermal and hydraulic performance required. Because the plates are made by stamping dies, the size of plate exchangers cannot be varied as easily as shell-and-tube exchangers.

As an alternative to plate mixing, a given die is modified to produce plates of two or three different lengths. By varying the length and number of plates, the required duty can be met within the allowed pressure drop, and possible problems of nonuniform heating or flow maldistribution can be alleviated.

The largest plate heat exchangers have 2.4 m × 1.2 m (8 ft × 4 ft) plates with a maximum surface area in an exchanger of about 2200 m² (24,000 ft²). The smallest plate exchanger has 250 mm × 100 mm (10 in. × 4 in.) plates with a minimum surface area in an exchanger of about 0.032 m² (50 in.²) [18].

Plates can be made from any material that can be pressed to the desired corrugation patterns—frequently from stainless steel and alloy materials. Recently, high alloy steel (254 SMO) has been used as a replacement for titanium in applications involving chlorides, sea water, etc. Tantalum has been introduced for very corrosive applications.

The gasket is the weakest component in the plate exchanger in high operating temperatures and pressures. Extensive research is under way to develop an ideal gasket material with a minimum of compression set and that resists many fluids and chemicals. Elastomer gaskets are limited to operating temperatures up to about 155°C (310°F); recently available Du Pont Kalrez elastomer, although expensive, can operate up to about 230°C (450°F). An asbestos gasket can be used with a few types of plates for operating temperatures up to 285°C (550°F).

In most cases cement is used between gasket and plate to hold the gasket in place and seal the unit. However, snap-on gaskets have been introduced in the past two years in the food industries, where daily manual cleaning is required and gasket replacement is high.

The frames of exchangers generally are very thick, in order to contain the common operating pressures of about 1.0 MPa (150 psig) (up to 2.5 MPa [360 psig] for some special exchangers). The weakest areas in the exchanger are the inlet and outlet ports. The recent trend for hygienic plate exchangers is to strengthen the port area with reinforcement braces and reduce the thickness of the frames.

EXTENDED SURFACE EXCHANGERS

In an extended surface exchanger, fins are used on or in plates or tubes; common types of such exchangers are known as plate-fin and tube-fin.

Plate-Fin Exchangers Much work has been done in header design for uniform flow distribution, new materials for corrosion and high-temperature applications, and fin geometries with a variety of interruptions. Other areas of research are formed tube designs for containing one-side fins and fluids, different plate-to-fin joining techniques, and designs to take care of thermal stresses. Major developments in plate-fin exchangers made of metal, ceramic, and paper are discussed here.

Advances in metal plate-fin exchang-

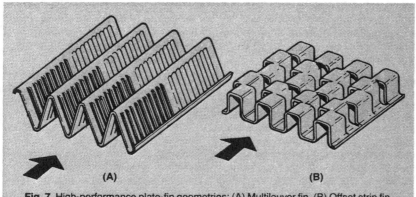

Fig. 7 High-performance plate-fin geometries: (A) Multilouver fin. (B) Offset strip fin.

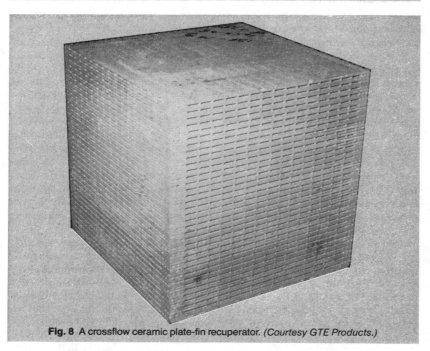

Fig. 8 A crossflow ceramic plate-fin recuperator. *(Courtesy GTE Products.)*

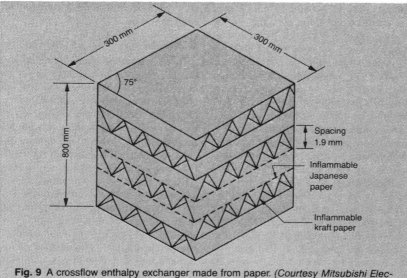

Fig. 9 A crossflow enthalpy exchanger made from paper. *(Courtesy Mitsubishi Electric Corp.)*

er manufacturing have been in two areas: very compact surfaces, and brazing techniques and materials for high-temperature applications. Two high-performance compact geometries are multilouver fins and offset strip fins (Fig. 7). Gränges Metallverken Co. of Sweden has developed a highly compact multilouver copper fin with a fin density of about 2000 fins/m (50.5 fins/in.) and corresponding heat transfer surface area density of 4400 m²/m³ (1341 ft²/ft³).

AiResearch has developed an aluminum offset strip fin that has fin density of about 1450 fins/m (36.85 fins/in.) and corresponding heat transfer surface area density of 5650 m²/m³ (1722 ft²/ft³).

Furnace brazing of relatively large size stainless steel compact plate-fin exchangers is another new technology. Plate-fin exchangers with 2.8 m × 1.0 m (110 in. × 40 in.) plates and a stack height of 1.5 m (5 ft) have been successfully brazed by the Harrison Radiator Division of GM. Furnace-brazed metal recuperators 56.5 cm × 43.9 cm × 28.2 cm (22.24 in. × 17.28 in. × 11.10 in.) for waste heat recovery at a continuous operating temperature of 816°C (1500°F) have been released by AiResearch Mfg. Co.

For high-temperature applications, in excess of 800°C (1472°F), GTE Products has developed a crossflow plate-fin recuperator made from magnesium aluminum silicate or cordierite (Fig. 8). The basic core measures 0.3 m × 0.3 m × 0.46 m (12 in. × 12 in. × 18 in.), has rectangular passages with an aspect ratio of 4:1, fin density of 60 fins/m (1.5 fins/in.), and surface area density of 440 m²/m³ (135 ft²/ft³). Other ceramic plate-fin exchangers that are not commercially available are described in [19]. A ceramic counterflow recuperator for Stirling and gas turbine engines is under development by the Coors Porcelain Co.

For energy recovery of ventilation air, plate-fin exchangers made from hygroscopic (permeable to humidity) paper have been developed by Mitsubishi Electric Corp. (Fig. 9). The paper is used as a prime surface; triangular fins, made of special kraft papers, are used as spacers. The maximum operating temperature is limited to 50°C (122°F).

Tube-Fin Exchangers There are two types of tube-fin exchangers: continuous fins on round or flat tubes, and individually finned tubes. Usually, liquid is on the tube side and gas (air) on the fin side.

The use of continuous fin-and-tube metal exchangers has become widespread for energy conservation. The bond between fin and tube is made by mechanically or hydraulically expanding the tube against the fin, and formation of this mechanical bond requires very small energy consumption compared to the energy required to solder, braze, or weld the fins to the tubes. Furthermore, the advanced tube-fin surfaces can offer reduced weight and envelope sizing. Some recent geometries are shown in Fig. 10.

A large number of interrupted fin geometries have been developed for individually finned metal tubes. These fins are usually tension wound, adhesive bonded, brazed, or welded. Tubes can be externally and internally finned. Figure 11 shows internally finned tubes that have been swaged or mechanically press-fit.

The art of extruding tubes with internal or external fins has been advanced considerably in the past decade. An extruded rectangular tube with integral skyve fins is shown in Fig. 12. When the tubes of individually finned tubes are replaced by heat pipes, and the fin side is divided into two compartments by a splitter plate, the resultant exchanger is referred to as a heat pipe exchanger. Applications of this exchanger have mushroomed in industrial waste heat recovery from exhaust gases [20].

Ceramic individually finned tubes have been developed for 1100 to 1200°C (2000–2200°F) operating temperatures. Tubes with both internal and external fins are available from Hague International, and tubes with only external fins are manufactured by the Norton Co. [3]. Ceramic heat pipe heat exchangers are also available.

ROTARY REGENERATORS

Rotary regenerators are used in gas-to-gas heat exchangers for waste heat recovery where there are relatively low operating pressure differentials.

Vehicular metal disk regenerators contain disks and seals that operate successfully at gas inlet temperatures of about 788°C (1450°F). Harrison Radiator has developed stainless steel disk regenerators that are durable for automotive life.

Ceramic disk regenerators have been developed by Corning Glass Works,

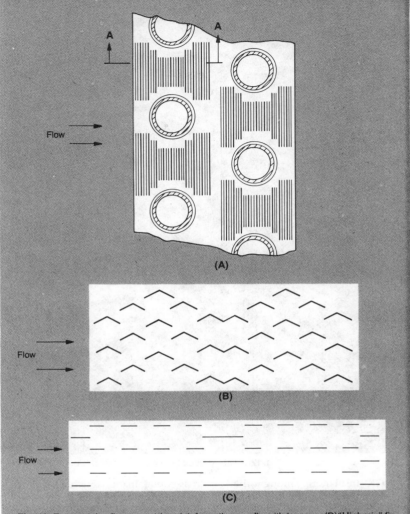

Fig. 10 Recent tube-fin geometries: (a) A continuous fin with louvers. (B)"Highmix" fin from Sundstrand Heat Transfer, Mich. (C) Strip fin from Mori and Nakayama of Japan *(From p. 8 of Ref. [1]).* (B) and (C) represent cross section A-A of (A).

Fig. 11 Internally finned tubes. *(Courtesy Forge-Fin Div., Noranda Metal Ind.)*

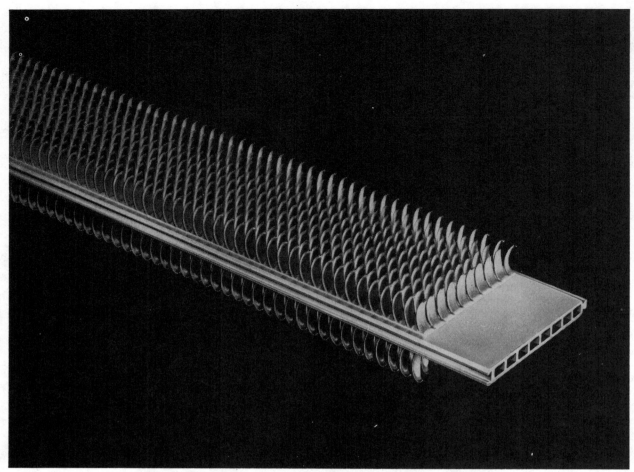

Fig. 12 A flat extruded tube with skyve fins. *(Courtesy Peerless of America Inc.)*

and NGK Insulators, among others. The Corning disk is made from aluminum silicate and has been successfully operated at 1000°C (1832°F) gas inlet temperature for over 3000 hours. The NGK disk is made from magnesium-aluminum-silicate (MAS) by an extrusion process that produces near-perfect triangular passages (Fig. 13). The thick wall (0.168 mm, 0.0066 in.) NGK disk has been operated at 1000°C (1832°F) gas inlet temperature for 236 hours [3].

The development of regenerators for ventilation air (heat wheels) was pioneered in Sweden about 20 years ago, principally by AB Carl Munters. The matrix form developed consists of alternate layers of corrugated and flat sheets wound to form a disk. Ventilation regenerators with such matrices are manufactured in several countries, including the U.S., U.K., and Japan.

In order to be cost effective at low temperature differentials, heat wheels must be efficient. Also, they must be fire resistant, nonbiodegradable, easily cleanable, and have very little cross-contamination. Such a regenerator, made up of a thin polyester film wound on spacers supported by channel section spokes, has been released by CSIRO, Australia (Fig. 14). Rotary regenerators for ventilation air are also made by Sharp, Hitachi and other Japanese firms. The Sharp unit has a form similar to CSIRO's with sheets made from textiles embedded in a plastic base. The sheets are charged with a humidity absorber and the regenerator acts as an enthalpy exchanger.

PLASTICS

The use of plastics for heat exchanger surfaces has increased substantially in areas involving corrosive and fouling fluids at moderate temperatures. All plastic exchangers on the market are prime surface (no fins) exchangers. Teflon tube exchangers are the most common, and are available from Du Pont in tube diameters of 2.5 to 6.3 mm (0.10 to 0.25 in.). The operating temperature is limited from 40°C (104°F) at 825 kPa (120 psig) to 150°C (300°F) at 300 kPa (44 psig). Because Teflon is inert to most corrosive fluids within these temperature and pressure limits, Teflon tube exchangers are generally used in place of exchangers with more exotic materials. A new material, Teflon PFA, extends the operating temperature to 200°C (392°F) at 350 kPa (50 psig).

Polypropylene tubes, used in tubular solar water heaters made by Fafco Inc., can operate at 70°C (158°F) at 200 kPa (30 psig). These have good resistance to sunlight and weather exposure. Polymeric capillary tubes with tube o.d. of 0.5 to 2.5 mm (0.02 to 0.10 in.) are made by Enka bv of Holland. Such tubes, made from a number of different polymers, have burst pressures of over 120 bars, operating temperatures up to 175°C (347°F), and good resistance to weather, salt solutions, and diluted acid or base solutions. These tubes are used in mat, wound package, bundle, and stack heat exchangers.

A new material, Ryton PPS, has been introduced by Phillips Petroleum Co.

Fig. 13 A section of NGK extruded MAS ceramic regenerator core.

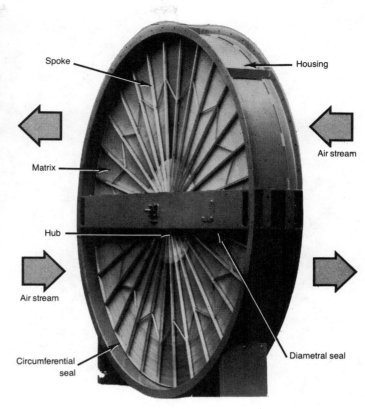

Spoke

Housing

Air stream

Matrix

Hub

Air stream

Diametral seal

Circumferential seal

Fig. 14 A rotary regenerator made from polyester film. *(Courtesy CSIRO, Australia.)*

The material has high strength, can be used at 180°C (356°F) at 1.0 MPa (150 psig), and is immune to most corrosive fluids except for concentrated nitric and hydrochloric acids. The material is being considered for plate heat exchangers.

Plastic tubes and plastic sheets that form narrow parallel passages are under study for dry cooling towers [21]. The durability of plastic-coated metal tubes against highly concentrated sulfuric acid has been demonstrated by Argonne National Laboratory [22]. Prototype testing and commercial application of these tubes are the next steps. ▣

REFERENCES

1 Shah, R.K., McDonald, C.F., and Howard, C.P., eds., *Compact Heat Exchangers,* Book No. G00183, HTD-Vol. 10. New York: ASME, 1980.

2 Taborek, J., "Evolution of Heat Exchanger Design Techniques," *Heat Transfer Engineering,* Vol. 1, No. 2, 1979, pp. 15–29.

3 Shah, R.K., "Advances in Compact Heat Exchanger Technology and Design Theory," *Heat Transfer 1982,* Vol. 1, 1982, pp. 123–42.

4 Tubular Exchanger Manufacturers Assn., *Standards of TEMA,* 6th ed., New York, 1978.

5 Cho, S.M., and Beaver, T.R., "LMFBR Inter-mediate Heat Exchanger Experience," ASME Paper No. 83-JPGC-NE-12, 1983.

6 Singh, K.P., and Soler, A.I., *Mechanical Design of Heat Exchangers and Pressure Vessel Components.* Cherry Hill, N.J.: Arcturus Publishers, 1984.

7 Adkins, C.R., Bongaards, D.J., and Smith, P.G., "Double-Wall Tube Steam Generator for Breeder Nuclear Plants," ASME Paper No. 81-JPGC-NE-2, 1981.

8 Gentry, C.C., Young, R.K., and Small, W.M., "RODbaffle Heat Exchanger Thermal-Hydraulic Predictive Methods," *Heat Transfer 1982,* Vol. 6, 1982, pp. 197–202.

9 Apblett, W.R., Jr., ed., *Shell and Tube Exchangers.* Metals Park, Ohio: American Society for Metals, 1982, pp. 239–49, 341–45, 389–409.

10 Boyer, R.C., and Pase, G.K., "The Energy-Saving NESTS Concept," *Heat Transfer Engineering,* Vol. 2, No. 1, 1980, pp. 19–27.

11 Halle, H., Chenoweth, J.M., and Wambsganss, M.W., "DOE/ANL/HTRI Heat Exchanger Tube Vibration Data Bank (Addendum 1)," Argonne Report ANL-CT-80-3, Jan. 1981, pp. 3–27.

12 Yokell, S., "Hydroexpanding: The Current State of the Art," ASME Paper No. 82-JPGC-Pwr-1, 1982.

13 Moore, J.A., "Fintubes Foil Fouling for Scaling Services," *Chemical Processing,* Aug. 1974, pp. 8–11.

14 Webb, R.L., "The Evolution of Enhanced Surface Geometries for Nucleate Boiling," *Heat Transfer Engineering,* Vol. 2, Nos. 3–4, 1981, pp. 46–69.

15 ———, "The Use of Enhanced Surface Geometries in Condenser: An Overview," in *Power Condenser Heat Transfer Technology,* P.J. Marto and R.H. Nunn, eds. Washington, D.C.: Hemisphere Pub. Corp., 1981, pp. 287–324.

16 ———, and Bergles, A.E., "Heat Transfer Enhancement: Second Generation Technology," *Mechanical Engineering,* June 1983, pp. 60–7.

17 Li, C.H., "Analytical Solution of the Heat Transfer Equation for a Bayonet Tube Heat Exchanger," ASME Paper No. 81-WA/NE-3, 1981.

18 Bell, K.J., "Plate Heat Exchangers," in *Heat Exchangers: Thermal-Hydraulic Fundamentals and Design,* S. Kakaç, A.E. Bergles, and F. Mayinger, eds. Washington, D.C.: Hemisphere Pub. Corp., 1981, pp. 165–75.

19 McDonald, C.F., "The Role of the Ceramic Heat Exchanger in Energy and Resource Conservation," *J. Engineering for Power,* Vol. 102, 1980, pp. 303–15.

20 Reay, D.A., *Heat Recovery Systems.* London: E. & F.N. Spon, Ltd., 1979.

21 Guyer, E.C., "Non-Metallic Heat Exchangers: A Survey of Current and Potential Designs for Dry Cooling," Final Report, EPRI Contract No. RP-1260-29, Dynatech R/D Co., Cambridge, Mass., June 1983.

22 Roach, P.D., and Holtz, R.E., "Using Plastics in Waste Heat Recovery," *Proceedings of the ASME-JSME Joint Thermal Engineering Conference,* Vol. 2, Book No. I00158. New York: ASME, 1983, pp. 409–12.

Pressure Vessels

Key elements in pressure-vessel design are specifying and locating the many nozzles through which process ingredients enter and resulting product is removed.

PRESSURE VESSEL PRIMER

Pressure vessels play a key role in a wide range of industrial activities from energy generation to the processing of minerals and chemicals. As these fields become more important, the "tight" design of vessels becomes more critical to process efficiency.

H. C. HOFFMAN
Staff Representative
Committee of Steel Plate Producers
American Iron and Steel Institute
Washington, D.C.

CALL them reactors, accumulators, surge tanks, autoclaves, or any of a dozen other names; they are all pressure vessels that contain air, gas, liquid, or a slurry for some processing operation. These components are used in such varied industries as food processing, mining, paper, petroleum, liquor, pharmaceutical, brewing, chemical, fertilizer, and power generating.

At one time, pressure-vessel design was concerned primarily with the simple containment of hydrostatic forces. Today, however, the economics of a growing number of industrial processes—particularly those associated with energy generation and production of critical materials—depend heavily on cost-effective design of pressure vessels. The feasibility of such processes as coal gasification, for example, hinges in large measure on the design of pressure-containing process equipment. Thermodynamics and design-to-cost now are ranked as high as stress analysis. What's more, because of the closer control demanded by many energy-related processes, pressure-vessel walls are more often penetrated by a variety of monitoring, inspection, and feed ports—all of

For pressure-vessel contents that require heating or cooling, a dimple jacket (left, above) and half-pipe coil (right, above) provide space between the jacket and vessel wall to hold steam, hot oil, or a cooling fluid. When external jackets are not sufficient for process heating or cooling needs, internal coils provide additional heat-transfer surface. Two commonly used arrangements are helical pipe coils (left, below) and plate-type coils (right, below). Both are used in conjunction with outer jackets to provide maximum heat transfer.

which bring complications to the design and analysis.

The design of a pressure vessel begins with basic steps: the determination of a capacity and shape, along with consideration of operating temperature and pressure and the service environment. From there, the design becomes more complex and deals with a multitude of factors such as selection of nozzles and access openings, instrumentation requirements, and provisions for heating or cooling the process ingredients.

Sizes and Shapes

Basically, a pressure vessel is a sealed cylindrical container for holding air, gas, or liquid pressures from full vacuum to 50,000 psi. Most vessels are fabricated from carbon or alloy-steel plate.

Custom-fabricated steel-plate vessels normally range in diameter from 25 in. up, for a prosaic reason: Standard steel pipe, which can be fabricated into a pressure vessel, is made in sizes up to 24 in. At the other extreme, most factory-built pressure vessels are limited to a 14-ft diameter because this size is the largest that can be shipped in one piece on a railroad flat car. Larger-diameter vessels are transported by barge or shipped in sections by rail, then assembled at the plant site.

The ratio of vessel length to diameter can vary from 1:1 to as great as 15:1. Typical vessels with high L/D ratios are main fractionators, stripper columns, scrubber columns, flash towers, distillers, and towers for drying and other treatments.

Wall thickness, of course, depends upon the pressure that must be contained. The thinnest allowable (by ASME Pressure Vessel Code) vessels, at 1/16 in. thick, are used as low-pressure accumulators and dryers. Maximum wall thickness is 14 in. for solid plate;

layered-construction vessels are made with wall thicknesses to 15 in. and greater. These heavy-wall vessels are usually used in processes having some reaction taking place within. Those with thicknesses of 15 in. and greater are often used in conjunction with presses requiring high pressure for effective operation. Vessels for coal gasification require walls 24 in. thick or more.

Volumetric capacity normally is established in multiples of 1,000 gallons. Vessels for holding liquid helium (usually made from 304 stainless steel) must withstand temperatures as low as −425°F. The highest temperature allowed by the ASME Code for 300 series steels is 1,500°F. The maximum for carbon steel is 1,000°F, and for Inconel 629, 1,600°F.

The cylindrical section of a pressure vessel is capped at both ends by various types of heads. The most commonly used shape is ellipsoidal, which is particularly suited for vessels whose contents require agitation. Thickness of ellipsoidal heads is usually the same as that of the shell. Toroconical heads are usually used on tank bottoms to facilitate drainage. Hemispherical heads provide a substantial increase in volume, making them frequent choices for storage vessels. They also can affect vessel cost favorably because, in most cases, their thickness can be one-half that of the shell.

Construction Materials

About three quarters of all pressure vessels are fabricated from carbon-steel plate ranging from 0.20 to 0.30% in carbon content to provide forming and welding ease. Yield strength of these steels ranges from 30,000 to 38,000 psi, and tensile strength ranges from 55,000 to 90,000 psi.

Where an inner vessel surface of carbon steel is not

suitable—for reasons of corrosion, product contamination, or temperature limitation— vessels can be made from clad plate. These laminated composite materials can consist of as much as 95% carbon steel (for structural integrity and economy), roll-bonded to stainless steel, copper, nickel, or other metal. Explosive bonding is another joining technique used to fuse steel to such metals as aluminum, tantalum, titanium, or zirconium.

Titanium provides excellent resistance to attack in most oxidizing, neutral, and reducing environments. It retains its corrosion resistance at high temperatures, for example, in hot salt, wet chlorine, and nitric and acetic acids. Tantalum requires special procedures in welding but is equivalent to glass in corrosion resistance. Zirconium resists most mineral and organic acids and is totally resistant to alkalies. These metals are expensive and are used only when carbon or stainless steel is inadequate for the job.

In addition to providing surface advantages, clad metals often improve mechanical or physical properties. For example, the metallurgical bond between clad metals ensures excellent thermal conductivity across the metals, which helps prevent thermal-shock damage.

Auxiliary Fittings

Most pressure vessels require a number of openings to accommodate the requirements of the process involved and to provide access to the interior for inspection, cleaning, and maintenance. These may consist of hinged closures, manways, inlet and outlet nozzles and fittings for piping-in and removing process ingredients, steam vents, drains, sight gages, and thermocouple provisions. Most of these are available as standard, code-approved hardware, ready to be welded into the vessel as needed.

Many processing systems require heat to be added to or removed from the vessel. In such cases, an external steel jacket is usually built around all or a portion of the vessel, leaving a space to contain steam, hot process oil, or a cooling medium. Three types of jackets are in common use:

• Conventional, where the space between vessel and jacket is not partitioned. This jacket construction is used principally for small vessels—to about 300 gal—and for vessels whose internal pressure is more than twice the jacket pressure.

• Half-pipe coil, where halves of pipe, in helical pattern, are welded to the exterior of the vessel. These jackets are used typically for large vessels, for high-temperature service, and for liquid heat-transfer media.

• Dimple, where the steel jacket is embossed so that the raised pattern holds the heating/cooling medium. This construction, also used for large vessels—to 45,000 gal—offers the advantages of an increase in rigidity from the embossing, which permits construction from a thinner steel.

In situations where an external jacket does not provide sufficient heat transfer, internal coils can be installed to accommodate additional heating or cooling fluid. Alternatively, a tube bundle can be installed through a nozzle to introduce or remove heat, as the process requires.

Preferred Practice

Because of the critical nature of the design, fabrication, and operation of pressure vessels, manufacturers have developed guidelines to maximize efficiency and safety. The design tips listed here embody a number of these recommendations.

• To ensure a sound, safe structure, provide the vessel fabricator with complete specifications—information beyond dimensions, materials, design pressure, and operating temperature. Often missing are the vital data "extras" that can mean the difference between success and failure. These include details on start-up operating conditions, special cleanliness or safety requirements, the heat-transfer medium, details on whether the process is continuous or batch, manner of field lifting, and any special considerations that could affect design, even down to the gasketing material that will seal the nozzles.

• Incorporate a corrosion allowance—an extra thickness of plate that can be sacrificed to corrosion. For example, if the intended service life of a steel vessel is ten years and anticipated annual corrosion is 0.012 in., specify an additional ⅛-in. plate thickness.

• Use particular care in specifying plate for vessels used in low-temperature service. Choice of plate is important to avoid brittle fracture, which can cause failure of vessels meant to function at −20°F and below. Basic carbon-steel plate, SA516-70 ksi, can operate to −20°F and, with special tests, can be taken to −50°F. SA203, Grade A or B, is the choice to −90°F; SA203, Grade D or E, to −150°F; SA553 to −275°F; and SA353 to −320°F. (Designations refer to ASME Pressure Vessel Code, Section VIII, Div. 1.) Other materials, such as nickel and stainless steels, can be operated at even lower temperatures.

• Use an adequate safety factor. The ASME code dictates using a design stress limit of one-fourth of the tensile strength of the steel plate.

• Use reliable instruments to sense and control pressure and temperature. Rupture discs are available to guard against a pressure excess. These devices incorporate thin membranes that burst under a designated overload.

• Provide for necessary functional vessel components. For example, a sparger, or lance, can inject an ingredient to a specific inner area; agitators such as paddles, turbines, blades, and sweeps can stir and disperse a vessel's contents; and baffles, usually in groups of four, can be attached to the inner wall to increase agitation. A system of perforated trays within a vessel usually meets fractionation needs or allows gas to bubble up through the process liquid. Support grates can hold catalysts and other substances used in the process.

• When possible, specify maximum length or width in steel plates to minimize the number of weld seams. Reducing the amount of welding cuts fabrication costs and also permits a significant cost saving in applications subject to rigid quality control standards. Fewer welds reduce the possibility of rejection and the ensuing need for repair work.

• Do not apply heavy external loads such as those from valves, pumps, blowers, or other rotating machinery mounted directly on a vessel unless the loads are provided for in the initial design. Where vessels are to operate outdoors, wind and snow loads should be considered as well as thermal loads from changes in ambient temperatures.

• Prescribe a service-prolonging maintenance program, starting with a periodic inspection of the interior, particularly if the process ingredients are corrosive. Maintenance procedures should be tailored to specific process and material requirements be they periodic repainting, cleaning of the inner liner, flushing of heat-exchanger tubes, or other procedures necessary to maintain proper operation. MD

A New Series of Advanced 3Cr-Mo-Ni Steels for Thick Section Pressure Vessels in High Temperature and Pressure Hydrogen Service

R. O. RITCHIE, E. R. PARKER, P. N. SPENCER, and J. A. TODD

A new series of 3Cr-Mo-Ni steels has been developed for use in thick section pressure vessels, specifically for coal conversion (high temperature and high pressure hydrogen) service. The new steels rely on minor alloy modifications to commercial 2.25Cr-1Mo (ASTM A387 Grade 22 Class 2) steel. Based on evaluations in relatively small heats (55 kg), the experimental alloys, which employ additions of Cr, Ni, Mo, and V, with Mn at 0.5 wt pct and C at 0.15 wt pct, display improved properties compared to commercial steels. Specifically, they show significantly improved hardenability (i.e., fully bainitic microstructures following normalizing of 400-mm (16-in.) plates), enhanced strength (i.e., yield strengths exceeding 600 MPa), far superior hydrogen attack resistance and better Charpy V-notch impact toughness, with comparable tensile ductility, creep rupture resistance and temper embrittlement resistance. The microstructural features contributing to these improved mechanical properties are briefly discussed.

INTRODUCTION

The development of second and third generation coal conversion systems, such as proposed large-scale coal liquefaction and gasification processes, has necessitated the design of large thick-section pressure vessels.[1,2,3] Materials requirements for such vessels include weldable steels with sufficient hardenability to maintain good mechanical properties throughout plate sections up to a maximum of roughly 400 mm (16 in.). In addition, yield strengths in excess of 350 MPa are required with sufficient toughness, creep rupture, fatigue and environmental degradation resistance to withstand mechanically and environmentally hostile environments, which in certain instances involve hydrogen plus hydrogen sulfide atmospheres at temperatures of ~550 °C and hydrogen gas pressures up to 20 MPa (3000 psi).[3–6]

Historically, the material favored for hydrogen service pressure vessel construction has been 2.25Cr-1Mo steel,[6,7] such as the normalized and tempered ASTM A 387 Grade 22 Class 2. Although this steel in general has ideal mechanical properties, it does not have sufficient hardenability to produce the fully bainitic microstructures necessary to provide the desired low and elevated temperature properties in normalized plate sections of the required 300 to 400 mm (12 to 16 in.) thicknesses.[8] Moreover, there is also some doubt as to the resistance of 2.25Cr-1Mo steel to hydrogen attack under the most severe in-service conditions,[9–13] where elevated temperature and pressure gaseous hydrogen environments can result in an internal reaction between ingressed hydrogen and carbides in the steel. This leads to decarburization, cavitation and, in extreme cases, fissuring from the formation and growth of methane bubbles at interfaces such as grain boundaries.[14,15,16]

Both laboratory research and in-service experience have shown that additions of certain alloying elements, specifically carbide stabilizing elements such as vanadium, niobium, and particularly chromium and molybdenum, can have a markedly beneficial effect on hydrogen attack resistance through the precipitation of stable carbides.[17,18] However, there is some concern over the elevated temper-

R. O. RITCHIE, E. R. PARKER, and P. N. SPENCER are Professor, Professor Emeritus, and Research Engineer, respectively, in the Department of Materials Science and Mineral Engineering, University of California, Berkeley, CA 94720. J. A. TODD is Assistant Professor, Department of Materials Science, University of Southern California, Los Angeles, CA 90089. Presented at American Society for Metals WESTEC '84, March 1984, Los Angeles, CA.

ature strength of such modified steels, due to their greater tendency to precipitate $M_{23}C_6$ carbides which tend to coarsen at service temperatures.[19,20] Furthermore, commercial 3Cr-1Mo compositions, such as ASTM A 387 Grade 21 Class 2, also have insufficient hardenability for section sizes in excess of about 200 mm.[21,22]

Over the past few years, several major alloy design programs have been instigated to develop superior alternatives to commercially available heavy section 2.25Cr-1Mo pressure vessel steel for elevated temperature hydrogen service.[5,19-25] Notable amongst these studies are the Japanese Steel Works 3Cr-1Mo steel microalloyed with Ti, V and B[24] and the Climax Molybdenum Company 3Cr-1.5Mo steels containing 0.1 pct V and 1 to 1.4 pct Mn with 0.12 pct C maximum.[19,21,23]

In the current alloy development program the objective was to design an improved hydrogen attack resistant steel of higher strength to permit applications with thinner section sizes, thereby saving in weight and cost. Similar additions of Mo and Cr were employed to improve elevated temperature strength, oxidation resistance and resistance to hydrogen damage. However, 0.5 to 1 pct Ni was also added for hardenability coupled with 0.2 pct V for creep resistance and grain refinement. In particular, the Mn content was limited to a nominal 0.5 pct in order to minimize potential problems, from temper embrittlement susceptibility,[19,26,27,28] excess retained austenite following slow cooling* and band-

---

*Both banding and the transformation of excess retained austenite on tempering can lead to unexpected local susceptibility to hydrogen attack due to non-uniform distributions of unstable alloy carbides.[12]

---

ing, which can be promoted by higher Mn contents. The specific compositions (Table I), consisting of two 3Cr-1Mo-1Ni steels, with and without 0.2 pct V (termed Steels C and B, respectively), and two 3Cr-1.5Mo-0.5Ni steels, with and without 0.2 pct V (termed Steels E and D, respectively), are compared with a conventional 2.25Cr-1Mo steel of similar purity and steelmaking practice (termed Steel A).

Such modified 3Cr-Mo-Ni steels are found to be far superior to commercial 2.25Cr-1Mo steels and to compare very favorably with other advanced 3Cr-Mo materials. The microstructural features contributing to the improved hydrogen attack resistance, as well as the superior hardenability strength and toughness, and comparable ductility, temper embrittlement resistance and creep rupture properties, are discussed in this paper.

EXPERIMENTAL PROCEDURES

The four experimental 3Cr-Mo steels and the reference 2.25Cr-1Mo steel were produced as 55 kg (125 lb) laboratory heats by Climax Molybdenum Company using vacuum induction melting and casting under argon atmospheres. Chemical compositions in wt pct are listed in Table I. The ingots were subsequently upset-forged and cross-rolled to approximately 30 mm thick plates before austenitizing for 1 hour at 1000 °C. Following austenitization, samples were either quenched into agitated oil (*i.e.*, cooling rate ~1200 °C/min) or subjected to slow continuous cooling, at 8 °C/min through the transformation range, using a programmable induction furnace. Based on cooling rate data supplied by Lukens Steel Company, these treatments simulated both the surface and quarter thickness (0.25 T) locations, respectively, of a 400-mm (16-in.) thick plate during accelerated surface cooling (herein referred to as normalizing). Tempering was carried out in neutral molten salt baths at temperatures of 650 °C and principally 700 °C, for a range of times varying from 1 to 1000 hours. In terms of the tempering parameter, P, sometimes referred to as the Larson Miller parameter and defined as $T[20+\log t] \times 10^{-3}$ where T and t are the tempering temperature (in Kelvin) and time (in hours), respectively, these treatments represent a variation in P between 19.0 and 22.4 (or between 34.2 and 40.3 for temperatures in Rankin).

Room temperature uniaxial tensile tests were conducted on 6.4-mm diameter specimens of 32-mm gauge length, machined in the longitudinal direction of the plate, according to ASTM Standard E8-69. A displacement rate of 0.5 mm/min was employed. Standard Charpy V-notch impact specimens were also prepared in the longitudinal direction, and tested according to ASTM Standard E23-72 over a temperature range from −196° to 160 °C. Constant load creep-rupture tests were performed in air using smooth round bar longitudinal specimens of initial diameter 6.4 mm and gauge length 25 mm, according to ASTM Standard E-139. Tests were carried out at 560 °C with engineering stresses between 138 and 345 MPa (20 to 50 ksi).

Susceptibility to temper embrittlement was evaluated by testing a further series of Charpy impact toughness speci-

Table I. Chemical Composition, in wt pct, of Reference 2.25Cr-1Mo and Experimental 3Cr-Mo Steels Tested

Designation	C	Mn	Si	Cr	Ni	Mo	V	P	S	Al
Steel A	0.15	0.48	0.21	2.24	—	1.01	—	0.006	0.0034	0.014
Steel B	0.15	0.50	0.23	3.00	0.98	1.03	—	0.008	0.0040	0.010
Steel C	0.14	0.49	0.24	2.99	0.98	1.02	0.21	0.006	0.0039	0.009
Steel D	0.15	0.47	0.22	3.00	0.50	1.50	—	0.006	0.0036	0.008
Steel E	0.14	0.49	0.22	2.95	0.50	1.50	0.21	0.006	0.0034	0.014

mens which had been step-cooled, rather than air cooled, following tempering at 700 °C. The step-cooling treatment[26] involved holding for progressively longer times at progressively decreasing temperatures from 595 °C to 470 °C, as illustrated schematically in Figure 1.

Susceptibility to hydrogen attack was evaluated with respect to both uniaxial tensile and impact toughness properties by prolonged exposure of oversize specimen blanks to gaseous hydrogen atmospheres at temperatures of 550° to 600 °C and at pressures of 14 to 18 MPa (2000 to 2800 psi) for times up to 1000 hours, prior to machining and testing.

Microstructures were examined with both optical and electron microscopy. Thin foils for transmission electron microscopy (TEM) were prepared from 0.6 mm steel slices by chemically thinning to 0.05 mm in a hydrofluoric acid/hydrogen peroxide solution before electropolishing at room temperature in chromium trioxide/acetic acid solution. Foils were examined using Phillips EM301 and 400 STEM electron microscopes at 100 kV. Further analysis of the carbide compositions was performed on the scanning transmission electron microscope (STEM) using extraction replicas. Quantitative estimates of the percentage of retained austenite were assessed using both X-ray and magnetic saturation techniques.

RESULTS

Strength and Ductility

The room temperature uniaxial tensile properties of the four 3Cr-Mo-Ni steels, after slow-cooling (8 °C/min) and tempering at 700 °C for various times between 1 and 1000 hours, are shown in Figure 2. These heat-treatments represent a range of tempering parameters (temperatures in Kelvin) between 19.5 and 22.4. It is apparent that all four

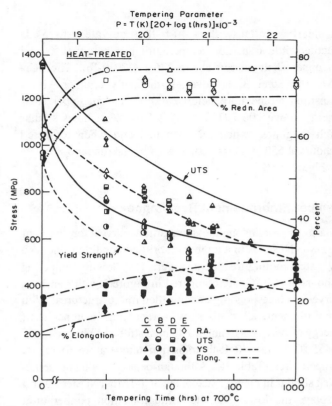

Fig. 2 — Room temperature uniaxial tensile properties for experimental 3Cr-Mo-Ni steels, slow-cooled and tempered at 700 °C, as a function of tempering time and tempering parameter (in Kelvin).

steels show similar mechanical properties (the V-containing C and E steels are marginally stronger), with ductilities fairly constant at pct RA between 70 and 80 pct, and with tensile strengths above 600 MPa (87 ksi), even at the longest tempering times.

Toughness

The Charpy V-notch impact toughness, as a function of test temperature for the four 3Cr-Mo-Ni steels, slow-cooled (8 °C/min) and tempered 4 hours at 700 °C (P = 20), is shown in Figure 3. The toughness of Steels C, D, and E

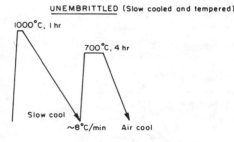

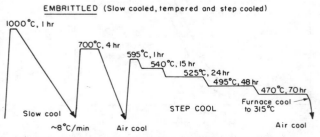

Fig. 1 — Schematic representation of the heat treatments used to assess susceptibility to temper embrittlement.

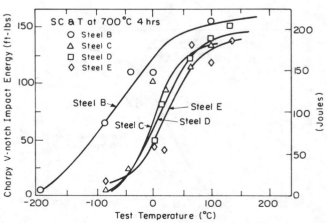

Fig. 3 — Charpy V-notch impact toughness transition curves for experimental 3Cr-Mo-Ni steels, slow-cooled and tempered (SC & T) at 700 °C for 4 h.

Source: Journal of Materials for Energy Systems, December 1984, 151-162

is clearly similar with a 40 ft-lb (equivalent to 54 J) ductile/brittle transition temperature just below 0 °C and an upper shelf energy above 176 J (130 ft-lb). The 3Cr-1Mo-1Ni Steel B, however, is distinctly tougher with a transition temperature below −100 °C and an upper shelf energy above 200 J (150 ft-lb).[25] This steel shows similar high toughness when oil quenched and when tempered 4 hours at 650 °C (P = 19), as seen by the (unexposed) data in Figure 4.[29]

Temper Embrittlement Resistance

Charpy V-notch toughness transition curves for the four 3Cr-Mo-Ni steels in a temper embrittled condition, induced by a step-cooling treatment (Figure 1) following tempering 4 hours at 700 °C, are compared in Figure 5 with transition curves of corresponding unembrittled microstructures. With embrittlement, all steels show a reduction in upper shelf energy of between 10 and 40 J and a shift (ΔT) between 15° to 55 °C in the 40 ft-lb transition temperature to higher temperatures (Table II). Although behavior is again somewhat similar in the four steels, the 3Cr-1Mo-1Ni Steel B still displays the lowest ductile/brittle transition temperature.

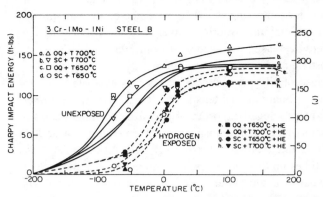

Fig. 4—Charpy V-notch impact toughness for experimental 3Cr-1Mo-1Ni Steel B, oil quenched (OQ) or slow-cooled (SC) and tempered (T) 4 h at 650 °C or 700 °C, in the prior hydrogen exposed (HE) condition (600 °C, 17 MPa pressure) and unexposed condition.[29]

Hydrogen Attack Resistance

In Figure 6, the room temperature uniaxial tensile data for the four 3Cr-Mo-Ni steels are given, as a function of tempering time at 700 °C following slow cooling from 1000 °C, for samples which have been previously exposed for 1000 hours to 14 MPa (2000 psi) hydrogen gas at 550 °C. By comparing with the scatter bands, which represent the

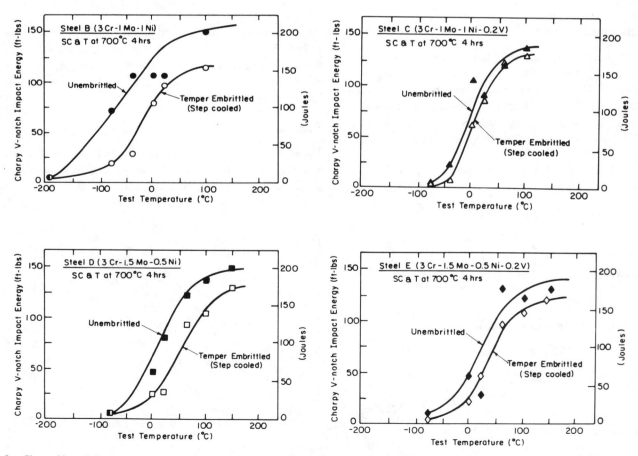

Fig. 5—Charpy V-notch impact toughness transition curves for experimental 3CR-Mo-Ni steels, slow-cooled and tempered (SC & T) 4 h at 700 °C, in the temper embrittled (step-cooled) and unembrittled conditions.

Table II. Effect of Temper Embrittlement on Charpy V-Notch Impact Toughness

| Steel* | 40 ft-lb (54 J) Transition Temp. | | Upper Shelf Energy | | Shift |
	Unembrittled	Temper Embrittled	Unembrittled	Temper Embrittled	ΔT
			(J)	(J)	
A	− 13 °C	—	122	—	—
B	−105 °C	−40 °C	206	165	65 °C
C	− 25 °C	−10 °C	186	175	15 °C
D	− 20 °C	25 °C	203	175	45 °C
E	− 10 °C	15 °C	176	163	25 °C

*Slow cooled (8 °C/min), tempered 4 hr at 700 °C.

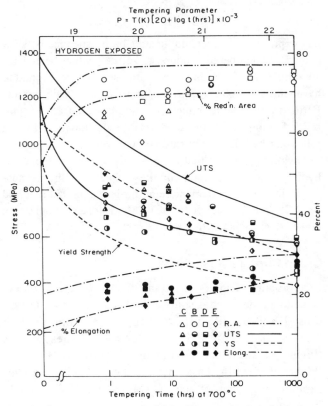

Fig. 6—Room temperature uniaxial tensile properties for prior hydrogen exposed 3Cr-Mo-Ni steels, slow-cooled and tempered at 700 °C, as a function of tempering time and tempering parameter. Hydrogen exposure for 1000 hr at 550 °C with 14 MPa pressure. Scatter bands represent bounds of data for corresponding unexposed conditions, shown in Figure 2.

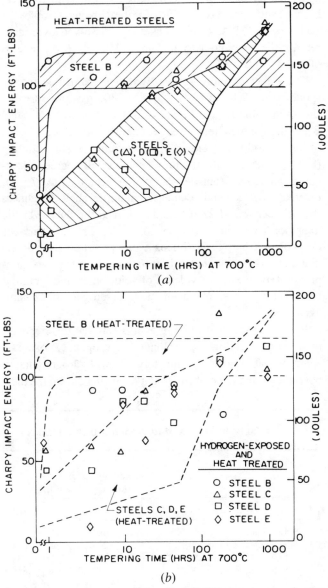

Fig. 7—Room temperature Charpy V-notch impact energy as a function of tempering time at 700 °C for slow-cooled 3Cr-Mo-Ni steels in the (a) heat-treated (unexposed) and (b) hydrogen-exposed conditions. Hydrogen exposure for 1000 h at 550 °C with 14 MPa pressure. Scatter bands represent bounds of data for unexposed conditions.

bounds of the corresponding data on unexposed samples from Figure 2, it can be seen that, apart from a slight decrease in pct RA in the higher strength conditions, there is little evidence of hydrogen attack damage in these steels (tempered at 700 °C) for this hydrogen exposure.

A similar result is obtained when comparing the effect of hydrogen attack damage on room temperature Charpy V-notch toughness (Figure 7). In Figure 7(a), the toughness of unexposed samples is plotted as a function of tempering

Source: Journal of Materials for Energy Systems, December 1984, 151-162

time at 700 °C following slow-cooling from 1000 °C. As noted above, Steel B is distinctly tougher and reaches a high room temperature toughness, corresponding to upper shelf ductile fracture by microvoid coalescence, after only short tempering times at 700 °C. The other steels show mixed mode transitional behavior, with consequent lower toughness, as their respective transition temperatures are closer to room temperature (c.f., Table II). Following prior 1000-hour exposure to 14 MPa hydrogen at 550 °C, the toughness in all steels is essentially unchanged [Figure 7(b)], with no apparent differences in fracture morphology compared to corresponding unexposed conditions. Steel B, however, does show a slight decrease in toughness. This may be more associated with entrapped hydrogen rather than hydrogen attack damage from internal voids and fissures, because subsequent re-heating of prior damaged samples for 80 hours at 275 °C (which bakes out dissolved hydrogen) returns the toughness to undamaged levels (i.e., excess of 135 J).[25]

The effect of more severe hydrogen exposures [corresponding to 17 to 18 MPa (2500 to 2800 psi) pressures] is shown in Table III and Figure 4 for the 3Cr-1Mo-1Ni Steel B, with samples tempered either at 650 °C or 700 °C (4 hours) following either oil quenching or slow-cooling (8 °C/min).[29] For all structures, some small degree of softening can be seen (tensile strengths reduced by 7 to 17 pct), together with a small decrease in upper shelf toughness (by between 2 and 22 pct) and a shift in the ductile/brittle transition temperature (by 65° to 97 °C). Fracture surfaces on broken Charpy specimens were predominately intergranular close to the notch in prior hydrogen exposed samples, but there was no evidence of voids on the intergranular facets which are so characteristic of hydrogen attack damaged structures.[12]

By comparison with Figure 5, it is felt that this decrease in toughness following prolonged hydrogen exposure is again probably not predominantly due to hydrogen attack, but rather to temper embrittlement occurring during the ther-

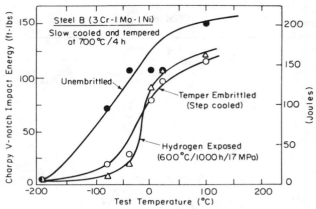

Fig. 8—Effect of prior temper embrittlement and hydrogen attack on Charpy V-notch impact toughness of experimental 3Cr-1Mo-1Ni Steel B, slow cooled and tempered 4 h at 700 °C.

mal exposure.[25] In fact, duplicate Charpy specimens, slow cooled and tempered 4 hours at 700 °C, when heated for 1000 hours at 550 °C in an inert atmosphere, were found to have similar toughnesses to the hydrogen exposed samples. Furthermore, the toughness of step-cooled (temper embrittled) samples in this steel was almost identical to that following hydrogen exposure (Figure 8).

Thus it is apparent that there is only a small influence of hydrogen attack damage in these steels, with the susceptibility to embrittlement being far less than that reported for 2.25Cr-1Mo steel or for 3Cr-1Mo-1Ni steels containing higher (i.e., 1 pct) Mn contents.[12]

Creep Rupture Resistance

Creep-rupture data for the four 3Cr-Mo-Ni steels (slow-cooled and tempered 4 hours at 700 °C) are shown in Figures 9 and 10. Results at 560 °C are compared to 2.25Cr-1Mo steel[20,30,31] in terms of applied stress vs life (Figure 9) and creep rupture ductility (pct RA) vs life (Figure 10). On the basis of these data, the elevated temperature properties of the 3Cr-Mo-Ni steels are clearly

Table III.   Effect of Hydrogen Attack on Uniaxial Tensile and Toughness Properties in 3Cr-1Mo-1Ni Steel B

Condition	Yield Stress (MPa)		U.T.S. (MPa)		pct Elong.*		pct Redn. Area		Upper Shelf Charpy Energy (J)		40 ft-lb (54 J) Transition Temp. (°C)	
	UE	HE	UE	HE	UE	HE	UE	HE	UE	HE	UE	HE
Slow-cooled, 4 hr at 650 °C	691	552	781	647	20	22	75	73	189	162	−100 °C	−23 °C
Slow-cooled, 4 hr at 700 °C	627	552	779	661	24	23	73	73	206	161	−105 °C	−40 °C
Oil Quenched, 4 hr at 650 °C	660	537	782	638	24	22	76	73	190	186	−115 °C	−50 °C
Oil Quenched, 4 hr at 700 °C	545	502	664	616	24	25	76	71	230	178	−120 °C	−23 °C

*On 32 mm gauge length.
Legend: UE = Unexposed, HE = Hydrogen Exposed (600 °C, 17 MPa)

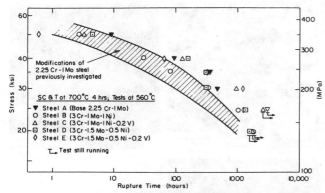

Fig. 9—Creep rupture properties at 560 °C of experimental 3Cr-Mo-Ni steels, slow-cooled and tempered 4 h at 700 °C. Results are compared with previous data[20,30,31] on 2.25Cr-1Mo steel.

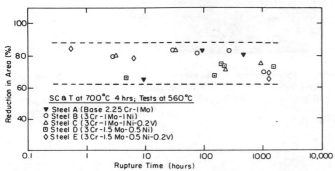

Fig. 10—Creep rupture ductility (pct RA) *vs* time to rupture for experimental 3Cr-Mo-Ni steels, slow-cooled and tempered 4 h at 700 °C, from constant load tests at 560 °C at stresses between 138 and 345 MPa.

comparable with those of 2.25Cr-1Mo steel. Furthermore, at longer lives (above 1000 hours), the 3Cr-Mo-Ni steels actually out perform commercial 2.25Cr-1Mo alloys, particularly in the V-containing C and E steels which tend to display the best performance at the lower applied stresses.

Hardenability

For cooling rates as slow as 8 °C/min, typical of the conditions at quarter thickness of a 400-mm thick plate during normalizing, commercial 2.25Cr-1Mo steel generally has inadequate hardenability to ensure fully bainitic microstructures, as evidenced by the substantial proportion (~40 pct) of the polygonal ferrite in as-cooled structures [Figure 11(a)]. The experimental 3Cr-Mo-Ni steels, conversely, with their increased Cr + Ni additions, show vastly enhanced hardenability and 100 pct granular bainitic structures after a similar slow cooling rate of 8 °C/min [Figure 11(b)].

Microstructures

As noted above, the microstructure of all four 3Cr-Mo-Ni steels was found to be 100 pct granular bainite, with a prior austenite grain size of approximately 50 μm (Figure 12). In the as-cooled untempered conditions, fine autotempered

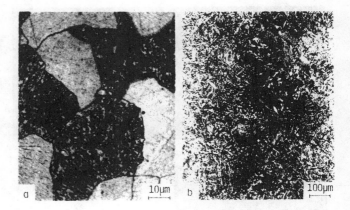

Fig. 11—Optical micrographs of microstructures after slow-cooling at 8 °C/min to simulate the 0.25T cooling rate in normalized 400 mm (16 inch) plate. (*a*) 2.25Cr-1Mo steel, showing bainite and 40 pct polygonal ferrite, and (*b*) 3Cr-1Mo-1Ni steel, showing 100 pct bainite (etched in nital).

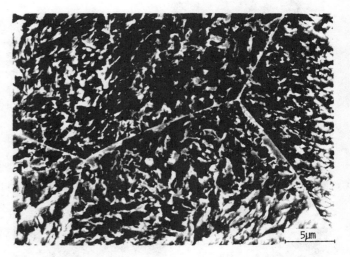

Fig. 12—Scanning electron micrograph of 3Cr-1.5Mo-0.5Ni Steel D in the slow-cooled and untempered condition, showing 100 pct granular bainitic microstructure (etched in hot picric acid and nital).

M_3C (cementite) precipitates (Figure 13) and films of retained austenite (Figure 14) are evident between bainitic laths. X-ray and magnetic saturation studies on Steel B have indicated that the proportion of such interlath austenite is approximately 17 pct in the as-cooled condition, compared to 5.5 to 7 pct in the as-quenched state.[29]

On tempering at 700 °C for 1 hour, fine accicular intralath carbides (~0.05 μm in size) together with coarser lath boundary carbides (~0.25 μm in size) were observed and identified, using selected area diffraction, as cubic $M_{23}C_6$ (Figure 15).[32] No evidence of the hexagonal M_2C or M_7C_3 or orthorhombic M_3C carbides could be detected. This is in stark contrast to 2.25Cr-1Mo steel where, at 700 °C, M_3C precipitation can persist up to tempering times of 30 hours (Figure 16).[33] After 1000 hours at 700 °C, $M_{23}C_6$ precipitates in the 3Cr-Mo-Ni steels are typically 0.5 to 1 μm in size (Figure 17) with additional evidence of the Mo-rich M_6C precipitates (Figure 18), and in the C and E steels VC

Source: Journal of Materials for Energy Systems, December 1984, 151-162

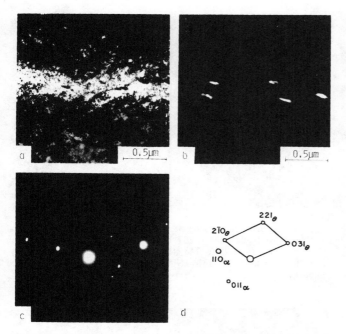

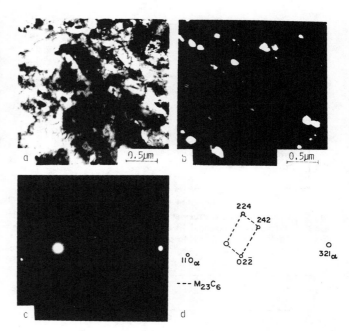

Fig. 13 — Transmission electron micrographs of slow-cooled 3Cr-1Mo-1Ni Steel B in the untempered condition, showing (a) bright field of ferrite (α) and fine autotempered cementite (θ), (b) dark field of cementite from $(2\bar{1}0)_\theta$ reflection of (c) selected area diffraction (SAD) pattern, and (d) interpretation of (c).[32]

Fig. 15 — Transmission electron micrographs of slow-cooled 3Cr-1.5Mo-0.5Ni Steel D after tempering at 700 °C for 1 h, showing (a) bright field of ferrite and $M_{23}C_6$ precipitates, (b) dark field of carbide from $(02\bar{2})$ $M_{23}C_6$ reflection of (c) SAD pattern, and (d) interpretation of (c).[32]

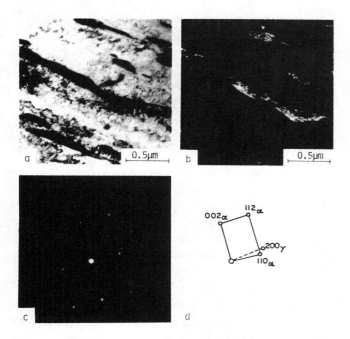

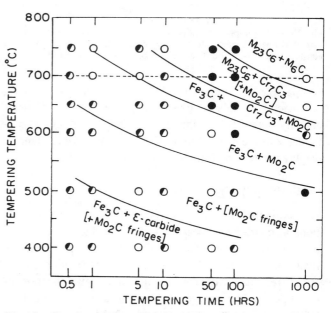

Fig. 16 — Experimental data of Baker and Nutting[33] showing the sequence of carbide formation during tempering of a normalized 2.25Cr-1Mo steel.

Fig. 14 — Transmission electron micrographs of slow-cooled (3Cr-1.5Mo-0.5Ni) Steel D in the untempered condition, showing (a) bright field of ferrite (α) and interlath retained austenite (γ), (b) dark field from $(200)_\gamma$ austenite reflection of (c) SAD pattern, and (d) interpretation of (c).[32]

precipitates. The types of carbide in these structures were confirmed with extraction replica studies in the STEM using analyses developed by Titchmarsh[34] and Shaw.[35] Carbide types in Steels B, D, and E are shown in Table IV together with their respective compositions in atomic percent.

DISCUSSION

On the basis of the TEM and STEM studies (Figures 15, 17, and 18), it is clear that the increased Cr and Ni content in the present 3Cr-Mo-Ni series of steels leads to a markedly accelerated tempering response compared to 2.25Cr-1Mo steel. In particular, the iron and chromium-rich $M_{23}C_6$ carbide replaces the less stable M_2C, M_3C and M_7C_3 carbides

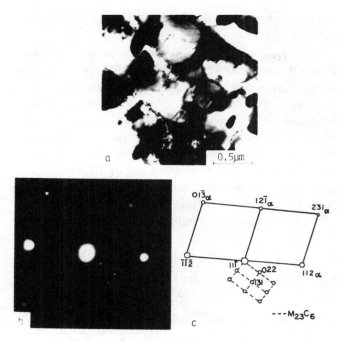

Fig. 17 — Transmission electron micrographs of slow-cooled 3Cr-1Mo-1Ni Steel B after tempering at 700 °C for 1000 h, showing (a) bright field of ferrite and $M_{23}C_6$ precipitates, (b) SAD pattern, and (c) interpretation of (b).[32]

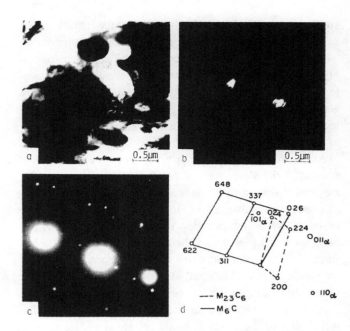

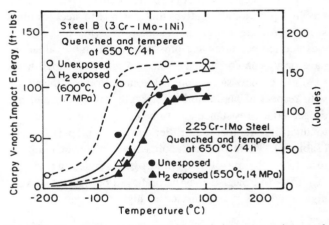

Fig. 18 — Transmission electron micrograph of slow-cooled 3Cr-1.5Mo-0.5Ni-0.2V Steel E after tempering at 700 °C for 1000 h, showing (a) bright field of ferrite, and $M_{23}C_6$ and M_6C carbides, (b) dark field of (026) M_6C reflection of (c) SAD pattern, and (d) interpretation of (c).[32]

Table IV. STEM Analysis of Carbide Precipitates in 3Cr-Mo-Ni Steels Tempered for 1000 hr at 700 °C

	Composition (Atomic Pct)				Type Carbide
	Pct Cr	Pct Mo	Pct Fe	Pct V	
Steel B					
Fine ppts (0.1 μm)	43.9	11.6	44.5	—	$M_{23}C_6$
Elongated ppts (1.0 μm)	24.7	32.7	42.6	—	M_6C
Massive ppts (1.0 μm)	37.7	11.2	51.1	—	$M_{23}C_6$
Steel D					
Fine ppts	17.2	36.5	46.3	—	M_6C
Elongated ppts	36.3	17.5	46.2	—	$M_{23}C_6$
Massive ppts	46.0	7.3	46.7	—	$M_{23}C_6$
Steel E					
Fine ppts	8.8	4.0	1.2	86.0	VC
Elongated ppts	10.6	45.8	43.6	—	M_6C
Massive ppts	11.9	43.1	45.0	—	M_6C

Fig. 19 — Comparison of Charpy V-notch impact toughness of 2.25Cr-1Mo steel and experimental 3Cr-1Mo-1Ni Steel B, both quenched and tempered 4 h at 650 °C, prior to, and after, 1000 h high temperature exposure to gaseous hydrogen (550 to 600 °C, 14 to 17 MPa pressure).

after only an hour of tempering at 700 °C, compared to approximately 400 hours in 2.25Cr-1Mo steel (Figure 16). It is felt that specifically the rapid elimination of Fe_3C (*i.e.,* within 1 hour at 700 °C compared to roughly 30 hours in 2.25Cr-1Mo) and the rapid precipitation of more stable alloy carbides on tempering is primarily responsible for the enhanced hydrogen attack resistance in the 3Cr-Mo-Ni steels. This enhanced resistance to hydrogen damage can be seen clearly in Figure 19 where the impact toughness 3Cr-1Mo-1Ni Steel B is compared to 2.25Cr-1Mo Steel A in the prior hydrogen exposed and unexposed conditions

(both steels were oil quenched and tempered 4 hours at 650 °C). Even though the 3Cr-1Mo-1Ni steel suffered a higher exposure (17 MPa hydrogen pressure at 600 °C compared to 14 MPa at 550 °C), its toughness remained superior to the 2.25Cr-1Mo steel, both in terms of a lower transition temperature and higher upper shelf energy.

The Cr + Ni combinations in the experimental steels also provide a potent effect in vastly increasing hardenability, as evident by the fully bainitic microstructures (Figure 11) following simulated normalizing of 400-mm thick plates (0.25T location). Mn contents, however, were deliberately limited to 0.5 pct as prior experience[12] with a 3Cr-1Mo-1N;

steel containing 1 pct Mn had indicated lower creep strength and problems with hydrogen attack susceptibility where, particularly after slow cooling, the transformation of high carbon retained austenite on tempering lead to a prolonged stability of Fe_3C carbides. Higher Mn contents are also known to promote banding, which again is detrimental to hydrogen attack resistance, and to increase susceptibility to temper embrittlement.[26,27,28]

Aside from generally improving temper embrittlement resistance, the Mo content of 1 to 1.5 pct provides for good elevated temperature strength and creep-rupture properties. This is further enhanced in Steels C and E by the precipitation of a vanadium carbide. Although in the past there has been some question about the creep resistance of 3Cr-Mo steels,[19,20] the present series of alloys compared very favorably with 2.25Cr-1Mo steel, at least over the testing range of temperatures and stresses (Figures 9 and 10).

Although not studied in the present investigation, the weldability of these steels should not cause problems with standard welding and pre-heat practices due to the relatively low carbon contents (*i.e.*, 0.15 max wt pct). In addition, preliminary data on fatigue crack propagation behavior indicate a crack growth resistance similar to 2.25Cr-1Mo steel.[36]

Thus, the present series of experimental 3Cr-Mo-Ni steels provides superior alternatives to 2.25Cr-1Mo steel for thick section pressure vessel applications involving hydrogen at elevated temperatures and pressures. Although evaluations have only been completed to date on relatively small heats (55 kg), the new steels appear to out perform 2.25Cr-1Mo with respect to hardenability, strength, toughness and hydrogen attack resistance and to be comparable with respect to ductility, temper embrittlement and creep resistance (Table V). There may be applications for lower temperature (*i.e.*, below 500 °C) hydrogen service where the 3Cr-1Mo-1Ni Steel B may be preferred because of its very high toughness.[25] For service at higher temperatures, the vanadium containing Steels C and E may be preferred because of their superior creep strength.

Finally, the present series of steels compare very favorably with other recently developed 3Cr-Mo alloys for thick section, hydrogen service pressure vessel application, namely the 3Cr-1.5Mo-0.1V steels developed by Climax Molybdenum[19,21,23] and the 3Cr-1Mo-0.25V, Ti, B steels developed by the Japan Steel Works.[24] As shown in Figures 20 to 22, all three series of steels out perform 2.25Cr-1Mo steels. However, whilst the creep rupture (Figure 20) and temper embrittlement resistance of the three 3Cr-Mo steels is similar, the current 3Cr-Mo-Ni alloys appear to offer the best combinations of strength and toughness (Figures 21 and 22).

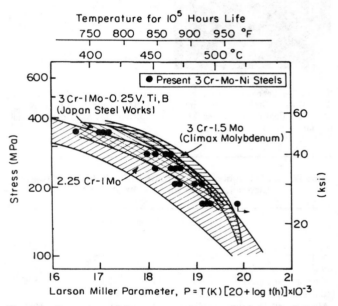

Fig. 20—Comparison of the creep rupture strength of slow cooled and tempered experimental 3Cr-Mo-Ni alloys with 2.25Cr-1Mo steel,[30,31] the Climax Molybdenum 3Cr-1.5Mo steels,[19,21] and the Japan Steel Works 3Cr-1Mo-0.25V, Ti, B steel.[24]

Table V. Comparison of Mechanical Properties of New 3Cr-Mo-Ni Steels with 2.5Cr-1Mo, Following Slow Cooling (8 °C/min) and Tempering at 700 °C (4 hr)*

Code	Alloy	Yield Stress	UTS	Elong.**	RA	Upper Shelf Charpy Energy	40 ft-lb Transition	Creep Rupture Life at 200 MPa (560 °C)
		MPa (ksi)	MPa (ksi)	Pct	Pct	J (ft-lb)	° C (°F)	hr
A	2.25Cr-1Mo	496 (72)	641 (93)	24	73	122 (90)	− 13 (9)	457
B	3Cr-1Mo-1Ni	627 (91)	779 (113)	24	73	206 (152)	−105 (−157)	269
C	3Cr-1Mo-1Ni-0.2V	593 (86)	772 (112)	26	75	186 (137)	− 25 (− 13)	947
D	3Cr-1.5Mo-0.5Ni	655 (95)	799 (116)	22	73	203 (150)	− 20 (− 4)	212
E	3Cr-1.5Mo-0.5Ni-0.2V	676 (98)	821 (119)	21	74	176 (130)	− 10 (14)	1253

*Tempering Parameter (K) = 20.04
**On 32 mm gauge length

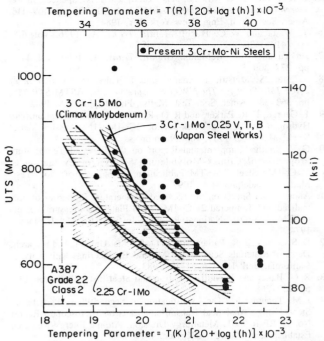

Fig. 21—Comparison of tensile strength of present experimental 3Cr-Mo-Ni alloys with 2.25Cr-1Mo steel,[37,38] the Climax Molybdenum 3Cr-1.5Mo steels,[19,21] and the Japan Steel Works 3Cr-1Mo-0.25V, Ti, B steel.[24]

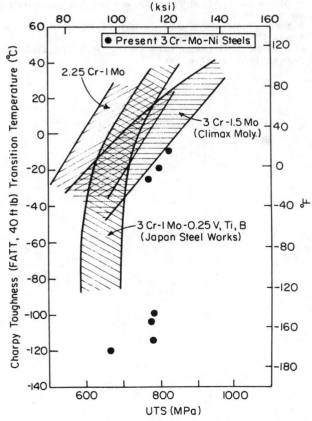

Fig. 22—Comparison of the tensile strength/toughness characteristics of present experimental 3Cr-Mo-Ni alloys with 2.25Cr-1Mo steel,[39] the Climax Molybdenum 3Cr-1.5Mo steels,[19,21] and the Japan Steel Works 3Cr-1Mo-0.25V, Ti, B steel.[24] Note lower ductile/brittle transition temperature implies higher toughness.

CONCLUSIONS

A new series of 3Cr-Mo-Ni steels has been developed to provide alternatives to 2.25Cr-1Mo steel for thick section pressure vessel applications involving high temperatures (up to 550 °C) and high hydrogen pressures (up to 18 MPa). The nominal compositions of these steels are 3Cr-1Mo-1Ni and 3Cr-1.5Mo-0.5Ni, with and without 0.2V. Based on evaluations on 55 kg heats the new steels are found to be superior to 2.25Cr-1Mo steel with respect to hardenability, room temperature strength, impact toughness and resistance to hydrogen attack, and to be similar with respect to room and elevated temperature ductility, creep strength and rupture ductility and temper embrittlement resistance. In addition, their mechanical properties compare very favorably to the recently developed 3Cr-1.5Mo-0.1V Climax Molybdenum and 3Cr-1Mo-0.25V, Ti, B Japan Steel Works steels, and actually are somewhat superior to these steels with respect to strength and toughness.

ACKNOWLEDGMENTS

This work was supported by the Department of Energy, Advanced Research and Development Fossil Energy Materials Program, through the Oak Ridge National Laboratory (ORNL), under Subcontract 7843 with the University of California, Berkeley, under Union Carbide Contract W-7605-eng-26 with the U.S. Department of Energy. The authors would like to thank R. W. Swindeman of ORNL for many helpful discussions, Dr. D. S. Sarma for transmission electron microscopy assistance, and T. George, R. K. Anders, R. I. Huntley, and L. -H. Chan for experimental help.

REFERENCES

1. D. A. Canonico, G. C. Robinson, and W. R. Martin: "Pressure Vessels for Coal Conversion Systems," Report No. ORNL/TM-5685, Oak Ridge National Laboratory, Oak Ridge, TN, September 1978.
2. W. Lochmann: *Proc. 4th Annual Conf. on Materials for Coal Conversion and Utilization,* Gaithersberg, MD, pp. 1-31, Dept. of Energy, Washington, DC, 1979.
3. T. E. Scott: in *Application of 2¼Cr-1Mo Steel for Thick-Wall Pressure Vessels,* ASTM STP 755, pp. 7–25, Amer. Soc. Test. Matls., Philadelphia, PA, 1982.
4. R. W. Swindeman, R. D. Thomas, R. K. Hanstad, and C. J. Long: "Assessment of Need for an Advanced High Strength Chromium-Molybdenum Steel for Construction of Third Generation Gasifier Pressure Vessels," Report No. ORNL-TM/8873, Oak Ridge National Laboratory, Oak Ridge, TN, 1983.
5. R. M. Horn, R. J. Kar, V. F. Zackay, and E. R. Parker: *J. Mat. for Energ. Syst.,* 1979, vol. 1, no. 1, pp. 77–92.
6. G. D. Nasman and W. P. Webb: in *Advanced Materials for Pressure Vessel Service with Hydrogen at High Temperatures and Pressures,* M. Semchyshen, ed., ASME Vol. MPC-18, pp. 1–23, Amer. Soc. Mech. Eng., New York, NY, 1982.
7. Y. Murakami, T. Nomura, and J. Watanabe: in *Application of 2¼Cr-1Mo Steel for Thick Wall Pressure Vessels,* ASTM STP 755, pp. 383–417, Amer. Soc. Test. Matls., Philadelphia, PA, 1982.
8. R. J. Kar and J. A. Todd: in *Application of 2¼Cr-1Mo Steel for Thick Wall Pressure Vessels,* ASTM STP 755, pp. 228–252.

9. G. A. Nelson: *Trans. ASME*, 1951, vol. 51, pp. 205–213.
10. J. Wanagel, T. Hakkarainer, and Che-Yu Li: in *Application of 2¼Cr-1Mo Steel for Thick Wall Pressure Vessels*, ASTM STP 755, pp. 93–108, Amer. Soc. Test. Matls., Philadelphia, PA, 1982.
11. P. G. Shewmon and Z. -S. Yu: in *Advanced Materials for Pressure Vessel Service with Hydrogen at High Temperatures and Pressures*, M. Semchyshen, ed., ASME Vol. MPC-18, pp. 85–92, Amer. Soc. Mech. Eng., New York, NY, 1982.
12. D. W. Chung, J. A. Todd, J. K. Youngs, and E. R. Parker: in *Advanced Materials for Pressure Vessel Service with Hydrogen at High Temperatures and Pressures*, ASME Vol. MPC-18, pp. 25–52.
13. A. R. Ciuffreda, N. B. Heckler, and E. B. Norris: in *Advanced Materials for Pressure Vessel Service with Hydrogen at High Temperatures and Pressures*, ASME Vol. MPC-18, pp. 53–68.
14. P. G. Shewmon: *Metall. Trans. A*, 1976, vol. 7A, pp. 509–516.
15. G. R. Odette and S. S. Vagarali: *Metall. Trans. A*, 1982, vol. 13A, pp. 299–303.
16. H. Natan and H. H. Johnson: *Metall. Trans. A*, 1983, vol. 14A, pp. 963–971.
17. Anon: *A Study of the Effects of High-Temperature High-Pressure Hydrogen on Low Alloy Steels*, API Publication 945, June 1975.
18. Anon: *Steels for Hydrogen Service at Temperatures and Pressures in Petroleum Refineries and Petrochemical Plants*, API Publication 941, 1977.
19. T. Wada, T. B. Cox, and F. B. Fletcher: *Proc. 7th Annual Conf. on Materials for Coal Conversion and Utilization*, Gaithersburg, MD, Department of Energy, Washington, DC, 1982.
20. J. A. Todd, D. W. Chung, and E. R. Parker: in *Advanced Materials for Pressure Vessel Service with Hydrogen at High Temperatures and Pressures*, M. Semchyshen, ed., ASME Vol. MPC-18, pp. 179–191, Amer. Soc. Mech. Eng., New York, NY, 1982.
21. T. Wada and T. B. Cox: in *Advanced Materials for Pressure Vessel Service with Hydrogen at High Temperatures and Pressures*, ASME Vol. MPC-18, pp. 111–121.
22. R. A. Swift: in *Application of 2¼Cr-1Mo Steel for Thick Wall Pressure Vessels*, ASTM STP 755, pp. 166–188, Amer. Soc. Test. Matls., Philadelphia, PA, 1982.
23. T. B. Cox and T. Wada: "Resistance of 2¼Cr-1Mo and 3Cr-Mo Steels to Hydrogen Attack," Report L-294-14, Climax Molybdenum Company, Ann Arbor, MI, November 1982.
24. Anon: "Technical Data Sheet of 3Cr-1Mo-¼V-Ti-B Steel for Elevated Temperature Applications," Report No. MR83-3, The Japan Steel Works, Muroran, Japan, August 1983.
25. E. R. Parker, R. O. Ritchie, J. A. Todd, and P. N. Spencer: in *Research on Chrome-Moly Steels*, ASME Vol. MPC-21, pp. 109–116, Amer. Soc. Mech. Eng., New York, NY, 1984.
26. T. Wada and W. C. Hagel: *Metall. Trans. A*, 1976, vol. 7A, pp. 1419–1426.
27. Jin Yu and C. J. McMahon: *Metall. Trans. A*, 1980, vol. 11A, pp. 277–300.
28. S. Sato, S. Matsui, T. Enami, and T. Tobe: in *Application of 2¼Cr-1Mo Steel for Thick Wall Pressure Vessels*, ASTM STP 755, pp. 363–382, Amer. Soc. Test. Matls., Philadelphia, PA, 1982.
29. T. George, E. R. Parker, and R. O. Ritchie: "On the Susceptibility to Hydrogen Attack of a Thick-Section 3Cr-1Mo-1Ni Pressure Vessel Steel," *Met. Sci.*, 1985, vol. 19, in press.
30. G. V. Smith: "Supplemental Report on the Elevated Temperature Properties of Climax-Molybdenum Steels (An Evaluation of 2¼Cr-1Mo Steel)," ASTM Publication DS 652, Amer. Soc. Test. Matls., Philadelphia, PA, 1097.
31. Anon: "Data Sheets on the Elevated-Temperature Properties of normalized and Tempered 2¼Cr-1Mo Steel Plates for Pressure Vessels (ASTM A387D)," National Research Institute for Metals, Japan, 1974.
32. D. S. Sarma, E. R. Parker, and R. O. Ritchie: Unpublished research, Dept. of Materials Science and Mineral Engineering, University of California, Berkeley, CA 94720, 1983.
33. R. G. Baker and J. Nutting: J. Iron and Steel Inst., 1959, vol. 192, pp. 257–268.
34. J. M. Titchmarsh: "The Identification of Second-Phase Particles in Steels Using a Analytical Transmission Electron Microscope," Report No. AERE-R 9661, Metallurgy Division, Atomic Energy Research Establishment, Harwell, UK, Dec. 1979.
35. B. J. Shaw: in *Research on Chrome-Moly Steels*, ASME Vol. MPC-21, pp. 117–128, Amer. Soc. Mech. Eng., New York, NY, 1984.
36. R. Pendse, E. R. Parker, and R. O. Ritchie: Unpublished research, Dept. of Materials Science and Mineral Engineering, University of California, Berkeley, CA 94720, 1984.
37. K. Miyano and T. Adachi: *Trans. Iron and Steel Inst. Japan*, 1971, vol. 11, p. 54.
38. S. Sawada, J. Matsumiya, and K. Nonaka: *Kawasaki Steel Technical Report*, 1974, vol. 6, p. 249.
39. C. D. Clauser, L. G. Emmer, and R. D. Stout: *Proc. API Division of Refining*, 1972, vol. 52, p. 790.

THE EFFECT OF THERMAL AND MECHANICAL TREATMENTS ON
THE PROPERTIES OF A572 GRADE 50 STEEL

Louis R. DePatto and Alan W. Pense
Department of Metallurgy and Materials Engineering
Lehigh University

ABSTRACT

This investigation was concerned with the effect of strain aging on the strength and toughness of ASTM A572 Grade 50 steel. The material was studied in the as-rolled and normalized conditions, with specimens taken in the transverse to the rolling direction orientation. The as-rolled plate was stronger and had less toughness than normalized plate. The difference in Charpy V-notch transition temperature measured at 345 (25 ft-lbs.) was approximately $60^{o}C$ ($108^{o}F$).

ASTM A572 Grade 50 steel is sensitive to the strain aging phenomenon. A plastic strain of 5% increased the yield strength by 96 to 152 MPA (14 to 22 Ksi) and the tensile strength by 28 to 38 MPA (4 - 5.5 Ksi). Subsequent aging at $370^{o}C$ ($700^{o}F$) further increased the yield and tensile strengths another 28 to 34 MPA (4 - 5 Ksi). The Charpy impact transition temperature increased with straining and aging, the maximum increase being about $44^{o}C$ ($78^{o}F$). Stress relieving at $620^{o}C$ ($1150^{o}F$) had little effect on the recovery of toughness. The most recovery was seen in the as-rolled condition with a short stress-relief cycle (2 hours).

INTRODUCTION

The Pressure Vessel Research Committee (PVRC) of the Welding Research Council (WRC), through the Pressure Vessel Steels Subcommittee of the Materials Division, has been supporting investigations to determine the basic characteristics of microalloyed HSLA steels. The program has completed its fourth year and four steels - A737 Grades B and C and A588 Grades A and B have been investigated. The program was intended to study the general toughness and strain aging effects on high strength, low alloy (HSLA) steels for pressure vessels and their supports.

ASTM A588 Grades A and B, which are structural steels, have been used frequently for their general strength characteristics as they have higher yield and tensile strength than A36 steel. In the pressure vessel industry they are used for vessel supports. ASTM A572 Grade 50 microalloyed steel is another structural steel used in this capacity, also for pressure vessel supports. Therefore, the study of strain aging response of A572 Grade 50 was undertaken this year to complement the work on A588 completed last year.[1]

The object of this investigation is the same as past investigations, to determine the effect of mechanical and thermal treatments on the properties of the steel. The work involved base plate in both as-rolled and normalized conditions and includes a series of straining, aging and stress relief cycles. Focus of the investigation was on toughness and the ability of the steel to return to its original toughness after receiving the thermo-mechanical cycle. Strength properties changes were also included in the study.

The theoretical considerations of strain aging were explored by Miguel Erazo in his study of A588. A588 and A572 have similar chemical compositions; A572 having less chromium and silicon. A572 would then have many of the same mechanisms of strain aging as A588. The basic phenomena in strain aging have been described by Erazo[1].

During fabrication and cold forming of pressure vessel supports, thermo-mechanical treatment gives rise to strain aging cycles. Generally, strain produces an increase in strength and decrease in notch toughness in the steel. Aging (long term holding at room temperature or service at high temperature) after strain increases the strength and reduces the toughness of the material still further. Therefore, the strain aging process, especially from the toughness standpoint, can become a significant factor since this embrittlement can be a major problem during service. The strain aging phenomenom is most significant when steels have been prestrained and a further aging heat treatment at high or low temperature takes place.

This phenomenon is due to the migration or diffusion of solute atoms, mostly carbon and nitrogen atoms, to the dislocations locking them and increasing the strength of the material.[2] In HSLA steels, the carbon content and the free nitrogen in solid solution are low, therefore, other alloy elements forming intermetallic precipitates lock the dislocations. As carbon and nitrogen content decrease, the strain aging effect is reduced and eventually should be eliminated.

Strain aging manifests itself chiefly by an increase in yield or flow stress on aging after or during straining. Aging after straining is classified as "static" aging and aging during straining, "dynamic" aging. Using a stress-strain diagram, as is shown in Figure 1[2], is the best way to see the strain aging effects on the mechanical properties of a steel[3]. Figure 1 shows a stress-strain curve for a mild normalized steel. If a specimen is strained to point A beyond the lower yield extension, unloaded, and then immediately retested, the stress-strain curve follows the same trace (a). At the initial yield point the curve is slightly rounded, and there is no evidence of the upper and lower yield points found when the steel yielded initially at B. However, if the specimen is unloaded at A and then allowed to age at room temperature or above, the discontinuous yielding behavior returns and the stress-strain curve follows a curve such as (b) in Figure 1. The yield point is now higher than the flow stress at the end of pre-straining. This increase in yield or flow stress on unloading and aging is the most universal indication of strain aging. There may also be an increase in ultimate tensile strength (UTS) and a decrease in elongation and reduction of area, but these do not always take place[2]. Toughness is usually decreased.

Others[4, 5], have proposed step models and mathematical models to explain the microstructural effects on the phenomenon and mechanical properties. Many variables effect the strain aging process and its effects. They are chemical composition, aging temperature, prior heat treatment, and type and extent of prestrain[1, 2]. The two kinds of alloying elements can be distinguished in connection with the strain aging process. First, there are solutes which lock dislocations and can diffuse with sufficient speed to produce strain aging (C, N). Second, there are elements which may affect the process by altering the concentration or mobility of the solute atoms producing strain aging[2], although they do not directly enter into it. These can be carbide or nitride formers.

The effect of aging temperatures has been investigated by the PVRC[1, 8] in the past, resulting in $500^{o}F$ ($260^{o}C$) being identified as a typical temperature with the maximum effect on strain aging. Some HSLA steels show the maximum strain aging effect at a higher temperature of $700^{o}F$ ($370^{o}C$)[1].

Cold forming is, in many cases, followed by a stress relief treatment. Four main reasons for such treatment are:[1, 6]

    a. to reduce the residual stresses which may cause dimensional instability and/or brittle fracture.

    b. to improve or recover the toughness of the material lost by the strain of cold forming.

    c. to remove aging effects on strength or ductility.

    d. to improve the resistance to stress corrosion.

It is important to realize that stress relief heat treatments may degrade the material's mechanical properties as well as enhance them.

This degradation may include a significant reduction of yield strength and ultimate tensile strength and also an increase in Charpy impact transition temperature. The embrittlement mechanism appears to be associated with carbides agglomeration and coarsening on the grain boundaries[6, 7] or with precipitation hardening. Thus, it is important to study the change in mechanical properties that occurs due to stress relief heat treatment[6].

The effect of this heat treatment had been already investigated by the PVRC[7, 8] for the A588 steels. The results of this research show that, in some cases, the stress relief restores the toughness of the material in some degree after the strain aging, but in others, not at all. Usually the final toughness level is lower than that in the original condition. On the other hand, for many other steels, this heat treatment restores the notch toughness to its original value and in some cases, a short term stress relief is beneficial when strain aging is a potential problem, from the notch toughness standpoint[10]. Therefore, it is clear that when a basic mechanical characterization of a HSLA steel is undertaken, the effect of stress relief heat treatment should be included and was included in this study.

MATERIALS AND PROCEDURES

Material

The material for this study was supplied by Lukens Steel Company, Coatsville, PA. The composition of this steel as supplied by Lukens is:

C	Mn	P	S	Si	Ni	Cr	Cu	V	Al
.16	1.19	.010	.023	.23	.19	.12	.27	.04	.04

The plate was received in the as-rolled condition and was 63.8 cm (25-1/8 inches) wide x 154.6 cm (60-7/8 inches) long x 4.8 cm (1-7/8 inches) thick. After an initial study of the general strength and toughness level in the transverse and longitudinal orientation to the rolling direction, the plate was sectioned at its centerline as shown in Figure 2. All specimens were machined adjacent to the centerline. This was to minimize through thickness property variations as well as to conserve material and reduce machining costs.

The material was tested in the as-rolled and normalized condition. The plate was received as rolled and as-rolled tests were in this condition. The normalizing heat treatment consisted of placing sections (full thickness) in a preheated, forced-air, Heavy-Duty furnace at 900°C (1600°F), holding for 2 1/2 hours, and air cooling.

Strain Aging - Stress Relief Sequence

The strain aging study followed the schedule shown in Table II. This schedule includes straining, aging and stress-relief cycles. The schedule was completed for both as-rolled and normalized conditions. The initial study of strength and toughness in the as-received condition showed anisotropic behavior for Charpy V-notch transition temperature between longitudinal and transverse orientation specimens. The transverse orientation was the worst case, therefore the study concentrated on this orientation.

Large tensile specimens transverse to rolling direction were machined from the as-received and heat-treated plates. The sampling method, orientation, and specimen size are shown in Figure 2. From each tensile specimen, 16 Charpy V-notch standard impact specimens and 12 standard button-head type (.252" dia. gage) tensile specimens were obtained.

The straining of the large tensile specimens was completed using a Tinius-Olsen 534 KN (120,000 lbf) testing machine with a constant crosshead speed of 1.27 cm/min (0.5 in/min) at room temperature. The specimens were strained to a nominal 5% strain. Past investigations have found fairly uniform distribution of strain and no significant mechanical property effects due to the resulting strain variations. To monitor the strain across the specimens, scribe marks every 2.5 cm (1") were made before testing. In addition, a 5 cm (2") gage extensometer was used to measure the strain while the test was in progress. The extensometer had a maximum extension of 4% strain, therefore the specimen required a halt to loading and reapplication of the extensometer at 4%. Variations in strain down the length of the specimen are shown in Figure 3.

After prestraining, Charpy V-notch and button-head tensile specimens were machined longitudinally to the strain direction as shown in Figure 2. One large strained tensile specimen was used for 12 button-head tensile specimens; the other large tensiles were used for Charpy V-notch impact specimens. Of the 12 button heads, three were used for each of the subsequent heat treatment conditions (no age - no stress relief; aged - no stress relief; aged - 2 hr stress relief; and aged - 10 hr stress relief). They were heat treated together with one large strained tensile specimen, used for Charpy V-notch specimens, for each of the heat treating conditions.

The post-strain heat treatments were performed in the forced-air, Heavy-Duty furnace. Aging was performed at 370°C (700°F) for 10 hours and air cooled. The stress relief cycle was performed at 620°C (1150°F) for 2 or 10 hours with a subsequent controlled cooling rate of approximately 25°C per hour down to 370°C (686°F) then air cooled. Two stress relief treatments were used, one for 2 hours, the other for 10 hours; both had the same cooling treatment. Charpy V-notch and button-head tensile specimens were machined following all heat treatments as in Figure 2.

Mechanical Testing

Standard 6.35 mm (.252") gage diameter button-head tensile specimens with 25.4 mm (1") gage length were machined transverse to the rolling direction and longitudinal to prestrain direction. ASTM E8 and A370 specifications were followed. The tests were performed on a 44.5 KN (10,000 lbf) Instron Universal testing machine at room temperature. Crosshead speed of 1.27 mm/min (.05 in/min) and a chart speed of 1.27 cm/mm (.5 in/min) were used as well as a 25.4 mm (1") extensometer on the gage length.

For each condition, three specimens were tested and .2% offset yield strength was determined from the chart using the extensometer. The extensometer was removed after passing .2% strain and testing resumed to determine maximum load for the ultimate tensile strength determination. Percent elongation and reduction in area were measured by fitting together the two tensile pieces and measuring final gage length and necked gage diameter.

Charpy specimens with standard 2 mm V-notch were machined transverse to rolling direction and longitudinal to prestrain direction. The notch was machined perpendicular to plate thickness (see Figure 2). A Satec SI-1D impact tester with a maximum impact energy of 325 J (240 ft-lb) was used. Testing was performed using a wide range of temperatures, -75°C to 97°C (-103°F to 207°F) to obtain the ductile-brittle transition temperatures. A methanol bath cooled by liquid nitrogen was used for temperatures below room temperature, and heat water baths were used for temperatures above room temperature. The testing procedure was done according to ASTM standards A370 and E23. The energy absorbed on impact was recorded from the machine and lateral expansion measured by placing a dial gauge along the side of the fractured impact bar. 34 J (25 ft-lb.) energy absorbed and 0.64 mm (25 mils) of lateral expansion were used as the transition temperature criteria.

RESULTS

Tensile Test Results

Room temperature mechanical properties are shown in Table I. The as-rolled plate is slightly stronger and harder than normalized plate, while the normalized plate is tougher and more ductile. The material meets ASTM A572 specifications for Grade 50 in both conditions, although this is not apparent from Table I, because a 50 Ksi yield point rather than a 50 Ksi yield strength is required. Anisotropy is noted most strongly in toughness properties.

Straining of the large tensile specimens was monitored using 25 mm scribe marks on the gage length. Strain was not totally uniform across the gage as is seen in Figure 3. The results are considered good in light of the fact that the specimens are relatively long. Actual plastic strain ranged from 4.0 to 5.5%. As discussed in the PROCEDURE, an extensometer was used as an auxillary measurement method.

The tensile test results are given in Table III and Figures 4 and 5 for the as-received (0% strain), strained, strain-aged, and strain-aged and stress-relieved conditions. As for the unstrained properties of as-received plate, as-rolled strengths are higher than normalized. Straining by itself produces the greatest increase in yield strength, between 96 and 152 MPA (14 and 22 Ksi), with a further increase occuring upon aging, 28 to 34 MPA (4 - 5

Ksi). Stress relieving lowered the strength, as expected, but not to pre-strain levels. The longer stress relieving time was virtually insignificant with respect to restoring tensile properties, as compared to shorter time stress relief.

Charpy Impact Results

Ductile-brittle transition curves for as-received and strain-aged material are illustrated in Figures 6 - 13. Table IV and Figures 6 and 7 summarize the results in terms of transition temperatures for an impact energy of 34 J (25 ft-lbs) and a lateral expansion of 0.64 mm (25 mils). Upper shelf energy is also recorded. The as-rolled plate had a significantly higher transition temperature ($\sim 60^{\circ}$C) than normalized plate. Moreover, the upper shelf energy is higher for the normalized plate as well. Accompanying the increase in strength by straining and subsequent aging is an increase in transition temperature of about 40°C. Stress relieving of as-rolled plate produced some toughness recovery but not full recovery to the unstrained state. Longer stress-relief cycles show no more, and perhaps less, recovery than short-term stress relieving. The normalized plate has negligible recovery of toughness upon stress relieving, regardless of time period.

DISCUSSION

It is clearly seen from the data presented that A572 Grade 50 micro-alloyed steel is susceptible to strain aging. This is illustrated by the fact that the yield and tensile strength increases following an aging treatment subsequent to straining. During the aging treatment, the solute atoms presumably migrate to, or precipitates (carbides or nitrides) interact with, dislocation regions thereby locking them. The flow stress is increased by the locked dislocations. The locking is probably due to carbon diffusion during the aging process since theoretical considerations dictate little nitrogen would be contained in ferrite (α) solid solution. The nitrogen should be tied up with Al and V as nitrides at the 0.04% Al level.

The impact transition temperature increased as the strength increased due to straining and aging, Figure 6. Others[1, 10], have seen this increase in transition temperature; however, Erazo[1] reports that, when prestressed at 5%, little increase occurs during aging for A588 A and B steel. A572 shows as great as, or greater, increase in transition temperature due to aging than the straining process.

As-rolled and normalized material behaved similarly in all processes and heat treatments except the stress relief cycles. In general, the stress relief cycles did not restore a sufficient amount of toughness to prove to be beneficial. However, there was an indication of recovery with short-term stress relief in the as-rolled steel, Figure 6. The stress relief cycles did effect the strengths of as-rolled and normalized steels similarly. The strengths decreased upon stress relieving with yield strength more markedly lowered. The stresses are removed rapidly, as indicated by the yield strength being the same for 2 and 10 hour heat treatments, Table III.

The practical implications of this investigation are quite similar to those that could be derived from the previous studies. The end result of a thermo-mechanical cycle involving straining and aging, with or without a stress relief, is an increase in steel transition temperature. Thus this loss must be accommodated for in the overall materials selection and processing sequence. This can be done by selecting a material of initial condition such that its initial transition temperature is low enough to permit a shift and still retain good toughness at service temperatures (for example, a normalized material), or by more effective post-strain treatments (for example, renormalizing after forming) to restore lost toughness.

CONCLUSIONS

1. The strain-aging behavior of As572 Grade 50 steel is similar to A588 Grades A and B.

2. There is a substantial effect of heat treatment and specimen orientation on the toughness of the steel. Strength is influenced to a much lesser extent. The 34J (25 ft-lb) transition temperature for the as-rolled plate was 60°C (108°F) higher than for the normalized plate.

3. The A572 Grade 50 steel is sensitive to strain aging in both the as-rolled and normalized condition. A plastic tensile strain of 5% produced a shift in 34J (25 ft-lb) transition temperature of about 40°C (72°F).

4. Stress-relief heat treatments had little beneficial effect on toughness after strain aging. Recovery of toughness was no greater for long time stress relief treatments (10 hours) than shorter ones (2 hours).

TABLE I

ROOM TEMPERATURE MECHANICAL PROPERTIES

	.2% Offset Yield Strength		Ultimate Tensile Strength		Elong. (%)	R.A. (%)	Charpy V-notch Energy		Hardness RB
	MPA	(Ksi)	MPA	(Ksi)			J	(ft-lb)	
Longitudinal - As Rolled	352	(51.1)	531	(77.1)	28.9	64.7	61	(45)	85
Transverse - As Rolled	370	(53.7)	532	(77.2)	29.8	64.0	39	(29)	85
Longitudinal - Normalized	341	(49.5)	484	(70.2)	34.4	65.6	128	(95)	77.5
Transverse - Normalized	341	(49.5)	487	(70.7)	34.4	64.5	88	(65)	77

TABLE II

MECHANICAL AND THERMAL TREATMENT SCHEDULE

Strain Level	Aging	Stress Relief
0%	-	-
5%	-	-
5%	370°C (700°F) - 10 Hr	-
5%	370°C (700°F) - 10 Hr	620°C (1150°F) - 2 Hr
5%	370°C (700°F) - 10 Hr	620°C (1150°F) - 10 Hr

TABLE III

STRAIN AGING DATA FOR ASTM A572 GRADE 50

TENSION TEST RESULTS

	Condition	.2% Offset Yield Strength MPA	(Ksi)	Ultimate Tensile Strength MPA	(Ksi)	% El (1" gage)	% R. A.
As-Rolled	Basic (0% Strain)	370	(53.7)	532	(77.2)	29.8	64.0
	5% Strain	522	(75.7)	568	(82.5)	22.0	62.4
	5% + Age	535	(77.6)	593	(86.0)	20.6	61.8
	5% + Age + SR [1]	427	(62.0)	560	(81.3)	23.9	61.4
	5% + Age + SR [2]	410	(59.5)	546	(79.2)	25.9	62.9
Normalized	Basic (0% Strain)	340	(49.3)	487	(70.7)	34.4	64.5
	5% Strain	442	(64.1)	515	(74.8)	28.0	67.2
	5% + Age	472	(68.5)	549	(79.7)	23.6	65.5
	5% + Age + SR [1]	381	(55.3)	510	(74.0)	29.3	66.8
	5% + Age + SR [2]	381	(55.3)	509	(73.9)	29.7	67.5

NOTES:

1) Specimen Orientation: Transverse to rolling direction
   Longitudinal to strain

2) Chart Speed: 12.5 mm/min (.5 inches/min)

3) Crosshead speed: 1.27 mm/min (.05 inches/min)

4) Heat Treatments:

    a. Age: 10 hours at 370°C (700°F)

    b. Stress Relief: SR[1] = 2 hours at 620°C (1150°F)

                    SR[2] = 10 hours at 620°C (1150°F)

TABLE IV

| Condition | Transition Temperature | | | | Upper Shelf Energy | |
| | 345 (25 ft-lb) | | 0.64 mm (25 mils) | | | |
	°C	°F	°C	°F	ft-lb	J
As-Rolled						
Basic (0% Strain)	9	48	19	66	52	71
5% Strain	31	88	30	86	47	63
5% + Age (1)	52	126	47	117	45	61
5% + Age + SR (2)	38	100	36	97	47	63
5% + Age + SR (3)	48	118	43	109	50	68
Normalized						
Basic (0% Strain)	-51	-60	-49	-56	67	90
5% Strain	-34	-29	-30	-22	63	85
5% + Age (1)	-8	18	-9	16	53	72
5% + Age + SR (2)	-10	14	-18	0	61	82
5% + Age + SR (3)	-10	14	-18	0	61	82

NOTES:

1) Specimen orientation: Transverse to rolling direction
   Longitudinal to Strain

2) Heat treatments:
   a. Age: 10 hours at 370°C (700°F)
   b. Stress Relief: 2 hours at 620°C (1150°F)
   c. Stress Relief: 10 hours at 620°C (1150°F)

Source: Welding Research Council Progress Report, November 1983, 2-31

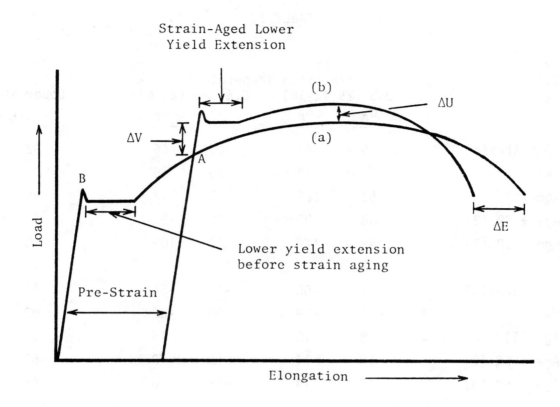

Strain-Aged Lower
Yield Extension

ΔV

B

Load

Pre-Strain

(b)

(a)

ΔU

ΔE

Lower yield extension
before strain aging

A

Elongation

Load/elongation curve for Low-Carbon Steel strained to point A,
unloaded, and then re-strained immediately (curve a) and after
aging (curve b).

ΔY = Change in yield stress due to strain-aging

ΔU = Change in UTS due to strain-aging

ΔE = Change in elongation due to strain-aging

Figure 1. -Strain Aging Effects on Mechanical Properties

Figure 2

SECTION OF PLATE FOR SPECIMEN LOCATION

TOP VIEW - $1\frac{7}{8}$" X $25\frac{1}{8}$" X $60\frac{7}{8}$" ROLLED PLATE

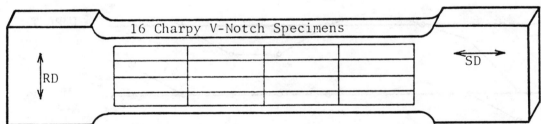

A. Preliminary long. and trans. properties - As Rolled and Normalized
B. 0% Strain Charpy V Specimens - Long. and trans. - As Rolled
C. 0% Strain Charpy Specimens - Long. and trans. - Normalized
D. 5% Strain Large Tensile Specimens - Trans. - Normalized
E. 5% Strain Large Tensile Specimens - Trans. - As Rolled

LOCATION OF SPECIMENS
RELATIVE TO PLATE THICKNESS

A. Button Head Tensile
   (.252 dia.)
B. Charpy V-Notch Impact
C. Large Tensile Strips

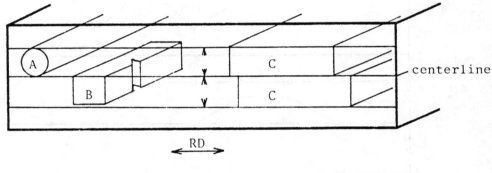

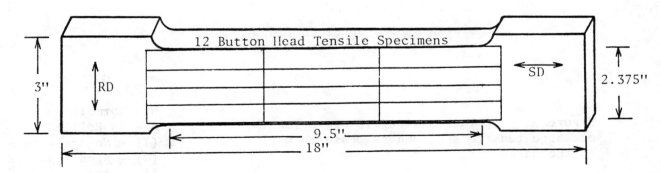

LARGE TENSILE STRIPS FOR STRAINING

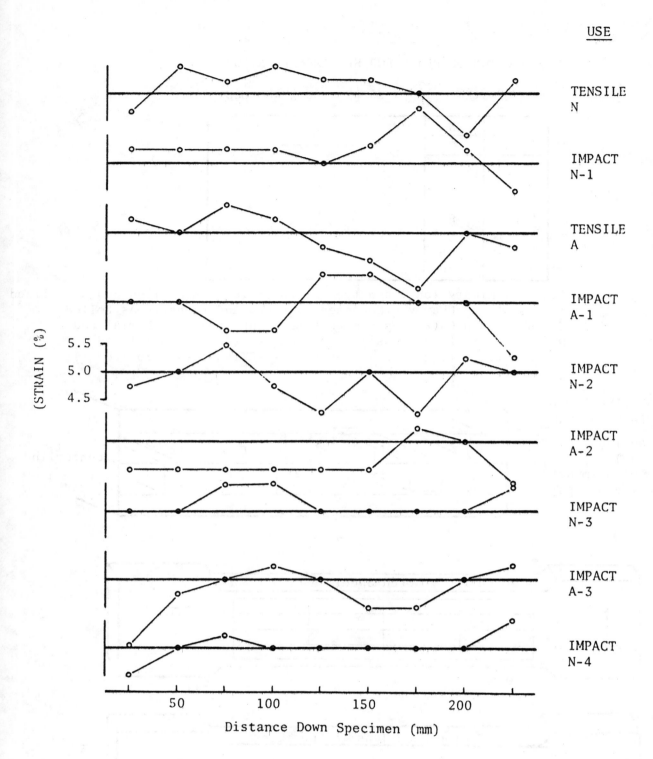

TENSILE
N

IMPACT
N-1

TENSILE
A

IMPACT
A-1

IMPACT
N-2

IMPACT
A-2

IMPACT
N-3

IMPACT
A-3

IMPACT
N-4

(STRAIN (%)

5.5
5.0
4.5

Distance Down Specimen (mm)

50   100   150   200

Figure 3. - Variation along 10" gage length
of large tensile specimen as measured by 1"
scribe marks

USE

N - Normalized
A - As Rolled
(1) - No Age - No S.R.
(2) - Aged - No S.R.
(3) - Aged - 2 Hr S.R.
(4) - Aged - 10 Hr S.R.

ASTM A572 GRADE 50

As-Rolled

Transverse to R.D.

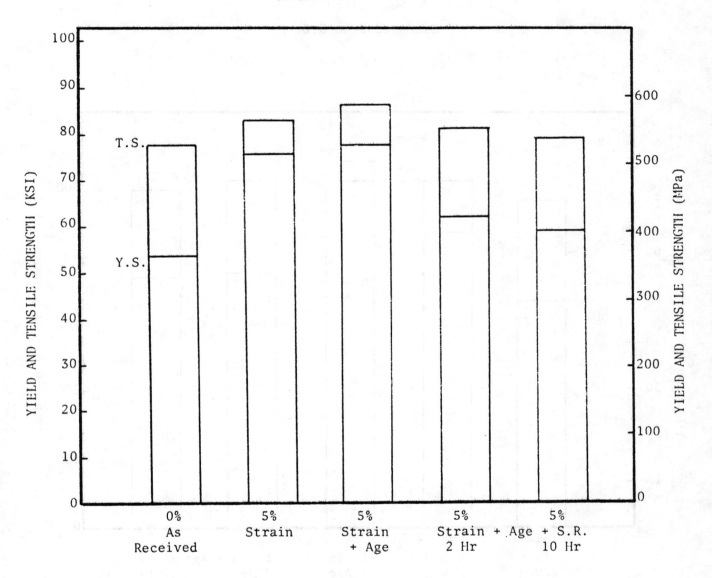

Figure 4. - Effect of strain aging and stress relief heat treatment on the yield and tensile strength

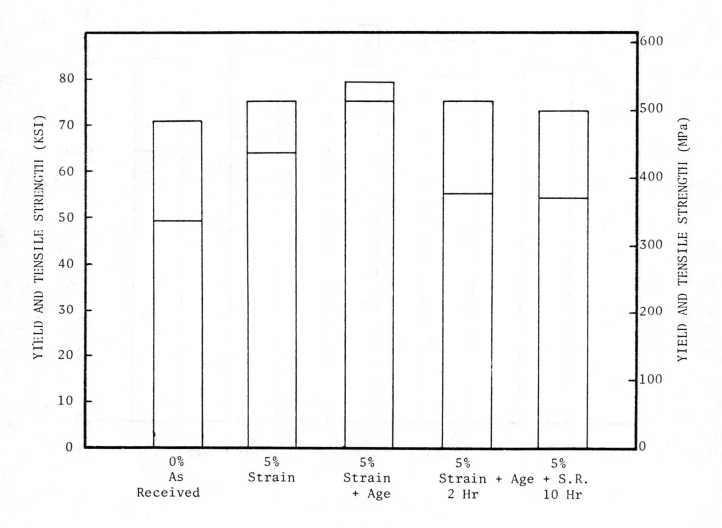

ASTM A572 GRADE 50

Normalized

Transverse to R.D.

Figure 5. - Effect of strain aging and stress relief heat treatment on the yield and tensile strength

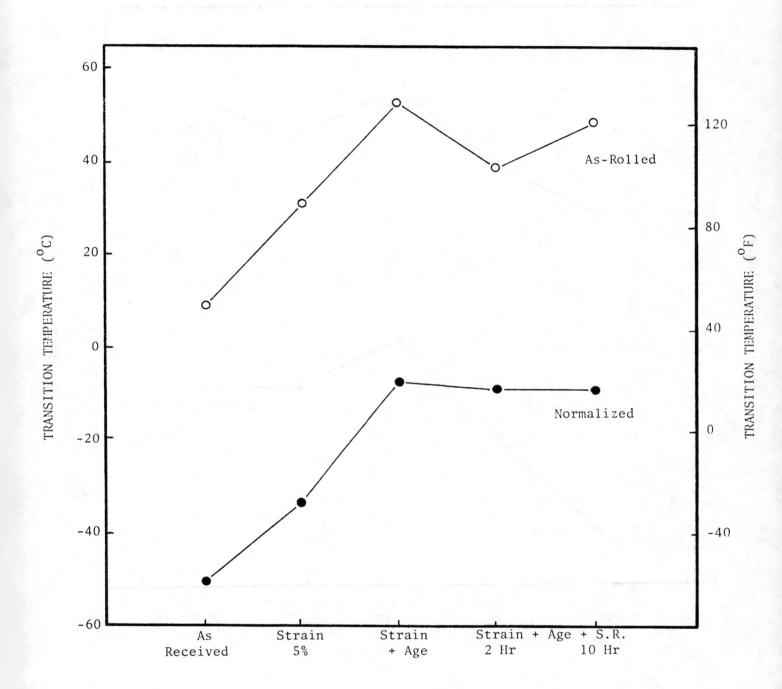

Figure 6. - Effect of strain aging and stress relief heat treatment on the impact transition temperature (measured at 25 ft-lbs)

Source: Welding Research Council Progress Report, November 1983, 2-31

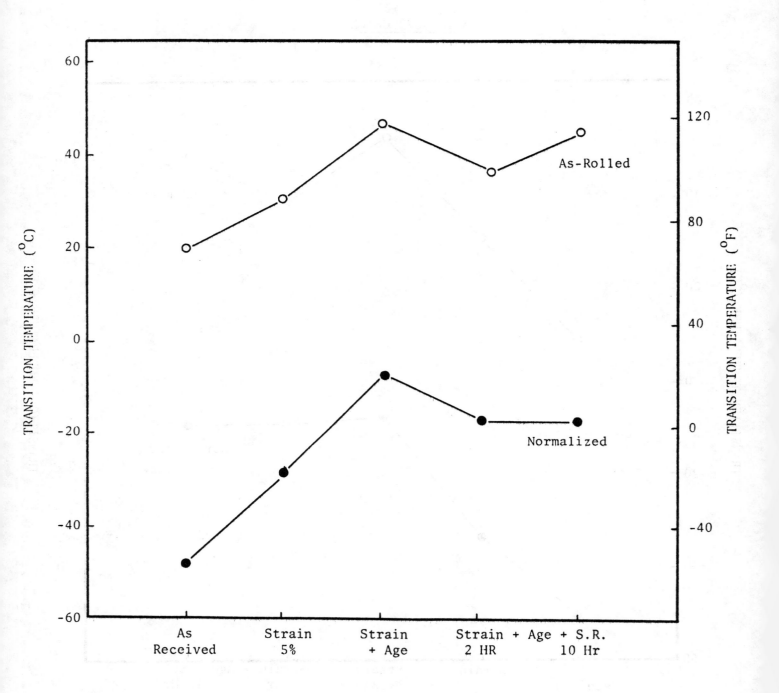

Figure 7. - Effect of strain aging and stress relief heat treatment on the impact transition temperature (measured at 25 mils)

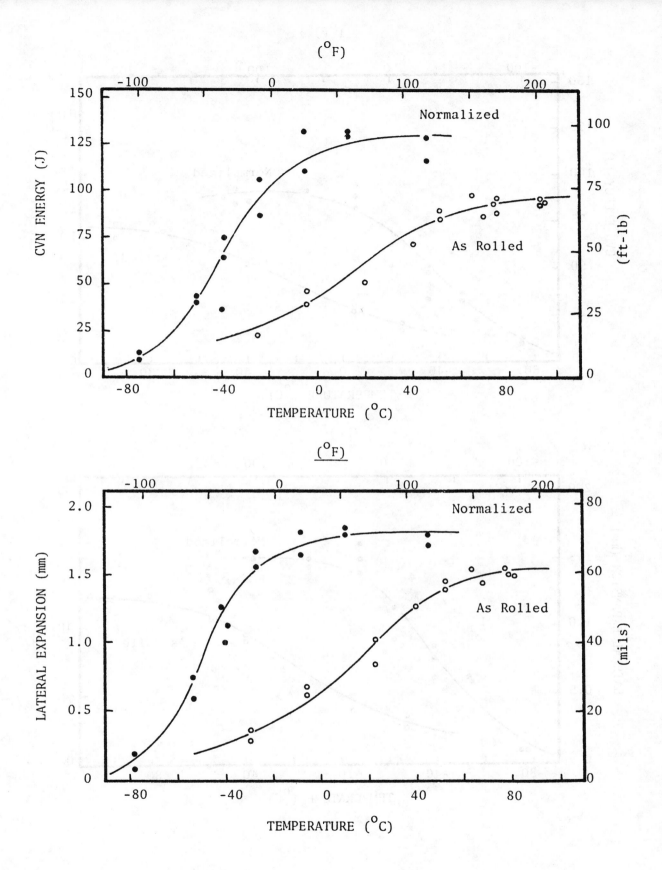

Figure 8. - Charpy transition curves for A572 Grade 50,
Strained 0%, longitudinal to R.D.

Source: Welding Research Council Progress Report, November 1983, 2-31

253

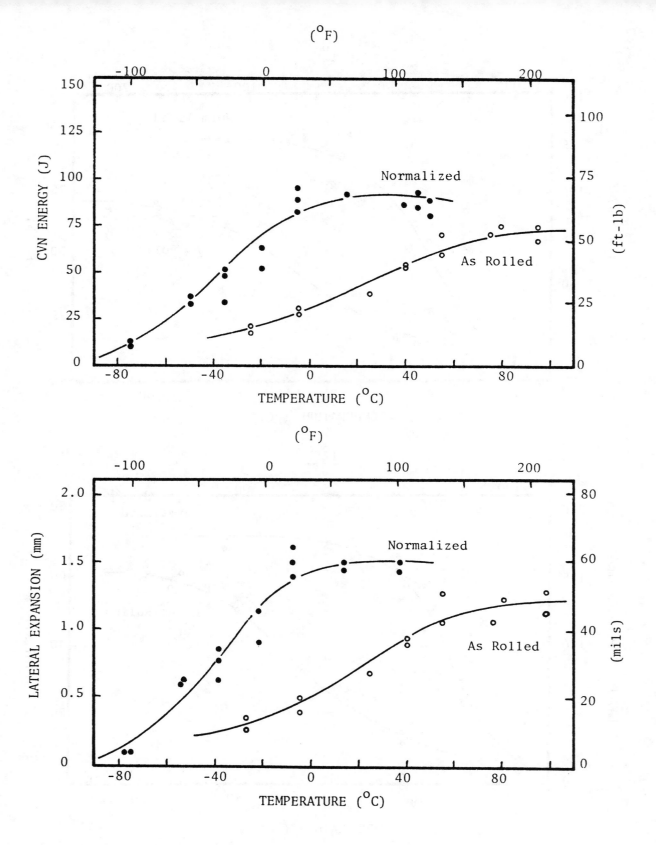

Figure 9. - Charpy transition curves for A572 Grade 50, strained 0%, transverse to R.D.

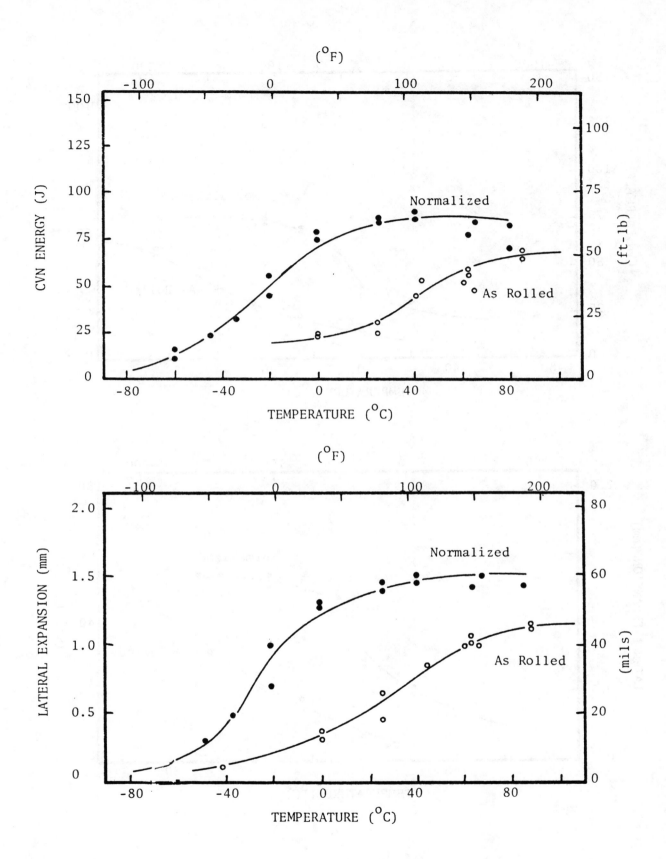

Figure 10. - Charpy transition curves for A572 Grade 50, strained 5%, transverse to R.D.

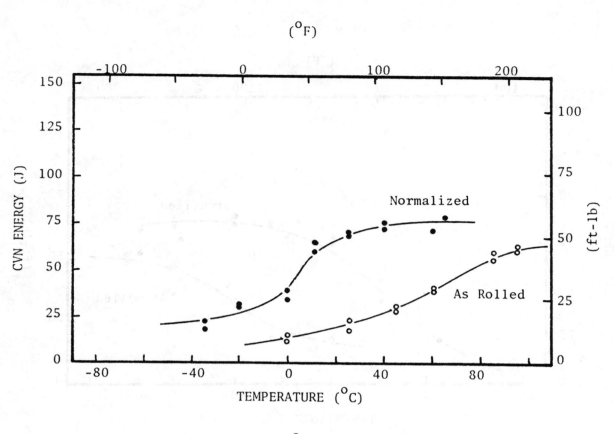

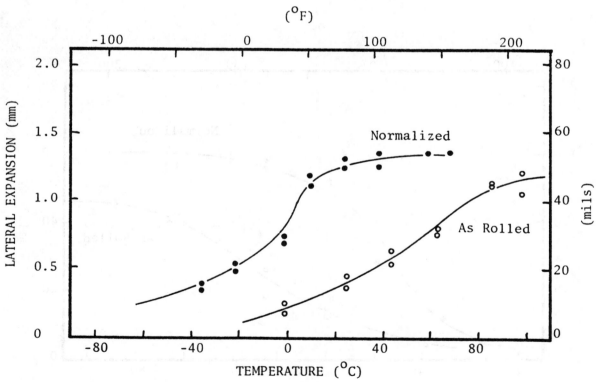

Figure 11. - Charpy transition curves for A572 Grade 50, strained 5%, aged, No S.R., transverse to R.D.

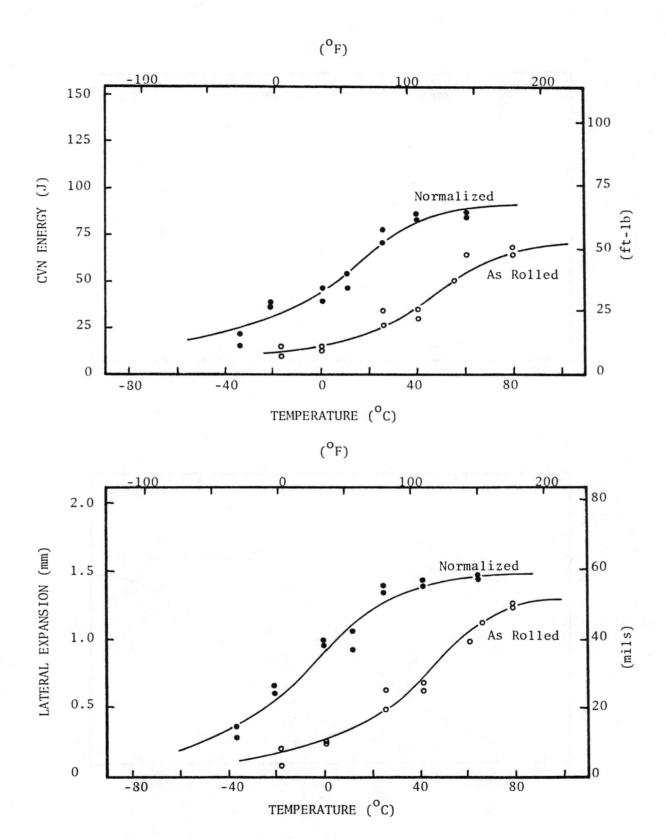

Figure 12. - Charpy transition curves for A572 Grade 50,
strained 5%, aged + 2 hr S.R. transverse to R.D.

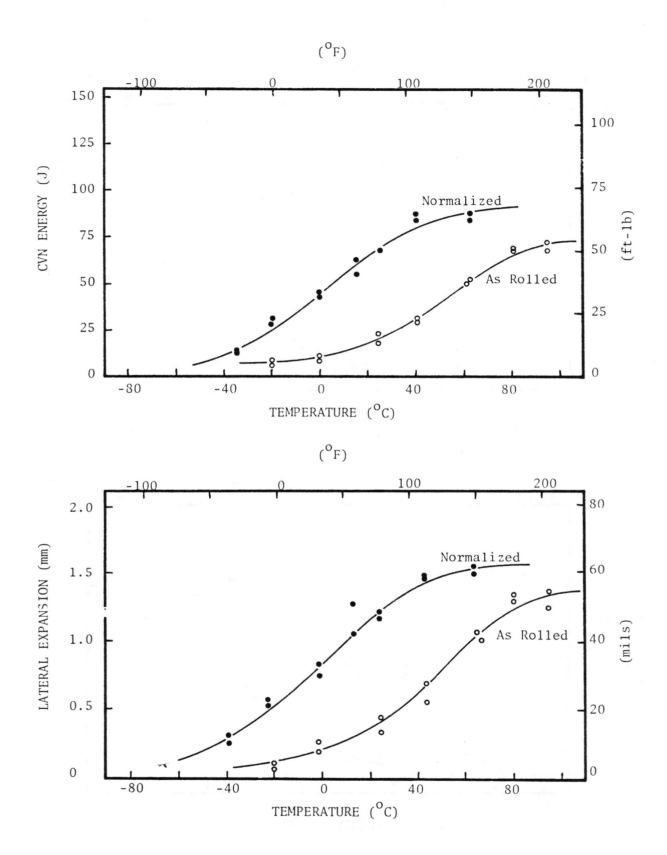

Figure 13. - Charpy transition curves for A572 Grade 50,
strained 5%, aged + 10 hr S.R., transverse to R.D.

REFERENCES

1. Erazo, M. A., Pense, A. W., "The Effect of Thermal and Mechanical Treatments on the Structure and Properties of A588 Grades A and B Steels." Department of Metallurgy and Materials Engineering, Lehigh University, April 1982.

2. Baird, J. D., "Strain Aging of Steel - A Critical Review," Journal of the Iron and Steel Institute, May - September 1963, pp. 186-192, 326-334, 368-374, 400-405, and 450-457.

3. Baird, J. D., "The Effects of Strain Aging Due to Interstitial Solutes on the Mechanical Properties of Metals," Metallurgical Reviews 149, Metals and Materials 5, 1971, pp. 1-8.

4. Wilson, D. V., Russell, B., "The Contribution of Atmosphere Locking to the Strain Aging of Low Carbon Steels," Acta Metallurgica, January 1960, Vol. 8, pp. 36-45.

5. Cottrell, A. H., Bilby, B. A., "Dislocation Theory of Yielding and Strain Aging of Iron," Proceedings of the Physics Society of London, Vol. 62, 1949, pp. 49-62.

6. Shinohe, N., "Stress Relief Effect in Niobium Containing Microalloyed Steel," Master's Thesis, Department of Metallurgy and Materials Engineering, Lehigh University, May 1979.

7. Pense, A. W., Stout, R. D., Kottcamp, E. H., "A Study of Subcritical Embrittlement in Pressure Vessel Steels," Welding Journal, December 1963, pp. 541-546.

8. Rubin, A. I., Gross, J. H., Stout, R. D., "Effect of Heat Treatment and Fabrication on Heavy Section Pressure Vessel Steels," Welding Journal, Research Supplement, April 1959, pp. 182-187.

9. Sekizawa, M., "The Effect of Strain Aging and Stress Relief Heat Treatment on A737 Grade C Pressure Vessel Steel," Master's Thesis, Department of Metallurgy and Materials Engineering, Lehigh University, 1981.

10. Herman, W. A., "The Effect of Strain Aging and Stress Relief Heat Treatment on A737 Grade B Pressure Vessel Steel," Master's Thesis, Department of Metallurgy and Materials Engineering, Lehigh University, 1981.

Pressure Vessel, Piping, and Welding Needs for Coal Conversion Systems

A. G. IMGRAM and R. A. SWIFT

The dissolution/hydrogenation areas of direct coal liquefaction processes, as well as some types of coal gasifiers, impose severe materials performance requirements due to the critical combination of high operating temperatures and high hydrogen partial pressures. Careful consideration must be given to selecting the materials of construction and fabrication procedures for the reactor vessels, heat exchangers, and transfer piping to assure reliability and safety of operation.

REACTOR (DISSOLVER) VESSELS

The materials and fabrication technology that were developed for the dissolver section of the Solvent Refined Coal II liquefaction process will serve as a typical example. Maximum design conditions were 15.5 MPa (2250 psi) at 482 °C (900 °F). A demonstration plant was designed to convert 6000 tons/day of coal into approximately 20,000 bbl/day of liquid fuel. Two very large, relatively complex reactor (dissolver) vessels were required for this size plant. Figure 1 is the drawing for one of the preliminary design configurations. Vessels have a 3.4-m (11-ft) inside diameter, 40-m (130-ft) tangent-to-tangent length, and 300-mm (12-in.) wall thickness and a total weight of approximately 1750 tons.

The material of construction was to be 2.25 Cr-1 Mo steel (SA-387, Gr. 22, Cl. 2). It was selected for resistance to high-temperature hydrogen attack based upon the "Nelson Curve" in API RP 941. Actually, the 15.5-MPa (2250-psi) pressure at 482 °C (900 °F) places 2.25 Cr-1 Mo steel above the limit established by the Curve, but since hydrogen is less than 90 pct of the gas phase, the limit is not exceeded.

DESIGN TEMPERATURE AND STRESS

These vessels could have been designed and fabricated with an extrapolation of the technology that has been successfully used by the petroleum refining industry[1,2] for hydrocracking and hydrodesulfurization (HDS) reactor vessels. This, however, would almost double the existing 900-ton maximum weight for a 2.25 Cr-1 Mo steel reactor vessel. To take advantage of more rigorous design analysis and thorough inspection requirements, Sect. VIII, Div. 2, of the ASME Code was used for the basis of design, but the lower design stresses given in Div. 1 were used. The reason for this decision is the higher design temperature of the SRC-II process combined with the effect of fabrication history upon the strength of 2.25 Cr-1 Mo steel. Maximum design temperature for HDS reactors is 454 °C (850 °F), whereas the SRC-II process requires design at 482 °C (900 °F).

The problem with the higher design temperature is shown in Figures 2 and 3.[3] The SA-387, GR. 22, Cl. 2 material is relatively unaffected by temperature up to 400 to 427 °C (750 to 800 °F). Above 482 °C (900 °F), the strength decreases rapidly with increasing temperature. When strength is expressed as the fraction of room temperature strength (Figure 3), these effects are more readily seen. Norris and Wylie[5] showed that the design criterion for 2.25 Cr-1 Mo steel changes from static (tensile governed) to time-dependent (stress-rupture governed) in the temperature range of 454 to 482 °C (850 to 900 °F), based upon Div. 1 and 2 criteria for establishing allowable design stresses

A. G. IMGRAM is with Gulf Research and Development, Pittsburgh, PA. R. A. SWIFT is with Lukens Steel Company, Coatesville, PA.

Reprinted from Journal of Materials for Energy Systems, December 1985, 212-221, © 1985 American Society for Metals

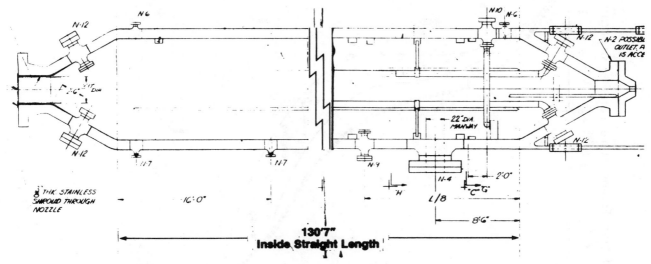

Fig. 1 — Preliminary design configuration for Solvent Refined Coal II dissolver vessel.

(Figure 4). The available stress-rupture data compiled by Smith[6] which are used to set the allowable design stresses are based largely upon data from lots 2.25 Cr-1 Mo steel having ultimate tensile strengths considerably above the 517 MPa (75 ksi) minimum allowed for SA-387, GR. 22, Cl. 2 material (Figure 5). For example, the rupture life data shown in Figure 6 demonstrate the ability of this material to meet the code requirements with room temperature strengths as low as 517 MPa (75 ksi).[3] In this regard, the post weld heat treatment (PWHT) required by the Code becomes a very important factor. A holding time of 6.75 hours at 677 °C (1250 °F) minimum is required for the 300-mm

(12-in.) wall thickness. Any weld seam, and adjacent base plate, could be subjected to up to four cycles (27 hours) PWHT during the reactor's service life. At least one weld must receive a local PWHT which would overlap with the initial heat treatment, because the reactors are too long to fit inside the various fabricators' largest heat treating furnaces. The fabricator may have to make a repair necessitating a third PWHT. Normally, provision is made for a fourth heat treatment to permit the operator to make a repair or modify the reactor after it has been placed in service. In this manner, the total possible temper parameter $[P = T(C + \log t) \times 10^{-3}]$ for PWHT exceeds 20.33 (36.6 for °R).

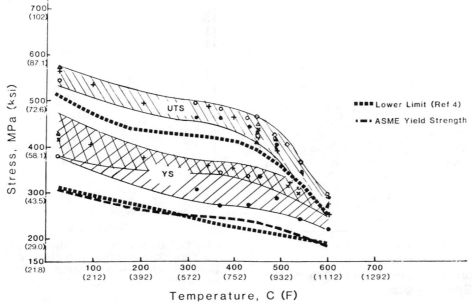

Fig. 2 — Elevated temperature strength of NT plate and all QT plates heat treated to SA 387-22 Class 2 and SA 542-3 properties.

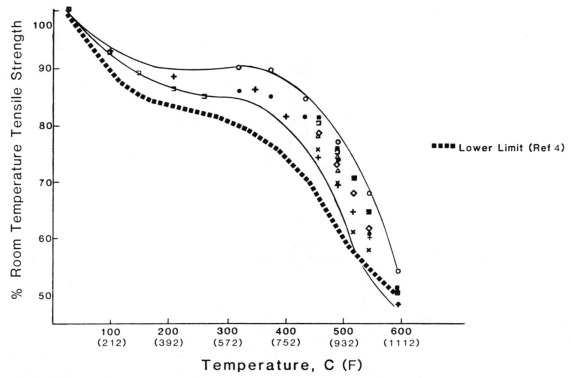

■■■■ Lower Limit (Ref 4)

Fig. 3 — Elevated tensile temperature strength as a percentage of room temperature tensile strength.

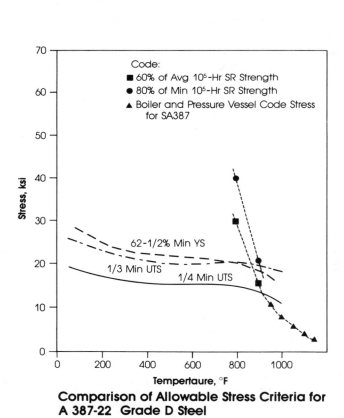

Comparison of Allowable Stress Criteria for A 387-22 Grade D Steel

Fig. 4 — Comparison of allowable stress criteria for A 387-22 grade D steel.

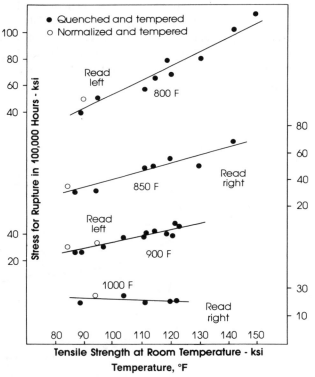

Effect of Room Temperature Tensile Strength on 100,000 Hour Rupture Strength of Quenched-and-Tempered and Normalized-and-Tempered 2¼ Cr-1 Mo Steel

Fig. 5 — Effect of room temperature tensile strength on 100,000 hour rupture strength of quenched-and-tempered and normalized-and-tempered 2.25 Cr-1 Mo steel.

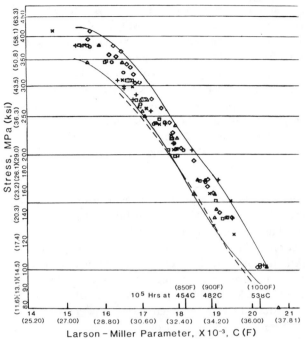

Fig. 6—Stress-rupture data plotted as a function of the Larson-Miller parameter.

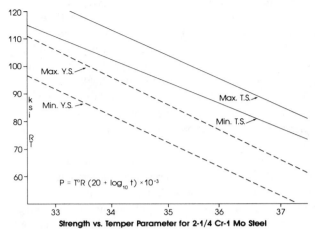

Fig. 7—Strength *vs* temper parameter for 2.25 Cr-1 Mo steel.

process fluid leading to spalling of the insulation and hot spots on the reactor shell. Internal attachments would provide a direct path for heat conduction to the shell, possibly resulting in hot spots. Any hot spots on the 2.25 Cr-1 Mo steel shell of the SRC-II reactor could be especially serious since it would raise the temperature from a relatively high stress, tensile limited design to the temperature regime where stress-rupture is the controlling failure mode at appreciably lower stresses.

An analysis of judiciously selected data concerning the effect of PWHT upon room temperature strength, summarized in Figure 7, indicates that the tensile strength could drop to 552 MPa (80 ksi) at this temper parameter.[5] In effect, strength decreases with increasing gage primarily because of the additional time required by the Code for PWHT. In actual practice, the control point for PWHT is usually set at 690 °C (1275 °F) to assure obtaining the required 677 °C (1250 °F) minimum. This corresponds to a temper parameter of 20.66 (37.2 for °R) for four PWHT cycles. Since a temper parameter of 20.49 (36.9) is reached after three cycles, tensile strengths below 552 MPa (80 ksi) can be expected even if the contingent fourth cycle of heat treatment is waved. An extrapolation of the data in Figure 5 to lower tensile strengths shows that the average stress-rupture strength for 2.25 Cr-1 Mo steel with a room temperature tensile of 552 MPa (80 ksi) or lower is not high enough to support the Div. 2 allowable design stress at 482 °C (900 °F) (Figure 8). However, the average stress-rupture strength for material with the minimum 517 MPa (75 ksi) tensile strength is ample to support the allowable design stress in Div. 1 for SA-387, Gr. 22, Cl. 1 (annealed) material at 482 °C (900 °F).

Consideration was given to design of an insulated, cold wall SRC-II reactor. It was rejected because the process fluid could infiltrate the insulation and impair its insulating properties. Subsequent cracks in the insulating properties. Subsequent cracks in the insulation can be penetrated by the

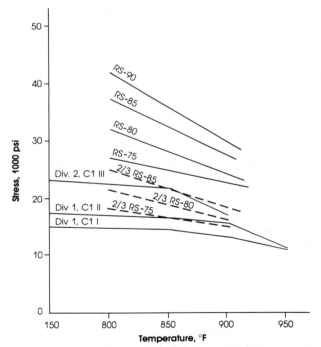

Rupture Strength (RS) and Allowable Design Stresses for 2-1/4 Cr-1 Mo Steel (A 387-22).

Fig. 8—Rupture strength (RS) and allowable design stresses for 2.25 Cr-1 Mo steel (A 387-22).

Source: Journal of Materials for Energy Systems, December 1985, 212-221

WELD OVERLAY

An austenitic stainless steel weld overlay is required to protect the reactor shell from high-temperature sulfide corrosion. A two-layer 309-347 system was selected for the SRC-II vessel. Both submerged-arc and electroslag welding procedures have been developed to apply this type of overlay. Both have been proven to provide reliable coatings in HDS reactors. The 309 layer is applied directly upon the 2.25 Cr-1 Mo base metal with a welding procedure designed to minimize dilution. The higher alloy content of the 309 stainless steel prevents formation of a hard, martensitic microstructure in the dilution zone at the interface with the base metal. This layer is susceptible to intergranular sensitization so that the sulfide scale that forms on the overlay during operation could lead to polythionic acid cracking during shutdowns when the reactor is opened for inspection and/or maintenance. To prevent this, a 347 stainless steel overlay is specified for the second layer that is exposed to the process fluids. This is a stainless steel that is Cb stabilized to impart a high resistance to intergranular sensitization. The chemical composition is balanced for the welding procedure employed to control the ferrite content to between 5 and 10 pct by microstructural point count. Ferrite contents below 5 pct can cause excessive hot cracking during solidification, whereas ferrite contents above 10 pct can result in hydrogen-induced cracking (HIC), especially in areas of high applied stress.

Some disbonding of austenitic stainless steel weld overlays has occurred in HDS reactors.[6] This disbonding has been detected primarily by UT inspection from the outside. There has been no blistering or cracking of the overlay cladding that has lead to corrosion, cracking, or other deterioration of the 2.25 Cr-1 Mo steel reactor shell. Furthermore, fracture mechanics analyses have indicated that discontinuities formed by disbonding do not directly endanger the reactor. The only structural requirement for the overlay is where internal attachments are welded to it. Repairs are generally recommended only at these locations. Despite the innocuous nature of the disbonding that has been detected, the major reactor fabricators have conducted research programs to investigate the phenomena. The disbonding has always occurred in the overlay at the first grain boundary from the interface with the base metal and has been related to the hydrogen that permeates the overlay during operation. Welding procedures have been developed that produce microstructures in the overlay, at the interface with the reactor shell, that are significantly less susceptible to hydrogen-induced disbonding.

INTERNAL ATTACHMENTS and FLANGE GASKET GROOVES

Cracks have developed in internal attachment fillet welds that have, on occasion, propagated through the overlay into the 2.25 Cr-1 Mo shell of HDS reactors requiring a difficult and costly repair.[7] The attachments were Type 321 stainless steel and were fillet welded to the two-layer 309-347 weld overlay with 347 filler metal. It was found that the delta ferrite in the 347 fillet weld partially converts to sigma phase during PWHT. The presence of sigma phase, in combination with the relatively high stress, makes the fillet weld susceptible to HIC when hydrogen permeates the weld during operation. It was observed that most of these cracks stopped propagating at the 347/309 interface in the two-layer weld overlay. There is less transformation of delta ferrite to sigma phase in 309 during PWHT and, therefore, it is less susceptible to HIC.

A fillet welding procedure has been developed for internal attachments that has proved to be resistant to this type of cracking. The internal attachment is fillet welded to the 309 layer of the weld overlay with 308 filler metal. Final PWHT is performed, and then the 308 and 309 are covered with 347. In this manner, sigma phase is not formed in the 347, and it is removed from the highest stressed regions of the attachment weld. A similar situation of HIC has occurred in highly stressed flange gasket grooves. This is now avoided by a three-layer 309-308-347 overlay, in the region of the grooves, where the 347 layer is applied after PWHT.

NOZZLES

All of the many nozzles are integrally reinforced for even thermal distribution and ease of inspection, regardless of permissible Code exemptions for the smaller sizes. The most critical were the liquid and gas quench nozzles, because the temperatures that the liquid and gas enter the reactor are below the shell temperature. An insulated nozzle design is employed to sufficiently reduce thermal gradients to avoid serious complications of a high thermal stress and/or thermal fatigue.

STEEL REQUIREMENTS

The 1750-ton reactor vessels for the demonstration plant could have been built from either plate or ring forgings, and either in the fabricator's shop or at the proposed plant site near Morgantown, West Virginia. Ring forgings eliminate the need for longitudinal welds, but longitudinal welds have given very reliable service in hydrocracker and HDS reactors for many years, providing proper welding procedures, and PWHT are specified. Very efficient welding procedures for longitudinal seams, such as electroslag (ESW) and more recently metal-inert-gas (MIG), generally make the cost of vessels fabricated from plate competitive with costs of vessels fabricated from ring forgings. The extra welding for plate fabrication can be offset by the higher cost per pound of the forgings. It is good practice to refine the grain size of the ESW by a renormalization, then temper to obtain acceptable toughness in the weld metal and heat affected zone (HAZ). Other welds require only a PWHT.

Both electric furnace and BOF steel-making practices are acceptable for producing 2.25 Cr-1 Mo steel for either plate or forgings. Because of the heavy gages required for these vessels, toughness is of utmost importance. The toughness levels attainable in SA-387, Gr. 22 are far superior to those believed possible in the 1960's. Presently 54 J (40 ft-lbs) at −40 °C (−40 °F) is a commonly specified level in gages to 300 mm (12 in.). This can be achieved by low sulfur, Ca treatment for shape control of inclusions and, less importantly, by microstructural control.[10-15] It has been shown that the desired toughness can be obtained with as much as 40 pct polygonal ferrite.[14] Figure 9 shows representative microstructures and the toughness for each structure. Fracture toughness tests of this same material substantiate these findings.[15]

Modern steel-making practices produce steels of sufficiently high purity that temper embrittlement can be virtually eliminated. Temper embrittlement susceptibility of 2.25 Cr-1 Mo steel can be judged by the J-factor.[16] This has been related to specific intentional additions and impurities by the equation

$$J = (\text{pct Si} + \text{pct Mn})(\text{pct P} + \text{pct Sn}) \times 10^4 \quad [1]$$

Good resistance to temper embrittlement is generally obtained with a J-factor below 200, but there is appreciable scatter in the correlation between the J-factor and the shift in the 54-J (40 ft-lb) CV transition temperature for individual heats.[17] It is thought that temper embrittlement is directly attributable to the segregation of P and/or Sn to prior austenitic grain boundaries during tempering at temperatures between 427 °C and 593 °C (800 °F and 1100 °F), or during prolonged service at temperatures as low as 371 °C (700 °F) with material that was post weld heat treated at 677 °C (1250 °F). The P and Sn segregation to the grain boundaries results in intergranular fracture and corresponding low toughness. The Si and Mn promote the segregation of P and Sn to the grain boundaries. Some steel suppliers prefer to reduce the susceptibility to temper embrittlement by obtaining a very low Si content,[10,11] whereas others indicate that this can adversely affect high-temperature strength[12,13] and, therefore, prefer to keep P and Sn contents very low.

Control of temper embrittlement in weld metal is considerably more difficult than with plate and forgings. Higher Si and Mn contents are necessary to sound weld metal. Basic fluxes generally provide the minimum susceptibility to temper embrittlement consistent with high-temperature strength requirements. Correlation of the J-factor with the temper embrittlement of weld metal is poor.[17] This is probably due to the significantly different chemical composition and microstructure of weld metal compared to the wrought product. Therefore, the susceptibility of weld metal to temper

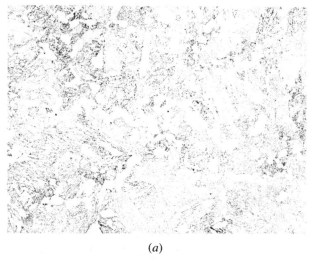

(a)

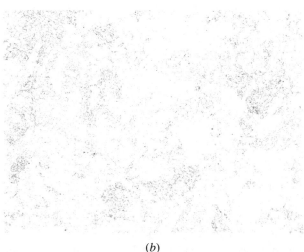

(b)

Fig. 9—Representative microstructures in heavy gage plate having in excess of 54J (40 ft-lb) Charpy V impact energy at −40 °C (−40 °F).

embrittlement is frequently determined by a direct measurement of C_v impact toughness using the equation

$$v\text{Tr }40 + 1.5\,\Delta_v\,\text{Tr }40 < 100\ °\text{F} \quad [2]$$

where vTr 40 is the transition temperature for unembrittled weld metal, and Δ_v Tr 40 is the change in transition temperature for weld metal that has been step cooled to cause temper embrittlement. It has been estimated that the temper embrittlement that can occur in 30 years of service is approximately three times greater than the embrittlement caused by step cooling, but there is considerable scatter in the data.[17]

It was decided not to overemphasize control of temper embrittlement in the specifications for the SRC-II reactor vessels, especially in view of the as-yet-unresolved discrepancies concerning the effects that some of the control methods may have upon high-temperature strength. Much of this concern has now been settled with the creep rupture data being generated on the state-of-the-art material[3] (see Figure 6). Furthermore, the greater susceptibility of weld metal to temper embrittlement makes highly restrictive specifications for plate and/or forgings, that can incur higher materials costs and possibly compromise strength, of questionable value.

It should also be emphasized that no reactor vessel has failed in service due to temper embrittlement. This is in spite of their construction from heats of 2.25 Cr-Mo steel that were state-of-the-art but now would be considered highly susceptible to temper embrittlement. It was considered advisable to adhere to the API guidelines that limit pressure to 25 pct of the design at temperatures below 130 °C (265 °F), regardless of temper embrittlement controls, in view of the wide scatter in temper embrittlement susceptibility data especially for weld metal. Limiting the applied stress at low temperatures in this manner essentially eliminates any possibility of brittle fracture.

COMMERCIAL PLANT REQUIREMENTS

It was desired to replace the two reactors in the SRC-II demonstration plant with one reactor containing twice the volume, for each 6000-ton/day module, in the eventual commercial plants. Economics in plant construction and simplified operation and maintenance would be achieved. The number of pumps and valves and the total length of piping would essentially be doubled for a two-train plant with two reactors. The complexity of operation and the cost of maintenance is more directly related to number of pieces of equipment rather than to the size.

A single reactor for a 6000-ton/day module would be approximately 4.6 m (15 ft) in diameter with a 400-mm (16-in.) wall thickness and would weigh approximately 3500 tons. A reactor this size would stretch pressure vessel technology beyond the present state-of-the-art. It is unlikely that the present SA-387, Gr. 22 has sufficient hardenability to develop a minimum tensile strength of 517 MPa (75 ksi) for Cl. 2 properties in this thickness. The cooling rate for water-quenched plate and forgings decreases sufficiently that polygonal ferrite can appear at the centerline of 300-mm (12-in.) thick material,[14] although the data and interpretation of microstructures vary.[19] Hardenability can be increased by increasing the austenitizing temperature from the customary 945 to 1052 °C (1750 to 1925 °F) or higher,[13] but there is some loss in C_v impact toughness due to a larger grain size. Vigorous research is now underway to develop modified Cr-Mo steels with increased hardenability and acceptable high-temperature strength in very thick gages.[12,20-22]

FABRICATION

Techniques were available to fabricate the 1750-ton reactors for the SRC-II demonstration either in a shop or at the plant site, with equal assurance of high integrity for safe operation.[21] The proposed plant site near Morgantown was on a navigable river, which avoided the severe problem of overland transportation for a vessel of this weight. The major trade-off between shop and field fabrication was the higher transportation and erection costs for shop fabrication vs the higher welding, heat treating, and inspection costs for field fabrication. The 3500-ton weight of the larger vessels for the eventual commercial plant would make field fabrication at the plant site almost mandatory. The preferred concept was to prefabricate ring course and head assemblies in the shop for final assembly in the field in a vertical position. All necessary longitudinal and nozzle welds would be made, post weld heat treated, and inspected in the fabrication shop prior to shipment to the field. Only girth welding would be performed in the field. Each girth weld would be locally post weld heat treated as it is completed. These girth welds could be made in a horizontal position with procedures and welding machines essentially identical to girth welding in a shop. However, stacking the ring courses vertically for girth welding would circumvent a very costly lift of the completed reactor.

Inspection of welds made in the field presents some difficult problems. Radiography is presently required by ASME Code, Sec. VIII. This can be accomplished with linear accelerators, but special licensing must be obtained.[21] A major difficulty, that incurs considerable extra cost, is shielding to protect job site personnel and surrounding areas against the radiation. This is particularly difficult if the girth welds must be inspected high above ground level as ring courses are stacked vertically. Furthermore, lengthy exposures are required to obtain satisfactory radiographs of 300-to-400-mm (12-to-16-in.) thick welds. A possibly greater cost penalty is the necessity to stop all other work on the reactor, and very likely all construction in adjacent areas of the plant, while the radiographic exposures are being made. Acceptance of the vessel based upon ultrasonic inspection would be highly desirable for these reasons.

WELDING

Development of high-speed welding processes and procedures could be of very significant benefit, especially for field fabrication. Shop-fabricated reactors are the longest lead time items and are usually ordered before plant site preparation is begun. They must be in place before the other equipment surrounding them can be installed. Therefore, field-fabricated vessels must be completed very quickly in order not to delay plant commissioning. Submerged-arc is presently the most widely used process for girth welds. It is an efficient process that uses a relatively narrow groove, but very careful selection and control of wire and flux is required to obtain good weld metal properties. Automatic, narrow groove MIG welding is being commercialized for vessel fabrication.[21] The groove is appreciably narrower than for submerged-arc welding, and no flux is required. Therefore, economies are realized through the reduction of welding time and labor, and in the cost of consumables. The most dramatic improvements could be realized by adaptation of the electron beam process for heavy wall vessel fabrication. The Department of Energy has sponsored one study,[22] and another private venture demonstrated its applicability for making heavy section girth welds with ring courses stacked vertically to simulate field fabrication.[23] Welding time is very short, it may be possible to eliminate preheat, and filler metal is not required. These very significant potential advantages must not be offset by the extra costs of the very precise joint preparation and fit-up required, to realize the great benefits of the process.[20] The weld and heat affected zones are very narrow compared to other processes, and the weldment can have strength and toughness properties that are essentially identical to the base metal.[20,23]

DOWNSTREAM VESSELS

Several other relatively large pressure vessels are required downstream of the reactors in the SRC-II plant. The first is the effluent separator, which is almost the same size as the dissolvers. Design temperature and pressure are the same as for the dissolver, and therefore, the materials and fabrication considerations are essentially identical. Effluent letdown drums are smaller vessels that operate at a slightly lower temperature. Nevertheless, 2.25 Cr-1 Mo steel is still required for resistance to high-temperature hydrogen attack. The 316 °C (600 °F) design temperature of the first high-pressure separator permits the use of 1.25 Cr-0.5 Mo steel. The second and third high-pressure separators have design temperatures of 204 °C (400 °F) or lower which permits the use of carbon steel. Similarly, all other low-pressure flash drums and separators are fabricated from carbon steel. Usually, SA-516 carbon steel is specified because of its good toughness. Post weld stress-relief is required for all carbon steel vessels, regardless of possible Code exemptions, to assure immunity to hydrogen stress cracking in H_2S service. Local regions of weld metals and heat affected zones can have a high enough hardness to risk failure if this precaution is not taken.

All vessels with a design temperature of at least 204 °C (400 °F), whether Cr-Mo or carbon steel, must be protected from high-temperature sulfur corrosion by a layer of austenitic stainless steel. Roll or explosion bonded Type 321 cladding is more economical for vessels with a wall thickness of 100 mm (4 in.) or less. Thicker walled vessels require the two-layer 309-347 weld overlay described for the reactor vessels. The rolling reduction for plates thicker than 100 mm (4 in.) does not always develop a strong enough

bond, and explosion cladding facilities are not available to economically handle 50- to 75-ton plates.

Carbon steel vessels, that operate at below 204 °C (400 °F) can be of multilayer construction. This can be more economical for vessels with a wall thickness of approximately 150 mm (6 in.) or greater such as the high-pressure separators. Multilayer carbon steel vessels have been used for many years in hydrodesulfurization plants and have given very reliable service.

All of the vessels downstream of the reactors, with the possible exception of the effluent separator, are comfortably within the pressure vessel technology that has been used very successfully for hydrocracker and hydrodesulfurization plants. Although advancements in technology are not necessary for their construction and successful operation in an SRC-II plant, there would be considerable economic benefit from some of the improvements in technology required to make the 3500-ton reactors for the commercial plant feasible.

PIPING

Transfer line piping materials selections for the SRC-II demonstration plant were basically Type 321 stainless steel for slurry and gas streams at 204 °C (400 °F) and above, 7 Cr-0.5 Mo steel for oil streams and 204 °C (400 °F), and carbon steel for oil streams below 204 °C (400 °F) based upon required corrosion resistance. The most demanding metallurgical requirements were for the stainless steel lines. Very large diameter, extra heavy wall pipe was required for critical, high-pressure service.

The slurry and gas streams contain H_2S that reacts with stainless steel at high temperature to form a sulfide film on the surface that protects against catastrophic sulfidation attack. However, this film can interact with moisture and oxygen in the air to form polythionic acid, when the stainless steel is at ambient temperature during shutdown periods. Polythionic acid can cause intergranular cracking of austenitic stainless steels if they are sensitized by the precipitation of chromium carbides in grain boundaries due to welding or other exposure to temperatures between approximately 427 °C and 816 °C (800 °F and 1500 °F). Type 321 stainless steel is stabilized with Ti and can be made immune to sensitization by a double heat treatment. A solution anneal at 1063 °C (1950 °F) followed by a rapid cool dissolves all metal carbides that might be present after hot working. A second stabilization anneal for 4 hours at 900 °C (1650 °F) causes the precipitation of titanium carbides. Once the titanium carbides are precipitated, there is insufficient carbon remaining in solution to permit chromium carbides to precipitate in the grain boundaries. The stabilization anneal, therefore, makes Type 321 essentially immune to sensitization. If the stabilization anneal is omitted, some chromium carbides will precipitate in the grain boundaries concurrent with the precipitation of titanium carbides upon exposure to temperatures in the sensitizing range, producing susceptibility to intergranular cracking in severe environments. It should be noted that the region of a heat affected zone that is heated above 1066 °C (1950 °F) during welding will, in effect, be re-solution annealed. Operation at temperatures in the sensitizing range could make this region of the heat affected zone susceptible to intergranular cracking. This is not an important consideration for transfer line piping, because it will operate at lower temperatures than required for sensitizing. However, it is of concern for the Type 321 heater tubes that can operate at 538 °C (1000 °F) or higher. For this reason, a local post weld stabilization anneal is desirable for all welds in heater tube coils.

Either seamless (A 312) or seam welded (A 358) Type 321 pipe can be used for the transfer lines. The seam welded pipe must be of either the Class 1 or Class 2 variety, that requires welding with the addition of filler metal and complete radiography, to obtain the highest assurance of weld integrity and reliability.

Any cold forming of the Type 321 pipe must be followed by the solution anneal and stabilization anneal heat treatment. Failures of small diameter stainless steel tubing during operation of one small SRC-II pilot plant revealed that cold-worked austenitic stainless steels are susceptible to hydrogen-induced cracking in the SRC-II process environment.

Drains and vents should be eliminated from the Type 321 transfer lines wherever possible. They act as traps for all of the contaminants in process streams that can cause stress-corrosion cracking of austenitic stainless steel during shutdown periods. Failures have occurred in hydrodesulfurization plants and a large SRC-II pilot plant because of this. This could be more of a problem in coal liquefaction plants due to the higher chloride content of most coals compared to crude oil.

In addition to the above heat treatment and fabrication requirements, specific procedures are recommended during shutdowns as a further precaution to prevent stress-corrosion cracking. They involve neutralization with a caustic soda wash or ammonia, and blanketing with dry nitrogen to prevent contact with oxygen and moisture condensation on the internal surfaces.

The only special requirements for the 7 Cr-0.5 Mo and carbon steel piping are that seal welds with austenitic stainless steel filler metal are prohibited. A martensitic microstructure is developed in the dilution zone of the weld that has a high hardness that cannot be lowered appreciably by post weld heat treatment. This hard, martensitic microstructure tends to be notch sensitive and, therefore, is conducive to crack nucleation at the toe of a seal weld. Mechanical and/or thermal fatigue are the predominant mechanisms of crack nucleation and propagation. Thermal fatigue is aggravated by the relatively large difference in thermal expansion between the austenitic filler metal and the ferritic base metal.

Seal welds should be made with a filler metal having a matching chemical composition. INCONEL* filler metals

---

*INCONEL is a trademark of the INCO family of companies.

---

would be less disadvantageous than austenitic stainless steel, with regard to developing an adverse microstructure and the mismatch in thermal expansion. However, they do not avoid the necessity for post weld heat treatment, and therefore, there is no advantage to using them.

NEEDED RESEARCH

Largely because of the concern about the stress-rupture strength of heavy gage 2.25 Cr-1 Mo material, the MPC, with API support, has embarked upon a program that should resolve the question. The first major goal is to increase the tensile strength of 2.25 Cr-1 Mo steel to 586 to 758 MPa (85 to 110 ksi) by reducing the minimum post weld heat treatment temperature to 621 °C (1150 °F). This will assure obtaining the average stress-rupture strength to sustain the Div. 2 allowable design stress at 482 °C (900 °F). In all probability, it will support even higher design stresses at 482 °C and 510 °C (900 °F and 950 °F). Tests will be conducted with hydrogen-charged specimens to verify that the reduced post weld heat treatment necessary to achieve the higher strengths for the base metal is adequate to temper the weld metal and heat affected zones sufficiently to prevent susceptibility to hydrogen stress cracking. In addition, stress-rupture tests will be conducted in a high-pressure hydrogen environment. There is some evidence that high applied stresses can reduce the temperature-pressure conditions at which high-temperature hydrogen attack can occur.[25] This requires better understanding before higher strength material and higher allowable design stresses can be used with confidence.

Modified 2.25 Cr and 3 Cr steels are being developed by several independent research efforts.[16,18,19] Alloying elements such as V, Cb, Ti, and B are added, in relatively small quantities, to increase the high-temperature strength. Some of these additions, especially B, also increase the hardenability, which is very important for obtaining high strengths in thick gages. The MPC is planning to select the most promising modifications, with high-temperature strengths significantly above conventional 2.25 Cr-1 Mo steel, and conduct sufficient additional high-temperature tensile and stress-rupture tests to develop an adequate data base for ASME Code approval. This will include tests of weldments and stress-rupture tests in a hydrogen environment.

Continued development of high-speed welding processes is important for the efficient, economic construction of the very large reactor vessels that will be required to commercialize coal liquefaction processes. The applicability of these processes for high-quality welding at the plant site is essential. Filler metals must be developed to match the chemical compositions and high-temperature strength of the modified 2.25 Cr and 3 Cr steels that are being developed. Testing of weldments will be part of the MPC program for the modified steels.

Improved inspection techniques and standards would be highly beneficial. The MPC is planning a closely related program to accomplish this in a joint effort with the Pressure Vessel Research Committee (PVRC). The first objective will be to demonstrate that modern ultrasonic inspection techniques can provide the same assurance of vessel integrity for safe operation as the radiography that is presently required by the ASME Code. The approach will be to expand upon PVRC's PISC I & II round-robin programs for nuclear vessels. A 300- to 350-mm (12- to 14-in.) thick 2.25 Cr-0.5 Mo weldment will be prepared with implanted defects to simulate all typical flaws that can be encountered in actual vessel construction. Ultrasonic inspection by all participating vessel fabricators and inspection agencies will reveal the accuracy and reliability of defect location and sizing that can be achieved.

The second effort will be to develop a fitness-for-service standard for vessel acceptance, in contrast to the present arbitrary, or workmanship, defect size limits in the ASME Code. A fracture mechanics type of analysis would be used to relate maximum defect size allowed to the operating stress and process environment. The governing criterion will be that no defect will be accepted in new construction that can be expected to grow to a critical size during the designed service life of the vessel. Considerably more fracture toughness and crack propagation data, under the operating conditions of the vessel, will be required than is presently available. A fitness-for-service acceptance standard will avoid the repair of innocuous fabrication defects without compromising safety of operation. Not only will the high cost of unnecessary repairs be saved, but repairs can create a metallurgical condition that is more deleterious than the minor defect.

Finally, a recommended practice for in-service inspection will be developed. Similar to the acceptance standard, it will be based upon a fracture mechanics type of analysis, but the governing criterion will be different. Presumably, a defect has been found during a shutdown that is larger than for initial vessel acceptance. Operation for the design life is no longer the primary concern. The questions that must be answered to assure continued safe operation now become: must an immediate repair be made, how much longer can the vessel be operated before repair, or when must the next

inspection be made? Crack propagation data and fracture mechanics methodology at high temperatures will be required to answer these questions.

To a large extent, the concerns and directions for research that are manifested in the MPC-sponsored work are parallel to those identified by the Department of Energy studies.[27] Materials requirements for other liquefaction processes were examined in connection with DOE-sponsored projects several years ago and found to be similar to those for SRC-II. The two-stage liquefaction process reduces some of the potential problems in the reactor vessel, but further treatment of the product is needed. Economics suggests that high-pressure, high-temperature conditions will be desired for further processing.

CONCLUSIONS

Adequate technology is presently available for pressure vessels and piping to construct and safely operate coal liquefaction plants. However, several advancements in technology can significantly improve the economics of commercial plant construction and thereby reduce capital investment to obtain the liquid fuel products at the lowest possible market price. Primary among these are higher strength pressure vessel steels compatible with the process environment, high-speed welding processes suitable for field fabrication, and fitness-for-service acceptance criteria using ultrasonic inspection.

REFERENCES

1. D. A. Canonico, et al.: "Assessment of Materials Technology for Pressure Vessel and Piping for Coal Conversion Systems," ORNL-5238, August 1978.
2. D. A. Canonico: "Structural Integrity of Vessels for Coal Conversion Systems," ORNL/TM-6969, September 1979.
3. J. A. Gulya and R. A. Swift: "Creep Rupture Properties of SA387-22 and SA542-3," ASME-MPC-21, 1984.
4. G. V. Smith: "Supplemental Report on the Elevated Temperature Properties of Chromium-Molybdenum Steels (An Evaluation of 2-1/4 Cr-1 Mo Steel)," ASTM DS6S2, 1971.
5. E. B. Norris and R. D. Wylie: "Analysis of Data from Symposium in Heat-Treated Steels for Elevated Temperature Service," ASME-MPC, 1967.
6. G. V. Smith: "The Strength of 2-1/4 Cr-1 Mo Steel at Elevated Temperatures," Symposium on 2-1/4 Cr-1 Mo Steel in Pressure Vessels and Piping, September 1970.
7. R. E. Lorentz et al.: "A Program to Study Materials for Pressure Vessel Service with Hydrogen at High Temperatures and Pressures," MPC, April 1982.
8. J. Watanabe et al.: "Hydrogen Induced Disbonding of Stainless Weld Overlay Found in Desulfurization Reactor," ASME, MPC-16, June 1981.
9. J. Watanabe et al.: "Field Inspection and Repairs of Heavy Wall Reactors," NACE, March 1978.
10. Y. Murakami et al.: "Heavy-Section 2-1/4 Cr-1 Mo Steel for Hydrogenation Reactors," ASTM STP 755, May 1980.
11. M. Kohn et al.: "Mechanical Properties of Vacuum Carbon-Deoxidized Thick-Wall 2-1/4 Cr-1 Mo Steel Forging," ASTM STP 755, May 1980.
12. J. A. Barthet et al.: "Data Obtained on Industrial Production Plates 150 to 500 mm Thick in Cr-Mo Steel," ASTM STP 755, May 1980.

SECTION VII
Shafts

The Crankshaft

Giovanni Riccio

Abstract
This paper examines the crankshaft, which can be considered the heart of the internal combustion engine, and the component which most reflects the development of engines as regards planning and progress in the choice of material. Spheroidal cast-iron is increasingly favoured by the design engineer. The importance of the discovery of this material can be compared only with that of malleable cast-iron, during the last century.

Riassunto
Viene preso in esame il componente che può essere considerato come il cuore del motore a combustione interna. Questo componente è anche quello che rispecchia in maniera più evidente l'evoluzione dei motori dal punto di vista della progettazione e del progresso realizzato nel campo dei materiali. Nella scelta che può essere fatta dal progettista è tenuta sempre più in considerazione la ghisa sferoidale, il materiale che ha rappresentato nella fonderia un fatto paragonabile soltanto alla scoperta della ghisa malleabile, avvenuta oltre un secolo prima.

What we now refer to as an engine may be said to have first come upon the scene when extensive use began to be made of the combination of a crank and a connecting rod to transform reciprocating motion into continuous, rotary motion.

From the crank handle employed by knife-grinders in the late Middle Ages, this system has wended its way down to machines designed for an infinity of applications, and mechanical devices of every kind. The conversion of the backward and forward movement of a steam engine into the circular movement of a wheel was the striking feature of the improvements introduced (1). The 19th century witnessed the birth of the internal combustion engine, at first complementary to, but eventually replacing the steam engine. This new system of propulsion continues to be the most widely employed. It has retained the crank and con-rod arrangement, and the crankshaft as its star performer (Fig. 1).

The crankshaft is one of the most important parts of the internal combustion engine and undergoes heavy duty. Therefore, it requires careful consideration when it comes to designing, choice of material, machining and finishing.

During the *design stage*, both the proportions and the strength of the crankshaft must be taken into account, that is, problems concerning systems of calculation and choice of material must be dealt with (2). These two aspects are linked, as calculation must be contained within the limits set by the material's specifications, while improved material specifications allow wider scope for calculations. As we have said, the geometry of

Fig. 1 - Crankshaft.

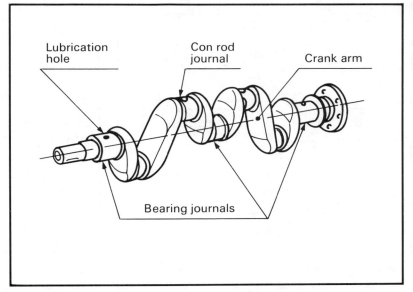

the part must take account of the material of which it is made. It is also subject to the constraints imposed by the specifications of the engine. The bore and stroke will be determined by the power rating, the number of cylinders, and the thermodynamic cycle employed, whereas the centre distances will depend on the way the cylinders are arranged, the type of cooling, and various parameters connected with heat.

Once both reciprocating masses and the thermodynamic cycle are known, crankpin loads and their dimensions are determined. Then main journal loads and their dimensions are defined. Finally, tensile conditions due to static and dynamic loads are analyzed in order to establish the critical point of fatigue strength and safety (Fig. 2). These calculations are based on simplified hypotheses and subsequent experimental examinations.

The prototypes for crankshafts usually undergo both static and dynamic tests (fatigue tests). The former are meant for the examination of form factors and maximum stresses while, at the same time, helping the planning; the latter for the control of the technological production process and evaluation of the product's reliability (Fig. 3).

Major innovations have recently been introduced, with regard to both calculation systems and planning, and these are being adopted in industrial practice. A calculation system which is based on modal analysis and direct integration has been set up. Thanks to this system, dubious aspects are removed and very significant mock-up crankshafts can be produced as early as the planning stage (3). The well-known and most up-to-date CAD/CAM systems are also applied, thus obtaining the optimum design of the part and engineering its manufacture in the most rational way.

As for the *material* used for making crankshafts for the automotive industry, the following ferrous alloys are considered:
— forged and cast steel;
— grey cast-iron;
— blackheart malleable cast-iron

Fig. 2 - Block diagram of current analytical and experimental processes applied in the designing of crankshafts.

Fig. 3 - S-N curves for hardened crankshafts subjected to alternating torsion stresses. (Wöhler).

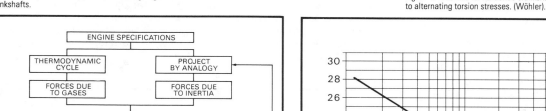

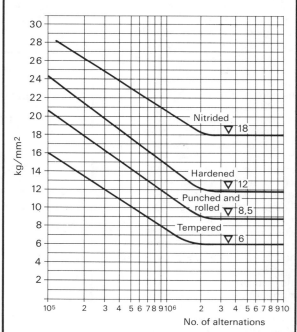

Fig. 4 - Hammer.

Fig. 5 - Forging press.

(American malleable iron);
— spheroidal cast-iron.

Forged steel was the first solution to the problem of the material for crankshafts, and still accounts for a large part of today's production (4), (5).
The forging process was enhanced by the perfecting of the steam engine during the first half of the 19th century, and by the invention of the Bourdon hammer at Creusot and the Nasmyth hammer in England. The forging press came shortly after, and the "closed-die" forging process was established. Since then forging has rapidly developed, thanks to new machines and equipment, and to better metallurgical knowledge of steel and of the laws regulating steel flow in the die (Figg. 4 and 5).

Fig. 6 - Die.

The closed-die forging process is used for crankshafts in the automotive industry. It confers the exact shape required on the steel once this has been heated to the point where it becomes plastic (Fig. 6). The forging blank (made of carbon or low-alloy steel) is cut to the length required to provide enough material to fill the die, plus an amount for the flashes, and (if necessary) a tail by which the piece can be held. Flashes represent the escape of surplus metal due to the pressure exerted inside the die. Quite complex and well-proportioned shapes can be obtained by means of "closed-die" forging which, combined with other processes, gives closer dimensional tolerances. Precision forging represents a recent improvement on the traditional process and gives very interesting results, especially as far as dimensions are concerned. The great variety of equipment available today for the forge shop

Fig. 7 - A forged and a cast version of the same crankshaft.

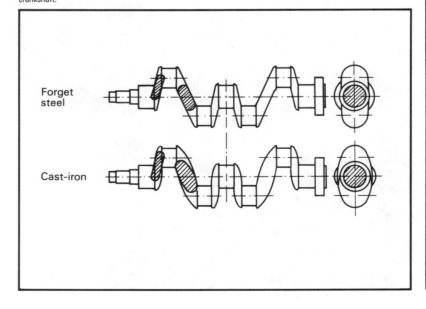

Forget steel

Cast-iron

through hot-forging machines manufacturers, offers a wide choice of possibilities, so as to guarantee the right choice for production requirements.

Depending on their weight and design, crankshafts can be either drop-forged or pressed.

Alternatively a combined process can be employed. The die must allow for the stock that will be removed during machining of the piece, and for shrinkage when the part cools down to the ambient temperature after removal from the die. Provision must also be made for suitable draft angles, fillets, joints at the edges, flash holes, etc.

The highly refined technology of the forge was tested during the development of the motor car industry. Today it bears considerable weight and can compete with other technologies for the production of crankshafts.

Cast steel has always gone side by side with forged steel as a solution to the problem of the material for crankshafts (Fig. 7).

The initial modest importance of the foundry increased as foundrymen improved their casting technology. It is worth mentioning, in this connection, the cast-iron crankshaft shown at the Exhibition held by the American Society for Metals in the United States of America in 1934. The machine was mounted on an engine used in a mine, and had been in operation since 1898 without showing any drawback. Researches started in 1925 by Ford Motor Co. led to the use, for crankshafts, of a foundry alloy having characteristics intermediate between those of cast iron and steel, and with the following composition:
C = 1.35-1.60%, Si = 0.85-1.1%, Mn = 0.6-0.8%, Cu = 1.5-2.0%, Cr = 0.4-0.5%, P max 0.1%, S max 0.06%.

This metal was obtained by melting a charge of scrap steel and cast-iron returns in electric furnace.

Crankshafts for tractors were cast into sand moulds; those for motor cars into sand core moulds bound with oil. The castings were annealed, and their structure was made of globular pearlite and a certain amount of graphite. They presented the following characteristics:
HB = 280-320, Rt = 70-80 kg/mm^2 and A% = 2.5-3%.

During the same years the so-called Esslinger Material was being developed in Germany. This material was made of cast-iron, melted in a cupola and subjected to a duplex process in an electric furnace; its carbon content was higher than that of the Ford material and it had small quantities of alloying elements.

Spheroidal cast-iron was already known when, at the end of the 'forties, the German metallurgist E. Piwowarsky considered heat-treated un-alloyed grey cast-iron to be the best suited material for crankshafts made in the foundry (6). But he also forsaw a future for Arma-Steel, i.e. American malleable iron with a pearlitic matrix, and for spheroidal cast-iron.

Of all foundry alloys, *cast steel* is the one with the best mechanical characteristics. It has a compact grain structure, high strength and toughness. Its impact strength is high and it can be bent and subjected to heavy loads before reaching its breaking point. However, its melting point is high, and it can be melted and cast into complex shapes only with difficulty.

Grey cast-iron (Fig. 8) has a relatively low melting point and can be melted

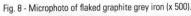
Fig. 8 - Microphoto of flaked graphite grey iron (x 500).

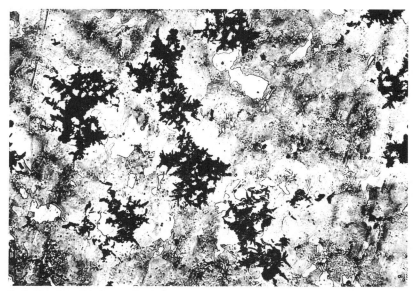

Fig. 9 - Microphoto of blackheart or American malleable cast iron (x 250).

and cast more easily than steel. The considerable carbon content of grey cast-iron (3-4%) causes the melting point to drop. During solidification, however, most of the carbon content splits as flaked graphite, thus breaking the continuity of the structure. Grey cast-iron is relatively brittle, and can be used only in the absence of heavy impact stresses. Malleable and spheroidal cast-iron present different features. The first *malleable cast-iron* was obtained in 1722 by the well-known French physicist Réaumur, and it was the source of the so-called whiteheart malleable cast-iron which is still produced in Europe.

White malleable cast-iron is prepared by melting a charge with a carbon content of 2.5-3%, i.e. half-way between that of grey iron and casting steel. After pouring, this solidifies to form what is known as a "white" structure, in other words, a hard, brittle material that can be made malleable by high-temperature decarburisation. This process was a major improvement in iron metallurgy because it used the carbon content in the liquid state to obtain fluidity, and then eliminated most of it during the subsequent heating treatment, so that the final product had features similar to those of steel.

Black-heart malleable cast-iron, invented in the United States of America around 1820 by S. Boyden, is more important (7), (Fig. 9). In an attempt to reproduce Réaumur's malleable cast-iron, Boyden used similar mixtures with intermediate carbon content, and obtained "white" structure castings with complex shape, which were subsequently annealed. With this process, however, the carbon was not eliminated, but precipitated into the metallic matrix in the form of dispersed flakes. The treatment was shorter than that for white-heart cast-iron. Both types of malleable cast-iron are tough and can be impact stressed, but they require long treatments and present difficulties when manufacturing very thick parts.

Spheroidal cast-iron was discovered in 1948 and is a kind of malleable cast-iron without treatment (8).

The addition of caesium and/or magnesium to liquid cast-iron results in an alloy which solidifies to form a spheroidal graphite which reduces continuity problems to a minimum, so much so that the tensile properties of the spheroidal iron thus obtained are only about 15% different from those of steel, including ultimate loads of 40 to over 100 k/mm^2, and elongations of 5 to 20% (Fig. 10).

The fatigue limit of spheroidal cast-iron is on a par with that of carbon steels. Its good wear resistance and high mechanical properties are points in its favour that are particularly appreciated by manufacturers of crankshafts.

Crankshafts followed the changes taking place in the manufacture of *vehicle engines* which have adjusted to the requirements of the market and the needs of the recent crisis (oil, raw materials, ecology, economy, politics).

For some time, during the development of the motor car industry, forged steel was the most used material in the manufacture of crankshafts.

Fig. 10 - Microphoto of spheroidal cast iron (x 250).

Fig. 11 - Cluster of spheroidal iron crankshafts for the Fiat 500.

Cast steel was adopted following the hard work of foundrymen who re-studied the preparation of the metal and its characteristics, shaping and finishing processes, testing, and, above all, production costs.

Towards the end of the 'fifties, black-heart malleable and spheroidal cast-iron were given special attention by those concerned with engine design because of the great scope they offered as regards design, machinability and savings achieved when casting crankshafts. FIAT showed the way with the spheroidal cast-iron crankshaft of its well-known Fiat 500 manufactured at its Mirafiori and Carmagnola Works (Fig. 11). This was followed by crankshafts made of pearlitic malleable cast-iron installed in cars and trucks. Then spheroidal cast-iron replaced malleable cast-iron (Fig. 12).

The great number of inspections carried out during manufacture and the performance of millions of vehicles on the road, has confirmed the use of cast-iron (9), (10), (Fig. 13).

Further rapid changes have been brought about in the engine manufacturing industry by the latest crisis and its related problems. In particular, the motor car industry has aimed at reducing both weights and costs, thus affecting both the foundry and the forge.

More particularly, the reduction in dimensions and machining tolerances, improved machining, the reduction or elimination of heat-treatment, and reduced machining

of blanks are worth mentioning. Engines for cars tend to be increasingly smaller and highly stressed. By manufacturing lighter

Fig. 12 - Spheroidal iron crankshafts for the Fiat 128.

cars the power to weight ratio becomes more favourable and fuel consumption is reduced. However, each component is optimised. The same objectives are pursued by the manufacturers of commercial vehicles, that is, reduction of weight and consumption, etc. Engines must be lighter, more reliable and less noisy.

With these objects in view, the designing of crankshafts for cars and major sectors of industrial vehicles tends towards foundry

Crankshaft N°	Cycles N°	F (kg/mm²)	HB
1	5.652.000	8,1	283
2	5.598.000	8,05	282
3	5.400.000	8	285
4	5.400.000	8	239
5	5.148.000	7,9	285
6	4.860.000	7,8	269
7	4.500.000	7,7	285
8	3.960.000	7,55	270
9	3.850.000	7,5	286
10	3.240.000	7,25	270
11	3.240.000	7,25	235
12	3.000.000	7,15	281
13	2.812.000	7,05	294
14	2.774.000	7	285
15	2.700.000	6,95	285
16	2.700.000	6,95	269
17	2.700.000	6,95	243
18	2.500.000	6,85	268
19	2.420.000	6,8	285
20	2.420.000	6,8	271
21	2.268.000	6,7	285
22	2.160.000	6,6	288
23	1.980.000	6,4	285
24	1.700.000	6,3	286
25	1.600.000	6,2	291
26	1.400.000	6	288
27	1.400.000	6	229
28	1.260.000	5,90	269
29	720.000	5,05	207
30	572.000	4,75	207

Fig. 13 - Endurance, fatigue strength and Brinell hardness ratings of experimental crankshafts for the Fiat 500.

solutions using spheroidal cast-iron. The main factors affecting the choice of cast-iron are the great flexibility allowed by the process as regards design, the wide possibility of reducing the weight and better stress distribution in the component (by using hollow shafts), the better machinability of the material and the potential production of big foundries, with up-to-date equipment and the ability to control the various technologies perfectly. As a matter of fact, spheroidal cast-iron is making constant progress in Italy and all over the world (Japan excluded) as far as the car manufacture is concerned, while gaining a good position in the

Fig. 14 - Hollow spheroidal iron crankshafts produced by the Policast process.

manufacture of commercial vehicles.

The Policast process (evaporable-pattern casting), which Teksid set up only very recently at its Carmagnola foundries, is bound to be a further technical and economic reason for choosing cast-iron in the future (Fig. 14).
At the same time, it must be mentioned that noteworthy progress has been made in the forge by exploiting the superior qualities of steel and achieving weights and sizes which can well compete with the foundry products. The various factors already mentioned, that is production cost, material machinability (bearing in mind that forged steels are either resulphurized or have improved machinability), crankshaft weight given the same engine performances, and the availability of an industrial structure which can comply with the chosen solution, are to be taken into account when choosing between cast-iron and forged steel or between foundry and forge. At any rate, the wide choice of possibilities offered by both the foundry and the forge can meet all the requirements of motor car engines planning as regards the production of crankshafts.

REFERENCES

(1) Boutan, A., and J. Ch. d'Almeida. **Physique**. Dunod, Paris, 1867, tome Ier.

(2) Giacosa, D. **Motori endotermici**. Hoepli, Milano, 1941.

(3) Bargis, E., A. Garro, and V. Vullo. Progettazione degli alberi a gomiti e verifica sperimentale. Il progettista industriale, 2 (1981).

(4) Bosco, G., and M. Solei. **Lo stampaggio a caldo degli acciai nell'industria moderna**. Fiat Automobili, Torino, 1965.

(5) Chamouard, A. **Estampage et Forge**. Compagnie Française d'Edition, Paris, 1970, tome III.

(6) Piwowarsky, E. **Hochwertiges Gusseisen**. Springer-Verlag, Berlin, 1951.

(7) Malleable Founders' Society (Ed.). **American malleable iron**. Malleable Founders' Society, Cleveland, 1944.

(8) Gagnebin, A.P. **I principi fondamentali della produzione di getti in ghisa e in acciaio**. The International Nickel Company, Milano, 1950.

(9) Locati, L. **La fatica dei materiali metallici**. Hoepli, Milano, 1946.

(10) Fortino, D. L'influenza della struttura micrografica sulla resistenza alla fatica degli alberi motore in ghisa sferoidale. Presented at the VIII Convegno Nazionale AIM, Torino, 1958.

Development of Sintered Integral Camshaft

The camshaft for an automobile engine is generally made of chilled cast iron. Due to increasing demand for higher performance, lower maintenance and better fuel economy, it is difficult to make the cast iron camshaft lighter and/or more durable. In order to overcome these problems, development of an integral camshaft comprising a sintered alloy cam piece for better wear resistance and steel tube for weight saving has been accomplished. C Thumuki, K Ueda, H Nakamura, K Kondo, and T Suganuma of the Toyoto Motor Corp in Japan describe the development of this new camshaft in the following report.

The camshaft for an automobile engine is generally made of chilled cast iron. Due to increasing demand for higher performance, lower maintenance and better fuel economy, it is difficult to make the cast iron camshaft lighter and/or more durable due to restrictions of its material, surface treatment, process, and production cost. In order to overcome these problems, development of an integral camshaft comprising a sintered alloy cam piece for better wear resistance and steel tube for weight saving has been accomplished. We have succeeded in practical application of the sintered integral camshaft and the following article describes the developmental procedures and advantages of this new sintered component.

DEVELOPMENT

It was suggested that an integral camshaft might be obtained by various methods, but the practical application had not been achieved. We carried out research on the possibility of an integral camshaft consisting of the best suitable materials for each part. From the result of this research, we started the development based upon the following two points.

(1) Cam piece : To adopt sintered material for freer selection of wear resistant material and to reduce the amount of machining.

(2) Shaft piece: To use tube for weight reduction and improved lubrication by using the inside for an oil channel.

We finished fundamental research on wear resistant sintered alloy and bonding method of cam pieces with shaft in 1978. We concluded that the sintered integral camshaft was better in wear resistance and lighter than chilled cast iron camshafts and started the study of its practical application to 1S engine. In 1981 Toyota Motor Corp succeeded in starting the mass production of the sintered integral camshaft for the new 1.8 litre 1S engine.

Appearance of the 1S engine for Celica, Corona and Carina is shown in Fig. 1. The 1S engine is a straight four cylinder with a single overhead camshaft the longitudinal section as shown in Fig. 2. Fig. 3 shows the valve system structure. The camshaft actuates alloy cast iron rocker arms. A hydraulic lash

FIG. 1 Appearance of 1S engine

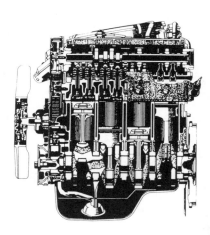

FIG. 2 Longitudinal section

adjuster provides the fulcrum for the rocker arm.

This is developed as a compact, light weight, high peformance and highly reliable engine which is Toyota's representative in the 1.8 to 2.0 litre class for front engine front wheel drive as well as front engine rear wheel drive vehicles. The sintered integral camshaft greatly contributes to these features.

SINTERED INTEGRAL CAMSHAFT CONFIGURATION

The sintered integral camshaft and its structural parts are shown in Fig. 4. Fig. 5 shows the configuration of the sintered integral camshaft. Each campiece, journal piece and so on, are produced separately by the best suitable method, and bonded together at a later stage. Fig. 6 shows an outline of the production process.

Cam Piece

The cam material requires superior wear, scuffing and pitting resistance against rocker arm material. Table 1 shows five sintered

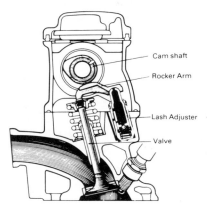

FIG. 3 Valve train structure

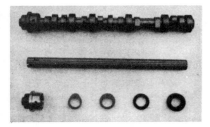

FIG. 4 Appearance of sintered integral camshaft and its structural parts

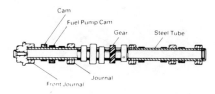

FIG. 5 Configuration of sintered integral camshaft

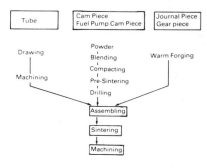

FIG. 6 Outline of the production process of sintered integral camshaft

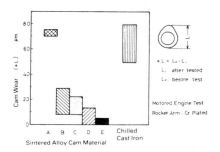

FIG. 7 Wear properties

chromium-molybdenum complex carbide are scattered uniformly in the bainite (partially martensite) matrix. Consequently, sintered alloy E has a relatively high hardness of Hv550 even under conventional sintering conditions without any additional work such as repress, resinter or hardening heat treatment.

For bonding the cam with the shaft, various methods-such as press-in, hydraulic expansion, welding, brazing, diffusion bonding and pin fitting were investigated from the view point of cost, reliability and productivity. The cam piece made of the wear resistant sintered material, which virtually allows no elongation, tends to crack due to internal stress when it is pressed-in or hydraulically expanded.

Therefore, the following specified bonding method was developed. Four pairs of V-shaped grooves are longitudinally provided on the outer periphery of shaft at the point of cam phase, and three V-shaped projections are provided on the inner periphery of cam piece. Therefore, the cam piece can be slid on

to the shaft, then the phase of cam piece is fixed. After that, the axial position of cam piece is fixed by making deformations on the groove adjacent to the projection at both sides of cam piece.

In the next stage, cam pieces and shaft are metallurgically bonded during the sintering process by shrinkage of cam piece and diffusion bonding by Fe-P-C liquid phase.

The sintering is carried out at 1110C in dissociated ammonia. Fig. 10 shows a diffusion bonded zone of cam piece with shaft.

Journal Piece

Low alloy steel (SAE 4140) is used for a journal material in order to obtain a hardness of Hv280 which is required for its wear resistance against the journal bearing material (SAE 306). The journal piece is produced by warm precision forging and air cooled. The journal piece is pressed on to the shaft and bonded by brazing.

Pure copper (AWS BCu-1) is used for a brazing material to suit the sintering

Sintered Alloy	Chemical Composition	Hardening	Notes
A	Fe-Cu-C with free Gr	Induction Hardening	
B	Fe-P-Cu-C with free Gr	↑	
C	Fe-Mo-Cu-C	Fe-Mo-C hard phase	Liquid phase sintering
D	Fe-Mo-P-Cu-C	↑	↑
E	Fe-Cr-Mo-P-Cu-C	Fe-Cr-Mo-C hard phase	↑

TABLE 1 Wear resistant sintered alloys for cam piece developed in this study

alloys for cam piece (A,B,C,D,E) developed in this study. The experimental results of screening test for wear resistance, which were carried out in engine test beds under the accelerated conditions, are shown in Fig. 7 and Fig. 8. According to these results, we selected sintered alloy E and made a further matching test of wear resistance against three rocker arm materials, which were high-chromium cast iron, chromium plated steel and soft-gas nitrided alloy steel. These results were compared with chilled cast iron. Sintered alloy E has better wear properties than chilled cast iron and also is less affected by varieties of rocker arm materials as shown in Table 2. The combination of sintered alloy E and high chromium cast iron was selected for cam and rocker arm materials for the 1S engine.

Table 3 shows a chemical composition of sintered alloy E, and a microstructure is shown in Fig. 9. Phosphorus activates the sintering by Fe-P-C liquid phase and lowers the sintering temperature. Therefore, the density of sintered alloy E reaches 7.6 g/cm³ even at 1110C.

Moreover, fine hard particles consisting of

Rocker Arm Material \ Cam Material	Result of Rocker Arm \ Result of cam	Sintered Alloy E		Chilled Cast Iron	
Alloy Cast Iron (High Chromium)		◎	◎	○	○
Chromium Plating		◎	◎	× (scuffing)	(scuffing) △
Alloy Steel (Soft Gas Nitrided)		○	◎	× (wear)	○

◎ Excellent, ○ Fair, △ Poor, × Bad

TABLE 2 Wear properties of cam materials against rocker arm materials

Elements	Cr	Mo	P	Cu	C	others	Fe
Weight %	5	1	0.5	2	2.5	⟨2	Bal

TABLE 3 Chemical composition of sintered alloy E

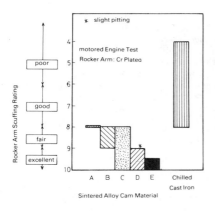

FIG. 8 Scuffing properties

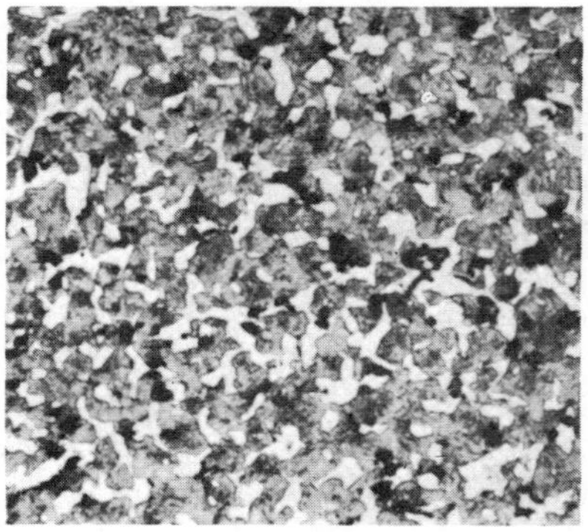

FIG. 9 Microstructure of sintered alloy E

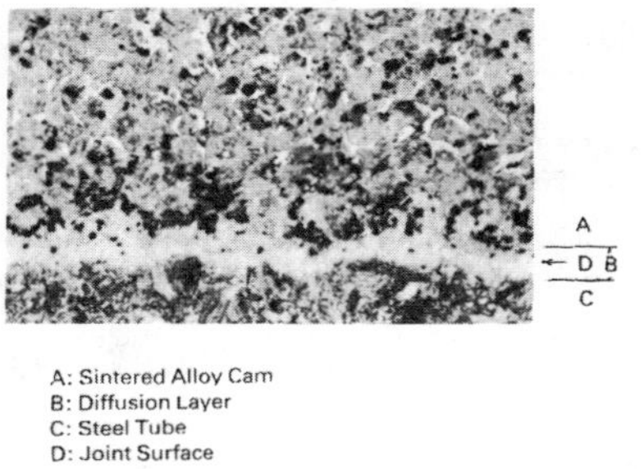

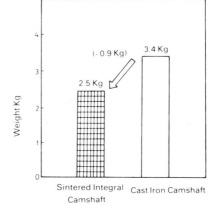

A: Sintered Alloy Cam
B: Diffusion Layer
C: Steel Tube
D: Joint Surface

FIG. 10 Microstructure of diffusion bonding zone

---

temperature of 1110C. This fact enables elimination of other brazing processes or equipment. The journal piece as well as cam piece are bonded with the shaft at the same time during the sintering process.

Shaft

For the purpose of insuring comparable strength or rigidity with the conventional cast iron camshaft, several kinds of materials and configurations were investigated. As the result, steel tube (ASTM A513-64 Grade 1030, 28mm of outer diameter and 4.5mm of thickness) was selected for the shaft. The V-shaped grooves are formed by cold drawing process.

ADVANTAGES OF SINTERED INTEGRAL CAMSHAFT

The advantages of the sintered integral camshaft applied to 1S engine are shown as follows.

Weight Savines

As shown in Fig. 11, a weight reduction of 26% (0.9 kg) is achieved by using a steel tube for shaft.

Excellent Wear Resistance

The newly developed wear resistant sintered alloy cam has excellent wear resistance compared with the conventional chilled cast iron under the accelerated fired engine test. These wear properties are shown in Fig. 12 and Fig. 13, which indicates that cam wear in the sintered alloy is only about 1/7th of chilled cast iron.

Improvement of Lubrication System

The hollow part of the steel tube is used as a main passage of engine oil. As shown in Fig. 14, engine oil from cylinder block is supplied to cams, journals, fuel pump cam, and distributor drive gear through the inside of the camshaft, and is also supplied to lash adjusters through journals. Thus, oil delivery tube becomes needless and the lubrication system is significantly simplified.

Futhermore, the voluminous oil passage inside the camshaft also satisfactorily separates air bubbles from the engine oil, as well as supplving a desirable oil pressure to

lash adjusters in the full operation range of low to high engine revolution.

Saving Machining Cost

The cam profile shape of sintered integral camshaft before grinding is exceedingly nearer to the finished shape than cast iron camshaft, therefore cam grinding machines are reduced by 20%. Also the drilling process for an oil passage, necessary for cast iron camshafts, is eliminated.

These advantages greatly saved the investment cost for machining process.

CONCLUSION

Development of a sintered integral camshaft was successfully achieved by bonding specified wear resistant sintered alloy cam pieces and steel journals and a gear piece with steel tube shaft. The first practical application of sintered integral camshaft for the 1S engine was quite successful and the camshaft is being mass produced today.

The significant advantages are as follows:

(1) Weight saving (75% of weight compared with cast iron camshaft)
(2) Excellent wear resistance (1/7th of wear)
(3) Simplified lubrication system
(4) Saving machining cost (camshaft grinding machines are reduced by 20%)

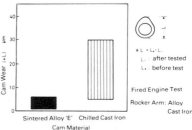

FIG. 11 Comparison of weight

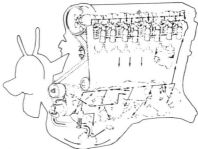

FIG. 12 Wear properties

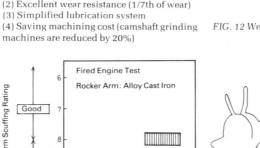

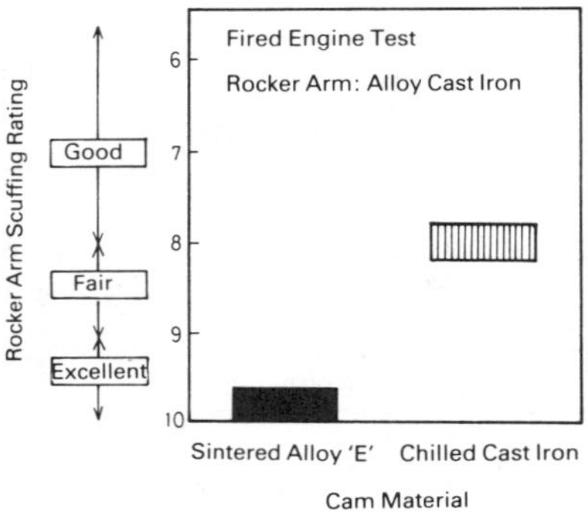

FIG. 13 Scuffing properties

FIG. 14 Lubrication system

SECTION VIII
Springs

Steel Springs

By James H. Maker
Chief Metallurgist
Associated Spring
Bristol Div. Barnes Group Inc.

STEEL SPRINGS are made in many types, shapes and sizes, ranging from delicate hairsprings for instrument meters to massive buffer springs for railroad equipment. The major portion of this article discusses those relatively small steel springs that are cold wound from wire. Relatively large, hot wound springs are quite different from cold wound springs in many respects and are treated in a separate section. Flat and leaf springs also are treated separately to the extent that they differ from wire springs in material and fabrication.

Wire springs are of four types: compression springs (including die springs), extension springs, torsion springs and wire forms. Compression springs are open wound with varying space between the coils and are provided with plain, plain and ground, squared, or squared and ground ends. Extension springs normally are close wound, usually with specified initial tension and, because they are used to resist pulling forces, are provided with hook or loop ends to fit the specific application. Ends may be integral parts of the spring, or specially inserted forms. Torsion springs are designed to exert force by unwinding—rarely by winding. Wire forms are made in a wide variety of shapes and sizes.

Flat springs usually are made by stamping and forming of strip material into shapes such as spring washers. However, there are other types, including motor springs (clock type), constant-force springs and volute springs, that are wound from strip.

Chemical composition, mechanical properties, surface quality, availability and cost are the principal factors to be considered in selecting steel for springs. Both carbon and alloy steels are used extensively.

Steels for cold wound springs differ from other constructional steels in four ways: they are cold worked more extensively; they are higher in carbon content; they can be furnished in the pre-tempered condition; and they have higher surface quality. The first three items increase the strength of the steel, and the last improves fatigue properties.

For flat cold formed springs made from steel strip or flat wire, narrower ranges of carbon and manganese are specified than for cold wound springs made from round or square wire.

Where special properties are required, spring wire or strip made of stainless steel, a heat-resistant alloy, or a nonferrous alloy may be substituted for the carbon or alloy steel, provided that the design of the spring is changed to compensate for the differences in properties between the materials. (See the Appendix on Design at the end of this article.

Table 1 lists grade, specification, chemical composition, properties, method of manufacture, and chief applications of the materials commonly used for cold formed springs. For further information on spring materials other than carbon and alloy steels, consult the appropriate volumes of this Handbook. Hot formed carbon and alloy steel springs are discussed in this article.

Mechanical Properties

Steels of the same chemical composition may perform differently because of different mechanical and metallurgical characteristics. These properties are developed by the steel producer through cold work and heat treatment, or by the spring manufacturer through heat treatment.

Selection of round wire for cold wound springs is based on minimum tensile strength for each wire size and grade (Fig. 1), and minimum reduction in area (45% for all sizes).

Rockwell hardness and tensile strength for any grade of spring steel depend on section thickness. The same properties in different section thicknesses may be obtained by specifying different carbon contents. The relation of thickness of spring steel strip containing 0.50 to 1.05% carbon and Rockwell hardness is shown in Fig. 2. The optimum hardness of a spring steel increases gradually with decreasing thickness.

The hardness scale that may be used for thin metal depends on the hardness and the thickness of the metal. (See the discussion and table on pages 7 and 8 in Volume 11, the 8th Edition, of this Handbook.) For testing spring steel strip, which has a minimum hardness of

Table 1 Common wire and strip materials used for cold formed springs (a)

Material type	Grade and specification	Nominal composition, %	Min tensile strength(b) MPa	ksi	Modulus of elasticity, E GPa	Million psi	Design stress, % of min tensile strength(c)	Modulus of rigidity, G GPa	Million psi	Hardness, HRC(d)	Max allowable temperature °C	°F	Method of mfg, chief applications, special properties
Cold Drawn Wire													
High-carbon steel	Music wire, ASTM A228	C 0.70-1.00 Mn 0.20-0.60	1590-2750	230-399	210	30	45	80	11.5	41-60	120	250	Drawn to high and uniform tensile strength. For high-quality springs and wire forms.
	Hard drawn, ASTM A227	C 0.45-0.85 Mn 0.30-1.30	Class I 1010-1950 Class II 1180-2230	Class I 147-283 Class II 171-324	210	30	40	80	11.5	31-52	120	250	For average-stress applications; lower-cost springs and wire forms.
	High-tensile hard drawn, ASTM A679	C 0.65-1.00 Mn 0.20-1.30	1640-2410	238-350	210	30	45	80	11.5	41-60	120	250	For higher-quality springs and wire forms.
	Oil tempered, ASTM A229	C 0.55-0.85 Mn 0.30-1.20	Class I 1140-2020 Class II 1320-2330	Class I 165-294 Class II 191-324	210	30	45	80	11.5	42-55	120	250	Heat treated before fabrication. For general-purpose springs.
	Carbon VSQ(e), ASTM A230	C 0.60-0.75 Mn 0.60-0.90	1480-1650	215-240	210	30	45	80	11.5	45-49	120	250	Heat treated before fabrication. Good surface condition and uniform tensile strength.
Alloy steel	Chromium vanadium, ASTM A231, A232(e)	C 0.48-0.53 Cr 0.80-1.10 V 0.15 min	1310-2070	190-300	210	30	45	80	11.5	41-55	220	425	Heat treated before fabrication. For shock loads and moderately elevated temperature.
	Chromium silicon, ASTM A401	C 0.51-0.59 Cr 0.60-0.80 Si 1.20-1.60	1620-2070	235-300	210	30	45	80	11.5	48-55	245	475	Heat treated before fabrication. For shock loads and moderately elevated temperature.
Stainless steel	Type 302(18-8), ASTM A313	Cr 17-19 Ni 8-10	860-2240	125-325	190	28	30-40	69	10.0	35-45	290	550	General-purpose corrosion and heat resistance. Magnetic in spring temper.
	Type 316, ASTM A313	Cr 16-18 Ni 10-14 Mo 2-3	760-1690	110-245	190	28	40	69	10.0	35-45	290	550	Good heat resistance; greater corrosion resistance than 302. Magnetic in spring temper.
	Type 631 (17-7 PH), ASTM A313	Cr 16-18 Ni 6.50-7.75 Al 0.75-1.50	Condition CH-900 1620-2310	Condition CH-900 235-335	200	29.5	45	76	11.0	38-57	340	650	Precipitation hardened after fabrication. High strength and general-purpose corrosion resistance. Magnetic in spring temper.
Nonferrous alloys	Copper alloy 510 (phosphor bronze A), ASTM B159	Cu 94-96 Sn 4.2-5.8	720-1000	105-145	100	15	40	43	6.25	98-104(f)	90	200	Good corrosion resistance and electrical conductivity.
	Copper alloy 170 (beryllium copper) ASTM B197	Cu 98 Be 1.8-2.0	1100-1590	160-230	130	18.5	45	50	7.0	35-42	200	400	May be mill hardened before fabrication. Good corrosion resistance and electrical conductivity; high mechanical properties.
	Monel 400, AMS 7233	Ni 66 Cu 31.5	1000-1240	145-180	180	26	40	65	9.5	23-32	230	450	Good corrosion resistance at moderately elevated temperature.
	Monel K-500, QQ-N-286(g)	Ni 65 Cu 29.5 Al 2.8	1100-1380	160-200	180	26	40	65	9.5	23-35	290	550	Excellent corrosion resistance at moderately elevated temperature.

Table 1 Common wire and strip materials used for cold formed springs (a) (contd.)

Material type	Grade and specification	Nominal composition, %	Tensile properties Min tensile strength(b) MPa	Tensile properties Min tensile strength(b) ksi	Modulus of elasticity, E GPa	Modulus of elasticity, E Million psi	Design stress, % of min tensile strength(c)	Torsion properties Modulus of rigidity, G GPa	Torsion properties Modulus of rigidity, G Million psi	Hardness, HRC(d)	Max allowable temperature °C	Max allowable temperature °F	Method of mfg, chief applications, special properties
High-temperature alloys	A-286 alloy	Fe 53, Ni 26, Cr 15	1100-1380	160-200	200	29	35	72	10.4	35-42	510	950	Precipitation hardened after fabrication. Good corrosion resistance at elevated temperature.
	Inconel 600, QQ-W-390(g)	Ni 76, Cr 15.8, Fe 7.2	1170-1590	170-230	215	31	40	76	11.0	35-45	370	700	Good corrosion resistance at elevated temperature.
	Inconel 718	Ni 52.5, Cr 18.6, Fe 18.5	1450-1720	210-250	200	29	40	77	11.2	45-50	590	1100	Precipitation hardened after fabrication. Good corrosion resistance at elevated temperature.
	Inconel X-750, AMS 5698, 5699	Ni 73, Cr 15, Fe 6.75	No. 1 temper 1070, Spring temper 1310-1590	155, 190-230	215	31	40	83	12.0	No. 1 34-39, Spring 42-48	400-600	750-1100	Precipitation hardened after fabrication. Good corrosion resistance at elevated temperature.
Cold Rolled Strip													
Carbon steel	Medium carbon (1050), ASTM A682	C 0.47-0.55, Mn 0.60-0.90	Tempered 1100-1930	160-280	210	30	…	…	…	Annealed 85 max(b) Tempered 38-50	120	250	General-purpose applications.
	"Regular" carbon (1074), ASTM A682	C 0.69-0.80, Mn 0.50-0.80	Tempered 1100-2210	160-320	210	30	…	…	…	Annealed 85 max(f) Tempered 38-50	120	250	Most popular material for flat springs.
	High carbon (1095), ASTM A682	C 0.90-1.04, Mn 0.30-0.50	Tempered 1240-2340	180-340	210	30	…	…	…	Annealed 88 max(f) Tempered 40-52	120	250	High-stress flat springs.
Alloy steel	Chromium vanadium, AMS 6455	C 0.48-0.53, Cr 0.80-1.10, V 0.15 min	1380-1720	200-250	210	30	…	…	…	42-48	220	425	Heat treated after fabrication. For shock loads and moderately elevated temperature.
	Chromium silicon, AISI 9254	C 0.51-0.59, Cr 0.60-0.80, Si 1.20-1.60	1720-2240	250-325	210	30	…	…	…	47-51	245	475	Heat treated after fabrication. For shock loads and moderately elevated temperature.
Stainless steel	Type 301	Cr 16-18, Ni 6-8	1655-2650	240-270	190	28	…	…	…	48-52	150	300	Rolled to high yield strength. Magnetic in spring temper.
	Type 302 (18-8)	Cr 17-19, Ni 8-10	1280-1590	185-230	190	28	…	…	…	42-48	290	550	General-purpose corrosion and heat resistance. Magnetic in spring temper.
	Type 316	Cr 16-18, Ni 10-14, Mo 2-3	1170-1590	170-230	190	28	…	…	…	38-48	290	550	Good heat resistance; greater corrosion resistance than 302. Magnetic in spring temper.
	Type 631 (17-7 PH, ASTM A693)	Cr 16-18, Ni 6.50-7.75, Al 0.75-1.50	Condition CH-900 1655	240	200	29	…	…	…	46 min	340	650	Precipitation hardened after fabrication. High strength and general-purpose corrosion resistance. Magnetic in spring temper.

Table 1 Common wire and strip materials used for cold formed springs (a)(contd.)

Material type	Grade and specification	Nominal composition, %	Tensile properties			Torsion properties	Hardness, HRC(d)	Max allowable temperature		Method of mfg, chief applications, special properties
			Min tensile strength(b)	Modulus of elasticity, E	Design stress, % of min tensile strength(c)	Modulus of rigidity, G				
			MPa / ksi	GPa / million psi		GPa / million psi		°C	°F	
Nonferrous alloys	Copper alloy 510 (phosphor bronze A), ASTM B103	Cu 94-96, Sn 4.2-5.8	650-750 / 95-110	100 / 15	...	... / ...	94-98(f)	90	200	Good corrosion resistance and electrical conductivity.
	Copper alloy 170 (beryllium copper), ASTM B194	Cu 98, Be 1.6-1.8	1240-1380 / 180-200	130 / 18.5	...	... / ...	39 min	200	400	May be mill hardened before fabrication. Good corrosion resistance and electrical conductivity; high mechanical properties.
	Monel 400, AMS 4544	Ni 66, Cu 31.5	690-970 / 100-140	180 / 26	...	... / ...	98 min(f)	230	450	Good corrosion resistance at moderately elevated temperature.
	Monel K-500, QQ-N-286(g)	Ni 65, Cu 29.5, Al 2.8	1170-1380 / 170-200	180 / 26	...	... / ...	34 min	290	550	Excellent corrosion resistance at moderately elevated temperature.
High-temperature alloys	A-286 alloy, AMS 5525	Fe 53, Ni 26, Cr 15	1100-1380 / 160-200	200 / 29	...	... / ...	30-40	510	950	Precipitation hardened after fabrication. Good corrosion resistance at elevated temperature.
	Inconel 600, ASTM B168, AMS 5540	Ni 76, Cr 15.8, Fe 7.2	1000-1170 / 145-170	215 / 31	...	... / ...	30 min	370	700	Good corrosion resistance at elevated temperature.
	Inconel 718, AMS 5596, AMS 5597	Ni 52.5, Cr 18.6, Fe 18.5	1240-1410 / 180-204	200 / 29	...	... / ...	36	590	1100	Precipitation hardened after fabrication. Good corrosion resistance at elevated temperature.
	Inconel X-750, AMS 5542	Ni 73, Cr 15, Fe 6.75	1030 / 150	215 / 31	...	... / ...	30 min	400-590	750-1100	Precipitation hardened after fabrication. Good corrosion resistance at elevated temperature.

(a) Based on a table in the Handbook of Spring Design (1977), published by the Spring Manufacturers Institute. (b) Maximum tensile strength generally is about 200 MPa (30 ksi) above the minimum tensile strength. (c) For helical compression or extension springs; design stress of torsion and flat springs taken as 75% of minimum tensile strength. (d) Correlation between hardness and tensile properties of wire is approximate only and should not be used for acceptance or rejection. (e) Valve-spring quality. (f) HRB values. (g) Federal specification.

38 HRC, the Rockwell C scale is used for metal more than 0.89 mm (0.035 in.) thick. For testing spring steel strip in thicknesses between 0.89 to 0.81 mm, or 0.035 to 0.032 in. (depending on the hardness), and 0.51 to 0.43 mm, or 0.020 to 0.017 in., the superficial Rockwell 30N scale is used. Between 0.51 to 0.43 and 0.36 to 0.28 mm (0.020 to 0.017 and 0.014 to 0.011 in.), the 15N scale is used. For spring steel strip thinner than 0.36 to 0.28 mm (0.014 to 0.011 in.), a microhardness tester should be used. It has been found that the readings obtained with the Vickers indenter are less subject to variation in industrial circumstances than those obtained with the Knoop indenter. The 500-g-load Vickers test is used for spring steel strip in thicknesses as low as 0.08 mm (0.003 in.).

If readings are made using the proper hardness scale for a given thickness and hardness, they may be converted to HRC values using charts like those on pages 425 to 427 in Volume 11, the 8th Edition, of this Handbook. Similar charts appear in ASTM A370 and in the cold rolled flat wire section of the Steel Products Manual of AISI. Chart No. 60 published by Wilson Instrument Div., American Chain & Cable Co., Inc., also can be used for this conversion. For specific steel springs, hardness can be held to within 3 or 4 points on the Rockwell C scale.

Note that in Table 1 and in the Appendix on Design, design-stress values are given as percentages of minimum tensile strength. These values apply to springs that are coiled or formed and then stress relieved, which are used in applications involving relatively few load cycles. If each spring is coiled or formed so as to allow for some set, and then deflected beyond the design requirements, higher design stresses can be used. This is discussed in the section on residual stresses in this article.

As a further aid in selecting steels for springs, Table 2 lists the suitable choices for cold wound helical springs in various combinations of size, stress and service. Each recommendation is the most economical steel that will perform satisfactorily under the designated conditions and that is commercially available in the specific size.

Fatigue strength is another important mechanical property of steel springs. However, this property is affected by many factors and, because of this complexity, fatigue is discussed in a separate section of this article.

Fig. 1 Minimum tensile strength of steel spring wire

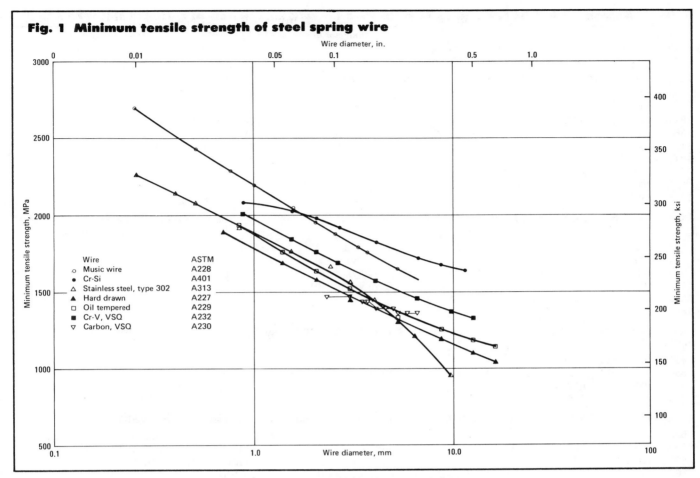

Wire	ASTM
○ Music wire	A228
● Cr-Si	A401
△ Stainless steel, type 302	A313
▲ Hard drawn	A227
□ Oil tempered	A229
■ Cr-V, VSQ	A232
▽ Carbon, VSQ	A230

Fig. 2 Effect of strip thickness on the optimum hardness of spring steel strip for high-stress use

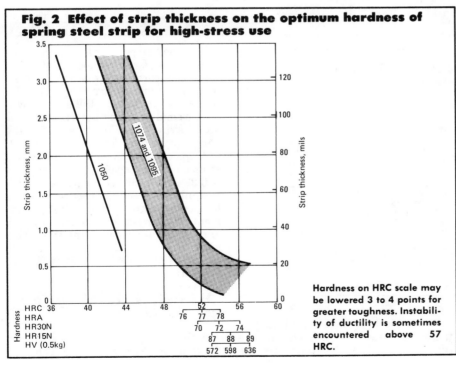

Hardness on HRC scale may be lowered 3 to 4 points for greater toughness. Instability of ductility is sometimes encountered above 57 HRC.

Flat Springs. Figure 3 illustrates the different working stresses allowable in flat and leaf springs of 1095 steel that are to be loaded in each of three different ways—statically, variably, and dynamically. (These three types of loading are dealt with separately in the selection table for cold wound springs, Table 2.) The stresses given in Fig. 3 are the maximum stresses expected in service. These data apply equally well to 1074 and 1050 steels if the stress values are lowered 10 and 20%, respectively. Except for motor or power springs and a few springs involving only moderate forming, most flat springs, because of complex forming requirements, are formed soft, then hardened and tempered.

For best combination of properties, hypereutectoid spring steel in coil form should be held at hardening temperature for the minimum period of time. The presence of undissolved carbides indicates proper heat treatment. Extent of decarburization can be determined by microscopic examination of transverse

Table 2 Recommended ASTM grades of steel wire for cold wound helical springs (a) (contd.)

Corrected max working stress MPa	ksi	Diameter of spring wire(b) 0.13 to 0.51 mm (0.005 to 0.020 in.)	0.51 to 0.89 mm (0.020 to 0.035 in.)	0.89 to 3.18 mm (0.035 to 0.125 in.)	3.18 to 6.35 mm (0.125 to 0.250 in.)	6.35 to 12.70 mm (0.250 to 0.500 in.)	12.70 to 15.88 mm (0.500 to 0.625 in.)
Compression Springs, Static Load (Set removed, springs stress relieved)(c)							
550	80	A228(d)	A227(d)	A227(d)	A227	A227	A227(e), A229
690	100	A228(d)	A227(d)	A227(d)	A227	A227(f), A229	A229
825	120	A228(d)	A227(d)	A227	A227(g), A229	A229(h)	...
965	140	A228	A227	A227(j), A229	A229	A401(h)	...
1100	160	A228	A227	A229	A229(k), A228	A401(m)(h)	...
1240	180	A228	A228	A228	A228(n)(h)	...	...
1380	200	A228	A228	A228(p)(h)	...	...	...
1515	220	A228	A228(q)(h)	...	...	...	...
1655	240	A228(r)(h)	...	...	...	...	...
Compression Springs, Variable Load, Designed for Minimum Life of 100 000 Cycles (Set removed, springs stress relieved)(s)							
550	80	A228(d)	A227(d)	A227(d)	A227(t), A229	A229(u), A401	A401
690	100	A228(d)	A227(d)	A229	A229(v), A401	A401	A401
825	120	A228(d)	A227	A227(w), A229	A229(x), A401	A401(h)	...
965	140	A228	A229	A229(y), A228	A228(h)	...	...
1100	160	A228	A228	A228(y)(h), A401	...	...	...
1240	180	A228	A228(z)(h)	...	...	...	...
1380	200	A228(aa)(h)	...	...	...	...	...
Compression Springs, Dynamic Load, Designed for Minimum Life of 10 Million Cycles (Set removed, springs stress relieved)(s)							
415	60	A228(d)	A227(d)	A227(d)	A227	A229(m)(h)	...
550	80	A228(d)	A227(d)	A227(bb), A229(w), A228	A230(h)	...	...
690	100	A228(d)	A228	A228(bb), A230	A230(cc)(h)	...	...
825	120	A228(d)	A228	A230(h)	...	...	...
Compression and Extension Springs, Static Load (Set not removed, compression springs stress relieved)(c)							
550	80	A228	A227	A227	A227	A227(dd), A229	A229
690	100	A228	A227	A227(w), A229	A401	A401(h)	...
825	120	A228	A227	A227(ee), A229	A401	A401(m)(h)	...
965	140	A228	A228	A228(ff), A401	A401(h)	...	...
1100	160	A228(h)	...	...	...	...	...
1240	180	A228(r)(h)	...	...	...	...	...

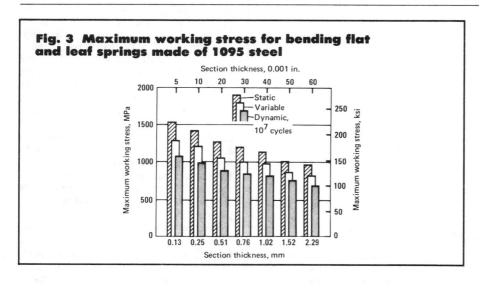

Fig. 3 Maximum working stress for bending flat and leaf springs made of 1095 steel

sections or by microhardness surveys using Vickers or Knoop indenters with light loads (usually 100 g).

Power (clock) springs made from pretempered stock have a longer service life if controlled heat treatment can produce a fine, tempered martensitic structure with uniform distribution of excess carbide. If carbides are absent and the tempered martensitic structure is relatively coarse grained, the springs will have a smaller maximum free diameter after having been tightly wound in the barrel or retainer for a long time.

A recent development for lower-stressed flat springs is a hardened, 0.04 to 0.22% plain carbon strip steel, which is blanked, formed and used with only a low-temperature stress-relief treatment. Thickness tolerances, however, are not as close as for spring steel. This material is available in tensile strengths of 900 to 1520 MPa (130 to 220 ksi).

Characteristics of Spring Steel Grades

Among the grades of steel wire used for cold formed springs (see Table 1),

Table 2 Recommended ASTM grades of steel wire for cold wound helical springs (a)(contd.)

Corrected max working stress MPa	ksi	0.13 to 0.51 mm (0.005 to 0.020 in.)	0.51 to 0.89 mm (0.020 to 0.035 in.)	0.89 to 3.18 mm (0.035 to 0.125 in.)	3.18 to 6.35 mm (0.125 to 0.250 in.)	6.35 to 12.70 mm (0.250 to 0.500 in.)	12.70 to 15.88 mm (0.500 to 0.625 in.)
				Diameter of spring wire(b)			

Compression and Extension Springs, Designed for Minimum Life of 100 000 Cycles (Set not removed, compression springs stress relieved)(s)

MPa	ksi	0.13 to 0.51 mm	0.51 to 0.89 mm	0.89 to 3.18 mm	3.18 to 6.35 mm	6.35 to 12.70 mm	12.70 to 15.88 mm
415	60	A228	A227	A227	A227	A227	A229
550	80	A228	A227	A227	A227(gg), A229	A229	A401
690	100	A228	A229	A229(hh), A228	A228(v), A401	A401(h)	...
825	120	A228	A228	A228(y)	A401(h)	...	...
965	140	A228	A228(h)	...	...	...	...
1100	160	A228(r)(h)	...	...	...	...	...

Compression and Extension Springs, Designed for Minimum Life of 10 Million Cycles (Set not removed, compression springs stress relieved)(s)

MPa	ksi	0.13 to 0.51 mm	0.51 to 0.89 mm	0.89 to 3.18 mm	3.18 to 6.35 mm	6.35 to 12.70 mm	12.70 to 15.88 mm
275	40	A228	A227	A227	A227	A227(m), A229	A229
415	60	A228	A227	A227(bb), A230	A230(cc)(h)	...	...
550	80	A228	A228	A228(bb)(h), A230	...	...	...
690	100	A228(r)	...	...	...	...	...

Torsion Springs (Springs not stress relieved)(s)

MPa	ksi	0.13 to 0.51 mm	0.51 to 0.89 mm	0.89 to 3.18 mm	3.18 to 6.35 mm	6.35 to 12.70 mm	12.70 to 15.88 mm
690	100	A228	A227	A227	A227(cc), A229	A229(jj), A401	...
825	120	A228	A227	A227(y), A229	A229(cc), A228	A401(u)(h)	...
965	140	A228	A229	A229(j), A228	A228(t)(h)	...	...
1100	160	A228	A228	A228(y)(h)	...	...	...
1240	180	A228	A228(h)	...	...	...	...
1380	200	A228(h)	...	...	...	...	...

(a) A227, hard drawn spring wire; A228, music wire; A229, oil-tempered wire; A230, carbon steel, valve-spring quality; A401, chromium-silicon steel spring wire. See Table 1 for compositions of these steels. (b) Where more than one steel is shown for an indicated range of wire diameter, the first is recommended up to the specific diameter listed in the footnote referred to; the last steel listed in any multiple choice is recommended for the remainder of the indicated wire diameter range. (c) Shot peening is not necessary for statically loaded springs. (d) Set removal not required in this range. (e) To 14.29 mm (0.563 in.). (f) To 10.32 mm (0.406 in.). (g) To 4.11 mm (0.162 in.). (h) Yielding likely to occur beyond this limit. (j) To 1.83 mm (0.072 in.). (k) To 3.81 mm (0.150 in.). (m) To 11.11 mm (0.437 in.). (n) To 5.33 mm (0.210 in.). (p) To 2.29 mm (0.090 in.). (q) To 0.81 mm (0.032 in.). (r) To 0.20 mm (0.008 in.). (s) Shot peening is recommended for wire diameter greater than 1.57 mm (0.062 in.) and smaller where obtainable. (t) To 3.94 mm (0.155 in.). (u) To 7.77 mm (0.306 in.). (v) To 4.76 mm (0.187 in.). (w) To 2.34 mm (0.092 in.). (x) To 3.76 mm (0.148 in.). (y) To 1.57 mm (0.062 in.). (z) To 0.74 mm (0.029 in.). (aa) To 0.36 mm (0.014 in.). (bb) To 1.37 mm (0.054 in.). (cc) To 5.26 mm (0.207 in.). (dd) To 7.19 mm (0.283 in.). (ee) To 1.12 mm (0.044 in.). (ff) To 2.69 mm (0.105 in.). (gg) To 4.50 mm (0.177 in.). (hh) To 2.03 mm (0.080 in.). (jj) To 9.19 mm (0.362 in.).

hard drawn spring wire is the least costly grade. Its surface quality is comparatively low with regard to such imperfections as hairline seams. This wire is used in applications involving low stresses or static conditions.

Oil-tempered wire is a general-purpose wire, although it is more susceptible to the embrittling effects of plating than hard drawn spring wire. Its spring properties are obtained by heat treatment. Oil-tempered wire is slightly more expensive than hard drawn wire; it is significantly superior in surface smoothness, but not necessarily in seam depth.

Music wire is the carbon steel wire used for small springs. It is the least subject to hydrogen embrittlement by electroplating (see the section on electroplating) and is comparable to valve-spring wire in surface quality.

These three types of wire (hard drawn spring wire, oil-tempered wire and music wire) are used in the greatest number of applications. Most cold wound automotive springs are made of oil-tempered wire, and only a small percentage are made of music wire and hard drawn spring wire.

High-tensile hard drawn wire fills the gap where high strength is needed but where the quality of music wire is not required.

Carbon steel spring wire is the least costly of the valve-spring quality wires. *Chromium-vanadium steel wire* of valve-spring quality (A232) is superior to the same quality of carbon steel wire (A230) for service at 120 °C (250 °F) and above. Springs of *chromium-silicon steel wire* (A401) can be used at temperatures as high as 230 °C (450 °F). All valve-spring wires have the highest surface quality attainable in commercial production.

Carbon steel wire of valve-spring quality, and chromium-vanadium steel wire of both spring and valve-spring quality, can be supplied in the annealed condition. This will permit severe forming of springs with a low spring index (ratio of mean coil diameter to wire diameter) and also will permit sharper bends in end hooks. (Although a sharp bend is never desired in any spring, it is sometimes unavoidable.)

Springs made from *annealed wire* can be quenched and tempered to spring hardness after they have been formed. However, without careful control of processing, such springs will have greater variations in dimensions and hardness. This method of making springs usually is used only for springs with special requirements, such as severe forming, or for small quantities, because springs made by this method may have less uniform properties than those of springs made from pretempered wire, and are higher in cost. The amount of cost increase depends largely on design and required tolerances, but the cost of heat treating (which often involves fixturing expense) and handling can increase total cost by more than 100%.

Cold drawn *type 302 stainless steel spring wire* (A313) is high in heat resistance and has the best corrosion resistance of all the steels in Table 1. The surface quality of *type 302 stainless steel spring wire,* which formerly was very good, has become variable, with a serious effect on fatigue resistance. Type 316 stainless is superior in corrosion resistance to type 302, particularly against pitting in salt water, but is more costly and is not considered a standard spring wire. Type 302 is readily available and has excellent spring properties in the full-hard or spring-temper condition. It is more expensive than any of the carbon steel wires for designs requiring a diameter larger than about 0.30 mm (0.012 in.), but less expensive than music wire for sizes under about 0.30 mm. In many applications, type 302 stainless can be substituted for music wire with only slight design changes to compensate for the decrease in modulus of rigidity.

For example, a design for a helical compression spring was based on the use of 0.25-mm-diam (0.010-in.-diam) music wire. The springs were cadmium plated to resist corrosion, but they tangled badly in the plating operation because of their proportions. A redesign substituted type 302 stainless steel wire of the same diameter for the music wire. Fewer coils were required because of the lower modulus of rigidity, and the springs did not require plating for corrosion resistance. The basic cost of this small-diameter stainless wire was, at the time, 20% less than the cost of the music wire. Elimination of plating and reduction of handling resulted in total savings of 25%.

Wire Quality

The steels shown in Table 1 exhibit much greater differences in performance in fatigue applications than in static applications. The similarity of composition of the carbon steels emphasizes that performance differences depend more on surface quality than on composition. As shown in Fig. 1, the minimum tensile strengths of these steels are not greatly different.

Specification requirements for these wires include twist, coiling, fracture, or reduction in area tests, in addition to dimensional limits and minimum tensile strength. Such tests ensure that the wire has not been overdrawn (which would produce internal splits or voids) and has the expected toughness.

Seams are evaluated visually, often after etching with hot 50% muriatic acid. The depth of surface metal removed can vary from 0.006 mm (¼ mil) to 1% of wire diameter. Examination of small-diameter etched wire requires a stereoscopic microscope, preferably of variable power so that the sizes of seams can be observed in relation to the diameter of the wire. The least expensive wires can have seams that are quite pronounced. Hard drawn and oil-tempered wires occasionally have seams as deep as 3.5% of wire diameter, but not often deeper than 0.25 mm (0.010 in.). On the other hand, wires of the highest quality (music wire, valve-spring wire) have only tiny scratches, generally not deeper than ½% of wire diameter. Some grades can be obtained at moderate cost with seam depth restricted to 1% of wire diameter.

There is no general numerical limit on decarburization, and phrases such as "held to a minimum consistent with commercial quality" are very elastic. It is usual for seams present during hot rolling to be partly decarburized to the full depth of the seam, or slightly deeper.

For valve-spring quality, most manufacturers require that wire conform to aircraft quality as defined by the AISI Steel Products Manual. Decarburization limits are more severe. Some manufacturers permit loss of surface carbon only if it does not drop below 0.40% for the first 0.025 mm (0.001 in.) and, within the succeeding 0.013 mm (0.0005 in.), becomes equal to the carbon content of the steel.

General decarburization can be detrimental to the ability to maintain load. For hot wound springs made directly from hot rolled bars, it is common practice to specify a torsional modulus of 72 GPa (10.5 million psi) instead of the 80 GPa (11.5 million psi) used for small spring wires. In part, this compensates for the low strength of the surface layer. Total loss of carbon from the surface during a heat treating process is infrequent in modern wiremill products. Partial decarburization of spring wire is often blamed for spring failures, but quench cracks and coiling-tool marks are more frequently the actual causes. In wires of valve-spring or aircraft quality, a decarburized ferritic ring around the wire circumference is a basis for rejection. The net effects of seams and decarburization are described in the section on fatigue in this article.

Inspection for seams and other imperfections in finished springs generally is carried out by magnetic-particle inspection. In its various forms, this inspection method has proved to be the most practical nondestructive method for inspection of springs that may affect human safety or that for other reasons must not fail as a result of surface imperfections. The inspection is always concentrated selectively on the inside of the coil, which is more highly stressed than the outside and is the most frequent location of start of failure.

Freedom from surface imperfections is of paramount importance in some applications of highly stressed springs for shock and fatigue loading, especially where replacement of a broken spring would be difficult and much more costly than the spring itself, or where spring failure could cause extensive damage to other components.

Residual Stresses

Residual stresses can increase or decrease the strength of a spring material, depending on their direction. For example, residual stresses induced by bending strengthen wire for deflection in the same direction while weakening it for deflection in the opposite direction. In practice, residual stresses are either removed by stress relieving or induced to the proper direction by cold setting and shot peening.

Compression springs, torsion springs, flat springs and retainer rings may be stress relieved and cold set. The treatment used depends on design and application requirements of the individual spring.

Many compression springs are preset for use at higher stress. They are then known as springs with set removed. Compression springs can be made to close solid, without permanent set and without presetting, if the shear stress is less than a specified proportion of the tensile strength of the wire in the fully compressed spring (about 45% for music wire), and if the springs are properly stress relieved. The maximum shear stress in a fully compressed preset spring is about 33% higher, or approximately 60% of the tensile strength of the wire. Hence, presetting to a maximum stress will permit the use of up to 40% less steel than is otherwise required, a savings greater than the cost of presetting for wire larger than about 3 mm (⅛ in.) in diameter. Also, the smaller, equally strong spring requires less space.

When the calculated uncorrected stress at solid height is greater than about 60% of the tensile strength (or, for cold set springs, greater than the proportional limit stress), the spring can be neither cold set nor compressed to its solid height without taking a permanent set. Several types of springs are in this category, where the maximum permissible deflection must be calculated and positive stops provided to avoid permanent set in service.

Compression springs, cold wound and cold formed from pretempered high-carbon spring wire, should always be stress relieved to remove residual stresses produced in coiling.

Extension springs usually are given a stress-relieving treatment to relieve stresses induced in forming hooks or other end configurations, but such treatment should allow retention of stresses induced for initial tension.

The treatment of wire retainer rings depends on whether the loading tends to increase or decrease the relaxed diameter of the spring. Most rings contain residual stresses in tension on the inside surface. For best performance, rings that are reduced in size in the application should not be stress relieved, while expanded rings should be. This consideration applies equally to torsion springs.

Stress relieving affects the tensile strength and elastic limit, particularly for springs made from music wire and hard drawn spring wire; properties of both types of wire are increased by heating in the range from 230 to 260 °C (450 to 500 °F). Oil-tempered spring wire, except for the chromium-silicon grade, shows little change in either tensile strength or elastic limit after stress relieving below 315 °C (600 °F). Both properties then drop because of temper softening. Wire of chromium-silicon steel temper softens only above about 425 °C (800 °F).

Properties of spring steels are not usually improved by stress relieving for more than 30 min at temperature, except for age-hardenable alloys like type 631 (17-7 PH) stainless steel, which requires about 1 h to attain maximum strength. Temperatures that produce optimum values of the torsional elastic limit for ferritic steels are shown in Table 3.

Table 3 Optimum stress-relieving temperatures for steel spring wire (a)(b)

Steel	Temperature	
	°C	°F
Music wire	230-260	450-500
Hard drawn spring wire	230-290	450-550
Oil-tempered spring wire	230-400	450-750(c)
Valve spring wire	315-345	600-650
Cr-V spring wire	315-370	600-700
Cr-Si spring wire	425-455	800-850
Type 302 stainless	425-480	800-900
Type 631 stainless	480 ± 6(d)	900 ± 10(d)

(a) Applicable only for stress relieving after coiling and not valid for stress relieving after shot peening. (b) Based on 30 min at temperature. (c) Temperature is not critical and may be varied over the range to accommodate problems of distortion, growth, and variation in wire size. (d) Based on 1 h at temperature.

When springs are to be used at elevated temperatures, the stress-relieving temperatures should be near the upper limit of the range to minimize relaxation in service. Otherwise, lower temperatures are better.

Plating of Springs

Steel springs often are electroplated with zinc or cadmium to protect them against corrosion and abrasion. In general, zinc has been found to give the best protection in atmospheric environments, but cadmium is better in marine and similar environments involving strong electrolytes. Electroplating increases the hazards of stress raisers and residual tensile stresses because hydrogen released at the surface during acid or cathodic electrocleaning or during plating can cause a time-dependent brittleness, which can act as though added tensile stress had been applied and can result in sudden fracture after minutes, hours or hundreds of hours. Unrelieved tensile stresses can result in fracture during plating. Such stresses occur most severely at the inside of small-radius bends. Parts with such bends should always be stress relieved before plating. However, because even large-index springs have been found to be cracked, general stress relief is always good practice.

Preparation for plating is also very important because hydrogen will evolve from any inorganic or organic material on the metal until the material is thoroughly covered. Such contaminants

may be scarcely noticeable before plating except by their somewhat dark appearance. Thorough sand blasting or tumbling may be required to remove such layers.

Hydrogen Relief Treatment. If stress relieving has been attended to, and the springs are truly clean before plating, then the usual baking treatment of around 200 °C (400 °F) for 4 h should lessen the small amount of hydrogen absorbed and redistribute it to give blister-free springs, which will not fail.

Mechanical Plating. Another technique that solves the hydrogen problem is mechanical plating, which involves cold welding particles of zinc or other soft-metal powder to an immersion copper flash plate on the spring. While some hydrogen may be absorbed during acid dipping before plating, it does not result in a time-dependent embrittlement because the plated layer is inherently porous, even though it has a shiny appearance. The hydrogen easily diffuses through the pores within 24 h, leaving the steel ductile.

Fatigue

For those springs that are dynamically loaded, it is common practice to obtain basic mechanical data from S-N fatigue curves. A typical S-N diagram is shown in Fig. 4(a). For each cycle of fatigue testing, the minimum stress is zero and the maximum stress is represented by a point on the chart. An alternative method of presenting data on fatigue life of springs is shown in Fig 4(b).

Stress Range. In most spring applications, the load varies between initial and final positive values. For example, an automotive valve spring is compressed initially during assembly, and during operation it is further compressed cyclically each time the valve opens.

The shear-stress range (that is, the difference between the maximum and minimum of the stress cycle to which a helical steel spring may be subjected without fatigue failure) decreases gradually as the mean stress of the loading cycle increases. The allowable maximum stress increases up to the point where permanent set occurs. At this point, the maximum stress is limited by the occurrence of excessive set.

Figure 5 shows a fatigue diagram for music wire springs of various wire diameters and indexes. This is a modified

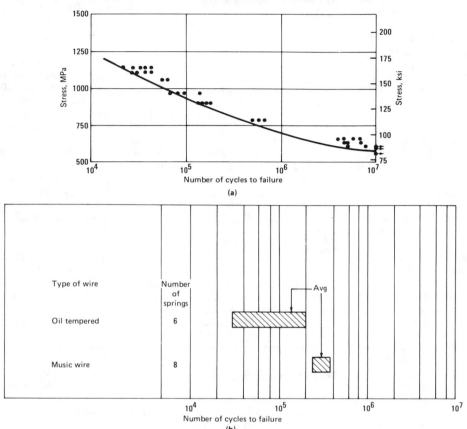

Fig. 4 Fatigue lives of compression coil springs made from various steels

(a)

(b)

(a) *S-N* diagram for springs made of minimum quality music wire 0.59 mm (0.022 in.) in diameter. Spring diameter was 5.21 mm (0.205 in.); *D/d* was 8.32. Minimum stress was zero. Stresses corrected by Wahl factor.

(b) Life of springs used in a hydraulic transmission. They were made of oil-tempered wire (ASTM A229) and music wire (A228). Wire diameter was 4.75 mm (0.187 in.), outside diameter of spring was 44.45 mm (1.750 in.), with 15 active coils in each spring. The springs were fatigue tested in a fixture at a stress of 605 MPa (88 ksi), corrected by the Wahl factor.

Goodman diagram and shows the results of many fatigue-limit tests on a single chart. In this diagram, the 45° line *OM* represents the minimum stress of the cycle, while the plotted points represent the fatigue limits for the respective minimum stresses used. The vertical distances between these points and the minimum-stress reference line represent the stress ranges for the music wire springs.

In fatigue testing, some scatter may be expected. The width of the band in Fig. 5 may be attributed partly to the normal changes in tensile strength with changes in wire diameter. There appears to be a trend toward higher fatigue limits for the smaller wire sizes. Line *UT* usually is drawn so as to intersect line *OM* at the average ultimate

shear strength of the various sizes of wire.

Modified Goodman diagrams for helical springs made of several steels are shown in Fig. 6. In all instances, the plotted stress values were corrected by the Wahl factor. The data were obtained from various sources, including controlled laboratory fatigue tests, spot tests on production lots of springs, and correlation between rotating-beam fatigue tests on wire and unidirectional-stress fatigue tests on compression and extension helical springs.

The points on the vertical axes are the stresses at 10 million cycles taken from *S-N* curves similar to those in Fig. 4(a) for each wire size, where the minimum stress is zero. In Fig. 6, the stress range is the vertical distance between the 45°

line and the lines for the several wire sizes. The allowable maximum stress increases to a point of permanent set, indicated by the horizontal sections of the lines on the diagrams. On the right of Fig. 6 are shown the allowable stresses for less than 10 million cycles. For equal wire sizes, these diagrams show that the most fatigue-resistant (music) wires have fatigue limits 50% greater than those of the least resistant (hard drawn) wires. This difference is largely maintained under high-stress, short-life conditions. These graphs represent normal quality for each grade. Due to variations in production conditions, however, quality is not constant.

Figure 7 shows the statistical range of fatigue life for five lots of music wire,

Fig. 5 Fatigue limits for compression coil springs made of music wire

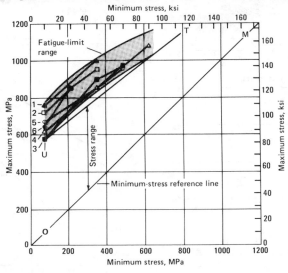

Spring No.	Wire diam		Spring OD		Spring index	Free length		Total turns	Active turns	Total tested
	mm	in.	mm	in.		mm	in.			
1 . . . . .0.81		0.032	9.52	0.375	10.7	22.10	0.87	6.0	4.2	16
2 . . . . .0.81		0.032	6.35	0.250	6.8	26.97	1.062	7.0	5.2	28
3 . . . . .1.22		0.048	15.88	0.625	12.0	44.45	1.75	7.0	5.2	38
4 . . . . .2.59		0.102	22.22	0.875	7.6	60.20	2.37	7.0	5.2	43
5 . . . . .3.07		0.121	22.22	0.875	6.2	57.15	2.25	7.5	5.7	35
6 . . . . .4.50		0.177	22.22	0.875	4.9	57.15	2.25	7.5	5.7	25

Data are average fatigue limits from *S-N* curves for 185 unpeened springs of various wire diameters run to 10 million cycles of stress. All stresses were corrected for curvature using the Wahl correction factor. The springs were automatically coiled, with one turn squared on each end, then baked at 260 °C (500 °F) for 1 h, after which the ends were ground perpendicular to the spring axis. The test load was applied statically to each spring and a check made for set three times before fatigue testing. The springs were all tested in groups of six on the same fatigue testing machine at ten cycles per second. After testing, the unbroken springs were again checked for set and recorded. Number 4 springs, tested at 1070 MPa (155 ksi) max stress, had undergone about 2½% set after 10 million stress cycles, but the stresses were not recalculated to take this into account. None of the other springs showed appreciable set. The tensile strengths of the wires were according to ASTM A228.

all 0.51 mm (0.020 in.) in diameter. Wire was tested on a rotating-beam machine at a maximum stress of 1170 MPa (170 ksi) and a mean stress of zero. Results were correlated with fatigue tests on torsion springs as follows: a minimum fatigue life of 50 000 cycles was required of each spring; a minimum life of 20 000 cycles for the wire in the rotating-beam machine at 1170 MPa (170 ksi) gave satisfactory correlation with the 50 000-cycle service life of springs made from the wire. Lot 5 in Fig. 7 was rejected because it failed to meet the fatigue requirement. Subsequent fatigue tests on a pilot lot of springs made from lot 5 wire confirmed the inability of these springs to meet the fatigue requirement of 50 000 cycles.

Shot peening of springs improves fatigue strength by prestressing the surface in compression. It can be ap-

plied to wire 1.6 mm (1/16 in.) or more in diameter, and slightly smaller wire using special techniques. The kind of shot used is important; better results are obtained with carefully graded shot having only a few broken, angular particles. Shot size may be optimum at roughly 20% of the wire diameter. However, for larger wire, it has been found that excessive roughening during peening with coarse shot lessens the benefits of peening, apparently by causing minute fissures. Also, peening too deeply leaves little material in residual tension in the core; this negates the beneficial effect of peening, which requires internal tensile stress to balance the surface compression.

Shot peening is effective in largely overcoming the stress-raising effects of shallow pits and seams. Proper peening intensity is an important factor, but

more important is the need for both the inside and outside surfaces of the spring to be thoroughly covered. An Almen test strip necessarily receives the same exposure as the outside of the spring, but to reach the inside, the shot must pass between the coils and is thereby much restricted. Thus, for springs with closely spaced coils, a coverage of 400% on the outside may be required to achieve 90% coverage on the inside.

Cold wound steel springs normally are stress relieved after peening to restore the yield point. A temperature of 230 °C (450 °F) is common because higher temperatures degrade or eliminate the improvement in fatigue strength.

The extent of improvement in fatigue strength to be gained by shot peening, according to one prominent manufacturer of cold wound springs, is shown in Fig. 8. The bending stresses apply to flat

Fig. 6 Modified Goodman diagrams for steel helical springs

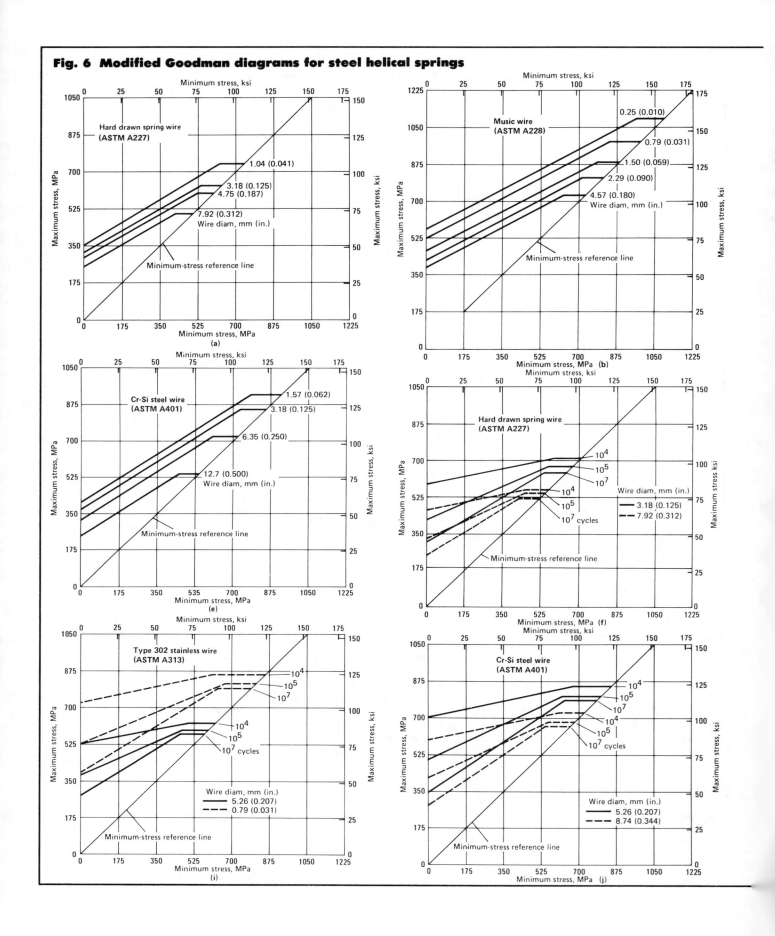

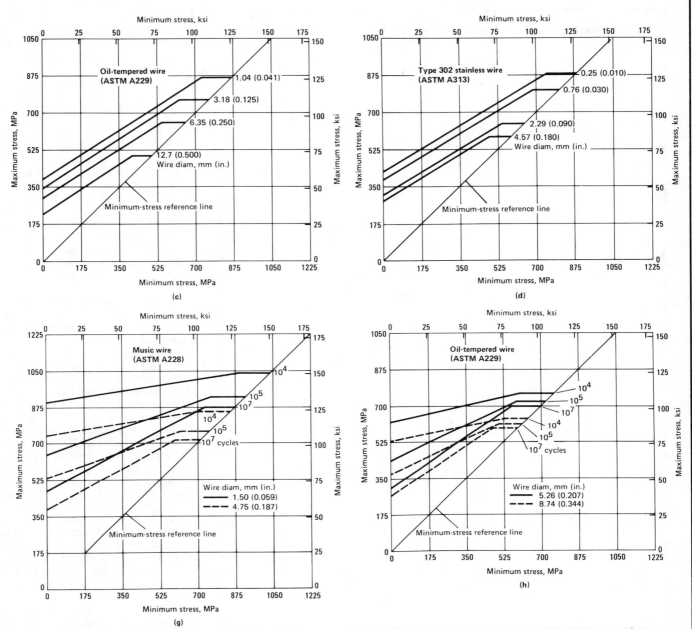

Stresses were corrected by the Wahl factor. Data for these curves were obtained from controlled laboratory fatigue tests, spot tests on production lots of springs, and correlation between rotating-beam fatigue tests and unidirectional-stress fatigue tests on compression and extension helical springs. Points on the vertical axis are the stresses at 10 million cycles taken from *S-N* curves similar to those in Fig. 4 (a) for each wire size, where the minimum stress is zero. The stress range is the difference between the 45° line and the lines for the several wire sizes indicated on the charts. (a) to (e) 10 million stress cycles, wire diameters as indicated. (f) to (j) two wire diameters, and number of stress cycles as indicated.

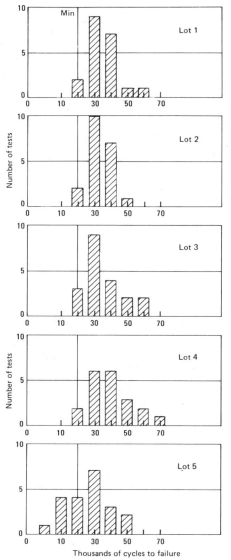

Fig. 7 Fatigue-life distribution for 0.51-mm (0.020-in.) diam music wire

Tested in a rotating-beam machine at a maximum stress of 1170 MPa (170 ksi) and a mean stress of zero.

tion calculated. For example, a music wire spring designed at a corrected stress of 690 MPa (100 ksi) will relax a maximum of 3.8% when held at 90 °C (200 °F) for 72 h.

Springs made from music wire (A228) are equal in performance to those made from oil-tempered wire (A229) at 90 °C (200 °F) but are inferior at 150 °C (300 °F) (Fig. 9a and b), and alloy steel springs at 200 °C (400 °F) are superior to the carbon steel springs at 150 °C (300 °F). Percentage of load loss increases with shear stresses.

The effect of time and temperature on relaxation of springs is shown in Fig. 9 (g) and (h). Rate of relaxation is greatest during the first 50 to 75 h. For longer periods of time, the rate is lower, decreasing as the logarithm of time when no structural change or softening occurs.

A significant effect of temperature on relaxation of piston rings is indicated in Fig. 9(i). The rings were confined in test cylinders to maintain the outside ring diameter and were exposed to test temperature for 3 to 4 h. The load required to deflect the ring to working diameter was measured before and after each test to calculate the amount of relaxation.

Tests on thousands of springs under various loads at elevated temperatures are summarized in Fig. 10. Plain carbon spring steels of valve-spring quality are reliable at stresses up to 550 MPa (80 ksi) (corrected) and temperatures no higher than 175 °C (350 °F), in wire sizes no greater than 9.5 mm (⅜ in.). Slightly more severe applications may be successful if springs are preset at the operating temperature with loads greater than those of the application. Plain carbon spring steels of valve-spring quality should not be used above 200 °C (400 °F) (see Table 1).

Except for high speed steel, these tests revealed no advantage in springs heat treated after coiling compared with those at the same hardness made of pretempered wire and properly stress relieved.

Deflection of a spring under load is inversely proportional to the modulus of rigidity, *G,* of the material. Variation with temperature is shown in Fig. 11.

A uniform deflection under load over a range of temperatures sometimes must be maintained. The instrument spring in Fig. 12 required a constant modulus up to 90 °C (200 °F) and, when made of music wire, drifted 5% in service. It was replaced with a satisfactory spring made of a nickel alloy with con-

springs, power springs and torsion springs; the torsional stresses apply to compression and extension springs.

Effect of Temperature

The effect of elevated temperatures on mechanical properties and performance of fabricated springs is shown in Fig. 9 and 10; effect is reported as amount of load loss (relaxation), which

is a function of chemical composition and maximum stress. Helical compression springs were tested to determine the maximum relaxation at a given static working stress, temperature, and time. A specific spring height was determined for a given corrected stress. The spring was clamped at this height and placed in a convection oven for 72 h. It was removed and cooled, and the new free height was measured. Load loss was determined and amount of relaxa-

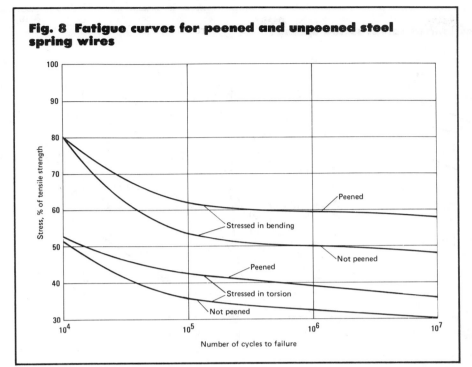

Fig. 8 Fatigue curves for peened and unpeened steel spring wires

Table 4 Comparison of two spring materials for elevated-temperature service (a)

| | Condition at elastic limit | | | |
| | Load | | Total deflection | |
Grade of wire	kg	lb	mm	in.
Oil-tempered (A229)	1227	270.5	19.3	0.76
Cr-Si Steel (A401)	1397	308.0	21.8	0.86

(a) Design data (the same for both types of wire): mean diameter, 15.88 mm (0.625 in.); inside diameter, 12.12 mm (0.477 in.); wire diameter; 3.76 mm (0.148 in.); spring index, 4.22; total number of coils, 10; number of active coils, 8; spring rate, 6.36 kg/mm (356 lb/in.); working load, 517 kg (114 lb); deflection at working load, 7.92 mm (0.312 in.); stress at working load, 383 MPa (55.6 ksi); solid height, 37.6 mm (1.48 in.) max; free height, 49.0 mm (1.93 in.) approx; set removed, plain finish, variable type of load, environment of about 90 °C (200 °F).

Table 5 Typical minimum hardness and hardenability for steel used for hot wound helical springs (a)

| Corrected max solid stress | | Hardness, HRC | |
MPa	ksi	At surface	At center
Static Load			
690	100	45	35
825	120	50	45
965	140	60	50
1100	160	60	50
1240	180	60	50
Variable Load, Designed for a Minimum Life of 50 000 Cycles (Set removed, 2.5% probability of failure, mean stress 515 MPa, or 75 ksi)			
690	100	45	35
825	120	50	45
965	140	60	50
1100	160	60	50
1240	180	60	50
Dynamic Load, Designed for a Minimum Life of 2 Million Cycles (Set removed, shot peened, 2.5% probability of failure, mean stress 515 MPa, or 75 ksi)			
690	100	45	35
825	120	60	50

a) As oil quenched, prior to tempering. Normal hardness, as tempered, 44 to 49 HRC at surface.

stant modulus (42 Ni, 5.4 Cr, 2.40 Ti, 0.60 Al, 0.45 Mn, 0.55 Si, 0.06 C, remainder Fe).

In another example, a spring originally was fabricated from oil-tempered wire (A229) and performed satisfactorily when tested at room temperature. However, in service it was immersed in oil that attained a temperature slightly above 90 °C (200 °F), which was high enough to cause excessive relaxation over a period of 2 to 3 h. Chromium-silicon steel spring wire (A401) was substituted for the oil-tempered wire, and at the identical operating temperature service was satisfactory. Design data for these springs are given in Table 4.

Hot Wound Springs

Although some hot wound springs are made of steels that are also used for cold wound springs, hot wound springs usually are much larger, which results in significant metallurgical differences.

Hardenability Requirements. Steels for hot wound springs are selected mainly on the basis of hardenability. Carbon steels with about 0.70 to 1.00% carbon (1070 to 1095) are suitable and widely used for statically and dynamically loaded springs in the smaller sizes. Carbon steels also are used for larger springs in the lower stress range, where some hardenability can be sacrificed safely. However, alloy steels usually are required for the larger sizes because of the need for hardenability. Most specifications for hot wound alloy steel springs require 0.50 to 0.65% carbon and a minimum hardness of 50 HRC at the center after oil quenching from about 815 °C (1500 °F) and before tempering. (The austenitizing temperature will vary, depending on the specific steel.)

Springs subjected to lower stress ranges may not require this high hardenability and thus can be made of the lower-priced carbon steels. Typical minimum hardness and hardenability values are given in Table 5 for hot wound springs used under specific conditions of stress. Here it can be seen that lower surface hardness and lower hardenability are permitted in the lower stress ranges for both statically and variably loaded springs. These requirements gradually increase as the stress increases, with 50 HRC min specified for the solid stress of 1240 MPa (180 ksi), and a solid stress of 825 MPa (120 ksi) for dynamically loaded springs.

Recommended steels for hot wound helical springs are given in Table 6, covering variations in stress range, type of loading, and wire size. Hardenability requirements increase as required strength and/or wire diameter increases.

The strengths obtained in bars with different hardenabilities are shown in Fig. 13. The band for 1095 steel demon-

Fig. 9 Relaxation curves for steel helical compression springs

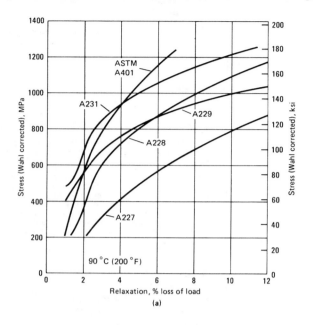

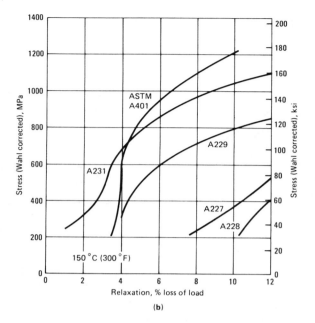

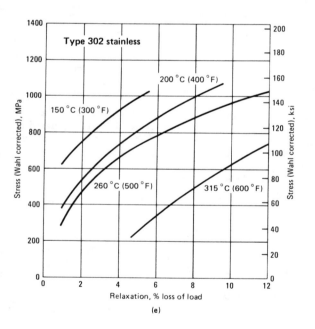

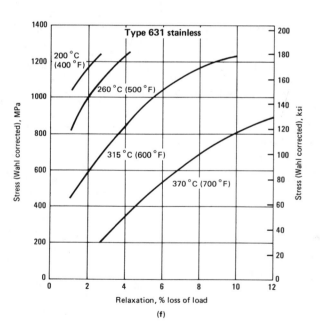

(a) to (f) Relaxation of helical springs after exposure for 72 h at indicated temperatures. A231 has same composition as A232; however, the latter grade is produced to valve spring quality.

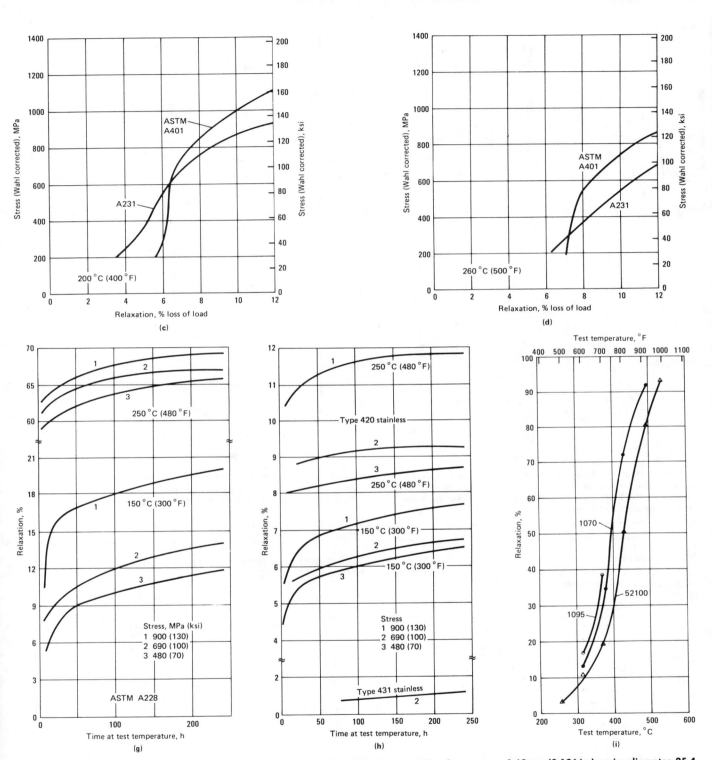

(g) and (h) Relationship of time and temperature on relaxation of ten-turn helical springs. Wire diameter was 2.69 mm (0.106 in.); spring diameter, 25.4 mm (1.00 in.); free length, 76.2 mm (3.00 in.); stresses were corrected by the Wahl factor.
(i) Relaxation of circular flat springs (piston rings) at elevated temperature. Spring hardness was HRC 35.

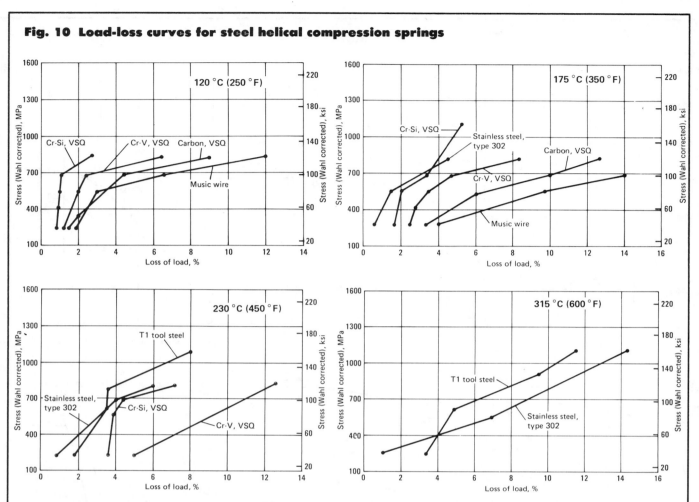

Fig. 10 Load-loss curves for steel helical compression springs

120 °C (250 °F)

175 °C (350 °F)

230 °C (450 °F)

315 °C (600 °F)

Based on tests of thousands of springs. The steels are: (ASTM) A228, music wire; A230, carbon steel, valve spring quality; A232, chromium-vanadium steel, valve spring quality; A313, Type 302 stainless steel; 9254, chromium-silicon steel, valve spring quality; T1, high speed tool steel. All stresses Wahl corrected. All springs were made of pretempered wire and were stress relieved after coiling, none were shot peened, and all were at the indicated temperatures at least 72 h.

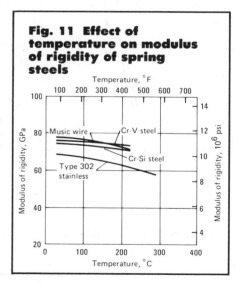

Fig. 11 Effect of temperature on modulus of rigidity of spring steels

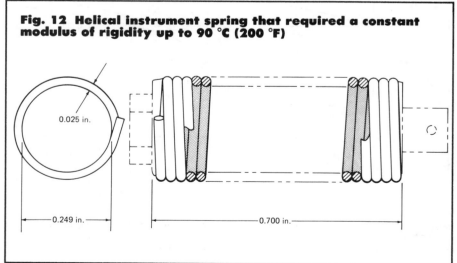

Fig. 12 Helical instrument spring that required a constant modulus of rigidity up to 90 °C (200 °F)

0.025 in.

0.249 in.

0.700 in.

Table 6 Recommended steels for hot wound helical springs (a)

Corrected max solid stress MPa	ksi	Diameter of spring wire(b)		
		9.5 to 25.4 mm (⅜ to 1 in.)	25.4 to 50.8 mm (1 to 2 in.)	50.8 to 76.2 mm (2 to 3 in.)
Static Load (Set removed, static stress up to 80% of max solid stress)				
825	120	1070 1095	1095	1095
965	140	1070 1095	51B60H 4161	4161
1100	160	5150H 5160H 50B60H	51B60H 4161	4161
1240	180	5150H 5160H 50B60H	51B60H 4161	4161
Variable Load, Designed for Minimum Life of 50 000 Cycles (Set removed, 2.5% probability of failure, operating stress range not over 50% of solid stress)				
690	100	1095	1095	1095
825	120	1095	51B60H 4161	4161
965	140	5150H 5160H 50B60H	51B60H 4161	4161
1100	160	5150H 5160H 50B60H	51B60H 4161	4161
1240	180	5150H 5160H 50B60H	51B60H 4161	4161
Dynamic Load, Designed for Minimum Life of 2 Million Cycles (Set removed, shot peened, 2.5% probability of failure, mean stress 515 MPa or 75 ksi, operating stress range not over 50% of solid stress)				
690	100	1095	1095	1095
825	120	1095	51B60H 4161	4161

(a) Where more than one steel is recommended for a specific set of conditions, they are arranged in the order of increasing hardenability. The first steel listed applies to the lower end of the designated wire diameter range and the last to the upper end of the range. (b) Hot rolled material.

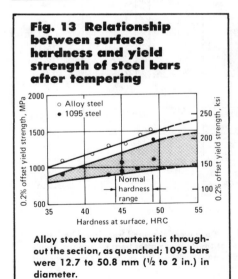

Fig. 13 Relationship between surface hardness and yield strength of steel bars after tempering

Alloy steels were martensitic throughout the section, as quenched; 1095 bars were 12.7 to 50.8 mm (½ to 2 in.) in diameter.

strates the variation in yield strength that results from variations in the thickness of the bar and the severity of quenching and from limited hardenability. Smaller bars, and those quenched more rapidly from the austenitizing temperature, follow the top of the band, while larger bars and those quenched less drastically fall in the lower half of the band. For example, a 12.7-mm (½-in.) round bar quenched in 5% caustic solution will have maximum properties, and a 50.8-mm (2-in.) round bar quenched in still oil will have yield strength near the minimum shown.

The scatter for alloy steels in Fig. 13 is much narrower because the points only represent steels with sufficient hardenability to have a martensitic structure throughout the bar section, as quenched.

Distribution of hardness-test results at surface and center, as quenched and as tempered, is shown in Fig. 14 for a multiplicity of heats of 1095 steel and five alloy steels commonly used for hot wound springs. The results of testing specimens 12 in. long correlate with those of testing production springs. For the alloy steels, a minimum of 50% martensite at the center of the quenched section was specified.

Table 7 lists steels that meet hardenability requirements for torsion-bar springs with section thicknesses from 29.21 to 57.15 mm (1.150 to 2.250 in.).

Surface Quality. Hot wound springs fabricated from bars of large diameter will normally show much deeper surface decarburization, in the range from 0.13 to 0.38 mm (0.005 to 0.015 in.), unless special material preparation and processing techniques are used. For hot wound springs, where design will permit, the use of the desirable stress pattern created by shallow hardening can be effective in overcoming effects of decarburization (Table 8).

Detrimental effects of decarburization are less noticeable on hot wound than on cold wound springs because other weaknesses, such as the surface imperfections and irregularities typical of a hot rolled surface, create additional focal points for fatigue cracks.

Specifications for hot wound springs and torsion bar springs usually include maximum seam depth. For example, the specifications used by a manufacturer of railway equipment allow seams with a maximum depth of 0.025 mm (0.001 in.) per 1.59 mm (¹⁄₁₆ in.) of wire diameter up to 0.41 mm (0.016 in.) for any bar size above 25.4 mm (1 in.).

Another manufacturer allows seam depths of 0.41 mm (0.016 in.) for bars 25.4 mm (1 in.) in diameter, 0.81 mm (0.032 in.) for bars 25.4 to 44.4 mm (1.00 to 1.75 in.) in diameter, and 1.22 mm (0.048 in.) for bars 44.4 to 63.5 mm (1.75 to 2.50 in.) in diameter.

Design Stress. It should be noted that some organizations specify a much more conservative approach to stress than that presented in Table 6. The Spring Manufacturers Institute has adopted a chart from the Manufacturers Standardization Society of the Valve and Fittings Industry, for essentially static service, that calls for an admittedly uncorrected stress for alloy steel of 760 MPa (110 ksi) for a bar diameter of 12.7 mm (½ in.), reducing to 590 MPa (86 ksi) for a bar diameter of 92.1 mm

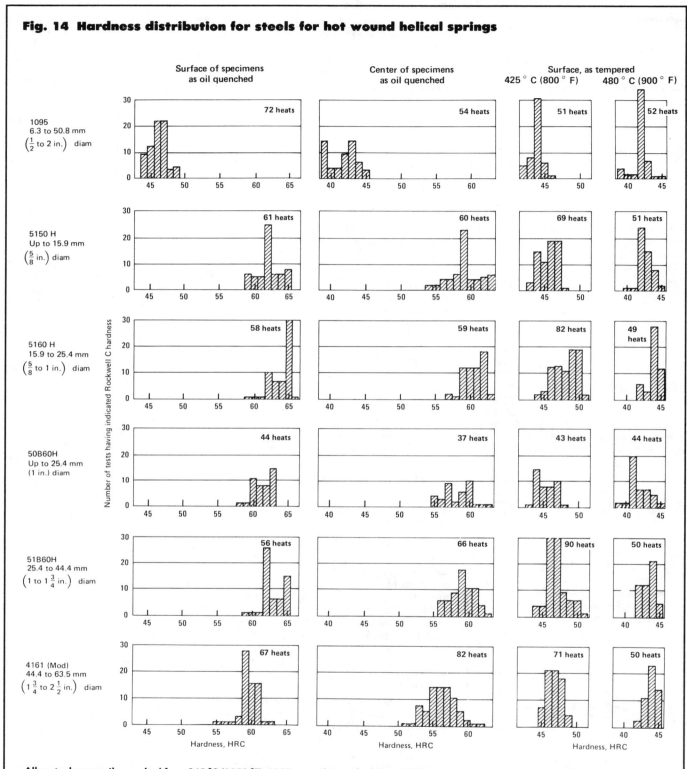

Fig. 14 Hardness distribution for steels for hot wound helical springs

Alloy steels were oil quenched from 845 °C (1550 °F); 1095 was oil quenched from 890 °C (1625 °F). Data were obtained from hot rolled, heat treated laboratory test coupons, 305 mm (12 in.) long. Specimens were sectioned from the center of the coupons after heat treatment. These results on bars correlate with those on production springs. For the alloy steels, a minimum of 50% martensite at the center of the quenched section was specified.

Table 7 Steels with sufficient hardenability for torsion bar springs

Thickness of section mm	in.	Minimum hardenability required	Steel
29.21	1.150	J50 at 8	8650H, 5152H
33.91	1.335	J50 at 9	8655H, 50B60H
36.32	1.430	J50 at 8½	50B60H
39.88	1.570	J50 at 10	51B60H
46.23	1.820	J50 at 11	8660H
48.26	1.900	J50 at 14	4150H
57.15	2.250	J50 at 22	9850H

(3⅝ in.). Their curve for 1095 steel is roughly 140 MPa (20 ksi) lower.

The desirability of conservative design in cyclical service is illustrated in Fig. 15, in which the minimum stress used was low. Such data on springs hot wound from bars with as-rolled surfaces are limited, and interpretation is therefore difficult. The value of peening, however, is made quite apparent. Surface imperfections can be removed by grinding, and this is normal practice, where the increased cost can be borne, in order to increase reliability at higher stresses.

With all of the limitations discussed above, there can be both cost and reliability advantages in using 1095 steel at low stresses. For example, railroad freight car springs, which are subject to severe corrosion pitting, are made of 1095 steel and designed for very low stress, and very little difficulty is encountered.

Costs

The relative costs of various spring steels in the form of round wire are given in Fig. 16. Base price may be outweighed by other costs, as indicated in Table 9, which shows possible extras for two alloy spring steels of about the same base price. Also, base price and other costs will vary somewhat with time.

The amount of material in a spring can be minimized by designing for the highest safe stress level. From the equation for volume of material:

$$V = E(4G/S^2) \qquad \text{(Eq 1)}$$

in which V is volume of material in the spring, in cubic inches; E is spring energy, in inch-pounds; G is torsion modulus of elasticity, in pounds per square inch; and S is stress, in pounds per square inch. A reduction of 5% in stress level increases required volume of material by 11%.

Cost of Springs. Tolerances often are the most important factors in spring cost, and should be selected with cost in mind. Tolerances superimposed on tolerances are especially costly. For example, if a free-length tolerance were specified in addition to load tolerances at each of two deflections, the cost of manufacturing the spring could be much higher than if free-length tolerance were unspecified, or indicated as approximate. Important dimensions sometimes can be held to close tolerances at no additional cost by allowing wider tolerances elsewhere. Tolerances on spring diameter and pitch are the least expensive to hold.

Appendix on Design

All spring design is based on Hooke's Law: when a material is loaded within its elastic limit, the resulting strain in the material is directly proportional to the stress produced by the load. For springs, this means that the amount of stretch or other deflection of the spring is directly proportional to the load or other force producing it.

Charts and formulas are available to aid in the design of springs. Charts that recommend categorically a design stress for a given spring steel can be reliable guides for springs used in static service only, unless otherwise qualified. Charts of any sort are of value only when it is known to which wire size(s) they apply, whether the data are for springs with or without presetting, and whether the cited nominal stresses include a "correction factor" or a "safety factor".

The spring-design formulas given in Table 10 are valid for extension and compression springs of round and square wire below the stress at which yielding begins. For rectangular wire, the numerical coefficients vary with the ratio of width to thickness. Detailed information on these calculations is given in the SAE Handbook, Supplement J795 on Spring Design, and in the Spring Manufacturers Institute Handbook on Spring Design. Much of the labor of computation in design can be eliminated by the use of "spring slide rules", which often are more effective than an unprogrammed electronic calculator.

Wahl Correction. In a straight torsion-bar spring of circular section, twisting produces a shear stress uniform at every point on the surface; but in a helical spring coiled from round wire, the stress at the inside of the coil is higher.

The helical spring formulas in Table 10 do not include additional stresses that exist because the load on the spring imposes bending and shear stresses, as well as torsion, in the wire. The magnitude of these stresses varies with the spring index D/d. A set of formulas developed by A. M. Wahl, involving a correction factor known as the Wahl factor, takes into account these two effects (Fig. 17).

Table 11 shows a typical comparison between uncorrected and corrected stresses for two spring indexes and five grades of spring wire. The values in Table 11 indicate that the stress in the spring is higher because of combined bending and torsion at the lower index. Thus, in the first line, if the elastic limit

Table 8 Properties of shallow-hardened and through-hardened hot wound helical compression springs (a)

Heat treatment	Hardness, HRC Distance below surface 1.59 mm (¹⁄₁₆ in.)	3.18 mm (⅛ in.)	6.35 mm (¼ in.)	Center	Fatigue tests(b) Millions of cycles to failure High	Low	Average
1045 Steel, Decarburized 0.38 mm (0.015 in.)							
As quenched	60	50	45	30	. . .	. . .	. . .
Tempered	54.5	49	37.5	25	1.28	0.46	0.772
8655H Steel, Decarburized 0.03 mm (0.001 in.)							
As quenched	60	60	60	60	. . .	. . .	. . .
Tempered	42	43	43	42	0.789	0.443	0.572

(a) Wound from bar 31.8 mm (1¼ in.) in diameter. (b) 485 MPa (70 ksi) stress range (corrected), 415 MPa (60 ksi) mean stress (corrected); eight springs tested in each group.

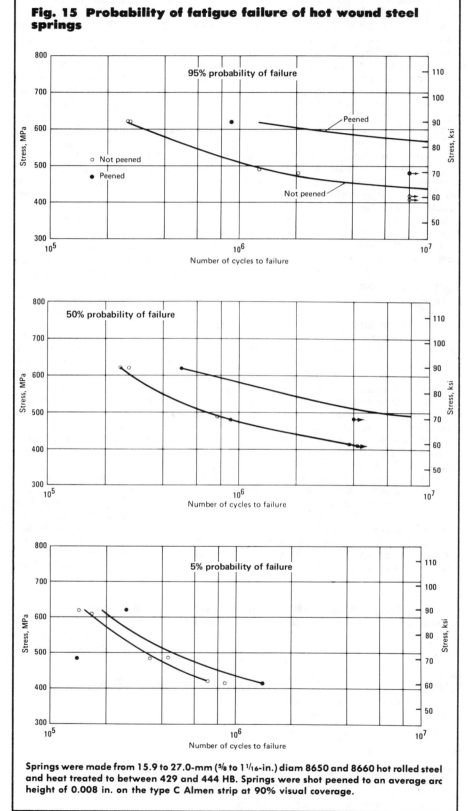

Fig. 15 Probability of fatigue failure of hot wound steel springs

Springs were made from 15.9 to 27.0-mm (⅝ to 1¹/₁₆-in.) diam 8650 and 8660 hot rolled steel and heat treated to between 429 and 444 HB. Springs were shot peened to an average arc height of 0.008 in. on the type C Almen strip at 90% visual coverage.

is not over 1080 MPa (157 ksi) the spring will take a permanent set.

For springs with an index greater than 12, the Wahl correction is small. Stresses for statically loaded springs that are not preset and that must not take a permanent set need not be computed with the correction factor, except for near-maximum stresses and springs of low index. Correction is needed for springs to be repeatedly loaded.

Stress Range. Safe maximum stresses for a minimum of 10 million cycles of stress will be somewhat lower than the fatigue limits shown in Goodman diagrams (Fig. 5 and 6), chiefly because of variations in the wire, especially with regard to surface condition, and variations in the spring itself and its manufacture.

Commercial tolerances on spring wire also will affect the selection of allowable stress for a production spring. For instance, the 0.013-mm (0.0005-in.) tolerance on 0.81-mm (0.032-in.) diam music wire gives rise to a stress variation of 4½%. In addition, stresses may vary proportionately with load tolerance from 7 to 15%, depending on the number of coils in the spring. Tolerances on number of coils, squareness, and spring diameter also permit stresses different from those calculated. For these reasons, stresses near the upper fatigue limits may be unsafe for some kinds of applications.

Life. Unless required, it is wasteful to design a spring for infinite life. A fatigue curve or S-N diagram provides approximations of allowable stresses for designing to a desired life. Springs made of hard drawn or oil-tempered wire, which may contain seams or scratches as deep as 3.5% of wire diameter or 0.25 mm (0.010 in.), should be limited to static applications or subjected to less than 10 000 loadings.

Ordnance springs typify long-stroke springs that have limited space and life requirements and that are subjected to shock loads. Stranded-wire springs sometimes have been used because of damping or design considerations. However, these springs are costly and chafe between coils. Stress level, stress range, and life (until set becomes excessive) of some springs used at high stresses, including some stranded-wire springs, are listed in Table 12. The life of these springs is not entirely dependent on maximum operating stress and stress range, but also is affected by the shock loading that is exerted. However, a fracture in one strand does not signif-

Fig. 16 Relative cost of spring steel wire

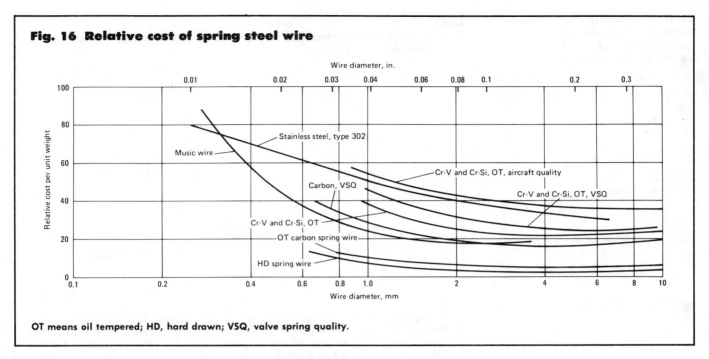

OT means oil tempered; HD, hard drawn; VSQ, valve spring quality.

Table 9 Pricing of automotive coiled spring steel

Item		Cost per 45.4 kg (100 lb)	Item		Cost per 45.4 kg (100 lb)	Item		Cost per 45.4 kg (100 lb) (a)
Alloy Steel 6150			**Alloy Steel 5160**				show heat number, weight, part number, grade, length and size	
Base price	Alloy steel	$16.80	Base price	Alloy steel	$16.80			
Grade	6150	4.10	Grade	5160, standard fine grained	1.55			
Quality	Electric furnace	1.25	Quality	Open hearth	...	Packaging	1.59 to 1.81 t (3500 to 4000 lb) lifts wired with 4 or 5 wires, wrapped in waterproof paper with three steel bands outside paper for magnet unloading	...
Restrictions	Restricted carbon and manganese	2.35	Restrictions	None, standard steel	...			
Size	12.7 mm (½ in.) diam	2.75	Size	16.48 mm (0.649 in.)	2.20			
Straightness	Special straightness	1.75		Close tolerance (½ standard)	.90			
Treatment	Hot rolled, precision ground	4.45	Straightness	½ standard tolerance, machine straightened	1.75			
Cleaning and coating	Pickled and oiled	2.75					1.59 to 1.81 t (3500 to 4000 lb) lifts	.10
Preparation	None	...	Treatment, coating and cleaning	Hot rolled, Pickled and oiled	2.70		Paper wrap	.15
Testing	Restricted hardenability	2.00				Loading	Gondola cars blocked to maintain straightness; not less than 18.1 t (20 tons) per car	...
Cutting	1.5 to 2.4 m (5 to 8 ft) abrasive	.50	Preparation	No surface conditioning allowed	...			
Length	4.6 m (15 ft) dead length	.40	Testing	Special decarburization, standard chemistry and hardenability	...			
Quantity	Less than 0.91 t (2000 lb)	8.65				Total		$27.05
Marking	Continuous line	2.25	Cutting	Machine cut	.50	(a) 1977 prices		
Packaging	Paper wrap	.15	Length	Dead length 3.63 m (143 in.)	.40			
	1.13 t (2500 lb) or less box with runners, bar 4.6 to 6.7 m (15 to 22 ft)	2.75	Quantity	Heat lots	...			
			Marking	Paint one end of the bundle green; attach metal tag;	...			
Loading	Box car loading	.10						
Total		$53.00						

Table 10 Design formulas for helical extension and compression springs

Round wire	Square wire

Stress for Statically Loaded Springs, psi

$$S = \frac{2.55\,PD}{d^3} \qquad \frac{2.4\,PD}{t^3}$$

$$S = \frac{FGd}{\pi D^2 N} \qquad \frac{FGh}{2.32\,D^2 N}$$

Stress for Round Wire Corrected by the Wahl Factor (Fig. 17), psi

$$S_K = \frac{2.55\,PD}{d^3}\left[\frac{4C-1}{4C-4} + \frac{0.615}{C}\right]$$

Stress for Square Wire Corrected by Wahl Factor (Fig. 17), psi

$$S_K = \frac{2.4\,PD}{t^3}\left[1 + \frac{1.2}{C} + \frac{0.56}{C^3} + \frac{0.50}{C^3}\right]$$

Wire Size for Given Load and Stress, in.

$$d^3 = \frac{2.55\,PD}{S}$$

$$t^3 = \frac{2.4\,PD}{S}$$

Deflection of the Spring Under Load, in.

$$F = \frac{8\,PD^3 N}{Gd^4} \qquad \frac{5.58\,PD^3 N}{Gt^4}$$

Spring Loading at a Given Deflection, lb

$$P = \frac{FGd^4}{8\,D^3 N} \qquad \frac{FGt^4}{5.58\,D^3 N}$$

$$P = \frac{Sd^3}{2.55\,D} \qquad \frac{St^3}{2.4\,D}$$

Note: S is nominal shear stress, in pounds per square inch; P is spring load or force, in pounds; D is mean diameter of the spring, in inches (outside diameter of the spring minus the wire diameter); d is diameter of the wire, in inches; t is the thickness of the square wire before coiling, in inches; F is the deflection, in inches, under load P; G is the modulus of elasticity in shear (modulus of rigidity), equal to 11.5 million psi for ordinary steels and 10 million psi for stainless steel near room temperature; N is the number of active coils; C is the spring index, D/d; and K is the Wahl correction factor. Similar formulas can be developed for metric units.

icantly affect the functioning of such a spring, which may develop several distributed breaks before becoming seriously impaired.

Compression Springs

The usual types of ends for compression springs are shown in Fig. 18, in

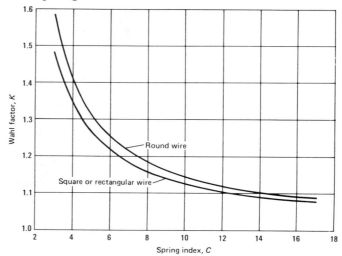

Fig. 17 Wahl correction factors for helical compression or extension springs

Spring index, C, is the mean diameter of the spring, D, divided by the diameter of the wire, d. When square wire, or rectangular wire coiled on the flat side, is used, the thickness of the wire, t, is substituted for d. When rectangular wire is coiled on edge, the width, b, is substituted for d. (From data in A. M. Wahl, *Mechanical Springs*, 2nd Edition, McGraw-Hill, 1963.)

Table 11 Effect of spring index on calculated spring working stress

Steel wire	Spring index	Wahl correction factor	Calculated spring working stress			
			Uncorrected		Corrected	
			MPa	ksi	MPa	ksi
Music	4.60	1.33	814	118	1082	157
	8.38	1.16	820	119	951	138
Hard drawn spring	4.60	1.33	634	92	841	122
	8.50	1.16	641	93	745	108
Oil-tempered spring ...	4.75	1.32	814	118	1075	156
	8.68	1.15	779	113	896	130
Cr-V spring	4.75	1.32	869	126	1145	166
	8.65	1.15	814	118	938	136
Cr-Si spring	4.75	1.32	1020	148	1344	195
	8.26	1.16	1076	156	1248	181

Table 12 Performance of highly stressed, shock-loaded music wire springs

Max stress		Stress range		Wire diam		Spring index(a),	Cycles to
MPa	ksi	MPa	ksi	mm	in.	D/d	excessive set
1103	160	862	125	2.29(b)	0.090(b)	3.9	2000
965	140	689	100	2.49	0.098	6.8	4000
1082	157	696	101	0.94(c)	0.037(c)	3.4	5000
1110	161	793	115	3.76	0.148	6.3	4000
1124	163	572	83	1.35	0.053	6.6	6000
1172	170	600	87	0.97	0.038	6.5	6000
945	137	558	81	1.57(b)	0.062(b)	4.8	7500
910	132	614	89	1.40	0.055	5.1	5000

(a) d, strand diameter. (b) Seven-strand wire. (c) Three-strand wire.

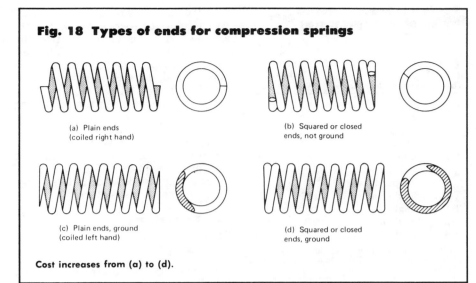

Fig. 18 Types of ends for compression springs

(a) Plain ends
(coiled right hand)

(b) Squared or closed
ends, not ground

(c) Plain ends, ground
(coiled left hand)

(d) Squared or closed
ends, ground

Cost increases from (a) to (d).

order of increasing cost. An additional increase in the cost of grinding may be anticipated with close tolerances on squareness and length.

When springs are designed to work at maximum stress, allowance should be made for the effect on stress of the specified type of end. For example, in a compression spring with squared, or squared and ground, ends, the number of active coils will vary throughout the entire deflection of the spring, and this results in loading rate and stress different from those originally calculated. This is especially important in springs having less than seven coils, because changes in the number of active coils result in high percentage changes in stress. The number of active coils changes because part of the active coil adjacent to the squared end closes down solid and becomes inactive as the spring approaches the fully compressed condition. In designs where this is objectionable, loads at two compressed lengths should be specified on the manufacturing print in order to ensure loading rates and stresses no higher than those calculated. Controlling loading in this way also controls the number of active coils and therefore the stresses at all positions of the spring during deflection.

For springs that will be operated at high stresses, tolerances that do not increase calculated stresses should be liberal so that tolerances on other dimensions that might increase stresses may be held closely. Thus, if the number of coils must be held closely in order to keep spring stresses in control, toler-

ance on free length with ends ground should be large.

Active Coils. The number of active coils equals the total number of coils less those that are inactive at each end. For plain grounds ends, the number of inactive coils depends on wire diameter and pitch of the spring. The number of active coils in a spring with plain, unclosed ends ground would approximately equal the number of turns of the wire untouched by grinding. One inactive coil on each end should be allowed in springs with squared ends or squared and ground ends. For springs of less than seven coils, this rule should be applied with care; in some springs, the degree of squareness should be considered.

Solid heights for the types of ends shown in Fig. 18(a) and (b) are computed by multiplying the wire diameter by the total number of coils plus one. Solid height for the type of end in Fig. 18(c) equals wire diameter times total number of coils plus one half the wire diameter, if grinding at each end removes at least one quarter of the wire thickness. For the end shown in Fig. 18(d), solid height is equal to the wire diameter times the total number of coils. A spring should be designed so that the coils do not touch or clash in service, because such contact will induce wear, fretting corrosion, and failure. The maximum solid height, specified primarily for inspection, should be determined after consideration of clearances and of tolerances on wire size, to allow spring manufacturers to meet economically the requirements estab-

lished for solid height.

Wire that is square or rectangular before coiling will upset at the inside of the coil and become trapezoidal in section during coiling. This limits deflection per coil, because solid height is predicated on wire thickness at the inside of the coil. An approximate formula for the upset thickness (t_1) at the inside compared to the original thickness (t) of the rectangular or square wire is:

$$t_1 = t(1 + k/c) \qquad \text{(Eq 2)}$$

where $c = (D_o + D_i)/(D_o - D_i)$; D_o is the outside diameter of the spring; D_i is the inside diameter of the spring; and k is 0.3 for cold wound springs, and 0.4 for hot wound springs and annealed materials.

With wire that is keystone shaped before coiling, this difficulty is not encountered, but such wire is costly and is not readily available—particularly in small lots.

Effect of Modulus Change. A change in the value of G (modulus of rigidity) resulting from a change in material must be compensated for by a change in wire diameter, which in turn changes spring rate. If the springs made from different steels must have precisely the same properties, the number of coils must differ. A typical example follows:

	Music wire	Type 302
Wire diam, in.	0.010	0.010
Coil diam, in. . . .	0.20	0.20
Number of coils . .	10	8.5
Spring rate, lb/in.	0.179	0.179

Extension Springs

Extension springs usually are close wound with some initial tension at the discretion of the springmaker, unless definite space between coils is specified; or close wound with no initial tension (difficult to accomplish); or close wound with a definite initial tension, usually within ±15%. In making stress calculations, initial tension is included in the load, P, and if P is measured, initial tension is included in the measurement obtained.

The necessity for fastening extension springs to some other part may require secondary operations or a costly fastening design. End hooks with sharp bends may have stress concentrations that increase stresses by a factor of two or three. There are many common designs for end hooks.

End Hooks. Stresses in end hooks are combinations of torsional and bending stresses. Tensile stresses diminish toward the body of the spring, where torsional stresses prevail. There is a region where both tensile and torsional stresses are present and are difficult to compute accurately. If the last coil of the spring adjacent to the end hook is reduced in diameter, the stress in the last coil and the end hook usually will be lower than that in the other coils.

The bending stresses in an end hook may be calculated from:

$$S_b = \frac{32\,PR}{\pi\,d^3}\,\frac{r_c}{r_i} \qquad \text{(Eq 3)}$$

where PR is the bending moment (load times moment arm from centerline of load application to centerline of wire), d is the diameter of the wire, r_c is the radius of the sharp bend measured to centerline of wire, and r_i is the inside radius of the sharp bend. The ratio r_c/r_i is an approximate evaluation of the stress-concentrating effect of the conformation these radii describe. The ratio indicates the destructive effect of a small radius, r_i. The smaller this radius, the higher the stress at the corner, other values being the same.

Because of the possibilities for over-extension and the stress concentrations in the ends, allowable stresses are more likely to be exceeded in extension springs than in compression springs. Therefore, allowable stresses that are 20% lower than those for compression springs are sometimes recommended for extension springs. However, when stresses in the ends are accurately known and the stroke of the spring is controlled in design of the spring or of the associated parts, the same values for allowable stresses can be used for extension springs as for compression springs that have not been preset.

Appendix on Leaf Springs*

Leaf springs, like all other springs, serve to absorb and store energy and then to release it. During this energy cycle, the stress in the spring must not exceed a certain maximum in order to avoid settling or premature failure. This consideration limits the amount of

*Abstracted by A. R. Shah, Senior Metallurgist, Automotive Operations, Rockwell International, from the "Manual on Design and Application of Leaf Springs", SAE Information Report J788a, 1970. Used with permission of the copyright holder, Society of Automotive Engineers, Inc.

Table 13 Energy stored by steel leaf springs(a)

Leaf type (b)	Spring design	Energy per unit weight of spring	
		J/kg	in.·lb/lb
F-1	Single leaf	43	173
	Multileaf, with all leaves full length	43	173
F-2	Multileaf, with properly stepped leaves	95	380
T-2	Single leaf	106	426
T-1	Single leaf	109	438
P-2	Single leaf	122	488
F-4	Single leaf	123	493

(a) Springs designed for maximum stress of 1100 MPa (160 ksi). (b) See Table 14 for descriptions of leaf types.

energy that can be stored in any spring. Table 13 lists values of energy that may be stored in the active parts of leaf springs designed for maximum stress of 1100 MPa (160 ksi). If consideration of the inactive part of the spring required for axle anchorage, spring eyes, etc., is included, the energy per pound of the total spring weight will be less than shown. For comparison, the stored energy in the active material of a helical spring made of round wire is 510 J/kg (2050 in·lb/lb) at 1100 MPa (160 ksi), and for a torsion bar of round cross section is 390 J/kg (1570 in.-lb/lb) at 965 MPa (140 ksi). This comparison shows that a leaf spring is inherently heavier than other types of springs. Balancing this weight disadvantage is the fact that leaf springs also can be used as attaching linkage or structural members. In order to be economically competitive, leaf springs must therefore be so designed that this advantage is fully utilized.

A leaf spring can be constructed of a single leaf, of several leaves of equal (full) length, or of several leaves of decreasing (stepped) lengths. The various types of leaves used in single-leaf springs are described in Table 14. Of course, flat-profile, or type F, leaves can be used to construct multileaf springs. A leaf spring made entirely of full-length leaves of constant thickness (see type F-1) is much heavier and less efficient than a leaf spring made of properly stepped leaves (see type F-2) or a single-leaf spring (see types F-4, P-2, T-1 and T-2).

The maximum permissible leaf thickness for a given deflection is propor-

tional to the square of the spring length. By choosing too short a length, the designer often makes it impractical for the spring maker to build a satisfactory spring, although the requirements for normal load, deflection and stress can be fulfilled.

When a stepped spring is made of type F-2 leaves, its length should be chosen so that the spring will have no less than three leaves. Springs with many leaves sometimes are used for heavy loads, but they are economical only where use of short springs leads to definite savings in the supporting structure.

Leaf Springs for Vehicle Suspension

Leaf springs are most frequently used in vehicle suspension. The characteristics of a suspension system are affected chiefly by the spring rate and the static deflection.

The "rate", or more accurately the "load rate", of a spring is the change of load per inch of deflection. This is not the same at all positions of the spring and is different for the spring alone and for the spring as installed. "Static deflection" of a spring equals the static load divided by the rate at static load; it determines the stiffness of the suspension and the ride frequency of the vehicle. In most instances, the static deflection differs from the actual deflection of the spring between zero load and static load due to influences of spring camber and shackle effect.

A soft ride generally calls for a large static deflection of the suspension. There are, however, other considerations and limits, among them the following:

1. A more flexible spring will have a larger total deflection and will be heavier.
2. In most applications, a more flexible spring will cause more severe striking through or will require a larger ride clearance (the spring travel on the vehicle from the design load position to the metal-to-metal contact position), disregarding rubber bumpers.
3. The change of standing height of the vehicle due to a variation of load is larger with a more flexible spring.

The static deflection to be used also depends on the available ride clearance. Further, the permissible static deflection depends on the size of the vehicle,

Table 14 Types of spring leaves

With Flat Profile: "F" Types

F-1: Rectangular Cantilever

Constant in thickness (t_0) and in width (w_0). Under load (P), the stress is greatest at line of encasement and decreases at a constant rate to zero at line of load application. The elastic curve in bending (from an intially flat spring) has its smallest radius at line of encasement, i.e. the rate of change of curvature is greatest at this line.

This design is inefficient.

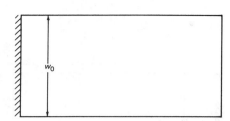
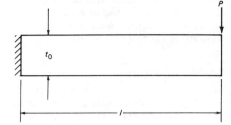

F-2: Trapezoidal Cantilever

Constant in thickness (t_0). Width decreases at a constant rate from w_0 at line of encasement to a specified dimension (w_e) at line of load application. Under load (P), the stress is greatest at line of encasement.

This design is more efficient than Type F-1.

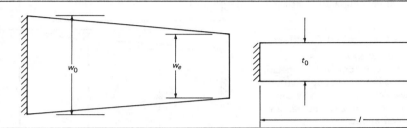

F-3: Triangular Cantilever

Constant in thickness (t_0). Width decreases at a constant rate from w_0 at line of encasement to zero at point of load application. Under load (P), the stress is constant throughout the length; the elastic curve in bending is circular, i.e. the rate of change of curvature is constant throughout the length.

Although this design is highly efficient, it is impractical as no material is provided for load application.

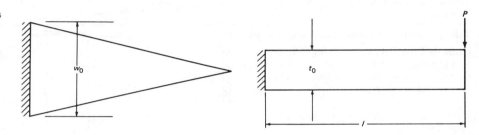

F-4: Modified Triangular Cantilever

Same as Type F-3 except for an end portion of length c with constant cross section ($t_0 \times w_c$) to facilitate load application.

This design is slightly less efficient than Type F-3.

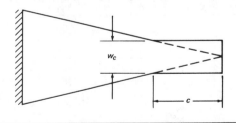
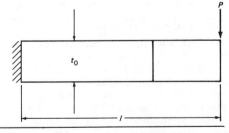

With Tapered Profile: "T" Types

T-1: Tapered Cantilever

Constant in width (w_0). Thickness decreases at a constant rate from t_0 at line of encasement to a specified dimension (t_e) at line of load application. Under load (P), the stress is greatest at line of encasement when the t_e/t_0 ratio is 0.50 or more. When the t_e/t_0 ratio is less (between 0.49 and 0.24), a higher degree of efficiency is obtained, with the line of peak stress some distance away from the line of encasement. The highest efficiency (approaching but not equaling that of the triangular F-3

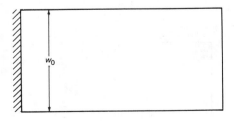
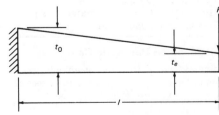

Source: Metals Handbook, Vol. 1, 9th Ed., 283-313

Table 14 Types of spring leaves (contd.)

and of the parabolic P-1 cantilevers) occurs when t_e/t_0 equals 0.357, with the peak stress (8.9% greater than at encasement) located at a distance from the line of encasement equal to 44.5% of the cantilever length (l).

T-2: Modified Tapered Cantilever

Same as Type T-1 except for an end portion of length c with constant cross section ($t_c \times w_0$) for material strength required for eye or load-bearing area.

This design is slightly less efficient than Type T-1.

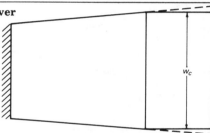
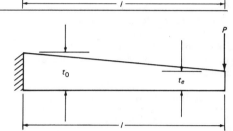

T-3: Tapered–Trapezoidal Cantilever

Thickness decreases at a constant rate from t_0 at line of encasement to a specified dimension (t_e) at line of load application. Width increases at a substantially constant rate from w_0 at line of encasement to w_e at line of load application.

This design approximates Type T-1 for efficiency.

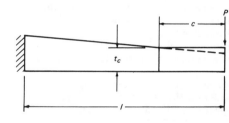

T-4: Modified Tapered–Trapezoidal Cantilever

Same as Type T-3 except for an end portion of length c with constant cross section ($t_c \times w_c$) for material strength required for eye or load-bearing area.

This design is slightly less efficient than Type T-3.

P-1: Parabolic Cantilever

Constant in width (w_0). Thickness decreases from t_0 at line of encasement in a parabolic profile that terminates in zero thickness at line of load application. Under load (P), the stress is constant throughout the length. The elastic curve in bending (from an initially flat spring) has its smallest radius at line of load application, i.e. the rate of change of curvature is greatest at this line.

Although this design is highly efficient, it is impractical because no material is provided for load application.

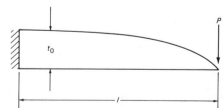

P-2: Modified Parabolic Cantilever

Same as Type P-1 except for an end portion of length c with constant cross section ($t_c \times w_0$) to facilitate load application.

This design is slightly less efficient than Type P-1.

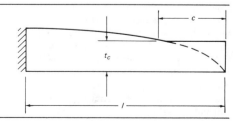

Table 14 Types of spring leaves (contd.)

With Flat Profile: "F" Types

P-3: Parabolic–Trapezoidal Cantilever

Thickness decreases from t_0 at line of encasement in an approximately parabolic profile that terminates in zero thickness at line of load application. Width increases at a substantially constant rate from w_0 at line of encasement to w_e at line of load application. Under load (P), the stress is constant throughout the length.

Although this design is highly efficient, it is impractical because no material is provided for load application.

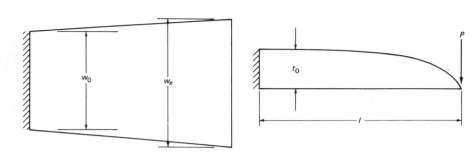

P-4: Modified Parabolic–Trapezoidal Cantilever

Same as Type P-3 except for an end portion of length c with constant cross section ($t_c \times w_c$) to facilitate load application.

This design is slightly less efficient than Type P-3.

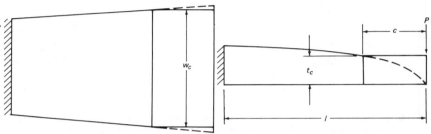

Table 15 Static deflections and ride clearances of various types of vehicles with steel leaf springs

Type of vehicle and load	Static deflection		Ride clearance	
	mm	in.	mm	in.
Passenger automobiles, at design load	100 to 300	4 to 12	75 to 125	3 to 5
Motor coaches, at maximum load	100 to 200	4 to 8	50 to 125	2 to 5
Trucks, at rated load: For highway operation	75 to 200	3 to 8	75 to 125	3 to 5
For "off-the-road" operation	25 to 175	1 to 7	50 to 125	2 to 5

because of considerations of stability in braking, accelerating, cornering, etc.

Table 15 shows typical static deflections and ride clearances for various types of vehicles. These values are approximate and are meant to be used only as a general indication of current practice in suspension system design.

The weight of a spring for a given maximum stress is determined by the energy that is to be stored; this energy is represented by the area under a load-deflection diagram. The effects of changes in rate and clearance on the weight of the spring can easily be seen in this type of diagram.

Figure 19 shows theoretical load-deflection diagrams for two springs designed for the same load and ride clearance; the spring represented by the solid line in Fig. 19(a) is stiffer than the optimum spring for these design conditions, whereas the spring represented by the solid line in Fig. 19(b) is too flexible. Each spring, when fully deflected, stores the same amount of energy (9000 in·lb), and the two springs will weigh almost the same if made of the same kind of material. The optimum spring (minimum energy and weight) for this design load and ride clearance will have stiffness intermediate between those of the two springs represented in Fig. 19. For this optimum spring, static deflection is equal to ride clearance, as indicated by the dashed lines in Fig. 19; the stored energy for this spring is only 8000 in·lb. Figure 19 also brings out the

fact that a change in ride clearance will affect the stored energy, and therefore the required weight, of the stiff spring much more than those of the flexible spring.

Steel Grades

The basic requirement of a leaf-spring steel is that it have sufficient hardenability relative to leaf thickness to ensure a fully martensitic structure throughout the entire cross section. Nonmartensitic transformation products detract from the fatigue properties.

Automotive chassis leaf springs have been made from various fine grained alloy steels such as grades 9260, 4068, 4161, 6150, 8660, 5160 and 51B60.

In the United States, almost all leaf springs are currently made of chromium steels such as 5160 or 51B60, or their H equivalents. With 5160, the chemical composition is specified as an independent variable (while the hardenability of the steel is a dependent variable that will vary with the composition); 5160H is essentially the same steel except that the hardenability is specified as an independent variable (while the composition is a dependent variable that may be adjusted to meet the hardenability-band requirement).

In general terms, higher alloy con-

Fig. 19 Theoretical load-deflection diagrams for two leaf springs

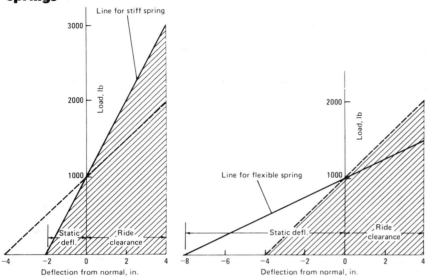

In each diagram, the dashed line represents the minimum-energy spring having the same design load and ride clearance as the spring represented by the solid line.

tent is necessary to ensure adequate hardenability when thicker leaf sections are used. When considering a grade of steel, it is recommended that hardenability be either calculated from chemical composition or (for the various H steels) determined from the hardenability-band charts published in this Handbook.

The following "rule of thumb" may be useful for correlating section size and steel grade:

Maximum section thickness		Steel
mm	in.	
8.2	0.323	5160
15.9	0.625	5160H
36.6	1.440	51B60H

Mechanical Properties

Steels of the same hardness in the tempered martensitic condition have approximately the same yield and tensile strengths. Ductility, as measured by elongation and reduction in area, is inversely proportional to hardness. Based on experience, the optimum mechanical properties for leaf-spring applications are obtained within the hardness range 388 to 461 HB. This range contains the six standard Brinell hardness numbers 388, 401, 415, 429, 444 and 461 (corresponding to the ball-indentation diameters 3.10, 3.05, 3.00, 2.95, 2.90 and 2.85 mm obtained with an applied load of 3000 kg). A specification for leaf springs usually consists of a range covered by four of these hardness numbers, such as 415 to 461 HB (for thin section sizes). Typical mechanical properties of leaf-spring steel are as follows:

Tensile strength .	1310 to 1690 MPa (190 to 245 ksi)
Yield strength (0.2% offset) . . .	1170 to 1550 MPa (170 to 225 ksi)
Elongation	7% min
Reduction in area	25% min
Hardness	42 to 49 HRC, or 338 to 461 HB for 3000-kg load (3.10 to 2.85-mm indentation diameter)

Mechanical Prestressing

Presetting, shot peening, and/or stress peening at ambient temperatures produces large increases in fatigue durability without increasing the size of the spring. These prestressing methods are more effective in increasing the fatigue properties of a spring than are changes in material.

When a load is applied to a leaf spring, the surface layers are subjected to the maximum bending stress. One surface of each leaf is in tension, and the other surface is in compression. The surfaces that are concave in the free position generally are tension surfaces under load, while the convex surfaces generally are in compression. Fatigue failures of the leaves start at or near the surface on the tension side. Because residual stresses are algebraically additive to load stresses, the introduction of residual compressive stresses in the tension surface by prestressing reduces operating stress level, thereby increasing fatigue life.

Presetting (synonymous terms: cold setting, bulldozing, setting-down, scragging) produces residual compressive stresses in the tension surface and residual tensile stresses in the compression surface by forcing the leaves to yield or take a permanent set in the direction of subsequent service loading. Although this operation is beneficial to fatigue life, its primary effect is the reduction of "settling" (load loss) in service. Presetting usually is done after assembly.

Shot peening introduces compressive residual stresses by subjecting the tension side of each individual leaf to a high-velocity stream of shot. The SAE Manual on Shot Peening, HS-84, deals with the control of process variables, while techniques for control of peening effectiveness and quality are explained in Procedures for Using Standard Shot Peening Test Strip, SAE J443. Cut wire shot, sizes CW-23 to CW-41, and cast steel shot, sizes S-230 to S-390, generally are used for this purpose. The intensity of shot peening applied to lightweight and medium-weight springs usually is in the range of 0.010 to 0.020 in. as measured by Almen "A" strip. For heavy springs, single-leaf springs and stress peened springs, the intensity is usually 0.006 to 0.014 in. as measured by Almen "C" strip. Coverage in both cases should be at least 90%.

Stress peening (strain peening) is a means of introducing higher residual compressive stresses than are possible with shot peening with the leaf in the free (unloaded) position. Stress peening is done by shot peening the leaf while it is loaded (under stress) in the direction of subsequent service loading.

Curvature of a leaf spring will be

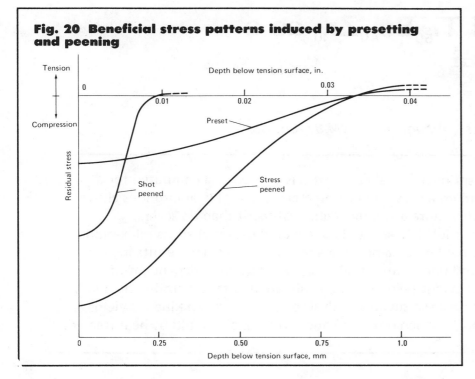

Fig. 20 Beneficial stress patterns induced by presetting and peening

changed by mechanical prestressing. The magnitude of the changes due to shot peening and presetting can be calculated using the formulas given in SAE Information Report J788a. The stress patterns induced by these processes are compared schematically in Fig. 20. The effects of presetting and shot peening are cumulative to some extent, but the results may be influenced by the sequence of operations.

Surface Finishes and Protective Coatings

Surface finish is defined as the surface condition of the spring leaves after the steel has been formed and heat treated, and prior to any subsequent coating treatment. Normally, automotive leaves are utilized "as heat treated" or in the "shot peened" condition. An as-heat-treated finish will be a tight oxide produced by the quenching and tempering operations and will exhibit a blue or blue-black appearance. A shot peened finish is the result of removing the blue or blue-black color by the peening operation and is characterized by a matte luster appearance.

A protective coating is a material added to the surfaces of individual leaves or to the exposed areas of leaf-spring assemblies. Its primary purpose is to prevent corrosion both in storage and in operational environments. All exposed surfaces to be coated must be free from loose scale and dirt. Peened surfaces should be coated as soon as possible to prevent the formation of corrosion. It is important that an enveloping coat is applied. An unprotected area or a break in the coating may contribute to localized corrosion and a reduction in fatigue life. Before a coating is specified, whether grease, oil, paint or plastic, it should be evaluated for effects that its application might have on the fatigue life of the spring steel. The thickness and adhesion characteristics of the coating must be within the tolerances that have been established for the type of material being used in order to provide adequate corrosion protection and ensure satisfactory performance.

On the use of $Ti_{50}Be_{40}Zr_{10}$ as a spring material

F. G. YOST
Sandia National Laboratories*, Albuquerque, NM 87185, USA

A useful figure-of-merit for spring material candidates is derived and computed for a select group of iron and titanium alloys. Two commercially available amorphous alloys are shown to have considerably more attractive figures-of-merit than do several crystalline alloys. Since springs left in a stressed condition must resist stress relaxation, tests of this type were performed on an amorphous titanium alloy. The results indicate that relaxation kinetics depend upon initial heat treatment but are independent of initial applied stress. This behaviour is shown to be consistent with hyperbolic flow in the limit of low stress. A simple least-squares analysis of the relaxation kinetics yields an activation energy of 67 kJ mol⁻¹. When used for springs, this alloy should be heat treated.

1. Introduction

Applications of metallic glass are increasing rapidly as more investigators become intrigued by their interesting properties. Magnetic components which rely upon the high permeability and saturation magnetization of commercially available metallic glass alloys have received careful attention [1–4]. For reasons of handling ease, economy and improved strength, amorphous brazing alloys are enjoying an expanding market [5]. Higher crystallization temperatures, strength and corrosion resistance promise many obvious and unforeseen applications of this interesting class of materials. Their practical use as spring materials apparently has not been recognized in the literature. Material used in a spring component must be fabricable, have adequate fatigue strength and acceptable corrosion resistance. In addition, since springs are frequently left in a stressed condition, they must resist low-temperature stress relaxation.

An elastic analysis is reported here that provides a useful figure-of-merit for spring material candidates called spring force effectiveness. This figure-of-merit was computed for a few select iron- and titanium-based alloys, often used for springs, and is compared to that computed for two com-

mercially available amorphous alloys. It will be seen that amorphous alloys are, indeed, a very attractive material for spring components. Also reported are coiled ribbon stress relaxation tests performed at three initial stress levels on as-received Metglas 2204† ribbons. Similar testing was carried out at the intermediate stress level on similar ribbons previously annealed at 523 K for 7.2 ksec under zero stress. The stress relaxation tests were carried out at five temperatures so that an Arrhenius analysis of the data will allow prediction of room-temperature behaviour of spring components. The results are presented and discussed in the light of previous stress relaxation tests of this type.

2. Experimental procedure

The stress relaxation test discussed by Luborsky *et al.* [6] and Graham *et al.* [7] has been used in these experiments; however, two test procedures were followed. Specimens in Group A were 0.05 mm thick by 1.00 mm wide Metglas 2204 ribbon in the as-received condition. Specimens in Group B were these same ribbons heat treated under zero stress at 523 K for 7.2 ksec prior to the stress relaxation tests. Aluminium blocks, 3.73 cm

---

*A U.S. Department of Energy Facility.
†This $Ti_{50}Be_{40}Zr_{10}$ alloy is produced by Allied Chemical, PO Box 1021R, Morristown, NJ 07960, USA.

long by 1.27 cm wide by 0.318 cm thick, containing holes bored at 0.476, 0.635 and 0.953 cm diameter were used to confine the ribbons during stress relaxation. The ribbons were cut to fit the circumference of the holes when coiled and inserted. The stress relaxation heat treatments were carried out at five temperatures ranging from 423 to 523 K (in 25 K increments) in stirred oil baths capable of maintaining ±1 K.

To determine how repeatable the experiment was, a separate specimen was used for each point in the Group A set of data. For the Group B data, each of the five specimens was re-used to obtain subsequent data points at longer stress relaxation times. After each anneal, the specimen was removed from the aluminium block and its radius of curvature was measured with an optical comparator which had a resolution of 0.25 mm. The data were reduced by calculating the ratio of the bending moment, $M(t)$, after anneal time, t, to that at $t = 0$, $M(0)$

$$\frac{M(t)}{M(0)} = \frac{\int_0^{d/2} \sigma(z,t)z\,dz}{\int_0^{d/2} \sigma(z,0)z\,dz}, \qquad (1)$$

where z is a co-ordinate perpendicular to and begiining at the neutral axis, d is the ribbon thickness and $\sigma(z,t)$ is the local stress. Later, this relationship will prove useful when certain flow models, $\sigma(z,t)$, are compared with the experimental data. The ratio, given in Equation 1 and designated R, is also given by

$$R = \frac{1 - R_h/R_t}{1 - R_h/R_s}, \qquad (2)$$

where R_h is the hole radius, R_s is the unconstrained radius at $t = 0$ and R_t is the unconstrained radius after anneal time t. While being stored on a spool, this material assumes a permanent set resulting in a finite R_s-value. Estimates of error in R may be calculated using the resolution 0.25 mm and allowing R_t to take on the values $R_t = \pm 0.25$ mm. Consequently this error becomes larger as R_t and R decrease.

3. Elastic analysis
Consider a flat spring element, having thickness d and width w, bent into a contour having radius of curvature, $C(s)$, where s is the arc length along the neutral axis. Then the elastic energy, U, in the spring is given by [8]

$$U = \oint \frac{M^2(s)\,ds}{2EI}, \qquad (3)$$

where $M(s)$ is the bending moment, E is the elastic modulus and

$$I = wd^3/12 \qquad (4)$$

is the moment of inertia. Substituting the definition of the bending moment

$$M(s) = \frac{EI}{d/2 + C(s)} \cong \frac{EI}{C(s)}, \qquad (5)$$

and the above expression for I into Equation 3 yields

$$U = \frac{EWd^3}{24} \oint \frac{ds}{C^2(s)}. \qquad (6)$$

The maximum strain in this spring element is

$$\epsilon_m = \frac{\sigma_m}{E} = \frac{d/2}{d/2 + C_{min}} \cong \frac{d}{2C_{min}}, \qquad (7)$$

where C_{min} is the smallest bending radius encountered along s. Consequently, to avoid plastic deformation and a permanent set, the maximum allowable thickness would be

$$d_m = \frac{2\sigma_m C_{min}}{E}, \qquad (8)$$

where σ_m is the elastic–plastic proportional limit which, for amorphous alloys, is approximately equal to the yield stress. Substituting Equation 8 into Equation 6 gives an expression for the upper bound of the stored elastic energy

$$U = \frac{\sigma_m^3}{E^2} \frac{WC_{min}^2}{3} \oint \frac{ds}{C^2(s)}. \qquad (9)$$

The quantity

$$F = \frac{\sigma_m^3}{E^2} \qquad (10)$$

has been called spring force effectiveness [9] and is the figure-of-merit referred to previously. The spring force effectiveness is a combination of material properties, whereas the remainder of Equation 9 is geometric in origin. It would be desirable to make F large so that the force per unit of volume exerted by the spring would be large. This quantity has been calculated for several spring steels and strong titanium alloys as well as for commercially available iron-based and titanium-based amorphous alloys; the results are shown in Table I. Notice that there is a decided difference between the iron-based crystalline alloys and the iron–boron amorphous alloy as well as between

TABLE I Spring force effectiveness for selected ferrous and titanium alloys

Alloy	Condition	Spring force effectiveness
15-5	H-900	0.037
4130	Quenched and tempered	0.082
440C	Hardness and stress relieved	0.159
52100	Quenched and tempered	0.908
Metglas 2605 ($Fe_{80}B_{20}$)	As-cast	1.649
Ti 5-2.5	Quenched	0.086
Ti 6-4	Aged	0.125
Ti 7-4	Aged and extruded	0.139
Ti 6-6-2	Extruded and aged	0.165
Metglas 2204 ($Ti_{50}Be_{40}Zr_{10}$)	As-cast	1.030

the titanium-based crystalline alloys and the titanium—beryllium—zirconium amorphous alloy. The spring force effectiveness for the glass alloy is approximately an order of magnitude larger than that for the crystalline alloys. Yield strengths and elastic moduli data for these calculations were taken from [10] and details regarding alloy chemistry and preparation can be found therein. Measurements have shown that the velocity of sound [11] in Metglas 2204 changes approximately 4% upon annealing at 523 K for 7.2 ksec under zero stress; therefore, its elastic modulus changes approximately 8% since negligible changes in density are expected. These structural modifications are not sufficient to make the spring force effectiveness of the crystalline alloys comparable with that of amorphous alloys.

4. Stress relaxation results and discussion

As mentioned in Section 2, separate as-received specimens were used for each data point in Group A to obtain some indication of the repeatability of the stress relaxation behaviour. As a result, the data are expected to be more scattered than that in Group B where fewer specimens were used. Group A data are plotted in Figs 1, 2 and 3 corresponding to the three increasing initial outer fibre stress levels. Each figure includes data from five stress relaxation temperatures and, except for the two highest temperatures, the data fall on remarkably smooth curves, suggesting a high degree of experimental repeatability. At all temperatures and stresses a rapid initial drop is observed in the stress relaxation parameter, R. Then a more gradual stress relaxation tendency is observed. The severity of the initial drop appears to increase in both extent and rate with increasing temperature.

This behaviour was also apparent in the data of Graham et al. [7] which was discussed and analysed by Ast and Krenitsky [12]. At any given temperature, data from Figs 1, 2 and 3 superimpose indicating that R is independent of initial outer fibre stress. This observation suggests that stress in the coiled ribbon should be of the form

$$\sigma(z,t) = \sigma_0 S(z,t), \qquad (11)$$

where $S(z,t)$ is independent of σ_0, the initial outer fibre stress. If this relationship for $\sigma(z,t)$ is substituted into Equation 1 it can be seen that $R(t)$ would be independent of σ_0. This apparent independence would not have been predicted from the analysis of Ast and Krenitsky [12] since in either of their derived relationships for stress

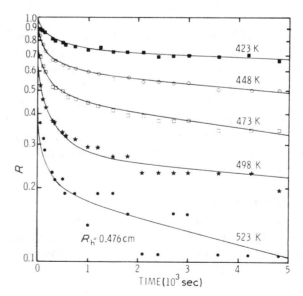

Figure 1 Stress relaxation of as-received Metglas 2204 ribbon (coil radius of 0.476 cm) against anneal time at five temperatures.

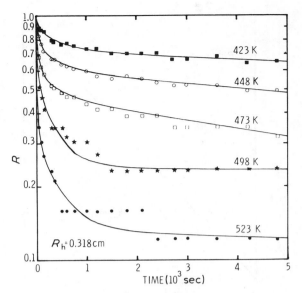

Figure 2 Stress relaxation of as-received Metglas 2204 ribbon (coil radius of 0.318 cm) against anneal time at five temperatures.

relaxation kinetics, in the form $M(t)/M(0) = R(t)$, there is a more complex dependence on σ_0. However the hyperbolic flow model analysed by Ast and Krenitsky [12] can be put into the form of Equation 11. When

$$\sigma(z, 0)v^* \ll 2kT,$$

expansion to first order yields

$$\sigma(z, t) = \frac{2z\sigma_0}{d} \cdot \exp(-t/\tau), \qquad (12)$$

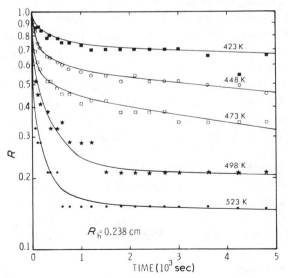

Figure 3 Stress relaxation of as-received Metglas 2204 ribbon (coil radius of 0.238 cm) against anneal time at five temperatures.

where [12]

$$\tau = \frac{kT \exp(\Delta H/kT)}{2E\dot{\epsilon}_0 v^*}, \qquad (13)$$

where $\dot{\epsilon}_0$ is a pre-exponential factor, v^* is an activation volume and ΔH is the activation energy for transient creep, which is the form of Equation 11. It is not apparent how one can eliminate the dependence of $M(t)/M(0)$ on σ_0 using the power-law creep model which is Equation 8 of Ast and Krenitsky [12].

The rapid initial drop and gradual relaxation behaviour observed in the Group A data and that of Graham *et al.* [7] are highly suggestive of two or more flow processes. This is especially significant when it is realized that the data are plotted on a logarithmic co-ordinate. Using numerical fits of their two flow models Ast and Krenitsky [12] concluded that one process, namely hyperbolic flow, is capable of describing the stress relaxation kinetics of as-received Metglas 2826B. As will be shown, the heat treatment of the Group B specimens appears to simplify the stress relaxation kinetics by suppressing one or more flow processes.

Group B data are obtained from specimens which were first annealed at 523 K for 7.2 ksec under zero stress, then stress relaxed. These data are plotted against annealing time in Fig. 4. The data obtained at 423 K are not plotted since relaxation at this temperature was so slow that crowded plotting symbols made the graph difficult to read. Quite an obvious change in relaxation behaviour has taken place. The non-steady state behaviour, characterized by a rapid initial drop, is no longer apparent. Consequently, while the rapid relaxation behaviour of as-received material might prevent its use as a spring material, a low-temperature heat treatment, prior to use, restores its potential.

The kinetics exhibited by the Group B specimens appear to be similar to the gradual second-stage kinetics of the Group A specimens. These more gradual kinetics have been called isoconfigurational or steady-state homogeneous flow [13, 14]. The almost linear behaviour suggests a simple exponential decay of the ratio of bending moments, R. By combining Equations 1, 2 and 12 one finds

$$\ln R = -\frac{t}{\tau} \qquad (14)$$

so a least-squares analysis of the data was performed in the natural logarithmic domain to ob-

Source: Journal of Materials Science, 1981, 3039-3044

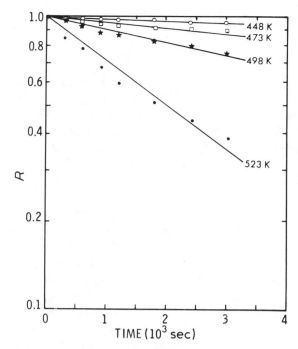

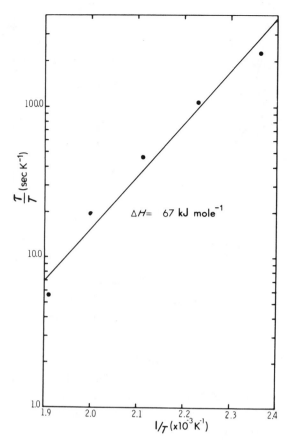

Figure 4 Stress relaxation of heat treated Metglas 2204 ribbon (coil radius of 0.318 cm) against anneal time at four temperatures.

Figure 5 Relaxation time τ divided by absolute temperature against reciprocal degrees Kelvin. An activation energy of 67 kJ mol^{-1} was calculated.

tain the straight lines in Fig. 4. The slopes of these lines are, according to Equation 14, equal to $-1/\tau$. The relaxation times, τ, were then divided by the test temperature in degrees Kelvin and plotted against reciprocal degrees Kelvin as suggested by Equation 13. The calculated relaxation times show a slight curvature, but of opposite sign to that expected for glasses [15]. The curvature may be due to temper embrittlement [16] or may simply be due to statistical scatter in the data. More work is necessary to resolve this question. A standard Arrhenius analysis yielded an activation energy of 67 kJ mol^{-1} which is much less than that calculated by Ast and Krenitsky [12] for Metglas 2826B. At the risk of oversimplification, it is suggested that this isoconfigurational flow can be described by Equation 12. In this way the stress relaxation parameter, R, is independent of σ_0 and decreases linearly in the natural logarithmic domain. To ensure the inequality used to obtain Equation 12 the volume of a flow unit, v^*, must be of the order of 10 Å^3. By way of contrast, Taub [14] quotes 48 Å^3 for $Pd_{82}Si_{18}$ and Ast and Krenitsky [12] calculated 400 Å^3 for Metglas 2826B.

5. Conclusion

An elastic analysis has shown that spring force effectiveness is a useful figure-of-merit for selecting spring materials. Two commercially available

amorphous alloys were shown to have an extremely large spring force effectiveness relative to that of typical crystalline alloys, making them excellent candidates for spring applications. As-received Metglas 2204 appears to stress relax in two stages. Stress relaxation kinetics for this alloy are independent of initial applied stress. Annealed material relaxes in simple exponential fashion with an activation energy of 67 kJ mol^{-1} and an activation volume of approximately 10 Å^3. If used as a spring material, this alloy must be annealed to avoid rapid initial relaxation.

Acknowledgements
The author wishes to thank K. S. Varga and J. H. Gieske for their technical assistance. This work was supported by the US Department of Energy (DOE) under Contract No. DE-ACO4-76-DP00789.

References
1. F. E. LUBORSKY, *IEEE Trans. Magn.* **14** (1978) 1008.
2. F. E. LUBORSKY, J. J. BECKER, P. G. FRISCH-

MANN and L. A. JOHNSON, *J. Appl. Phys.* **49** (1978) 1769.

3. F. E. LUBORSKY, P. G. FRISCHMANN and L. A. JOHNSON, *J. Magn. Magn. Mater.* **8** (1978) 318.

4. L. I. MENDELSOHN, E. A. NESBITT and G. R. BRETTS, *IEEE Trans. Magn.* **12** (1976) 924.

5. N. DeCRISTOFARO and C. HENSCHEL, *Welding J.* **57** (1978) 33.

6. F. E. LUBORSKY, J. J. BECKER and R. O. McCARY, *IEEE Trans. Magn.* **11** (1975) 1644.

7. C. D. GRAHAM, JR, T. EGAMI, R. S. WILLIAMS and Y. TAKEI, "Magnetism and Magnetic Materials", edited by J. J. Becker, G. H. Lander and J. J. Rhyne, (American Institute of Physics, New York, 1976) p. 218.

8. S. TIMOSHENKO and D. H. YOUNG "Elements of Strength Materials" 5th edn. (D. VanNostrand Company, Inc, New York (1978) p. 220.

9. M. S. WALMER, "First Symposium on Rolamite", (University of New Mexico Press, Albuquerque, 1969) p. 65.

10. "Aerospace Structural Metals Handbook", Mechanical Properties Data Center, Battelle Columbus Laboratories (1981).

11. J. H. GIESKE, unpublished work 1980.

12. D. G. AST and D. J. KRENITSKY, *J. Mater. Sci.* **14** (1979) 287.

13. A. S. ARGON, *Acta Metall.* **27** (1979) 47.

14. A. I. TAUB, *ibid.* **28** (1980) 633.

15. D. R. UHLMANN and R. W. HOPPER, "Metallic Glasses", edited by J. J. Gilman and H. J. Leamy, (American Society for Metals, Metals Park, 1978) p. 128.

16. L. A. DAVIS, *ibid.* p. 190.

Titanium: tomorrow's spring material?

Substitution of titanium alloy for steel can save 60-65% of spring weight, but calls for careful design.

Stiffness considerations usually limit direct substitutions of lighter materials for steel in automotive applications. Elastic modulus differences make possible weight savings even greater than those indicated by relative material densities in the case of titanium springs. Careful design can be shown, in the laboratory, to provide comparable properties at nearly 60-65% weight saving. Alloy development work and treatment optimization, as well as field testing, should indicate the real potential of this Ford Motor study.

One of the popular means of reducing automobile weight to conserve fuel has been substitution of lighter weight materials for conventional ones. Titanium, since it is available in as wide a range of strengths as steel yet is only 56% as dense, is such a substance. A substitution for steel on an equal strength basis could save 44% in weight. However, in most structural applications stiffness is also important and, since titanium's elastic moduli are half those of steel, a gauge-for-gauge substitution realizing this full weight saving is impossible. An exception to this is the case of springs, for which lower elastic properties increase potential weight savings.

Spring design

The equation governing coil

spring design relates its load response to material properties and dimensions. For coil springs the spring rate, R, is given by:

$$R = \frac{P}{\delta} = \frac{Gd^4}{8D^3} \tag{1}$$

in which G = shear modulus,
d, D = bar and coil diameters,
P = load
δ = deflection, and
n = number of active coils.

The shear stress is given by:

$$\tau = \frac{8PD}{\pi d^3} \tag{2}$$

While equation [2] is not corrected for the effect of spring curvature (which raises shear stress at the coil's interior) it is used to simplify this derivation. The corrected formula is used in later portions of this study.

The weight of such a spring, neglecting inactive coils at the ends) is:

$$\text{weight} = \frac{\pi^2 d^2 Dn\rho}{4} \tag{3}$$

in which ρ is the material density. Rearranging and substituting in equation [3] to eliminate terms involving spring dimensions d, D, and n results in the expression:

$$\text{weight} = \frac{2G\rho}{\tau^2} \cdot \frac{P^2}{R} \tag{4}$$

in which material properties G, ρ, and τ (the stress level sustainable for the required fatigue life) are separated from functional or design requirements P and R. Thus, equation [4] shows the effect on spring weight of a material substitution while keeping functional characteristics constant. Analogous expressions may be derived for leaf springs.

In all cases it is clear that the desirable material characteristics for lightweight springs are high strength, low density, and low elastic moduli. Titanium alloys are available which have strength nearly equivalent to that of heat treated steel commonly used in springs. Thus, because of lower density (56%) and elastic moduli, (50%) titanium springs can potentially weigh only 28% as much as steel springs having the same load carrying characteristics (P and R).

To realize the available weight savings while maintaining these characteristics will require that a titanium spring have a slightly different design from the corresponding steel part. For coil springs, wire and coil diameters can remain essentially fixed while the number of active coils is reduced in proportion to the lower shear modulus. This means that a titanium coil spring can directly replace a steel spring of identical performance. Design differences required to maintain constant P and R are more extensive for a leaf spring, for which new values of leaf length, width, and/or height must be chosen. For example, a single cantilever titanium leaf spring might have the same thickness but only be 71% as wide and long as a steel spring with the same characteristics. Thus, while titanium leaf springs offer large potential weight savings, they cannot be directly substituted for steel leaf springs. For new designs, or where redesigns to accommodate springs of different dimensions are feasible, however,

titanium leaf springs can be considered.

Since suspension springs are the largest springs in cars, their replacement by lighter substitutes offers significant weight saving potential. Titanium coil springs can be designed to replace steel ones directly, so it was decided to design, build, and test such springs to determine how much of the calculated potential weight saving could be realized in practice.

Design and fabrication

A typical front suspension spring for a full-size automobile is shown at the left of both the lead and cover photos. It has a spring rate of 59.5 N/mm (340 lb$_f$/in), and a design load of 8927 N (2007 lb$_f$) at its design height of 265 mm. These requirements dictate that the spring's free height be 415 mm and that the loads at jounce and rebound heights (204 and 305 mm) be 12,583 N (2829 lb$_f$) and 6534 N (1469 lb$_f$), respectively. When loaded between these limits the spring should exhibit a minimum fatigue life of 77,000 cycles. Besides matching these performance criteria, a titanium spring designed to replace the steel spring must fit the suspension system geometry. The steel spring is made of 16.8 mm diameter AISI 5160 steel wire heat treated to a 444-495 Brinell hardness and shot peened to an Almen "A" intensity of 0.43 to 0.53 mm arc height.

There are several titanium alloys which attain strength levels approaching those of spring steel. These β-titanium alloys are quite ductile when their structure consists of the metastable β (body centered cubic) phase. An aging treatment causes the α (hexagonal close packed) phase to precipitate and the two-phase structure thus produced can have yield strengths above 1400 MPa (203 ksi). Thus β-titanium alloys are chosen for spring applications because they can be formed while in the soft condition and subsequently age-hardened to suitably high strengths.

To design a titanium unit to replace the steel suspension spring, an equation relating shear stress to wire diameter

[5]

$$\tau' = \frac{8PD}{\pi d^3} \cdot K$$

is used to determine the necessary wire diameter. To make the titanium spring fit into the existing suspension system it is necessary that its inside diameter be the same as that of the steel spring. Therefore, the coil diameter is D = I.D.+d where I.D. = 102 mm. In this calculation,

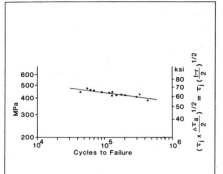

Fig. 1—Stress parameter vs fatigue life plot for titanium suspension springs permits correlation of data otherwise unsuitable for usual S-N curve.

τ' represents the cyclic shear stress level for the specified spring fatigue life. Although some torsional fatigue measurements have been made on β-titanium alloys, sufficient information to permit selection of an optimum stress level was unavailable. It was thus decided to fabricate titanium springs with three wire diameters designed to operate at three different jounce (maximum) shear stress levels. Stress level of the intermediate wire size spring closely matched that of the steel spring, 974 MPa (141 ksi); other springs bracketed this value. Using three wire sizes, the number of turns (n) for each spring design was determined by solving equation [1] for n and substituting values for R, d, D, and G (40 GPa or 5.8×10^6 psi). Titanium spring weights were estimated using equation [3] to calculate the active coil weight and then adding the weight of 1.18 inactive coils at the spring ends. Parameter values describing the three titanium spring variations are listed in Table 1 together with steel spring values. Note that weight savings are projected to be 50, 63, and 70% for the three designs.

The major difference between steel and titanium designs is the number of coils. Because of their fewer coils, titanium springs fabricated to the same free length as steel will have a higher coil pitch. With a constant coil spacing, titanium spring tangent tails will not seat properly in the pocket formed into the lower suspension arm, which is designed to accommodate the much closer steel spring coil spacing. Consequently, it is necessary to alter the titanium spring design so that final tangent tail coil spacing is such that proper seating in the lower arm will occur at full jounce. To calculate required spacing of the final coil it is assumed that as the spring is compressed, deflections of all coils are equal. Thus, deflection of the last coil at full jounce is the jounce deflection divided by the number of active coils. When added to the pitch built into the lower suspension arm, the result represents the proper pitch for the final coil. Since this is a lower pitch than a uniform spacing would dictate, the spring's remaining coils will have a slightly higher pitch to maintain constant free height. It is recognized that a pitch transition zone near the spring's tangent tail end introduces bending stresses which increase the probability of fatigue failure occurring at this location. But the opposite, ground end, already has a more severe transition, in effect to zero coil spacing. Details of the pitch spacings are listed in Table 1 for the three spring designs and are shown at right of lead and cover figures.

Titanium alloy 38644 (3 Al- 8 V- 6 Cr- 4 Mo- 4 Zr) was selected for the springs based on its favorable mechanical properties and availability in suitable form. Wires were produced by hot forging bars at 925°C to 21 mm diameter followed by solution annealing 30 minutes at 815°C and air cooling. Diameters called for in the spring designs were obtained by centerless grinding and pickling. Bars were subjected to ultrasonic inspection at this point to detect those with defects such as seams. A few bars had defects traceable to the forging process. In large volume production, wire drawing would be used, making this type of flaw unlikely.

Production of springs from the wire proceeded as follows:
• Wind on a mandrel of suitable diameter in a lathe. Pitch change is introduced by varying tool carrier speed as the wire wraps around the mandrel.
• Clean in a basic solution to eliminate organic contamination such as oil, followed by cleaning in an acid solution to remove any steel particles possibly picked up during winding.
• Age harden at 496° for 16 hours in an air furnace controlled to ±6°C.
• Grind spring's closed end.
• Dry grit blast clean.
• Dye penetrant inspect.

Table 2
Results of Spring Tests

Test	Spring Design	Maximum Shear Stress, MPa	r Value	Spring Rate, N/mm	Fatigue Life, cycles
1	M	840	0.51	53.9	150,000
2	O	821	0.42	54.3	130,000
3	C	805	0.49	62.3	299,000
4	M	841	0.51	53.2	130,000
5	O	814	0.43	53.1	118,000
6	M	945	0.53	NA	62,000
7	O	917	0.51	NA	70,000
8	M	852	0.51	54.1	179,000
9	O	809	0.41	54.5	91,000
10	M	848	0.51	53.8	202,000
11	M	857	0.51	54.6	333,000
12	O	822	0.43	53.9	44,000
13	O	765	0.50	54.8	432,000
14	M	878	0.42	55.2	56,000

NA = not measured
M, O, C = moderate, optimistic and conservative

• Shot peen to Almen "A" intensity of 0.51-0.61 mm.
• Measure spring dimensions.

This heat treatment was chosen, on the basis of results from variously aged tensile bars, to produce the best combination of high strength and ductility. The shot peening specification called for a higher intensity than that used on steel springs. Because of its lower elastic modulus titanium, shot peened to the same intensity as steel, will have lower surface residual stresses. Thus, to produce residual stresses equivalent to those in steel springs, a titanium spring must be shot peened to a greater intensity, since intensity is measured using a standard steel test bar. Therefore, the intensity specified was the greatest of which the available equipment was capable, though even greater intensities could prove desirable for producing longer fatigue lives.

Fatigue testing

Springs were fatigue tested using closed loop electrohydraulic testing machines, with either load or deflection the controlled parameter. In some tests springs were cycled between jounce and rebound limits while others involved cycling between other limits to acquire data about relationships between shear stress level and life. Prior to, and at intervals during, fatigue testing the spring rates were determined by measuring loads at several deflections. This information enabled jounce and rebound loads to be calculated for those tests run in the deflection controlled mode. Equation [5] was used to calculate shear stress at each test's limits. It was found that wire diameters had been reduced by ~0.13 mm during processing, which was taken into account in applying the equation.

Spring test results are shown in Table 2. Because r values (ratios of minimum to maximum stress) varied from test to test, the results do not determine a conventional stress vs life curve. However, a plot of a modified stress parameter vs life, suggested in the literature for comparing fatigue tests at different r values, results in a good log-log linear plot (Fig. 1). The stress parameter used takes into account both maximum shear stress and stress amplitude, hence the r value. This curve thus provides the relationship between design, load conditions, and fatigue life for springs made and processed in a particular way. A different curve would result if a different alloy, heat treatment, wire preparation method, or shot peening intensity were used. However, for the material used here, a stress parameter value of 407 MPa (59.0 ksi) can be selected from the curve for a life of 300,000 cycles, which would provide a safety factor of about four over the minimum required 77,000-cycle life. This stress level corresponds to a maximum shear stress of 822 MPa (119.2 ksi) under the required loading conditions of r = 0.51. A spring designed to this stress level would offer a 53% weight saving over the steel spring it replaces.

It is useful to compare these results with torsional fatigue results reported in the literature. It must be noted, however, that this is not a direct comparison since the present data were acquired through testing actual springs in which loading conditions differ from those of straight wires subjected to torsional stress. The latter data are shown in Fig. 2 as bands representing results from several sizes of cold worked, centerless ground, and aged β-titanium alloy wires. It is evident that the fatigue performance of the 38644 and 13 V-11 Cr-3 Al alloys in the as-aged condition is substantially similar. Shot peening is seen to improve markedly the fatigue life of the 13-11-3 alloy, though the literature did not include results of shot peened 38644 material, particularly at longer lives where the band's slope is flatter. Current test results for 38644 springs parallel those of the shot peened 13-11-3 but are about 65 MPa (9.5 ksi) lower. These observations provide an in-

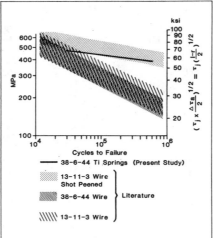

Fig. 2—Comparison of fatigue testing of springs with fatigue tests on β-titanium wires in torsion.

dication that the use of cold worked, e.g. cold drawn rather than forged, wire could result in a higher maximum stress level for a given r value and required fatigue life.

Potential improvements

Despite the potential 53% weight saving shown above, current springs do not represent the optimum achieveable with further development. There are at least three areas in which significant improvements leading to a larger weight reduction and lower cost can be identified:

• — Cold work improves torsional fatigue performance of titanium wires. This is consistent with other observations of cold work's boosting strength in 38644 titanium alloy. Consequently, springs made of cold drawn, rather than forged, β-titanium alloy can be operated at a higher cyclic stress amplitude for a given fatigue life requirement, increasing weight saving potential. Another advantage of using cold drawn wire would be a decrease in required aging time, perhaps to as little as one-fourth that needed with forged wire, reducing manufacturing cost.

• — The shot peening treatment used for these springs was probably not optimum for titanium units of this size. To produce a given residual stress distribution in a titanium spring, it would have to be subjected to a greater shot peening intensity than a steel one. Further development will be needed to determine an optimum treatment, but such treatment is likely to improve fatigue performance — affording greater weight reduction.

• — The β-titanium alloys currently used in spring production contain expensive alloy additions such as vanadium and chromium. Development of alloys tailored to spring applications, using more economical alloy additions such a iron, could result in appreciable cost savings.

Thus, with further development, weight savings potential of titanium springs can be increased to perhaps 60-65% for springs such as those considered here. Such springs' cost would simultaneously be reduced, lowering the cost penalty per unit weight saved relative to steel springs.

While it has been demonstrated that titanium springs offer an attractive level of weight savings with a potential for future improvements, much work remains before such springs can be considered for vehicle production use. None of the exhaustive vehicle durability testing necessary for suspension components was carried out, nor were detailed manufacturing feasibility or cost studies conducted. Results of such investigations and analyses are needed to assess potential technical and economic barriers and to assure satisfactory performance with any new concept. ▲

Fasteners

Threaded Steel Fasteners

By the ASM Committee on
Carbon and Alloy Steels

THREADED FASTENERS for service between -50 and $+200$ °C (-65 and $+400$ °F) may be made from several different grades of steel, as long as the finished fastener meets the specified strength requirements. This article discusses the properties of the carbon and alloy constructional steels containing a maximum of 0.55% carbon that are used to produce fasteners intended for use under these service conditions.

Guidelines for the selection of steels for bolts (including cap screws), studs (including U-bolts) and nuts for service at this temperature range and also for service between 200 and 370 °C (400 and 700 °F) are also discussed. Threaded fasteners for service above 370 °C (700 °F) and below -50 °C (-65 °F) will be discussed in subsequent volumes of the Metals Handbook.

The purchaser of steel bolts, studs and nuts usually selects the desired strength level by specifying a grade or class in the widely used SAE, ASTM, IFI or ISO specifications. The producer then selects a particular steel from the broad chemical composition ranges in these specifications. This allows the producers freedom to use the most economical material consistent with their equipment and production procedures to meet the specified mechanical properties. This situation has forced producers to adopt substantially the same manufacturing process for a given class of product, which has resulted in a certain degree of steel standardization.

Strength Grades and Property Classes

The strength level of a bolt, stud or nut is designated by its strength grade or property class number—the greater the number, the higher the strength level. A second number, following a decimal point, is sometimes added to represent a variation of the product within the general strength level. SAE strength grade numbers are often used for mechanical fasteners made to the United States system of inch dimensions, while the property class numbers defined in ISO Recommendation R898 are used for metric fasteners. Strength and property designations of bolts and studs are based on tensile (breaking) strength, while those of nuts are based on proof stress.

The commonly used strength grades and property classes of steel threaded fasteners are shown in Tables 1 and 2, along with the mechanical properties associated with those grades and classes. As may be seen in these tables, the strength grade numbers cannot be directly converted to a specific strength level. The property class numbers, however, indicate the general level of tensile strength or proof stress in MPa; for example, ISO class 9 nuts have a proof stress ranging from 900 to 990 MPa. The number following the decimal point in a class number for a bolt or stud indicates the ratio of the yield strength to the tensile strength; for example, an ISO class 5.8 stud has a tensile strength of 520 MPa and a yield strength of 420 MPa. The mechanical properties of bolt and stud classes are compared in Table

3 to those of bolt and stud grades. The mechanical properties of these various grades and classes of steel are discussed in greater detail in a subsequent section of this article.

Steels for Threaded Fasteners

Many different low-carbon, medium-carbon and alloy constructional steels are used to make all of the various strength grades and property classes of threaded steel fasteners suitable for service between -50 and $+200$ °C (-65 and $+400$ °F). The chemical compositions of those steels used for the grades and classes of threaded steel fasteners listed in Tables 1 and 2 are given in Tables 4 and 5. The following sections discuss the selection and processing of these steels for each type of end product: bolt, stud or nut.

Bolt Steels. As previously noted, the producer of bolts is free to use any steel within the grade and class limitations of Table 4 to attain the properties of the specified grade or class in Table 1. As strength requirements and section size increase, hardenability becomes the most important factor.

Sometimes, specific applications require closer control, and the purchaser will consequently specify the steel composition. However, except where a particular steel is absolutely necessary, this practice is losing favor. A specific steel may not be well-suited to the fastener producer's processing facilities; specification of such a steel may result in unnecessarily high cost to the purchaser.

Table 1 Mechanical properties of steel bolts and studs (a)(b)

Strength grade or property class	Nominal diameter	Proof stress (c) MPa	Proof stress (c) ksi	Min tensile strength (d) MPa	Min tensile strength (d) ksi	Min yield strength (e)(f) MPa	Min yield strength (e)(f) ksi	Min elongation, %(e)	Rockwell hardness Surface 30N, min	Rockwell hardness Core min	Rockwell hardness Core max
SAE Strength Grades (g)											
1 . . . . ¼-1½ in.		225	33	415	60	250(h)	36(h)	18	· · ·	B70	B100
2 . . . . ¼-¾ in. (j)		380	55	510	74	395	57	18	· · ·	B80	B100
>¾-1½ in.		225	33	415	60	250(h)	36(h)	18	· · ·	B70	B100
4 (k) . . . . ¼-1½ in.		· · ·	· · ·	795	115	690	100	10	· · ·	C22	C32
5 . . . . ¼-1 in.		585	85	830	120	635	92	14	54	C25	C34
>1-1½ in		510	74	725	105	560	81	14	50	C19	C30
5.2 (m) . . . . . . ¼-1 in.		585	85	830	120	635	92	14	56	C26	C36
7 (m) (n) . . . . . ¼-1½ in.		725	105	915	133	795	115	12	54	C28	C34
8 . . . . ¼-1½ in.		830	120	1035	150	895	130	12	58.6	C33	C39
8.1 (k) . . . . . . ¼-1½ in.		830	120	1035	150	895	130	10	· · ·	C32	C38
8.2 (m) . . . . . . ¼-1 in.		830	120	1035	150	895	130	10	61	C35	C42
ISO Property Classes (p)											
4.6 . . . . .5-36 mm		225	33	400	58	240(h)	35(h)	22	· · ·	B67	B100
4.8 . . . . .1.6-16 mm		310	45	420	61	340	49	14	· · ·	B71	B100
5.8 . . . . .5-24 mm(g)		380	55	520	75	420	61	10	· · ·	B82	B100
8.8 . . . . .16-36 mm		600	87	830	120	660	96	12	54	C24	C34
9.8 . . . . .1.6-16 mm		650	94	900	131	720	104	10	56	C27	C36
10.9 . . . . .5-36 mm		830	120	1040	151	940	136	9	59	C33	C39
12.9 . . . . .1.6-36 mm		970	141	1220	177	1100	160	8	63	C39	C44

(a) Including cap screws and U-bolts. (b) The minimum reduction of area for specimens machined from all grades and classes of fasteners listed is 35%. (c) Determined on full size fasteners. (d) Determined on both full size fasteners and specimens machined from fasteners. (e) Determined on specimens machined from fasteners. (f) Yield strength is stress to produce a permanent set of 0.2%. (g) Data from SAE Standard J429. (h) Yield strength instead is stress at 0.2%. (j) For bolts and screws longer than 6 in., grade 1 requirements apply. (k) Studs only. (m) Bolts and screws only. (n) Roll threaded after heat treatment. (p) Data from IFI Standard 501. Values for fasteners with coarse threads. (q) Requirements apply to bolts 150 mm long and shorter, and to studs of all lengths.

Most bolts are made by cold or hot heading. Resulfurized steels are not suitable for heading because they will split. However, if splitting did not occur, their high cost could not be justified against the small amount of machining necessary to produce a headed bolt. Only a few bolts are machined from bars; these usually are of special design or the required quantities are extremely small. For such bolts, the extra cost of resulfurized grades of steel may be justified. For example, 1541 steel might be selected to make headed bolts of a specific size. If the same bolts were to be machined from bars, 1141 steel would be selected because of its superior machinability. Special bolts can usually be made more economically by machining from oversize upset blanks instead of from bars.

Stud Steels. The chemical compositions of studs (and U-bolts, which are basically studs formed into a U-shape) are given in Table 4; special modifications that apply to studs may be found in the footnotes. Because studs (and U-bolts) are not headed, it is not essential to restrict sulfur. It may be noted that grade 2 and class 5.8 permit 0.33% maximum sulfur, while grade 5 and classes 8.8 and 9.8 permit 0.13% maximum sulfur.

Table 2 Mechanical properties of steel nuts (a)

Strength grade or property class	Nominal diameter	Proof stress (b) MPa	Proof stress (b) ksi	Hardness, HRC min	Hardness, HRC max
SAE Strength Grades (c)					
2 (d) . . . . ¼-1½ in.		620	90	· · ·	32
5 . . . . ¼-1 in.		830	120(e)	· · ·	32
		750	109(f)	· · ·	32
>1-1½ in.		725	105(e)	· · ·	32
		650	94(f)	· · ·	32
8 . . . . ¼-⅝ in.		1035	150	24	32
>⅝-1 in.		1035	150	26	34
>1-1½ in.		1035	150	26	36
ISO Property Classes (g)					
5 (h) . . . . .5-36 mm		570	83	· · ·	30
9 (h) . . . . .1.6-4 mm		900	131	· · ·	30
4-16 mm		990	144	· · ·	30
20-36 mm		910	132	· · ·	30
10 (h) . . . . .5-36 mm		1040	151	26	36

(a) Not normally including jam, slotted, castle, heavy or thick nuts. (b) Determined on full size nuts. (c) Data from SAE Standard J995. (d) Normally applicable only to square nuts, which are normally available only in grade 2. (e) For UNC, 9 UN thread series. (f) For UNF, 12 UN threaded series and finer. (g) Data from IFI Standard 508. Values for fasteners with coarse threads. (h) For hex nuts only.

Stud (or U-bolt) threads, however, are not necessarily cut, but may be rolled for economy and good thread shape. A smaller diameter rod must be used to roll a specific thread size than to cut the same thread size from rod. For example, a ½–13 thread could be cut from a rod 0.500 in. in diameter; a smaller diameter rod would be used to roll the same size threads. Grades 4 and 8.1 are made from a medium-carbon steel and obtain their mechanical properties not from quenching and tempering, but from being drawn through a die with special processes. They are particularly suitable for studs, for these materials can-

Table 3 Corresponding property classes and strength grades of steel bolts and studs

ISO property class	Corresponding SAE strength grade	Tensile strength			Yield strength	
		Nominal MPa	Min, MPa	Min, ksi(a)	Min, MPa	Min, ksi(a)
3.6	...	300	330	48	190	28
4.6	1	400	400	58	240	35
5.6	2	500	500	73	300	48
5.8	...	500	520	75	420	61
6.8	3	600	600	87	480	70
8.8	5	800	830	120	660	96
10.9	8	1000	1040	151	940	136
12.9	...	1200	1220	177	1100	160

(a) Converted values

Table 4 Chemical compositions of steel bolts and studs (a)

Strength grade or property class	Material and treatment	Composition, % (b)			
		C	P	S	Others
SAE Strength Grades (c)					
1	Low- or medium-carbon steel	0.55	0.048	0.058	...
2	Low- or medium-carbon steel	0.28(d)	0.048	0.058(e)	...
4	Medium-carbon cold drawn steel	0.55	0.048	0.058	...
5	Medium-carbon steel, quenched and tempered	0.28-0.55	0.048	0.058(f)	...
5.2	Low-carbon martensitic steel, fully killed, fine grain, quenched and tempered	0.15-0.25	0.048	0.058	(g)
7	Medium-carbon alloy steel, quenched and tempered (h, j)	0.28-0.55	0.040	0.045	...
8	Medium-carbon alloy steel, quenched and tempered (h, j)	0.28-0.55	0.040	0.045	...
8.1	Drawn steel for elevated-temperature service: medium-carbon alloy steel or 1541 steel	0.28-0.55	0.048	0.058	...
8.2	Low-carbon martensitic steel, fully killed, fine grain, quenched and tempered (k)	0.15-0.25	0.048	0.058	(g)
ISO Property Classes (m)					
4.6	Low- or medium-carbon steel	0.55	0.048	0.058	...
4.8	Low- or medium-carbon steel, partially or fully annealed as required	0.55	0.048	0.058	...
5.8	Low- or medium-carbon steel, cold worked	0.13-0.55	0.048	0.058(e)	...
8.8	Medium-carbon steel, quenched and tempered(n) (p)	0.28-0.55	0.048	0.058(f)	...
9.8	Medium-carbon steel, quenched and tempered(n)	0.28-0.55	0.048	0.058(f)	...
10.9	Medium-carbon alloy steel, quenched and tempered(h,q) (k,n)	0.28-0.55	0.040	0.045	...
12.9	Alloy steel, quenched and tempered	0.31-0.65	0.045	0.045	(r)

(a) Including cap screws and U-bolts. (b) All values are for product analysis; where a single value is shown, it is a maximum. Unless otherwise noted, manganese contents of the steels in this table are not specified. (c) Data from SAE Standard J429. (d) Carbon may by 0.55% max for all sizes and lengths of studs and for bolts larger than ¾ in. diameter and/or longer than 6 in. (e) For studs only, sulfur may be 0.33% max. (f) For studs only, sulfur may be 0.13% max. (g) 0.74 min Mn and 0.0005 min B. (h) Fine grain steel with hardenability that will produce 47 HRC min at the center of a transverse section one diameter from the threaded end of the fastener after oil quenching (see SAE J407). (j) For diameters of ¼ through ¾ in., carbon steel may be used by agreement. At producer's option, 1541 steel, oil quenched and tempered, may be used for diameters through 7/16 in. (k) Steel with hardenability that will produce 38 HRC min at the center of a transverse section one diameter from the threaded end of the fastener after quenching. (m) Data from IFI Standard 501. (n) For diameters through 24 mm, unless otherwise specified by the customer, the producer may use a low-carbon martensitic steel with 0.15 to 0.40 C, 0.74 min Mn, 0.048 max P, 0.058 max S and 0.0005 min B. (p) At producer's option, medium-carbon alloy steel may be used for diameters over 24 mm. (q) For diameters through 20 mm, carbon steel may be used by agreement. At producer's option, 1541 steel, oil quenched and tempered, may be used for diameters through 12 mm. (r) One or more of the alloying elements chromium, nickel, molybdenum or vanadium shall be present in the steel in sufficient quantity to assure that the specified mechanical properties are met after oil quenching and tempering.

not readily be formed into bolts.

Selection of Steel for Bolts and Studs. The following guidelines should be consulted before selecting steel for bolts and studs (including cap screws and U-bolts):

1 Bolts over 150 mm (6 in.) long or over 19 mm. (¾ in.) in diameter are usually hot headed.
2 Strength requirements for steels for grade 1 bolts can be met with hot rolled low-carbon steels.
3 The strength requirements for steels for grade 2 bolts 19 mm (¾ in.) and less in diameter can be met with cold drawn low-carbon steels; sizes over 19 mm (¾ in.) in diameter require hot rolled low-carbon steel only, but may be made of cold finished material.
4 Grade 4 fasteners (studs only) require a cold finished medium-carbon steel, specially processed to obtain higher than normal strength. Resulfurized steels are acceptable.
5 Grade 5 bolts and studs require quenched and tempered steel. The choice among carbon, 1541 and alloy steel will vary with the hardenability of the material, the size of the fastener, and the quench employed. Cost dictates the use of carbon steel, including 1541; however, the threading practice (before or after hardening) determines the severity of quench that can be used if quench cracks in the threads are to be avoided. Figure 1 shows cost-hardenability relationships for both oil- and water-quenched steels. An increase in hardenability does not necessarily mean an increase in cost per pound. Figure 1 is not intended to prescribe or imply the use of water quenching for alloy steels that are normally oil quenched. These data are presented only to show the economic advantages of water quenching when it can be properly and successfully applied to the product being heat treated.
6 Fasteners made to grade 7 and 8 specifications normally require alloy steel. This steel is selected on a hardenability basis so a minimum of 90% martensite exists at the center after oil quenching. This requirement ensures fasteners of the highest quality. Grade 7 bolts must be roll threaded after heat treatment. This practice substantially increases the fatigue limit in the threaded section, as shown in Fig. 2. Other factors being equal, a bolt with threads properly rolled after heat treatment—that is, free from mechanical imper-

Fig. 1 Cost and hardenability relations for oil-quenched and water-quenched steels for cold heading

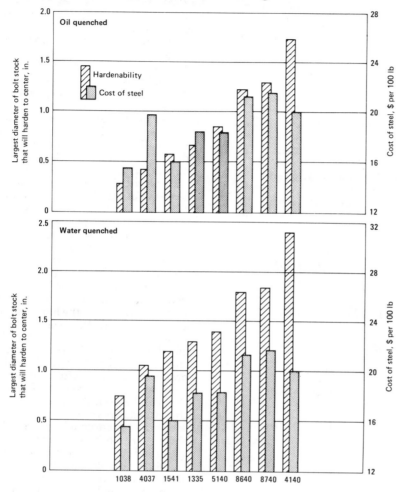

Cost data were based on hot rolled rounds in coils, and include base price and grade extra as of June 1977. Hardenability for both oil-quenched and water-quenched steels was based on hardness of 42 HRC min at the center of the as-quenched diameter.

Fig. 2 Fatigue limits for roll threaded steel bolts

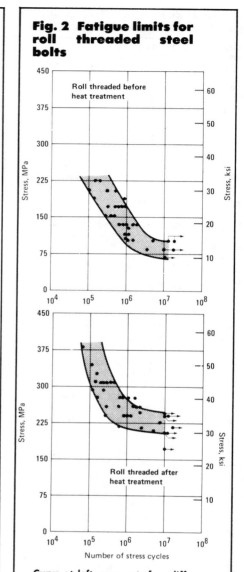

Curve at left represents four different lots of bolts that were roll threaded, then heat treated to average hardness of 22.7, 26.6, 27.6 and 32.6 HRC. Curve at right represents five different lots that were heat treated to average hardnesses of 23.3, 27.4, 29.6, 31.7 and 33.0 HRC, then roll threaded. Bolts having higher hardnesses in each category had higher fatigue strengths.

fections—has a higher fatigue limit than one with cut threads. This is true for any strength category. The cold work of rolling increases the strength at the weakest section (the thread root) and imparts residual compressive stresses, similar to those imparted by shot peening. The larger and smoother root radius of the rolled thread also contributes to its superiority.

7 Fasteners of grades 5.2 and 8.2 are made from the low-carbon boron steels now used for many bolts. These steels are readily formed because of the low carbon content, yet the boron gives them relatively high harden-

ability. Fasteners of these grades are hardened in oil or water, then tempered at minimum temperatures of 425 °C (800 °F) and 340 °C (650 °F), respectively. Grades 5.2 and 8.2 offer the same tensile properties as the corresponding non-boron grades 5 and 8, but grades 5.2 and 8.2 have slightly higher hardnesses due to their lower tempering temperatures.

8 For service temperatures of 200 to 370 °C (400 to 700 °F), specific bolt steels are recommended (see Table 6) because relaxation is an influencing factor at these temperatures. Although other steels will fulfill re-

quirements for the tabulated conditions, those listed are the lowest cost grades. Only medium-carbon steels are recommended; in all instances, they should be quenched and tempered.

Nut Steels. The selection of steel for nuts is less critical than for bolts. The nut is usually not made from the same

Table 5 Chemical compositions of steel nuts

Strength grade or property class	Composition, % (a)			
	C max	Mn min	P max	S max
SAE Strength Grades (b)				
2	0.47	...	0.12	0.15(c)(d)
5	0.55	0.30	0.05(e)(f)	0.15(d)(f)
8	0.55	0.30	0.04	0.05(g)
ISO Property Classes (h)				
5(j)	0.55	...	0.12	0.15(c)
9	0.55	0.30	0.05(f)	0.15(f)
10	0.55	0.30	0.04	0.05(g)

(a) All values are for heat analysis. (b) Data from SAE Standard J995. (c) Resulfurized and rephosphorized material is not subject to rejection based on product analysis for sulfur. (d) If agreed, sulfur may be 0.23% max. (e) For acid bessemer steel, phosphorus may be 0.13% max. (f) If agreed, phosphorus may be 0.12% max and sulfur may be 0.35% max, provided manganese is 0.70% min. (g) If agreed, sulfur may be 0.33% max, provided manganese is 1.35% min. (h) Data from IFI Standard 508. (j) If agreed, free cutting steel having maximums of 0.34 S, 0.12 P, and 0.35 Pb may be used.

Fig. 3 Variation of breaking strength with number of exposed threads for ¾-in.-diam SAE grade 5 bolts

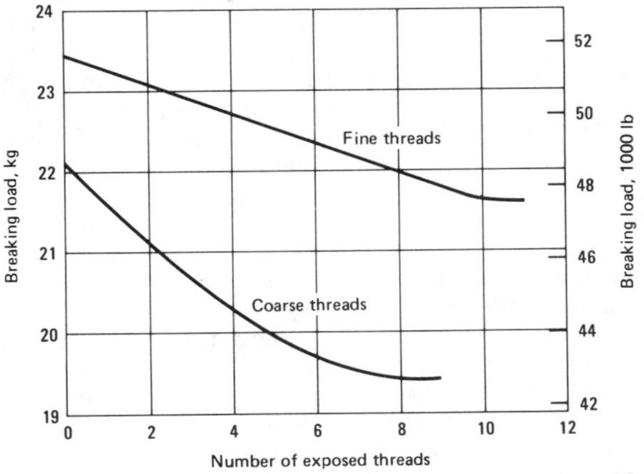

material as the bolt. Table 5 gives the chemical composition requirements for each property grade and class of steel nut shown in Table 2.

Lower strength nuts (such as grades 2 and 5) are not heat treated. However, higher grade nuts (such as grade 8) may be heat treated to attain specified hardness.

Nuts are machined from bar stock, cold formed or hot formed, depending on configuration and production requirements. Size and configuration are usually more important than the material from which the nuts are made.

The bolt is normally intended to break before the nut threads strip. Regular hex nut dimensions are such that the shear area of the threads is greater than the tensile stress area of the bolt by more than 100%. Consequently, low-carbon steel nuts are customarily used even when the bolts are made of much higher strength material.

Low-carbon steel nuts are usually heat treated to provide mar resistance to the corners of the head or to the clamping face. Light case carburizing or carbonitriding is often employed to improve mar resistance.

When nuts are to be quenched and tempered, the steel must have the appropriate hardenability. Increasing the amounts of carbon and manganese or adding other alloying elements to provide increased hardenability all decrease the suitability of the material for cold forming. For this reason, low-car-

bon boron steels are widely used for quenched and tempered high strength nuts. The low carbon content permits easy cold forming, while the boron enhances hardenability. Threading may be done before or after heat treatment, depending on the class of thread fit required and the hardness of the heat treated nut.

Because the selection of steel for nuts is not critical, practice varies considerably. A common practice is to use steels such as 1108, 1109, 1110, 1113 or 1115, cold formed or machined from cold drawn bars, for grade 2 nuts. Grade 5 nuts are commonly made from 1035 or 1038 steels, cold formed from annealed bars, cold drawn and stress relieved, or quenched and tempered. Grade 8 nuts are formed from low-carbon boron steels, then quenched and tempered.

Fastener Tests

The most widely accepted method for determining the strength of full size bolts, studs and nuts is a test for proof stress. For bolts, a wedge tensile test for tensile (breaking) strength is also commonly made on the fastener, following the proof-stress test.

The wedge test simulates conditions where bolts are assembled under misalignment; thus, it is also the most widely used quality control test for head ductility. The test is performed by placing the wedge under the bolt head and, by means of suitable fixtures, stressing the bolt to fracture in a tensile-testing machine. To meet requirements of the

test, the tensile fracture must occur in the body, or threaded section, with no fracture at the junction of the body and head. In addition, the breaking strength should meet specified strength requirements. Details of this test are given in ASTM A370 and SAE J429.

The number of exposed threads between the bolt shank and the beginning of the nut influences the recorded tensile strength for both coarse-thread and fine-thread bolts. A typical variation of breaking strength with number of exposed threads is plotted in Fig. 3. The data are for ¾-in.-diam. SAE grade 5 bolts. Because of this variation, the number of exposed threads should be specified for wedge tensile testing of bolts and other threaded fasteners; generally, three exposed threads are specified.

The proof stress of a bolt or stud is a specified stress which the bolt or stud must withstand without detectable permanent set. For purposes of this test, a bolt or stud is deemed to have incurred no permanent set if the over-all length after application and release of the proof stress is within ±0.013 mm (0.0005 in.) of its original length. Length measurements are ordinarily made to the nearest 0.0025 mm (0.0001 in.). Because bolts and studs are manufactured in specific sizes, the proof stress values are commonly converted to equivalent proof load values and it is the latter that are actually used in testing full-size fasteners. To compute proof load, the stressed area must first be determined. Because the smallest cross-sectional area is in the threads, the stressed area is computed by the following formula:

$$A_s \text{ (in.}^2\text{)} = 0.7854 \left(D - \frac{0.9743}{N} \right)^2 \quad \text{(Eq 1)}$$

where A_s is the mean equivalent stress area, in.^{2}; D is the nominal diameter, in.; N is the number of threads per inch. The equivalent formula for metric threads is

$$A_s \text{(mm}^2\text{)} = 0.7854(D - 0.9382P)^2 \quad \text{(Eq 2)}$$

where A_s is the mean equivalent stress area, mm^2; D is the nominal diameter, mm; P is the thread pitch, mm.

For example, ½-in. diam bolts made to grade 5 requirements have values of mean equivalent stress area for coarse threads (13 threads per inch) and fine threads (20 threads per inch) equal to 91.5 mm^2 (0.1419 in.^{2}) and 103.2 mm^2 (0.1599 in.^{2}), respectively, according to the above formulas. These bolts must withstand a proof stress of 585 MPa (85 ksi) without a detectable difference between the initial length and the length after the proof load has been applied and released. For coarse threads, this requires a proof load of 53.6 kN (12 060 lb); for fine threads, the proof load is 60.5 kN (13 600 lb).

The proof stress of a nut is determined by assembling it on a hardened and threaded mandrel or on a test bolt conforming to the particular specification. The specified proof load for the nut is determined by converting the specified proof stress, using the mean equivalent stress area calculated for

the mandrel or test bolt. This proof load is applied axially to the nut by a hardened plate as shown in Fig. 4. The thickness of the plate is at least equal to the diameter of the mandrel or test bolt. The diameter of the hole in the plate is a specified small amount greater than that of the mandrel or test bolt diameter. To demonstrate acceptable proof stress, the nut must resist the specified proof load without failure by stripping or rupture, and must be removable from the mandrel or test bolt by hand after initial loosening.

For details relating to the prescribed proof-stress tests and other requirements for threaded steel fasteners of various sizes in both coarse and fine threads, the reader should consult SAE J429 and J995 for fasteners made to SAE strength grades or ISO 898 for fasteners made to ISO property classes. Also see ASTM A370.

Mechanical Properties

The mechanical properties of the commonly used SAE strength grades and ISO property classes of steel bolts, studs and nuts are listed in Tables 1 and 2. Further information on the mechanical properties of threaded steel fasteners can be found in this section, and in ASTM A307, A325, A354 and A449. Grade 1038 steel is one of the most widely used steels for threaded fasteners up to the level of combined size and proof stress at which inadequate hardenability precludes further use. This steel has achieved its popularity because of excellent cold heading properties, low cost and availability. Typical distributions of tensile properties for bolts and cap screws made from 1038 steel, as evaluated by the wedge tensile test, are shown in Fig. 5. These data were obtained from one plant and represent tests from random lots of grade 5 fasteners. The three histograms in Fig. 5 show three distributions typical of grade 5 fasteners. No significance should be attached to the apparent difference between average values, especially for the two hex head bolts of different lengths. Specifications require only that bolt strength exceed a specified minimum value, not that bolts of different sizes have statistically equivalent average strengths.

Grade 1541 steel is used extensively for applications requiring hardenability greater than that of 1038 steel, but less than that of alloy steel. The depth of

hardening of both steels is compared in Fig. 6.

Hardness-tensile relations, showing the scatter of results when bolts are tested in each of three different ways, are given in Fig. 7. Scatter is much greater when the hardness tests are made on the upset head, probably because of decarburization. The difference in section size between the head and the threaded section may also contribute to greater scatter, because of the difference in quenching rate. Scatter is greater at higher hardness levels.

The bolts in Fig. 7 (grade 5, ¾ in. in diameter both coarse and fine thread) were made from one heat of 1038 steel of the following composition: 0.38 C, 0.74 Mn, 0.08 Si, 0.025 P, 0.040 S, 0.08 Cr, 0.07 Ni and 0.12 Cu. The bolts were quenched in water from 850 °C (1550 °F) and tempered at different temperatures in the range 360 to 600 °C (690 to 1125 °F) to produce a range of hardness of 20 to 40 HRC. Heat treating was done in one plant, but the hardness tests were made in five different laboratories.

Figure 9 shows the results of a similar cooperative study in which eight laboratories tested a large number of bolts made from a single heat of 1038 steel. There were eight different lots of bolts, each heat treated to a different hardness level. Each of the eight laboratories tested bolts from all eight lots. Hardness readings were taken on a transverse section through the threaded portion of the bolt at the mid-radius of the bolt. The transverse section was located one diameter from the threaded end of the bolt.

Fatigue Strength. If bolts made of two different steels have equivalent hardnesses throughout identical sections, their fatigue strengths will be similar, (see Fig. 8) as long as other factors such as mean stress, stress range and surface condition are the same. If the results of fatigue tests on standard test specimens were interpreted literally, high-carbon steels would be selected for bolts. Actually, steels of high carbon content (more than 0.55% carbon) are unsuitable because they are notch sensitive.

The principal design feature of a bolt is the threaded section, which establishes a notch pattern inherent in the part because of its design. The form of the threads, plus any mechanical or metallurgical condition that also creates a surface notch, is much more important than steel composition in determining

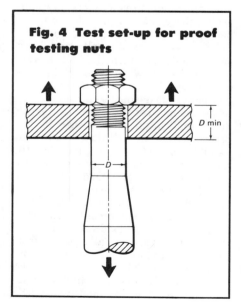

Fig. 4 Test set-up for proof testing nuts

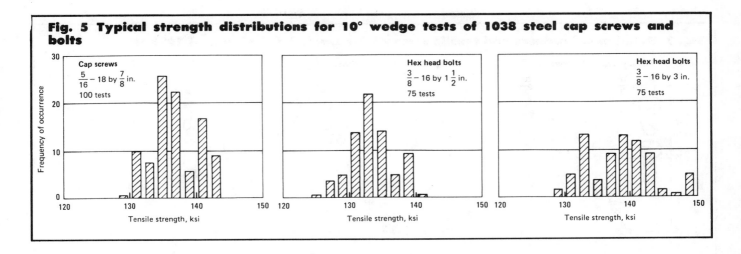

Fig. 5 Typical strength distributions for 10° wedge tests of 1038 steel cap screws and bolts

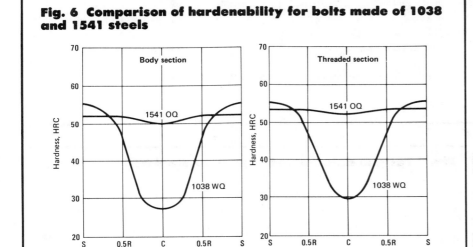

Fig. 6 Comparison of hardenability for bolts made of 1038 and 1541 steels

Curves represents average as-quenched hardnesses of fifteen ¾-in.-diam bolts from one heat of each grade. C is center of the bolt, 0.5R is mid-radius, S is surface. The 1038 steel bolts were water quenched; the 1541 steel bolts, oil quenched.

the fatigue resistance of a particular lot of bolts.

The method of forming the thread is an important factor influencing fatigue strength of bolts. Threads may be formed by cutting, grinding or rolling. There is no marked difference in fatigue strength between cut and ground threads. However, there is a marked improvement when threads are rolled rather than either cut or ground, particularly when the threads are rolled after the bolt has been heat treated. (See item 6 in the section on Selection of Steel for Bolts and Studs).

Light cases, such as from carburizing or carbonitriding, are rarely recommended and should not be used for criti-cal externally threaded fasteners, such as bolts, studs or U-bolts. The cases are quite brittle and crack when the fasteners are tightened or bent in assembly or service. These cracks may then lead to fatigue cracking and possible fracture.

Chromium and nickel platings decrease the fatigue strength of threaded sections and should not be used except in a few applications, such as automobile bumper studs or similar fasteners that operate under conditions of low stress and require platings for appearance. Cadmium and zinc have little effect on fatigue strength; these softer platings do not improve fatigue strength but may have a slightly adverse effect on it.

When placed into service, bolts are most likely to fracture in fatigue if the assemblies involve soft gaskets or flanges, or if the bolts are not properly aligned and tightened. Fatigue resistance also is related to clamping force. In many assemblies a certain minimum clamping force is required to ensure both proper alignment of the bolt in relation to other components of the assembly and proper preload on the bolt. The former ensures that the bolt will not be subject to undue eccentric loading, and the latter that the correct mean stress is established for the application. In some instances, clamping stresses that exceed the yield strength may be desirable; recent experiments showed bolts clamped beyond the yield point to have better fatigue resistance than bolts clamped below the yield point.

Fatigue failures in bolts usually occur in the threaded section at or near the first thread inside the nut. Other locations of possible fatigue failure of a bolt under tensile loading are the thread runout and the head-to-shank fillet. All three locations are areas of stress concentration. Any measures that decrease stress concentration can lead to improved fatigue life. Typical examples of such measures are use of UNJ increased root radius threads (see MIL-S-8879A) and use of nut designs that distribute the load uniformly over a large number of bolt threads. Shape and size of the head-to-shank fillet are important, as is a generous radius from the thread runout to the shank.

Decreasing bolt stiffness can also reduce cyclic stresses. A method commonly used is reduction of the cross-sectional area of the shank to form a "waisted" shank.

Source: Metals Handbook, Vol. 1, 9th Ed., 273-282

Fig. 7 Relation of hardness and tensile strength for grade 5 bolts made of 1038 steel

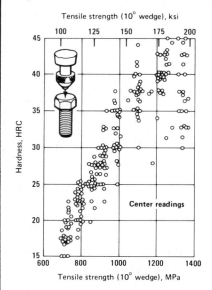

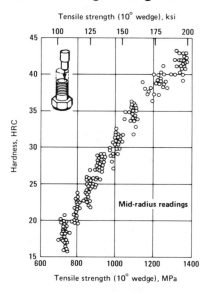

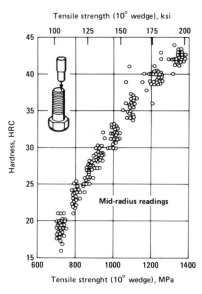

Bolts, ¾ in. in diameter, were all made from one heat of steel. Hardness tests were made in three different locations, as shown by the insets in the three graphs. See text for discussion.

Table 6 Steels recommended for bolts to be used at elevated temperatures (a) (b)

Bolt diameter, in.	Steel recommended for proof stress (c)		
	75 ksi	100 ksi	125 ksi
¼ to ¾	1038	4037	4037
¾ to 1¼	1038	4140	4140
1¼ to 2	4140	4140	4145

(a) Between 200 and 370 °C (400 and 700 °F). (b) All selections are based on a minimum tempering temperature of 455 °C (850 °F). (c) At room temperature.

Fabrication

Most bolts are made by cold heading. Other cold forming methods, including cold extrusion, are employed for making threaded fasteners, although these methods are primarily confined to the lower carbon and softer steels used for low proof stress applications.

Hot heading is required for forming heads that are large in relation to shank diameter and is usually used when the part cannot be cold headed in one or two blows. Hot heading is also more practical for bolts with diameters larger than about 32 mm (1¼ in.), because of equipment limitations and increased proba-

Fig. 8 Fatigue data for 1040 and 4037 steel bolts

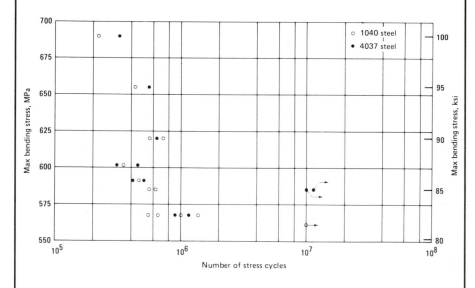

The bolts (⅜ by 2 in., 16 threads to the inch) had a hardness of 35 HRC. Tensile properties of the 1040 steel at three-thread exposure were: yield strength, 1060 MPa (154 ksi); tensile strength (axial), 1200 MPa (175 ksi); tensile strength (wedge), 1190 MPa (173 ksi). For the 4037 steel: yield strength, 1110 MPa (161 ksi); tensile strength (axial), 1250 MPa (182 ksi); tensile strength (wedge), 1250 MPa (182 ksi).

Fig. 9 Hardness distributions for eight lots of 1038 steel bolts

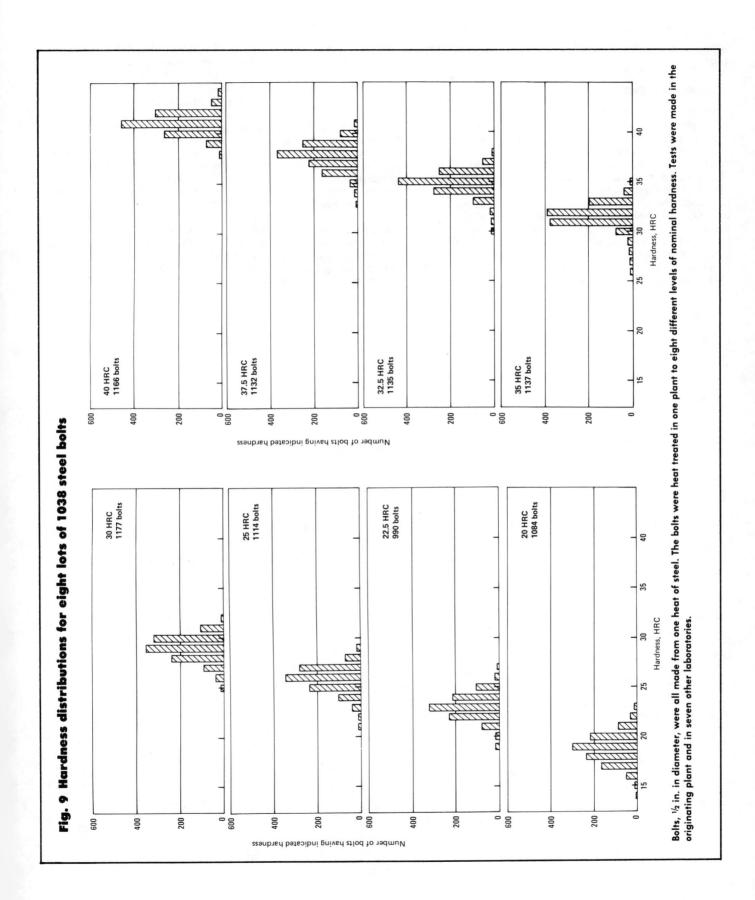

Bolts, ½ in. in diameter, were all made from one heat of steel. The bolts were heat treated in one plant to eight different levels of nominal hardness. Tests were made in the originating plant and in seven other laboratories.

bility of tool failures with cold heading.

Clamping Forces

To operate effectively and economically, threaded fasteners should be designed to be torqued near the proof stress, as dictated by bolt diameter and desired clamping force. The actual clamping force attained in any assembly will be influenced by such factors as (a) roughness of the faying surfaces, (b) coatings, contaminants or lubricants on the faying surfaces and (c) platings or lubricants on the threads. Typically, torque values are established to result in a clamping force equal to about 75% of the proof load. For some applications, bolts are torqued beyond their proof stress with no detrimental results, provided they are permanent fasteners.

Because it is difficult to measure bolt tension (clamping force) in production installations, torque values are commonly specified. The clamping forces generated at given torques are very dependent on the coefficient of friction at the threads and at the bearing face, and thus they are highly dependent on fastener coatings. Common fastener coatings are zinc, cadmium, and phosphate and oil.

The maximum clamping force that can be effectively employed in any bolt is often limited by the compressive strength of the materials being bolted. If this value is exceeded, the bolt head or nut will be pulled into the parts being bolted, with a subsequent reduction in clamping force. The assembly then becomes loose and the bolt is susceptible to fatigue failure. If high tensile bolts are necessary to join low compressive strength materials, hardened washers should be used under the head of the bolt and under the nut to distribute bearing pressure more evenly and avoid the condition described above.

The value of high clamping forces, apart from lessening the possibility of the nut loosening is that the working stresses (against solid abutments) are always less than the clamping forces induced in a properly selected bolt. This ensures against cyclic stress and possible fatigue failure.

Materials for
Corrosion-Resistant Fasteners

By Joseph S. Orlando and William Ballantine, ITT Harper

Condensed from Metals Handbook, Ninth Edition, Volume 3, pages 183 to 185.

CORROSION-RESISTANT metallic fasteners are those made of stainless steels and nonferrous alloys. This broad definition could include hundreds of alloys, but in practice the materials actually used are limited to several stainless steels and several copper alloys, plus a few nickel, aluminum and titanium alloys. Fasteners can and have been made from unusual materials (tantalum, for example), but this discussion is primarily limited to those corrosion-resistant materials used in commercial fasteners that are readily available as standard (see Table 1).

STAINLESS STEELS

Over half of all industrial fasteners classified as corrosion resistant are made of stainless steels. This general designation covers austenitic, martensitic and ferritic stainless steels. Of all stainless steels, the 300 series austenitic types are the most popular for fastener use. Austenitic stainless steels are not hardenable by heat treatment and are nonmagnetic for all practical purposes. All alloys in this group have at least 8% nickel in addition to chromium. They offer a greater degree of corrosion resistance than martensitic and ferritic types, but offer a lesser degree of resistance to chloride stress-corrosion cracking.

Martensitic and ferritic stainless steels contain at least 12% chromium, but contain little or no nickel because it stabilizes austenite. Martensitic grades, such as types 410 and 416, are magnetic and can be hardened by heat treatment. Ferritic alloys, such as type 430, are also magnetic but cannot be hardened by heat treatment.

The fastener industry generally markets fasteners made of types 302, 303, 304 and 305 stainless steels as "18-8." These four alloys are similar in both corrosion resistance and mechanical properties. From the manufacturer's point of view, the choice of alloy depends on method of fastener production, which in turn depends on type and size of fastener and, to some extent, on production volume. Because no two manufacturers have identical equipment, the alloy selected for a given fastener will vary; as an indication, however, the alloys that a major fastener producer uses on orders for 18-8 are as follows:

- *Type 302* is used for machine and tapping screws.
- *Type 303* is used to make nuts machined from bar. It contains a small amount of sulfur, for improved machinability.
- *Type 304* is used for hot heading (examples: long bolts or large-diameter bolts beyond the range of cold heading equipment).
- *Type 305* is used for cold heading (examples: hex-head bolts and cold formed nuts).

Other 300-series stainless steels used in fasteners include the following:

Table 1. Standard corrosion-resistant fastener alloys

Commercial name	UNS No.	ASTM specifications
Stainless steels		
17-4 PH	S17400	
Type 302	S30200	
Type 303	S30300	
Type 304	S30400	
Type 305	S30500	F593: stainless steel bolts, hex cap screws, and studs
Type 309	S30900	
Type 310	S31000	
Type 316	S31600	
Type 317	S31700	F594: stainless steel nuts
Type 321	S32100	
Type 347	S34700	
Type 410	S41000	
Type 416	S41600	
Type 430	S43000	
Copper alloys		
ETP copper	C11000	
Yellow brass	C27000	
High leaded brass	C34200	
Free-cutting brass	C36000	
Naval brass, 63½%	C46200	
Naval brass, uninhibited	C46400	
Si-bearing aluminum bronze	C64200	
Low-silicon bronze B	C65100	
High-silicon bronze A	C65500	
Nickel alloys		F468: nonferrous bolts, hex cap screws, and studs for general use
Monel 400	N04400	
Monel 405	N04405	
Monel K-500	N05500	F467: nonferrous nuts for general use
Inconel 600	N06600	
Aluminum alloys		
Aluminum 1100	A91100	
Alloy 2024	A92024	
Alloy 6061	A96061	
Titanium alloys		
Commercial-purity titanium:		
ASTM grade 1	R50250	
ASTM grade 2	R50400	
ASTM grade 4	R50700	
Ti-6Al-4V (ASTM grade 5)	R56400	
Ti-0.2Pd (ASTM grade 7)	R52400	

- *Types 309 and 310* are higher in both nickel and chromium than the standard 18-8 alloys, and are used for high-temperature applications.
- *Types 316 and 317,* because they contain molybdenum, have better elevated-temperature strength and better resistance to pitting than 18-8 alloys.
- *Types 321 and 347* are similar to 18-8 alloys but are stabilized by addition of titanium (type 321) or niobium (type 347) to increase resistance to intergranular corrosion.

Ferritic and martensitic stainless steels for fasteners are largely limited to:

- *Types 410 and 416* are general-purpose corrosion and heat-resistant alloys; they are hardenable by heat treatment.
- *Type 430* has better corrosion- and heat-resistant qualities than type 410; it is not hardenable by heat treatment.

COPPER ALLOYS

Silicon bronzes have tensile strengths higher than those of low-carbon steels and are resistant to corrosion by the atmosphere, by fresh water and seawater, and by gases and sewage. Silicon bronzes are the copper alloys most commonly used for fasteners. They are nonmagnetic and have excellent machining and forming characteristics. C65100 is a low-silicon alloy suitable for cold heading; C65500 is suitable for hot forged fasteners.

Aluminum bronzes have better mechanical properties than silicon bronzes, but are much less frequently used in fasteners. Because of its good machinability, C64200 is the aluminum bronze most often used for fasteners.

Brasses, once the most commonly used materials for corrosion-resistant fasteners, now are specified less frequently than steels and silicon bronzes, which have higher mechanical properties. Brasses are still used in various applications, including electrical communications equipment, builders' hardware and many other consumer and industrial products. C27000 is used for cold headed fasteners, and C36000 is used for fasteners milled from bar.

Naval brasses are copper-zinc alloys containing small amounts of tin, which give them higher resistance to salt water and atmospheric corrosion. C46200 is used for cold headed fasteners, and C46400 is used for hot forged fasteners and for fasteners milled from bar.

NICKEL ALLOYS

Nickel-base alloys are characterized by good strength and good resistance to heat and corrosion. They are often specified for marine and chemical-plant uses.

- *Monel 400* is used for fasteners more often than any other nickel-base alloy.
- *Monel K-500* is heat treatable and, in effect, is a high-strength version (900-MPa, or 130-ksi, minimum tensile strength) of Monel 400.
- *Inconel 600* is used for fasteners that must retain both high strength and resistance to oxidation at temperatures as high as 870 °C (1600 °F).

ALUMINUM ALLOYS

Some aluminum alloys are used for industrial fasteners. They have good corrosion resistance and low weight. Typically, aluminum fasteners are used to join aluminum components.

- *2024-T4* is a heat treated alloy usually used for cold headed fasteners; its tensile strength is

above 425 MPa (62 ksi).
- *6061-T6* is used for some nuts, both cold formed and machined.
- *1100* (commercial-purity aluminum) is used for some washers and rivets.

TITANIUM

Titanium and its alloys have excellent corrosion resistance and maintain their strength at moderately high temperatures. Most industrial titanium fasteners are made from commercial-purity titanium, and are used in chemical-equipment applications. Titanium aircraft fasteners, many of which are of proprietary design, are produced from titanium alloys of much higher strength.

INDUSTRY STANDARDS

The American Society for Testing and Materials (ASTM) has a working committee, "F16-Fasteners," with a series of subcommittees each of which deals with development of specific fastener standards. Subcommittee 4 works with nonferrous and stainless steel fastener standards. ASTM specifications in Table 1 (ASTM F467, F468, F593 and F594) are the four standards initially created by Subcommittee 4. They can be referred to for design criteria applicable to corrosion-resistant fasteners.

NONSTANDARD FASTENER ALLOYS

Previous sections of this article have dealt with those corrosion-resistant alloys that are used most often for fasteners whose designs are recognized as standard by the American National Standards Institute (ANSI). Not surprisingly, there are numerous other materials that are used, either for standard fasteners or for special parts, when dictated by strength considerations, corrosive conditions or temperature requirements. Some of these more specialized alloys are listed below.

- *Precipitation-hardening stainless steels*, such as 17-4 PH, 17-7 PH, PH 15-7 Mo, Custom 450 and Custom 455, are used to obtain higher strength than that available from 18-8 stainless steels.
- *Martensitic stainless steels* such as type 416 and type 420 are used to obtain better mechanical properties than can be achieved with types 410 and 430.
- *Carpenter 20Cb-3* is specified when greater corrosion resistance is required than can be offered by 18-8 stainless steels, such as for equipment handling hot sulfuric acid.
- *A-286*, a nonstandard stainless steel that has greater corrosion resistance than the 18-8 types, as well as good mechanical properties at elevated temperatures, has been used in applications requiring resistance to both heat and a corrosive substance, such as in specialized chemical-plant or petroleum-refinery applications.

7050 RIVET EXPERIENCE

Ronald E. Wood
Materials and Process Engineering
Lockheed-California Company
Burbank, California

Abstract

Development at the Lockheed-California Company to qualify 7050 aluminum alloy rivets is described. Program impetus was to replace 2024 aluminum protruding head rivets in commercial applications. Handling and installation problems for 2024 "ice box" rivets are discussed. Testing for mechanical properties and corrosion resistance as well as installation studies is described. The approved 7050 rivet necessitated changes to the heat treat procedure, rivet geometry and dimensions, rivet set size, as well as increased upset and shank diameters and tighter hole tolerances. Successful implementation into commercial production use of the 7050 protruding head rivet along with a very substantial decrease in the rejection rate for installed rivets is discussed. Other attributes of the new rivet are described including no "ice box" holding, freedom from clogging automatic machine hoppers and the ability to be restruck at a later date. Approval of a second 7050 rivet (flush head) is discussed along with continuing development towards the long-term goal of a single flush and protruding head 7050 aluminum rivet for all aluminum rivet applications.

"Keywords:" Rivets, Aluminum, 7050, Heat Treat, Fastener Installation.

1. INTRODUCTION

For the past 40 years the principal aluminum rivet for the 3/16 inch and above diameter used at the Lockheed-California Company has been the 2024-T31 or DD rivet. If handled properly, the rivet has excellent driving and installation properties. However, because undriven 2024 rivets must be held at sub-zero temperatures to resist natural aging, many problems exist for the rivet. If the cold cycle is not maintained, the rivets will fail or crack in the upsets when driven. If these chilled rivets are installed by automatic machines, the frost created as the rivets approach room temperature clogs the machine hopper. If a rivet, subsequent to driving, requires an additional "hit" to correct gaps or upset dimensions, the impact causes the rivet or upset to crack. In contrast, rivets made from 7050 aluminum are driven in the final T73 temper and do not require low temperature holding. These rivets are virtually impossible to flaw in the upset both at time of driving or at a later date if a rehit is required. The 7050 rivets also have mechanical and corrosion properties equivalent to 2024 aluminum rivets. The 7050 aluminum alloy was introduced in the 1970s and work started immediately to develop rivets from the alloy. Lockheed participated in the early military sponsored programs and since 1979 has been developing a 7050 rivet to replace the 2024 "ice box" rivet. Working with the basic 7050 chemistry, refinements were made to heat treat procedures, rivet configurations and dimensions, hole size tolerances and rivet set geometry. This effort has produced a flush and protruding head style 7050 aluminum rivet that is approved for use in Lockheed commercial aircraft. Ongoing work is developing 7050 rivets to replace MS style rivets.

2. PROCEDURE

Ten pounds of LS 13971E6 (7050 aluminum) rivets (Figure 1) were purchased. Rivet material was to the standard 7050 chemistry and the rivets were processed to the standard T73 temper. Baseline rivets were LS 13971DD6 (2024-T31 aluminum) reference Figure 1. Both rivets were the 3/16 inch nominal diameter size. Initial screening tests were conducted for hole fill and hole size dimensions, upset dimensions and rivet set geometry and size. Where required, constant amplitude fatigue tests were run. Variations to the standard T73 heat treat procedure were explored to obtain optimum installation and driven rivet properties. Rivets were installed by both hand bucking and machine squeezing.

Three stress corrosion blocks were fabricated from 7075-T6 aluminum extrusion. Alloy 7050-T73 rivets were installed in the three grain directions in each block. The block with dry installed rivets was exposed at the sea coast for one year. The other blocks with wet and dry installed rivets were given 60-day salt water alternate immersion testing. Twelve panels were prepared for galvanic corrosion testing. Specimen material was 7075-T6 bare sheet and 2024-T3 clad sheet with wet and dry installed 7050-T73 and 2024-T31 rivets. Two panels were exposed at the sea coast for five months, examined and then returned for further exposure. The remaining panels received the 60-day salt water alternate immersion test.

Three test specimen geometry styles, (Figures 2, 3, 4), were prepared for fatigue testing. The

100 percent load transfer joints were assembled with "wet" rivets and sealed faying surfaces. Rivet installation was both hand bucking and machine squeezed. Specimens were fabricated from 7075-T6 and 2024-T3 clad (both sides) aluminum. Each of the six built up panels (Figure 4) were measured for flatness prior to testing. In addition, seven process control specimens representing the critical material stackups for each of the built up panels were fabricated. The control specimens were fabricated in exactly the same manner as the panels. Each specimen had five holes, three filled and two open. After measuring the open hole diameter and the installed rivet upset dimensions, one rivet was removed and measured for expansion along with its hole. A second rivet was sectioned and photomacrographed. Fatigue loading for the specimens was primarily flight-by-flight spectrum consisting of flight and ground loads for the specimens shown in Figures 2 and 4 and constant amplitude for the Figure 3 specimens. Transport fuselage longitudinal loadings were selected for the spectrum loadings. The spectrum was modified by increasing the severity of the compressive taxi stresses, providing a flight-by-flight loading stress ratio of -0.35 and thereby a more severe stress range for the small lap joint specimens (Figure 2). The basic unmodified spectrum was used for the built up panels (Figure 4). The basic and modified spectra were 16 and 15 cycles per flight, respectively.

Engineering material and process specifications were prepared or updated for the 7050 rivet including a manufacturing installation standard. A typical production part (fuselage frame assembly) was selected and assembled with both hand bucked and machine driven 7050 rivets. After approval of the first rivet (Figure 5), an initial order of production rivets was received; and they were used to assemble large panels on automatic riveting equipment. In anticipation of production hand bucked installation with the 7050 rivet, modified rivet sets were procured.

A flush head 7050 aluminum rivet was developed concurrently with the protruding head style. The rivet was based on a geometry currently in use at Lockheed with refinements made to head geometry and dimensions. Figure 6 shows the flush rivet developed for 7050. Evaluation testing was the same as for the protruding head rivet except that static tension and shear specimens were made and tested. For the static tests, the rivets were wet installed in material thicknesses from 0.050 to 0.250 inch. Specimen material was 7075-T6 and 2024-T3 clad aluminum.

3. RESULTS AND DISCUSSION

The 7050 rivets heat treated to the standard T73 procedure had shear values closer to the maximums allowed (41 to 46 KSI) and did not exhibit good hole fill. To obtain better installation properties, a Lockheed T73 heat treat procedure was established:

o Solution Treatment
   890± 10°F for 30-60 minutes
o Double Age
   250± 5°F for 8-10 hours
   355± 5°F for 12-14 hours

The change in the procedure was to lengthen the times for each of the double age steps. This improved the ductility of the material and increased the hole fill. However, results from constant amplitude fatigue testing of small single lap specimens (.05 + .05 to .100 inch, see Figure 2) showed the need for additional changes, in this case the rivet set size and geometry, when the rivets in the initial specimens failed prematurely in the middle sheet because of inadequate shank expansion. Necessary shank expansion was obtained by using a rivet set with a concave radius greater than the radius of the rivet head. This resulted in using a one size larger rivet set than is used for the same size and type 2024-T31 rivet. The larger rivet set required modification by removing 0.013 inch from the end of the set so that it would not mark the skin material adjacent to the rivet head. The screening tests also demonstrated the need to tighten up the tolerance of the rivet hole (#10, 0.1935 inch) and increase the diameter of the rivet upset for adequate fatigue life results when compared to the baseline rivet. Figures 7 and 8,

which show the results of constant amplitude fatigue tests conducted on small joint specimens, illustrate the effect of upset diameter on fatigue life for hand bucked and machine squeezed rivets. The final upset diameter interval used in the testing was on the low side of the overall allowable range established for the 7050-T73 rivets and on the high side (just below the diameter which would result in cracking of bucktails) for the 2024-T31 rivets. Assembly parameters were then established for the balance of the program:

o ACV13 Rivet Gun, 3.5 pound bucking bar.
o 7050E6 rivet, -6 modified rivet set.
o 2024DD6 rivet, -5 rivet set.
o Rivets installed wet, faying surfaces sealed.
o Hole size 0.192 to 0.196 inch diameter.
o 7050E6 rivet, .280 to .290 inch diameter upset.
o 2024DD6 rivet, .270 to .280 inch diameter upset.
o Both rivets, upset height, 0.078 to 0.125 inch.

After one year of exposure at the sea coast, the stress corrosion block with dry installed 7050-T73 rivets showed no evidence of cracking. The exposure is continuing. The two other blocks with the wet installed rivets given the 60-day salt water alternate immersion test showed slight pitting on the upsets. This pitting is attributed to removal of anodize when the rivets were trimmed to length before installation. Two galvanic corrosion test panels exposed at the sea coast for five months showed no corrosion and have been returned for an additional six months exposure. The other panels given the 60-day salt water alternate immersion test showed slight corrosion attack on the 7050 rivets and was less evident on the 2024 baseline rivets. It was felt that the difference in corrosion was insignificant provided other rivet property comparisons were equal.

The balance of the fatigue testing was completed with the specimens assembled by the developed assembly parameters. Flatness measurements made before testing for the six built up panels (Figure 4) ranged from 0.010 inches to 0.035 inches out of flatness which was considered excellent for sheet metal fabrication. Product control specimens for the same panels were satisfactory and equivalent between the two rivet materials. Spectrum fatigue test results for both the small lap joint specimens (Figure 2) and the structural panels (Figure 4) are shown respectively in Figures 9 and 10. The data show the two rivet materials were basically equivalent for both hand bucked and machine assembled specimens or panels.

The fuselage frame assembly was trial assembled at a feeder plant. Shop personnel experience no difference in the effort required for hand bucking the 7050 rivet versus the usual "ice box" rivet and all installations including the machine squeezed rivets were satisfactory. The new or changed specifications were issued and the rivet, Figure 5, was released for production use. The new rivet basically followed the form of the 2024 rivet except for tighter dimensional tolerances and a 0.003 inch increase in the nominal diameter.

An initial purchase of 600 pounds of the new rivets in six grip lengths was made and the rivets have been used for machine installation on large structural panels. Through October 1982 approximately 55 panels have been assembled with in excess of 2,000 rivets per panel. Removal or rejection rate for the new rivet is 0.05 percent in contrast to a 9 to 11 percent rate when 2024 "ice box" rivets were used. A small amount of experience has been gained using the modified rivet sets for hand bucking of the new rivets; however, the majority of the rivets have been machine installed.

The development effort for a flush 7050 rivet was concurrent with the protruding head rivet effort and followed the same test plan. For the static tension and shear specimen tests the data showed similar results for the 7050 rivet (Figure 6) compared to the 2024 baseline rivet (LS 10052DD6). Fatigue testing with the various types of fatigue specimens also showed either equivalent properties for the 7050 rivet or an increase in fatigue life. The data are shown in Figures 11 and 12. Figure 11 shows the importance of the larger size upset for the 7050 rivet in order to attain fatigue life equal to or greater than the 2024 baseline rivet. Galvanic corrosion panel specimens with the flush rivets were tested by both 60-day salt water alternate immersion and sea coast exposure. For all specimen conditions, wet and dry rivet installation in 7075-T6 bare and 2024-T3 clad sheet, corrosion susceptibility of test and baseline rivets was equivalent and minimal. In May 1982 the flush 7050 aluminum rivet, Figure 6, was approved for production use.

4. CONCLUSIONS

o Rivets made from 7050-T73 aluminum alloy can be successfully installed by either hand bucking or automatic machine methods.

o The Lockheed developed T73 heat treat procedure reduced the shear strength of the rivet towards the low end of the range allowed, while increasing the malleability of the rivet.

o Changes to both the shape of the rivet protruding head and the rivet sets enhanced hole fill.

o Increasing the shank and upset diameters of the 7050 rivet gives fatigue strength equivalent to the 2024 rivet.

o Installed 7050 and 2024 rivets have similar corrosion resistance.

o Rivets made from 7050-T73 aluminum can be restruck after initial driving.

o Rivets made from 7050-T73 aluminum alloy result in substantial cost savings over 2024-T31 rivets because of a large reduction in the rejection rate for installed rivets of 0.05 percent compared to 9 to 11 percent for the 2024 rivets.

5. ONGOING WORK

A second protruding head rivet is in development to replace the MS 20470DD rivet. Rivets have been obtained from two sources and are undergoing static and fatigue testing. This rivet has an increased head height similar to the MS style but without the large diameter. The rivet can be installed with standard unmodified rivet sets. A second flush rivet is also in development to replace the MS 20426DD rivet. Rivets have been procured for test purposes. The rivet is distinctive in that the included angle of the flush head is 104°/105°. It is planned to install the rivet in the normal 100° countersink; it is hoped the angle difference will result in a tight rivet to countersink interface for the installed rivet.

The long term goal is to have a single flush and protruding head rivet style from 7050 aluminum for all aluminum rivet applications at the Lockheed-California Company.

6. ACKNOWLEDGEMENT

Development of 7050 aluminum alloy rivets at the Lockheed-California Company has been a group effort. The author, therefore, gratefully acknowledges the participation of Leon Bakow and David Richardson, Advanced Structures Department; Andy Incardona and Ken Sparling, Manufacturing Research Department; and Garth Kikendall, Parts and Equipment Department, of the Lockheed-California Company for this effort.

7. BIOGRAPHY

Ronald Wood, Design Specialist, Materials and Processes Department, has 15 years of engineering experience at the Lockheed-California Company. Mr. Wood has the design and methodology responsibility for mechanical fastening and machining and associated cutting methods for all materials. He is responsible for the preparation and maintenance of engineering process specifications and the Design Handbook. He provides technical assistance to design projects and manufacturing. For the past eight years he has been involved in IRAD programs for fastener development. Mr. Wood received a Bachelor of Science degree from the University of California, Los Angeles, in 1967 and is a registered engineer in California.

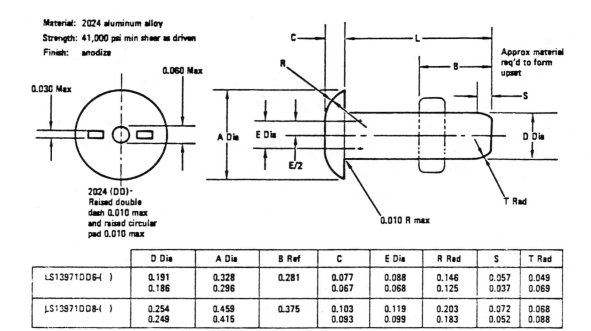

Material: 2024 aluminum alloy
Strength: 41,000 psi min shear as driven
Finish: anodize

0.030 Max

0.060 Max

2024 (DD)-
Raised double
dash 0.010 max
and raised circular
pad 0.010 max

C

R

L

B

Approx material
req'd to form
upset

S

A Dia

E Dia

E/2

D Dia

T Rad

0.010 R max

	D Dia	A Dia	B Ref	C	E Dia	R Rad	S	T Rad
LS13971DD6-( )	0.191 0.186	0.328 0.296	0.281	0.077 0.067	0.088 0.068	0.146 0.125	0.057 0.037	0.049 0.069
LS13971DD8-( )	0.254 0.249	0.459 0.415	0.375	0.103 0.093	0.119 0.099	0.203 0.183	0.072 0.052	0.068 0.088

Figure 1 – LS 13971 Rivet Geometry (DD shown) (E same dimensions)

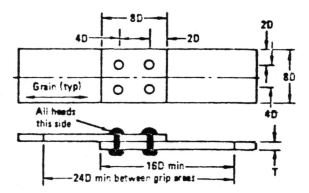

Figure 2 – Lap joint fatigue test specimen geometry

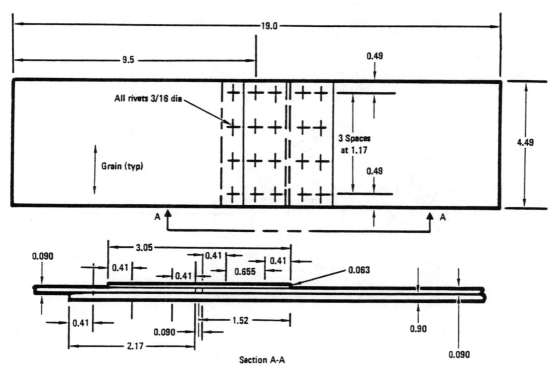

All rivets 3/16 dia

Grain (typ)

Section A-A

Figure 3 - LAP Splice Specimen

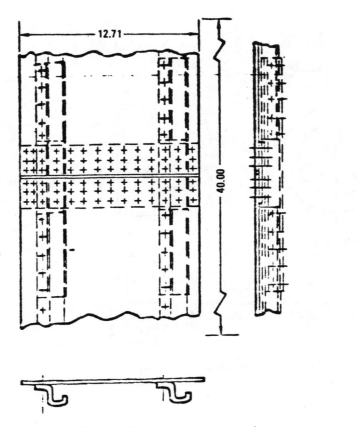

Figure 4 - Built-up fatigue test panel geometry

Source: 28th National SAMPE Symposium, April 12-14, 1983, 1202-1213

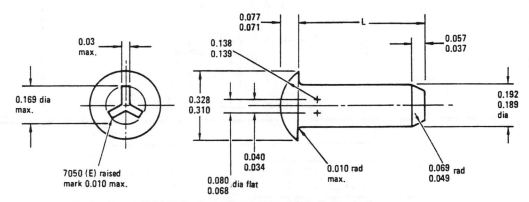

Figure 5 - LS15838 rivet geometry (E shown)

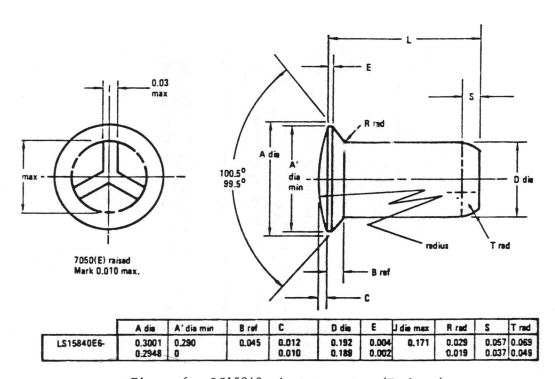

	A dia	A' dia min	B ref	C	D dia	E	J dia max	R rad	S	T rad
LS15840E6-	0.3001 0.2948	0.290 0	0.045	0.012 0.010	0.192 0.189	0.004 0.002	0.171	0.029 0.019	0.057 0.037	0.069 0.049

Figure 6 - LS15840 rivet geometry (E shown)

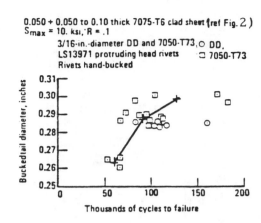

0.050 + 0.050 to 0.10 thick 7075-T6 clad sheet (ref Fig. 2)
S_{max} = 10. ksi, R = .1

3/16-in.-diameter DD and 7050-T73, ○ DD.
LS13971 protruding head rivets   ▭ 7050-T73
Rivets hand-bucked

Figure 7 – S-N fatigue test data for small riveted (handbucked) lap joints

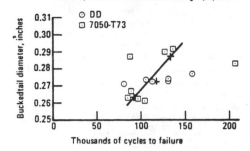

0.050 + 0.050 to 0.10 thick 7075-T6 clad sheet (ref. fig. 2)
S_{max} = 10. ksi, R = 0.1

3/16-in.-diameter DD and 7050-T73 LS 13971
protruding head rivets
Rivets squeezed on automatic riveting equipment

○ DD
▭ 7050-T73

Figure 8 – S-N fatigue test data for small riveted (squeezed) lap joints

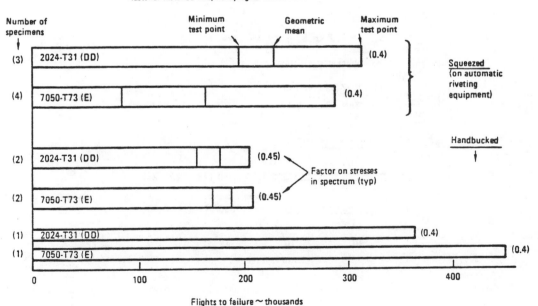

3/16-inch-diameter LS 13971 protruding head rivets
0.050 riveted to 0.050, 7075-T6 clad sheet (ref. Fig. 2)
fastener installed wet, w/faying surface sealant

Figure 9 – Spectrum fatigue test data for small riveted lap joints

Source: 28th National SAMPE Symposium, April 12-14, 1983, 1202-1213

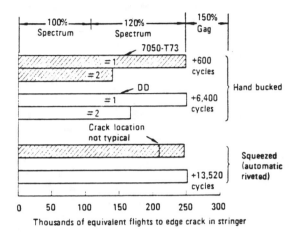

7050-T73 vs DD rivets, 3/16 diameter (LS13971) protruding head
2024-T3 clad skin & 7075-T6 clad stringers (ref. Fig. 4)

Figure 10 – Fatigue test results for
built-up riveted panels

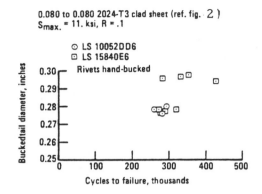

0.080 to 0.080 2024-T3 clad sheet (ref. fig. 2)
$S_{max.}$ = 11. ksi, R = .1

Figure 11 – S-N fatigue test data
for small riveted
(handbucked) lap joints

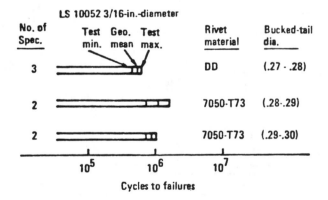

$S_{max.}$ = 11.8 ksi, $S_{min.}$ = 0.1 ksi, spec. unsupported

Hand-bucked rivets (specimen per Fig. 3)

Figure 12 – S-N fatigue test data for small riveted (squeezed) lap
joints

Blind rivets features and applications

M Zielonka, Technical and Production Director,
Ornit Blind Rivets, Kibutz or Haner, Israel

For the last seven years, Ornit has been Israel's sole manufacturer in the design and production of blind rivets for industrial and commercial purposes.

The comprehensive range of rivets in sizes, styles and alloy combinations of steel-steel, aluminium-steel and sealed rivets, is produced by the best cold-forging techniques.

High quality raw materials and strict quality control guarantee a consistently strong clenching action, high shear strength, vibration resistance, positive tightly retained and sealed rivet stem, factors that make a blind rivet so useful.

Mandrel production line

What is it?

Blind fastener. A blind fastener is a mechanical device which has the capability to join component parts in an assembly where access for fastener installation and activation is available from one side only.

Blind rivet. A blind rivet is a blind fastener which has a self-contained mechanical, chemical or other feature which permits the formation of an upset on the blind end of the rivet and expansion of the rivet shank during rivet setting to join the component parts of an assembly.

Blind rivet production line

Classification of blind rivets

Pull mandrel. A pull mandrel blind rivet is a multiple piece assembly consisting of at least a rivet body and a mandrel. In the setting operation the rivet is inserted into the components to be joined, the mandrel is gripped, pulled axially, and its head upsets the rivet body forming a blind head. Pull mandrel blind rivets are further classified into pull through mandrel, break mandrel, and non-break mandrel types.

Break mandrel. A break mandrel rivet is a pull mandrel type of blind rivet where during the setting operation the mandrel is pulled into or against the rivet body and breaks.

Non-break mandrel. A non-break mandrel is a pull mandrel type of blind rivet, where during the setting operation the mandrel is pulled into or against the rivet body, but does not break. This type requires the mandrel to be dressed in a subsequent operation.

Body styles

Closed end. The end of the rivet, as manufactured, is solid and remains closed on the blind side after setting.

Open end. The end of the rivet, as manufactured, is open, ie, the rivet body is hollow through its full length.

Core styles

Filled. After setting, the rivet body still retains enough of the mandrel or pin so that the break point of the mandrel, or the end of the pin, is essentially flush with the top of the rivet head.

Semi-filled. During rivet setting the mandrel breaks within the rivet body so that a short length of the mandrel is retained within the rivet body.

Hollow. After rivet setting the core of the rivet body is completely empty.

Characteristics and features of blind rivets

Rivet body. The rivet body is that part of a blind rivet assembly which incorporates the manufactured head, the shank, and the end. The cross section of the rivet body is usually round, and its outside diameter establishes the rivet size.

Head. The head is the manufactured upset portion of the rivet body. After setting the head is always located on the access side of the joint. There are many available head styles, eg, round, truss, countersunk, brazier, domed, etc. Flush heads are those styles where, after setting, the top of the rivet head is essentially flush with the surface of the assembly. Protruding heads are those where, after setting, the rivet head projects beyond the surface of the assembly.

Shank. The shank is that part of the rivet body, extending from the underside of the head to the extreme end.

End. The end is that part of the rivet body located at the extremity of the shank opposite the head, and may be closed, open or split.

Mandrel. The mandrel is that portion of a pull mandrel or threaded type blind rivet that is preassembled in the rivet body. During rivet setting the pulling or torquing of the mandrel is the action which forms the blind head. Mandrels usually are upset and may be smooth, serrated, or threaded.

Rivet length. Rivet length is the distance, measured parallel to the axis of the rivet, from the largest diameter of the bearing surface of the head to the extreme end of the rivet.

Blind side protrusion. Blind side protrusion is the distance, measured parallel to the axis of the rivet, from the largest diameter of the bearing surface of the head to the top of the upset on the mandrel.

Electroplating line for mandrels

Mechanical and performance properties of blind rivets

Ultimate tensile strength. The ultimate tensile strength of a blind rivet is the maximum tensile load in pounds which the rivet is capable of sustaining prior to failure.

Pull-together. Pull-together is the ability of the rivet to close a gap between the components to be joined while working against the rivet setting load.

Clamping force. Clamping force is the compressive load applied to the joint by a blind rivet after setting.

Sealing. Sealing is the ability of the rivet to prevent the escape of gas, liquid, or solid after rivet setting.

Blind rivet assembly line

Rivet setting

Setting. Setting is the operation of placing and activating the rivet in the members to be joined.

Hole fill. Hole fill is the lateral expansion of the rivet shank during rivet setting. Hole fill can be controlled or free.

Dress. Dress is the secondary operation of clipping the end of a mandrel projecting beyond the head of the rivet on the access side following rivet setting. If a flush surface is required a further operation may be necessary.

The blind rivet unit is supplied assembled. It fastens metal to metal, wood to metal, plastic to metal, etc. Blind rivets can be used with just about any combination of materials since threads and tapping are eliminated.

Blind rivets cut costs just where fastening expenses accumulate, during installation. Simply drill, insert and set. It is riveted all from one side, by one operator, using a lightweight tool, manually or power operated.

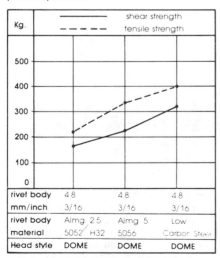
Fig 1 Strengths of different material combinations of materials

Four basic factors are to be considered in selecting the right rivet for any particular application.

Strength

Determine the shear and tensile strengths required for the application; then select a rivet that meets those requirements according to the manufacturer's technical information. Rivet strengths are a combination of rivet style, alloy and diameter. Thus the stronger the rivet the fewer (or smaller the sizes) are required.

The graph Figure 1 compares strengths available in different material combinations of non-structural Ornit blind rivets.

Rivet alloy

Rivet alloy should be compatible with the materials to be joined to avoid galvanic corrosion. Other considerations in selecting the proper rivet alloy may be temperature requirements. Table 1 gives some guidelines for rivet alloy selection.

Table 1
Guidelines for rivet alloy selection

GRADE DESIGNATION	RIVET BODY MATERIAL ★ melting temperature	MANDREL MATERIAL
18	Aluminium alloy 5052 ★ 607 - 649 C	Carbon Steel
19	Aluminium alloy 5056 ★ 571 - 638 C	Carbon Steel
16	Aluminium alloy 5154 ★ 593 - 643 C	Carbon Steel
30	Low Carbon Steel ★ 1518 C	Carbon Steel
40	Nickel-Copper alloy (Monel) ★ 1299 - 1349 C	Carbon Steel

Rivet head style

There are three different main head styles, Figure 2. The regular dome head is used for most industrial applications. A large head is used for soft or brittle top sheets requiring a wide bearing surface. The countersunk head is used to obtain a completely flush surface on the finished assembly.

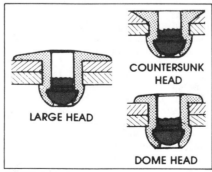
Fig 2 Rivet head styles

Rivet size

Most Ornit rivets are available in five diameters shown in Figure 3. Total thickness of the material to be joined is

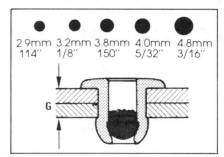
Fig 3 Rivet sizes available

the rivet's grip length (G). Each rivet's length (ie the shank length underhead) is suitable to a certain range of the rivet's grip length commonly called the rivet's grip range.

Blind rivet users should take into consideration this important point and choose the correct length which fulfils the requirements of the total material thickness to be joined.

Aluminium-steel rivets are manufactured in five diameters and cover a wide range of lengths, available in three different head styles and others upon special request. Likewise, they are produced from different aluminium alloys in order to meet customer's technical demands.

Steel-steel rivets are available in three diameters and two different head styles — covering most popular lengths.

Lost mandrel rivets, also known as Break Head mandrel rivets (hollow core), are available in three different diameters. This type of rivet is commonly used where the finished product's weight is of important consideration, also where a temporary assembly and a posterior easy rivet removal is required, etc.

Ornit, Unifix rivets are designed to provide maximum clamping action over a full range of material thickness, using the same length rivet. Since one length can cover several assembly requirements, production costs can be cut and inventories simplified.

Sealnit, Ornit sealed type rivets are the solution where a completely sealed and pressure-tight joint is necessary. The Sealnit rivets consist of a tubular rivet with a sealed end containing a steel mandrel. Its riveting sequence is similar to that of the standard open rivet, but the sealed type has an additional setting, the rivet is stressed beyond its elastic limit. This ensures that in addition to the normal clenching action, the rivet expands radially in setting. This, combined with the sealed end of the rivet, ensures a joint which is both air and water-tight.

Ornit's research division are aware of their responsibility as manufacturers, therefore, they are constantly looking for product's improvements, as well as for new products.

Surgical staples:
unique uses for unique fasteners

*First used in the early part of this century,
refined in the 1950s and perfected in the 1970s,
surgical staples are changing operating room procedure.*

One of the more unique uses of wire products is taking place in the operating rooms of hospitals throughout the world where fasteners, in the form of staples, are being used not only to close surgical incisions, but to rejoin internal organs.

This isn't the first time that products of the wire industry have found their way into operating rooms. A variety of metal threads—including silver, copper and brass—has been available for more than 100 years. Today, silver wire and various types of stainless steel wire are almost the only metal suture material being employed.

Surgical stapling is a concept that permits the performance of many operations safely and accurately with results that are at least equal to those obtained by manual suturing techniques.

A typical stainless steel surgical staple measures 0.157 in. by 0.217 in. and is produced on machines specially designed and built for this purpose. Sterilization, with cobalt, takes place after the staples and injection instrument are manufactured, packaged and placed in shipping crates. Because of the proprietary nature of surgical staple manufacturing, companies are unwilling to give more information about production, leaving those outside the industry to speculate on the advanced nature and capabilities of the machinery.

"We'd all like to know what's happening on the production line of the competition," says one manufacturer. "Just seeing one picture would help."

History and development

While manufacturers are not talking too much about the present, the history of surgical staples is clearly defined. And although the technique is just now gaining wide acceptance for specific surgical procedures, it is interesting to note that the concept was first envisioned more than 150 years ago.

The initial use of an instrument to implant surgical staples took place in Budapest, Hungary, in May 1908 where a physician by the name of Humer Hultl used a device designed by his brother and produced by a manufacturer of surgical instruments.

From the beginning, the new device employed a basic fabricating principle. The instrument had two long jaws, one containing the slots for the insertion of four rows of round, steel wire staples and the other containing the curved depressions, or anvils, for forming the staples. The staples were fabricated in a "B" shape, which is basic to modern stapling instruments.

A major step in the development of surgical stapling came at the Scientific Research Institute for Experimental Surgical Apparatus and Instruments in Moscow in the 1950s where scientists began a systematic program to develop stapling instruments for every conceivable surgical application. The instruments that were developed employed very fine wire staples which were at first made of tantalum and later of stainless steel.

American designers and inventors, once they got hold of the Russian instruments, began making modifications that resulted in practical, easy-to-use instruments with disposable anvils and cartridges.

Novel applications

Today, surgical staples are used in gynecology, urology and specialized surgical procedures including closure of pulmonary arteries and veins. The techniques described in medical literature, and frequently used by surgeons, are not a true indication of the potential combinations possible with staple systems. In fact, surgeons are reporting novel applications not originally envisioned. Some believe that the concept awaits only the ingenuity of the surgeon to develop new applications and technical uses.

The most practical use is skin closing. In this procedure, which is now commonplace, a disposable device containing 15, 35 or 55 staples is used to implant square-shaped stainless steel staples.

Surgeons are viewing these instruments and the techniques from more than one angle. Aside from the medical aspect, there is a financial savings to the surgeon because less operating room time is needed to complete a procedure. Since this is usually an hourly charge, stapling is viewed as a means to significantly reduce this cost.

The wire industry, particularly fastener makers, have made many contributions in diverse areas. Apparently, they haven't stopped making contributions. ∎

SECTION X
Tools

TOOL MATERIALS

Reviewed and revised by Neil J. Culp, Carpenter Technology Corp.; Dennis D. Huffman, Timken Research; and R. J. Henry, University of Pittsburgh–Johnstown

This section was condensed from Metals Handbook, Ninth Edition, Volume 3, Properties and Selection: Stainless Steels, Tool Materials and Special-Purpose Metals, pages 419 to 559. For more detailed information on the topics covered in this section, the reader is referred to the larger work.

_____ Introduction and Overview _____

_____ Tool Steels _____

A TOOL STEEL is any steel used to make tools for cutting, forming or otherwise shaping a material into a part or component adapted to a definite use. The earliest tool steels were simple, plain carbon steels, but beginning in 1868, and to a greater extent early in the 20th century, many complex, highly alloyed tool steels were developed. Although plain carbon tool steels were first used and still are employed occasionally, it is the alloy tool steels containing, among other elements, relatively large amounts of tungsten, molybdenum, manganese, vanadium and chromium which have made it possible to meet increasingly severe service demands and to provide greater dimensional control and freedom from cracking during heat treatment. Many alloy tool steels are also widely used for machinery components and structural applications where particularly severe requirements must be met.

In service, most tools are subjected to extremely high loads that are applied rapidly. They must withstand these loads a great number of times without breaking or undergoing excessive wear or deformation. In many applications, tool steels must provide this capability under conditions that develop high temperatures in the tool. No single tool material combines maximum levels of wear resistance, toughness, and resistance to softening at elevated temperatures. Consequently, selection of the proper tool material for a given application often requires a trade-off to achieve the optimum combination of properties.

Most tool steels are wrought products, but precision castings can be used to advantage in some applications. The powder metallurgy (P/M) process also is used in making tool steels, in both mill forms and near-net shapes. P/M tool steels may provide (*a*) more uniform carbide size and distribution in large sections and (*b*) special compositions that are difficult or impossible to produce by melting and casting and then mechanically working the cast product.

Tool steels are generally melted in small-tonnage electric-arc furnaces to economically achieve composition tolerances, good cleanness and precise control of melting conditions. Special refining and secondary remelting processes have been introduced to satisfy particularly difficult demands regarding tool steel quality and performance. Tool steels must have minimal decarburization held within carefully controlled limits. This requires that annealing be done by special procedures under closely controlled conditions.

The performance of a tool in service depends on proper design of the tool, accuracy with which the tool is made, selection of the proper tool steel and application of the proper heat treatment. A tool can perform successfully in service only when all four of these requirements have been fulfilled.

With few exceptions, all tool steels must be heat treated to develop specific combinations of wear resistance, resistance to deformation or breaking under high loads, and resistance to softening at elevated temperatures.

CLASSIFICATION AND CHARACTERISTICS

Table 1 gives composition limits for the tool steels most commonly used today. Each group of tool steels of similar composition, application or mode of quenching is identified by a capital letter; within each group, individual tool steel types are assigned code numbers.

High Speed Steels

High speed steels are tool materials developed largely for use in high speed cutting-tool applications. There are two classifications of high speed steels: molybdenum high speed steels (group M) and tungsten high speed steels (group T). Group M steels constitute about 95% of all high speed steel produced in the United States.

Group M and group T high speed steels are equivalent in performance; the main advantage of group M steels is lower initial cost (approximately 40% lower than that of similar group T steels).

Molybdenum high speed steels and tungsten high speed steels are similar in many other respects, including hardenability. Typical applications for group M and group T steels include cutting tools of all kinds. Some grades are satisfactory for cold work applications, such as cold-header die inserts, thread-rolling dies, punches and blanking dies. Steels of the M40 series are used to make cutting tools for machining modern, very tough, high-strength steels.

For die inserts and punches, high speed steels sometimes are underhardened — that is, quenched from austenitizing temperatures lower than those recommended for cutting-tool applications — as a means of increasing toughness.

Molybdenum high speed steels contain molybdenum, tungsten, chromium, vanadium, cobalt and carbon as principal alloying elements. Group M steels have slightly greater toughness than group T steels at the same hardness. Otherwise, mechanical properties of the two groups are similar.

Increasing the carbon and vanadium contents of group M steels increases wear resistance; increasing the cobalt content improves red hard-

Table 1. Composition limits of principal types of tool steels

	Designations						Composition(a), %				
AISI	SAE	UNS	C	Mn	Si	Cr	Ni	Mo	W	V	Co
Molybdenum high speed steels											
M1	M1	T11301	0.78-0.88	0.15-0.40	0.20-0.50	3.50-4.00	0.30 max	8.20-9.20	1.40-2.10	1.00-1.35	...
M2	M2	T11302	0.78-0.88; 0.95-1.05	0.15-0.40	0.20-0.45	3.75-4.50	0.30 max	4.50-5.50	5.50-6.75	1.75-2.20	...
M3, class 1	M3	T11313	1.00-1.10	0.15-0.40	0.20-0.45	3.75-4.50	0.30 max	4.75-6.50	5.00-6.75	2.25-2.75	...
M3, class 2	M3	T11323	1.15-1.25	0.15-0.40	0.20-0.45	3.75-4.50	0.30 max	4.75-6.50	5.00-6.75	2.75-3.75	...
M4	M4	T11304	1.25-1.40	0.15-0.40	0.20-0.45	3.75-4.75	0.30 max	4.25-5.50	5.25-6.50	3.75-4.50	...
M6	...	T11306	0.75-0.85	0.15-0.40	0.20-0.45	3.75-4.50	0.30 max	4.50-5.50	3.75-4.75	1.30-1.70	11.00-13.00
M7	...	T11307	0.97-1.05	0.15-0.40	0.20-0.55	3.50-4.00	0.30 max	8.20-9.20	1.40-2.10	1.75-2.25	...
M10	...	T11310	0.84-0.94; 0.95-1.05	0.10-0.40	0.20-0.45	3.75-4.50	0.30 max	7.75-8.50	...	1.80-2.20	...
M30	...	T11330	0.75-0.85	0.15-0.40	0.20-0.45	3.50-4.25	0.30 max	7.75-9.00	1.30-2.30	1.00-1.40	4.50-5.50
M33	...	T11333	0.85-0.92	0.15-0.40	0.15-0.50	3.50-4.00	0.30 max	9.00-10.00	1.30-2.10	1.00-1.35	7.75-8.75
M34	...	T11334	0.85-0.92	0.15-0.40	0.20-0.45	3.50-4.00	0.30 max	7.75-9.20	1.40-2.10	1.90-2.30	7.75-8.75
M36	...	T11336	0.80-0.90	0.15-0.40	0.20-0.45	3.75-4.50	0.30 max	4.50-5.50	5.50-6.50	1.75-2.25	7.75-8.75
M41	...	T11341	1.05-1.15	0.20-0.60	0.15-0.50	3.75-4.50	0.30 max	3.25-4.25	6.25-7.00	1.75-2.25	4.75-5.75
M42	...	T11342	1.05-1.15	0.15-0.40	0.15-0.65	3.50-4.25	0.30 max	9.00-10.00	1.15-1.85	0.95-1.35	7.75-8.75
M43	...	T11343	1.15-1.25	0.20-0.40	0.15-0.65	3.50-4.25	0.30 max	7.50-8.50	2.25-3.00	1.50-1.75	7.75-8.75
M44	...	T11344	1.10-1.20	0.20-0.40	0.30-0.55	4.00-4.75	0.30 max	6.00-7.00	5.00-5.75	1.85-2.20	11.00-12.25
M46	...	T11346	1.22-1.30	0.20-0.40	0.40-0.65	3.70-4.20	0.30 max	8.00-8.50	1.90-2.20	3.00-3.30	7.80-8.80
M47	...	T11347	1.05-1.15	0.15-0.40	0.20-0.45	3.50-4.00	0.30 max	9.25-10.00	1.30-1.80	1.15-1.35	4.75-5.25
Tungsten high speed steels											
T1	T1	T12001	0.65-0.80	0.10-0.40	0.20-0.40	3.75-4.00	0.30 max	...	17.25-18.75	0.90-1.30	...
T2	T2	T12002	0.80-0.90	0.20-0.40	0.20-0.40	3.75-4.50	0.30 max	1.00 max	17.50-19.00	1.80-2.40	...
T4	T4	T12004	0.70-0.80	0.10-0.40	0.20-0.40	3.75-4.50	0.30 max	0.40-1.00	17.50-19.00	0.80-1.20	4.25-5.75
T5	T5	T12005	0.75-0.85	0.20-0.40	0.20-0.40	3.75-5.00	0.30 max	0.50-1.25	17.50-19.00	1.80-2.40	7.00-9.50
T6	...	T12006	0.75-0.85	0.20-0.40	0.20-0.40	4.00-4.75	0.30 max	0.40-1.00	18.50-21.00	1.50-2.10	11.00-13.00
T8	T8	T12008	0.75-0.85	0.20-0.40	0.20-0.40	3.75-4.50	0.30 max	0.40-1.00	13.25-14.75	1.80-2.40	4.25-5.75
T15	...	T12015	1.50-1.60	0.15-0.40	0.15-0.40	3.75-5.00	0.30 max	1.00 max	11.75-13.00	4.50-5.25	4.75-5.25
Chromium hot work steels											
H10	...	T20810	0.35-0.45	0.25-0.70	0.80-1.20	3.00-3.75	0.30 max	2.00-3.00	...	0.25-0.75	...
H11	H11	T20811	0.33-0.43	0.20-0.50	0.80-1.20	4.75-5.50	0.30 max	1.10-1.60	...	0.30-0.60	...
H12	H12	T20812	0.30-0.40	0.20-0.50	0.80-1.20	4.75-5.50	0.30 max	1.25-1.75	1.00-1.70	0.50 max	...
H13	H13	T20813	0.32-0.45	0.20-0.50	0.80-1.20	4.75-5.50	0.30 max	1.10-1.75	...	0.80-1.20	...
H14	...	T20814	0.35-0.45	0.20-0.50	0.80-1.20	4.75-5.50	0.30 max	...	4.00-5.25	...	...
H19	...	T20819	0.32-0.45	0.20-0.50	0.20-0.50	4.00-4.75	0.30 max	0.30-0.55	3.75-4.50	1.75-2.20	4.00-4.50
Tungsten hot work steels											
H21	H21	T20821	0.26-0.36	0.15-0.40	0.15-0.50	3.00-3.75	0.30 max	...	8.50-10.00	0.30-0.60	...
H22	...	T20822	0.30-0.40	0.15-0.40	0.15-0.40	1.75-3.75	0.30 max	...	10.00-11.75	0.25-0.50	...
H23	...	T20823	0.25-0.35	0.15-0.40	0.15-0.60	11.00-12.75	0.30 max	...	11.00-12.75	0.75-1.25	...
H24	...	T20824	0.42-0.53	0.15-0.40	0.15-0.40	2.50-3.50	0.30 max	...	14.00-16.00	0.40-0.60	...
H25	...	T20825	0.22-0.32	0.15-0.40	0.15-0.40	3.75-4.50	0.30 max	...	14.00-16.00	0.40-0.60	...
H26	...	T20826	0.45-0.55(b)	0.15-0.40	0.15-0.40	3.75-4.50	0.30 max	...	17.25-19.00	0.75-1.25	...
Molybdenum hot work steel											
H42	...	T20842	0.55-0.70(b)	0.15-0.40	...	3.75-4.50	0.30 max	4.50-5.50	5.50-6.75	1.75-2.20	...

(continued)

ness but concurrently lowers toughness. High speed steels have unusually high resistance to softening at elevated temperatures (see Fig. 1) when compared with other tool steels; this is a result of their very high alloy contents.

Because group M steels readily decarburize and are easily damaged due to overheating under adverse austenitizing environments, they are more sensitive than group T steels to hardening conditions—particularly austenitizing temperature and atmosphere. This is especially true of the high-molybdenum, low-tungsten compositions.

Group M high speed steels are deep hardening. They must be austenitized at temperatures lower than those used for hardening group T steels, to avoid incipient melting. Group M high speed steels can develop full hardness when quenched from temperatures of 1175 to 1230 °C (2150 to 2250 °F). Type M10, which usually has slightly lower hardenability than other molybdenum high speed steels, must be oil quenched if section size is larger than about 40 to 50 mm (about 1 1/2 to 2 in.).

The maximum hardness that can be obtained in group M tool steels varies with composition and section size. For those with lower carbon

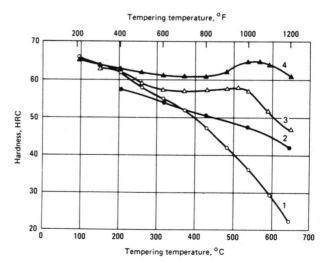

Curves are for 1 h at temperature. Curve 1 illustrates low resistance to softening at elevated temperature, such as is exhibited by group W and group O tool steels. Curve 2 illustrates medium resistance to softening, such as is exhibited by type S1 tool steel. Curves 3 and 4 illustrate high and very high resistance to softening, respectively, such as are exhibited by the secondary-hardening tool steels A2 and M2.

Fig. 1. Variation of hardness with tempering temperature for four typical tool steels

Table 1 (continued)

AISI	SAE	UNS	C	Mn	Si	Cr	Ni	Mo	W	V	Co
	Designations						Composition(a), %				
Air-hardening medium-alloy cold work steels											
A2	A2	T30102	0.95-1.05	1.00 max	0.50 max	4.75-5.50	0.30 max	0.90-1.40	...	0.15-0.50	...
A3	...	T30103	1.20-1.30	0.40-0.60	0.50 max	4.75-5.50	0.30 max	0.90-1.40	...	0.80-1.40	...
A4	...	T30104	0.95-1.05	1.80-2.20	0.50 max	0.90-2.20	0.30 max	0.90-1.40	...	...	...
A6	...	T30106	0.65-0.75	1.80-2.50	0.50 max	0.90-1.20	0.30 max	0.90-1.40	...	...	...
A7	...	T30107	2.00-2.85	0.80 max	0.50 max	5.00-5.75	0.30 max	0.90-1.40	0.50-1.50	3.90-5.15	...
A8	...	T30108	0.50-0.60	0.50 max	0.75-1.10	4.75-5.50	0.30 max	1.15-1.65	1.00-1.50	...	...
A9	...	T30109	0.45-0.55	0.50 max	0.95-1.15	4.75-5.50	1.25-1.75	1.30-1.80	...	0.80-1.40	...
A10	...	T30110	1.25-1.50(c)	1.60-2.10	1.00-1.50	...	1.55-2.05	1.25-1.75	...	...	...
High-carbon, high-chromium cold work steels											
D2	D2	T30402	1.40-1.60	0.60 max	0.60 max	11.00-13.00	0.30 max	0.70-1.20	...	1.10 max	1.00 max
D3	D3	T30403	2.00-2.35	0.60 max	0.60 max	11.00-13.50	0.30 max	...	1.00 max	1.00 max	...
D4	...	T30404	2.05-2.40	0.60 max	0.60 max	11.00-13.00	0.30 max	0.70-1.20	...	1.00 max	...
D5	D5	T30405	1.40-1.60	0.60 max	0.60 max	11.00-13.00	0.30 max	0.70-1.20	...	1.00 max	2.50-3.50
D7	D7	T30407	2.15-2.50	0.60 max	0.60 max	11.50-13.50	0.30 max	0.70-1.20	...	3.80-4.40	...
Oil-hardening cold work steels											
O1	O1	T31501	0.85-1.00	1.00-1.40	0.50 max	0.40-0.60	0.30 max	...	0.40-0.60	0.30 max	...
O2	O2	T31502	0.85-0.95	1.40-1.80	0.50 max	0.35 max	0.30 max	0.30 max	...	0.30 max	...
O6	O6	T31506	1.25-1.55(c)	0.30-1.10	0.55-1.50	0.30 max	0.30 max	0.20-0.30	...	...	...
O7	...	T31507	1.10-1.30	1.00 max	0.60 max	0.35-0.85	0.30 max	0.30 max	1.00-2.00	0.40 max	...
Shock-resisting steels											
S1	S1	T41901	0.40-0.55	0.10-0.40	0.15-1.20	1.00-1.80	0.30 max	0.50 max	1.50-3.00	0.15-0.30	...
S2	S2	T41902	0.40-0.55	0.30-0.50	0.90-1.20	...	0.30 max	0.30-0.60	...	0.50 max	...
S5	S5	T41905	0.50-0.65	0.60-1.00	1.75-2.25	0.35 max	...	0.20-1.35	...	0.35 max	...
S6	...	T41906	0.40-0.50	1.20-1.50	2.00-2.50	1.20-1.50	...	0.30-0.50	...	0.20-0.40	...
S7	...	T41907	0.45-0.55	0.20-0.80	0.20-1.00	3.00-3.50	...	1.30-1.80	...	0.20-0.30(d)	...
Low-alloy special-purpose tool steels											
L2	...	T61202	0.45-1.00(b)	0.10-0.90	0.50 max	0.70-1.20	...	0.25 max	...	0.10-0.30	...
L6	L6	T61206	0.65-0.75	0.25-0.80	0.50 max	0.60-1.20	1.25-2.00	0.50 max	...	0.20-0.30(d)	...
Low-carbon mold steels											
P2	...	T51602	0.10 max	0.10-0.40	0.10-0.40	0.75-1.25	0.10-0.50	0.15-0.40	...	...	...
P3	...	T51603	0.10 max	0.20-0.60	0.40 max	0.40-0.75	1.00-1.50	...	...	...	...
P4	...	T51604	0.12 max	0.20-0.60	0.10-0.40	4.00-5.25	...	0.40-1.00	...	...	...
P5	...	T51605	0.10 max	0.20-0.60	0.40 max	2.00-2.50	0.35 max	...	...	...	...
P6	...	T51606	0.05-0.15	0.35-0.70	0.10-0.40	1.25-1.75	3.25-3.75	...	...	...	...
P20	...	T51620	0.28-0.40	0.60-1.00	0.20-0.80	1.40-2.00	...	0.30-0.55	...	...	...
P21	...	T51621	0.18-0.22	0.20-0.40	0.20-0.40	0.20-0.30	3.90-4.25	...	...	0.15-0.25	1.05-1.25A1
Water-hardening tool steels											
W1	W108,W109, W110,W112	T72301	0.70-1.50(e)	0.10-0.40	0.10-0.40	0.15 max	0.20 max	0.10 max	0.15 max	0.10 max	...
W2	W209,210	T72302	0.85-1.50(e)	0.10-0.40	0.10-0.40	0.15 max	0.20 max	0.10 max	0.15 max	0.15-0.35	...
W5	...	T72305	1.05-1.15	0.10-0.40	0.10-0.40	0.40-0.60	0.20 max	0.10 max	0.15 max	0.10 max	...

(a) All steels except group W contain 0.25 max Cu, 0.03 max P and 0.03 max S; group W steels contain 0.20 max Cu, 0.025 max P and 0.025 max S. Where specified, sulfur may be increased to 0.06 to 0.15% to improve machinability of group H, M and T steels. (b) Available in several carbon ranges. (c) Contains free graphite in the microstructure. (d) Optional. (e) Specified carbon ranges are designated by suffix numbers.

contents—types M1, M2, M10 (low-carbon composition), M30, M33, M34 and M36—maximum hardness in typical tool cross sections is usually 65 HRC. For higher carbon contents—including types M3, M4 and M7—maximum hardness is about 66 HRC. A hardness of 66 HRC also can be developed in the lower-carbon, high-cobalt type M6. Maximum hardness of the higher-carbon cobalt-containing steels—types M41, M42, M43, M44 and M46—is 69 to 70 HRC.

Tungsten high speed steels contain tungsten, chromium, vanadium, cobalt and carbon as the principal alloying elements. Type T1 was developed partly as a result of the work of Taylor and White, who in the early 1900's found that certain steels with over 14% W, about 4% Cr and about 0.3% V resisted softening at temperatures high enough to cause the steel to emit radiation in the red part of the visible spectra—or in other words, they exhibited red hardness. In its earliest form, type T1 contained about 0.68 C, 18 W, 4 Cr and 0.3 V. By 1920, the vanadium content had been increased to about 1.0%. Carbon content was gradually increased over a 30-year period to its present level of 0.75%.

Group T tool steels are used primarily for cutting tools such as bits, drills, reamers, taps, broaches, milling cutters, and hobs. These steels are also used for making dies, punches, and high-load, high-temperature structural components such as aircraft bearings and pump parts. Type T15 is the most wear-resistant steel of this group.

Group T tool steels are characterized by high red hardness and wear resistance. They are so deep hardening that sections up to 75 mm (3 in.) in thickness or diameter can be hardened to 65 HRC or more by quenching in oil or molten salt.

Group T tool steels are all deep hardening when quenched from their recommended hardening temperatures of 1200 to 1300 °C (2200 to 2375 °F). They are seldom used to make hardened tools with section sizes greater than 75 mm (3 in.). Even very large cutting tools, such as drills 75 and 100 mm (3 and 4 in.) in diameter, have relatively small effective sections for hardening because metal has been removed to form the flutes. Some large-diameter solid tools are made from group T high speed steels; these include broaches and cold extrusion punches as large as 100 to 125 mm (4 to 5 in.) in diameter. For such tools, surface hardness is of primary importance.

The difference between surface hardness and center hardness varies with bar size. Section size and total mass of a given tool often have an effect on its response to a given hardening treatment that is equal to or greater than the effect of the grade of tool steel selected. For tools of extremely large diameter or heavy section, it is relatively common practice to use an accelerated oil quench to provide full hardness. This practice may yield values of Rockwell C hardness only one or two points higher than those obtainable through hot-salt quenching or air cooling, which ordinarily produce full hardness in tools smaller than about 75 mm (3 in.), but at such high hardnesses a one- or two-point increase in Rockwell hardness may prove quite significant.

Maximum hardness of tungsten high speed steels varies with carbon content, and to a lesser degree with alloy content. A hardness of at least 64.5 HRC can be developed in any high speed steel. Those types that have high carbon contents and hard carbides, such as T15, may be hardened to 67 HRC.

Hot Work Steels

Many manufacturing operations involve punching, shearing or forming of metals at high temperatures. Hot work steels (group H) have been developed to withstand the combinations of heat, pressure and abrasion associated with such operations.

Generally, group H tool steels have medium carbon contents (0.35 to 0.45%), and chromium, tungsten, molybdenum and vanadium contents totaling 6 to 25%. They are divided into three subgroups: chromium hot work steels (types H10 to H19), tungsten hot work steels (types H21 to H26) and molybdenum hot work steel (type H42).

Chromium hot work steels (types H10 to H19) have good resistance to heat softening because of their medium chromium contents and additions of carbide-forming elements such as molybdenum, tungsten and vanadium. The low carbon and low total alloy contents promote toughness at the normal working hardnesses of 40 to 55 HRC. Higher tungsten and molybdenum contents increase hot strength but slightly reduce toughness. Vanadium is added to increase resistance to washing (erosive wear) at high temperatures. An increase in silicon content improves oxidation resistance at temperatures up to 800 °C (1475 °F). The most widely used types in this group are H13, H12, H11 and, to a lesser extent, H19.

All of the chromium hot work steels are deep hardening. H11, H12 and H13 may be air hardened to full working hardness in section sizes up to 150 mm (6 in.); other group H steels may be air hardened in section sizes up to 300 mm (12 in.). The air-hardening qualities and balanced alloy contents of these steels result in low distortion during hardening. Chromium hot work steels are especially well adapted to hot die work of all kinds, particularly dies for extrusion of aluminum and magnesium, as well as die-casting dies, forging dies, mandrels and hot shears. Most of these steels have alloy and carbon contents low enough that tools made from them can be water cooled in service without cracking.

H11 tool steel is used to make certain highly stressed structural parts, particularly in aerospace technology. Material for such demanding applications is produced by vacuum-arc remelting of air-melted electrodes. Vacuum-arc remelting provides extremely low residual gas contents, excellent microcleanness and a high degree of structural homogeneity.

The chief advantage of H11 over conventional high-strength steels is its ability to resist softening during continued exposure to temperatures up to 540 °C (1000 °F) and at the same time provide moderate toughness and ductility at room-temperature tensile strengths of 1720 to 2070 MPa (250 to 300 ksi). In addition, because of its secondary hardening characteristic, H11 can be tempered at high temperatures, resulting in nearly complete relief of residual hardening stresses, which is necessary for maximum toughness at high strength levels. Other important advantages of H11, H12 and H13 steels for structural and hot work applications include ease of forming and working, good weldability, relatively low coefficient of thermal expansion, acceptable thermal conductivity and above-average resistance to oxidation and corrosion.

Tungsten Hot Work Steels. The principal alloying elements of tungsten hot work steels (types H21 to H26) are carbon, tungsten, chromium and vanadium. The higher alloy contents of these steels make them more resistant to high-temperature softening and washing than H11 and H13 hot work steels. However, high alloy content also makes them more prone to brittleness at normal working hardnesses (45 to 55 HRC) and makes it difficult for them to be safely water cooled in service.

Although tungsten hot work steels can be air hardened, they are usually quenched in oil or hot salt to minimize scaling. When air hardened, they exhibit low distortion. Tungsten hot work steels require higher hardening temperatures than chromium hot work steels, which makes them more likely to scale when heated in an oxidizing atmosphere.

Although these steels have much greater toughness, in many characteristics they are similar to high speed steels; in fact, type H26 is a low-carbon version of T1 high speed steel. If tungsten hot work steels are preheated to operating temperature before use, breakage can be minimized. These steels have been used to make mandrels and extrusion dies for high-temperature applications such as extrusion of brass, nickel alloys and steel, and are also suitable for use in hot forging dies of rugged design.

Molybdenum Hot Work Steel. There is only one active molybdenum hot work steel: type H42. This alloy contains molybdenum, chromium, vanadium and carbon, with varying amounts of tungsten. It is similar to tungsten hot work steels, having almost identical characteristics and uses. Although its composition resembles those of various molybdenum high speed steels, type H42 has a low carbon content and greater toughness. The principal advantage of type H42 over tungsten hot work steels is its lower initial cost. Type H42 is more resistant to heat checking than tungsten hot work steels but, in common with all high-molybdenum steels, requires greater care in heat treatment — particularly with regard to decarburization and control of austenitizing temperature.

Cold Work Steels

Because resistance to elevated temperatures (above 260 °C, or 500 °F) is not required for cold work tooling applications, the cold work die steels have alloy contents designed to provide good wear resistance and toughness in various combinations. There are three categories of cold work steels: air-hardening (group A); high-carbon, high-chromium die steels (group D); and oil-hardening steels (group O).

Air-hardening medium-alloy cold work steels (group A) contain enough alloying elements to enable them to achieve full hardness in sections up to about 100 mm (4 in.) in diameter on air cooling from the austenitizing temperature. (Type A6 through hardens in sections as large as a cube 175 mm, or 7 in., on a side.) Because they are air hardening, group A tool steels exhibit minimum distortion and the highest safety (least tendency to crack) in hardening. Manganese, chromium and molybdenum are the principal alloying elements used to provide this deep hardening. Types A2, A3, A7, A8 and A9 contain a high percentage of chromium (5%), which provides moderate resistance to softening at elevated temperatures (see curve 3 in Fig. 1 for a plot of hardness versus tempering temperature for type A2).

Types A4, A6 and A10 are lower in chromium content and higher in manganese content. They can be hardened from temperatures about 100 °C (about 200 °F) lower than those required for the high-chromium types, further reducing distortion and undesirable surface reactions during heat treatment at the expense of slightly lower abrasion resistance.

To improve toughness, silicon is added to type A8 and both silicon and nickel are added to types A9 and A10. Because of the high carbon and silicon contents of type A10, graphite is formed in

the microstructure; as a result, A10 has much better machinability when in the annealed condition, and somewhat better resistance to galling and seizing when in the fully hardened condition, than other group A tool steels.

Typical applications for group A tool steels include shear knives, punches, blanking and trimming dies, forming dies and coining dies. The inherent dimensional stability of these steels makes them suitable for gages and precision measuring tools. In addition, the extreme abrasion resistance of type A7 makes it suitable for brick molds, ceramic molds and other highly abrasive applications.

The complex chromium or chromium-vanadium carbides in group A tool steels enhance the wear resistance provided by the martensitic matrix. Therefore, these steels perform well under abrasive conditions at less than full hardness. Although cooling in still air is adequate to produce full hardness in most tools, very massive sections should be hardened by cooling in an air blast or by interrupted quenching in hot oil.

High-carbon, high-chromium cold work steels (group D) contain 1.50 to 2.35% carbon and 12% chromium; with the exception of type D3, they also contain 1% molybdenum. All group D tool steels except type D3 are air hardening, and attain full hardness when cooled in still air. Type D3 is almost always quenched in oil (small parts can be austenitized in vacuum and then gas quenched); therefore, tools made of D3 are more susceptible to distortion and are more likely to crack during hardening.

Group D steels have high resistance to softening at elevated temperatures. These steels also exhibit excellent resistance to wear — especially type D7, which has the highest carbon and vanadium contents. All group D steels — particularly the higher-carbon types D3, D4 and D7 — contain massive carbides that make them susceptible to edge brittleness.

Typical applications for group D steels include long-run dies for blanking, forming, thread rolling and deep drawing; dies for cutting laminations; brick molds; gages; burnishing tools; rolls; and shear and slitter knives.

Oil-hardening cold work steels (group O) have high carbon contents, plus enough other alloying elements so that small to moderate sections can attain full hardness when quenched in oil from the austenitizing temperature. Group O tool steels vary in type of alloy, as well as in alloy content, even though they are similar in general characteristics and are used for similar applications. Type O1 contains manganese, chromium and tungsten. Type O2 is alloyed primarily with manganese. Type O6 contains silicon, manganese and molybdenum; it has a high total carbon content that includes free carbon as well as sufficient combined carbon to enable the steel to achieve maximum as-quenched hardness. Type O7 contains manganese and chromium, and has a tungsten content higher than that of type O1.

The most important service-related property of group O steels is high resistance to wear at normal temperatures, a result of high carbon content. On the other hand, group O steels have low resistance to softening at elevated temperatures.

The ability of group O steels to harden fully on relatively slow quenching yields lower distortion and greater safety (less tendency to crack) in hardening than is characteristic of the water-hardening tool steels. Tools made from these steels can be successfully repaired or renovated by welding if proper procedures are followed. In addition, graphite in the microstructure of type O6 greatly improves the machinability of annealed stock and helps reduce galling and seizing of fully hardened stock.

Group O steels are used extensively in dies and punches for blanking, trimming, drawing, flanging and forming. Surface hardnesses of 56 to 62 HRC, obtained through oil quenching followed by tempering at 175 to 315 °C (350 to 600 °F), provide a suitable combination of mechanical properties for most dies made from type O1, O2 or O6. Type O7, which has lower hardenability but better general wear resistance than any other group O tool steel, is more often used for tools requiring keen cutting edges. Oil-hardening tool steels are also used for machinery components (such as cams, bushings and guides) and for gages (where good dimensional stability and wear properties are needed).

The hardenability of group O steels can be measured effectively by the Jominy end-quench test. Hardenability bands for group O steels are shown in Fig. 2.

At normal hardening temperatures, group O steels retain greater amounts of undissolved carbides and thus do not harden as deeply as do steels that are lower in carbon but similar in alloy content. On the other hand, group O steels attain higher surface hardness. Raising the hardening temperature increases grain size, increases solution of alloying elements and dissolves more of the excess carbide, thereby increasing hardenability. However, raising the hardening temperature can have an adverse effect on certain mechanical properties—most notably ductility and toughness—and also can increase the likelihood of cracking during hardening.

Shock-Resisting Steels

The principal alloying elements in shock-resisting (group S) steels are manganese, silicon, chromium, tungsten and molybdenum, in various combinations. Carbon content is about 0.50% for all group S steels, which produces a combination of high strength, high toughness and low-to-medium wear resistance. Group S steels are used primarily for chisels, rivet sets, punches, driver bits and other applications requiring high toughness and resistance to shock loading. Types S1 and S7 are also used for hot punching and shearing, which require some heat resistance.

Group S steels vary in hardenability from shallow hardening (S2) to deep hardening (S7). In these steels of intermediate alloy content, hardenability is controlled to a greater extent by actual composition than by the incidental effects of grain size and melting practice, which are so important for group W steels. Group S steels require relatively high austenitizing temperatures to achieve optimum hardness; consequently, undissolved carbides are not a factor in control of hardenability. Type S2 is normally water quenched; types S1, S5 and S6 are oil quenched; type S7 is normally cooled in air, except for large sections, which are oil quenched.

Because group S steels exhibit excellent toughness at high strength levels, they often are considered for nontooling or structural applications.

Low-Alloy, Special-Purpose Steels

The low-alloy, special-purpose (group L) tool steels contain small amounts of chromium, vanadium, nickel and molybdenum. At one time, seven

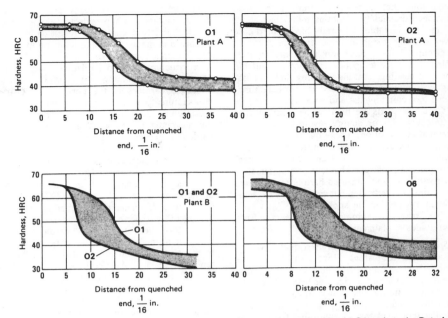

Hardenability bands from plant B represent the data from five heats each for O1 and O2 tool steels. Data from plant A were determined only on the basis of average hardness, not as hardenability bands. Data for O6 is for a spheroidized prior structure. O1 and O6 steels were quenched from 815 °C (1500 °F); O2 from 790 °C (1450 °F).

Fig. 2. End-quench hardenability bands for group O tool steels

steels were listed in this group, but because of falling demand, only types L2 and L6 remain. Type L2 is available in several carbon contents from 0.50 to 1.10%; its principal alloying elements are chromium and vanadium, which make it an oil-hardening steel of fine grain size. Type L6 contains small amounts of chromium and molybdenum, plus 1.50% nickel for increased toughness.

Although both L2 and L6 are considered oil-hardening steels, large sections of L2 are often quenched in water. A type L2 steel containing 0.50% carbon is capable of attaining about 57 HRC as oil quenched, but it will not through harden in sections more than about 13 mm (0.5 in.) thick. Type L6, which contains 0.70% carbon, has an as-quenched hardness of about 64 HRC; it can maintain a hardness above 60 HRC throughout sections 75 mm (3 in.) thick.

Group L steels generally are used for machine parts such as arbors, cams, chucks and collets, and for other special applications requiring good strength and toughness.

Mold Steels

Mold steels (group P) contain chromium and nickel as principal alloying elements. Types P2 to P6 are carburizing steels produced to tool steel quality standards. They have very low hardness and low resistance to work hardening in the annealed condition. These factors make it possible to produce a mold impression by cold hobbing. After the impression is formed, the mold is carburized, hardened and tempered to a surface hardness of about 58 HRC. Types P4 and P6 are deep hardening. In type P4, full hardness in the carburized case can be achieved by cooling in air.

Types P20 and P21 normally are supplied heat treated to 30 to 36 HRC—a condition in which they can be machined readily into large, intricate dies and molds. Because these steels are prehard-

ened, no subsequent high-temperature heat treatment is required, and distortion and size changes are avoided. However, when used for plastic molds, type P20 sometimes is carburized and hardened after the impression has been machined. Type P21 is an aluminum-containing precipitation-hardening steel and is supplied prehardened to 32 to 36 HRC. After machining and low-temperature aging, type P21 can reach 38 to 40 HRC in sections as large as it is practical to produce.

Nearly all group P steels have low resistance to softening at elevated temperatures, except for P4 and P21, which have medium resistance. Group P steels are used almost exclusively in low-temperature die-casting dies and in molds for injection or compression molding of plastics. Plastic molds often require very massive steel blocks up to 750 mm (30 in.) thick and weighing as much as 9 Mg (10 tons). Because these large die blocks must meet stringent requirements for soundness, cleanness and hardenability, electric-furnace melting, vacuum degassing and special deoxidation treatments have become standard practices in the production of group P tool steels. In addition, ingot casting and forging practices have been refined to achieve a high degree of homogeneity.

Water-Hardening Tool Steels

Water-hardening (group W) tool steels contain carbon as the principal alloying element. Small amounts of chromium and vanadium are added to most of the group W steels—chromium to increase hardenability and wear resistance, and vanadium to maintain fine grain size and thus enhance toughness. Group W tool steels are made with various nominal carbon contents (from about 0.60 to 1.40%); the most popular grades contain approximately 1.00% carbon.

Group W tool steels are very shallow hardening, and consequently develop a fully hard-

Source: Metals Handbook Desk Edition, 1985, 18.1-18.36

ened zone that is relatively thin, even when quenched drastically. Sections more than about 13 mm ($\frac{1}{2}$ in.) thick generally have a hard case over a strong, tough and resilient core.

Group W steels have low resistance to softening at elevated temperatures. They are suitable for cold heading, striking, coining and embossing tools; woodworking tools; wear-resistant machine-tool components; and cutlery.

Group W steels are made in as many as four different grades or quality levels for the same nominal composition. These quality levels have been given various names by different manufacturers, and range from a clean carbon tool steel with precisely controlled hardenability, grain size, microstructure and annealed hardness to a grade less carefully controlled but satisfactory for noncritical low-production applications.

The Society of Automotive Engineers defines four grades of plain carbon tool steels as follows:

- *Special* (*grade 1*) is the highest-quality water-hardening tool steel. Hardenability is controlled, and composition held to close limits. Bars are subjected to rigorous testing to ensure maximum uniformity in performance.
- *Extra* (*grade 2*) is a high-quality water-hardening tool steel that is controlled for hardenability and is subjected to tests that ensure good performance in general applications.
- *Standard* (*grade 3*) is a good-quality water-hardening tool steel that is not controlled for hardenability and that is recommended for applications where some latitude in uniformity can be tolerated.
- *Commercial* (*grade 4*) is a commercial-quality water-hardening tool steel that is neither controlled for hardenability nor subjected to special tests.

Limits on manganese, silicon and chromium generally are not required for "special" and "extra" grades. The following Shepherd hardenability limits are prescribed instead:

Hardenability classification	Radial depth of hardening (P), $\frac{1}{64}$ in.	Minimum fracture grain size (F)
Carbon content, 0.70 to 0.95%		
Shallow	10 max	8
Regular	9 to 13	8
Deep	12 min	8
Carbon content, 0.95 to 1.30%		
Shallow	8 max	9
Regular	7 to 11	9
Deep	10 to 16	8

The combined manganese, silicon and chromium contents of standard and commercial grades should not exceed 0.75%. Generally, both manganese and silicon are limited to 0.35% max in all standard and commercial grades; chromium is limited to 0.15% max in standard grades and 0.20% max in commercial grades.

The ability of a group W tool steel to perform satisfactorily in many applications depends on the depth of the hardened zone. Depth of hardening in these steels is controlled mainly by austenitic grain size, melting practice, alloy content, amount of excess carbide present at the quenching temperature and, to a lesser extent, initial structure of the steel prior to austenitizing for hardening. Typical results in the Shepherd PF test indicate an increase in P value of 0.80 mm ($\frac{2}{64}$ in.) for every increase in austenitic grain size of one ASTM number for the same grade. Increased amounts of undissolved carbides at the hardening

temperature will reduce hardenability. This is doubly important in hypereutectoid grades, which are deliberately quenched to retain carbides undissolved at the austenitizing temperature, in order to increase wear resistance. A fine lamellar microstructure prior to hardening, such as that obtained by normalizing, will result in fewer undissolved carbides at the normal austenitizing temperature than will a prior spheroidized microstructure. The presence of fewer carbides at the austenitizing temperature promotes deeper hardening because more carbon is dissolved in the austenite and because there are fewer carbides to act as nucleation sites for nonmartensitic transformation products. Thus, normalized bars have deeper hardenability than spheroidized bars of the same grade.

Addition of vanadium frequently decreases hardenability under normal hardening conditions due to formation of many fine carbides that not only act as nucleation sites for nonmartensitic transformation products, but also refine the austenitic grain size. Austenitizing at higher-than-normal temperatures will dissolve these excess carbides and thus increase the hardenability.

Group W steels with carbon contents lower than that of the eutectoid composition often have greater hardenability than hypereutectoid grades. Grain coarsening resulting from the higher austenitizing temperatures used for hypoeutectoid grades is one cause of this, but the main cause is the absence of excess carbides at the austenitizing temperature.

Figure 3 shows a typical relationship between bar diameter and case depth (60 HRC or above) for three W1 tool steels that are equal in carbon content (1% C) but differ in hardenability. Hardenability is varied by adjusting manganese and silicon contents and altering deoxidation procedure. This relationship illustrates the need for precise specification of hardenability in the selection of these grades: group W tool steels purchased without hardenability requirements could vary widely enough in this property to cause severe processing difficulties or actual tool failures.

With the very high cooling rates required for hardening of the W grades, there is a greater chance that the tool will crack during hardening. Consequently, most manufacturers prefer to use tool steels that can be hardened satisfactorily by

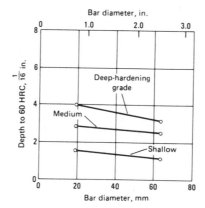

Fig. 3. Relation of bar diameter and depth of hardened zone for shallow-, medium- and deep-hardening grades of W1 tool steel containing 1% C

quenching in oil or cooling in air in order to attempt to avoid the expense involved if a tool cracks during heat treatment.

TYPICAL HEAT TREATMENTS AND PROPERTIES

Information on processing and service characteristics of tool steels is essential in understanding the problems involved in selection, processing and application of tool steels. A general guideline for these characteristics is presented in Table 2.

Technical representatives of tool steel producers can supply more specific information on the properties developed by specific heat treatments in the steels produced by their companies. They should be consulted as to the type of steel and heat treatment best suited to meet all service requirements at the least over-all cost.

The basic properties of tool steels that determine their performance in service are resistance to wear, deformation and breakage; toughness; and, in many instances, resistance to softening at elevated temperatures. Often, these characteristics can be measured by, or inferred from, direct measurement of hardness. Hardness of tool steels is most commonly measured and reported on the Rockwell C scale (HRC) in the United States and on the Vickers scale (diamond pyramid hardness, or HV) in the United Kingdom and Europe. It is significant that conversion from HRC to HV, or vice versa, is not linear (see Fig. 4).

For a given tool steel at a given hardness, wear resistance may vary widely depending on the wear mechanism involved and the heat treatment used. It is important to note also that among tool steels with widely differing compositions but identical hardnesses, wear resistance may vary widely under identical wear conditions.

The ability of a tool steel to withstand rapid application of high loads without breaking increases with decreasing hardness. With hardness held constant, wide differences can be observed among tool steels of different compositions, or among steels of the same nominal composition made by different melting practices or heat treated according to different schedules.

The ability of a tool steel to resist softening at elevated temperatures is related to (*a*) its ability to develop secondary hardening and (*b*) the amount of special phases, such as excess alloy carbides, in the microstructure. Useful information on the ability of tool steels to resist softening at elevated temperatures can be obtained from tempering curves such as those in Fig. 1.

Fabrication

The properties that influence the fabricability of tool steels include: machinability; grindability; weldability; hardenability; and extent of distortion, safety (freedom from cracking) and tendency to decarburize during heat treatment.

Machinability of tool steels can be measured by the usual methods applied to constructional steels. Results are reported as percentages of the machinability of water-hardening tool steels (see Table 3); 100% machinability in tool steels is equivalent to about 30% machinability in constructional steels, for which 100% machinability would be that of a free-machining constructional steel such as B1112.

Improving the machinability of a tool steel by altering either composition or preliminary heat treatment can be very important if a large amount

Table 2. Processing and service characteristics of tool steels

Adapted from "Tool Steels" (a Steel Products Manual): AISI, March 1978

AISI designation	Resistance to decarburization	Hardening response	Amount of distortion(a)	Resistance to cracking	Approximate hardness(b), HRC	Machinability	Toughness	Resistance to softening	Resistance to wear
Molybdenum high speed steels									
M1	Low	Deep	A or S, low; O, medium	Medium	60-65	Medium	Low	Very high	Very high
M2	Medium	Deep	A or S, low; O, medium	Medium	60-65	Medium	Low	Very high	Very high
M3 (class 1 and class 2) ...	Medium	Deep	A or S, low; O, medium	Medium	61-66	Medium	Low	Very high	Very high
M4	Medium	Deep	A or S, low; O, medium	Medium	61-66	Low to medium	Low	Very high	Highest
M6	Low	Deep	A or S, low; O, medium	Medium	61-66	Medium	Low	Highest	Very high
M7	Low	Deep	A or S, low; O, medium	Medium	61-66	Medium	Low	Very high	Very high
M10 ...	Low	Deep	A or S, low; O, medium	Medium	60-65	Medium	Low	Very high	Very high
M30 ...	Low	Deep	A or S, low; O, medium	Medium	60-65	Medium	Low	Highest	Very high
M33 ...	Low	Deep	A or S, low; O, medium	Medium	60-65	Medium	Low	Highest	Very high
M34 ...	Low	Deep	A or S, low; O, medium	Medium	60-65	Medium	Low	Highest	Very high
M36 ...	Low	Deep	A or S, low; O, medium	Medium	60-65	Medium	Low	Highest	Very high
M41 ...	Low	Deep	A or S, low; O, medium	Medium	65-70	Medium	Low	Highest	Very high
M42 ...	Low	Deep	S1, or S, low; O, medium	Medium	65-70	Medium	Low	Highest	Very high
M43 ...	Low	Deep	A or S, low; O, medium	Medium	65-70	Medium	Low	Highest	Very high
M44 ...	Low	Deep	A or S, low; O, medium	Medium	62-70	Medium	Low	Highest	Very high
M46 ...	Low	Deep	A or S, low; O, medium	Medium	67-69	Medium	Low	Highest	Very high
M47 ...	Low	Deep	A or S, low; O, medium	Medium	65-70	Medium	Low	Highest	Very high
Tungsten high speed steels									
T1	High	Deep	A or S, low; O, medium	High	60-65	Medium	Low	Very high	Very high
T2	High	Deep	A or S, low; O, medium	High	61-66	Medium	Low	Very high	Very high
T4	Medium	Deep	A or S, low O, medium	Medium	62-66	Medium	Low	Highest	Very high
T5	Low	Deep	A or S, low; O, medium	Medium	60-65	Medium	Low	Highest	Very high
T6	Low	Deep	A or S, low; O, medium	Medium	60-65	Low to medium	Low	Highest	Very high
T8	Medium	Deep	A or S, low; O, medium	Medium	60-65	Medium	Low	Highest	Very high
T15	Medium	Deep	A or S, low; O, medium	Medium	63-68	Low to medium	Low	Highest	Highest
Shock-resisting steels									
S1	Medium	Medium	Medium	High	40-58	Medium	Very high	Medium	Low to medium
S2	Low	Medium	High	Low	50-60	Medium to high	Highest	Low	Low to medium
S5	Low	Medium	Medium	High	50-60	Medium to high	Highest	Low	Low to medium
S6	Low	Medium	Medium	High	54-56	Medium	Very high	Low	Low to medium
S7	Medium	Deep	A, lowest; O, low	A, highest; O, high	45-57	Medium	Very high	High	Low to medium
Low-alloy special-purpose steels									
L2	High	Medium	W, low; O, medium	W, high; O, medium	45-63	High	Very high(c)	Low	Low to medium
L6	High	Medium	Low	High	45-62	Medium	Very high	Low	Medium
Low-carbon mold steels									
P2	High	Medium	Low	High	58-64(c)	Medium to high	High	Low	Medium
P3	High	Medium	Low	High	58-64(c)	Medium	High	Low	Medium
P4	High	High	Very low	High	58-64(c)	Low to medium	High	Medium	High

(continued on the next page)

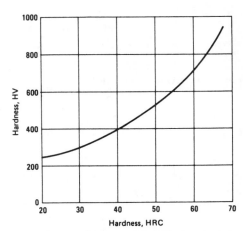

Fig. 4. Relation between Vickers and Rockwell C hardness scales

Table 3. Approximate machinability ratings for annealed tool steels

Type	Machinability rating
O6	125
W1, W2, W5	100(a)
A10	90
P2, P3, P4, P5, P6	75 to 90
P20, P21	65 to 80
L2, L6	65 to 75
S1, S2, S5, S6, S7	60 to 70
H10, H11, H13, H14, H19	60 to 70(b)
O1, O2, O7	45 to 60
A2, A3, A4, A6, A8, A9	45 to 60
H21, H22, H24, H25, H26, H42	45 to 55(b)
T1	40 to 50
M2	40 to 50
T4	35 to 40
M3 (class 1)	35 to 40
D2, D3, D4, D5, D7, A7	30 to 40
T15	25 to 30
M15	25 to 30

(a) Equivalent to approximately 30% of the machinability of B1112.
(b) For hardness range 150 to 200 HB.

of machining is required to form the tool and a large number of tools is to be made.

Grindability. One measure of grindability is the ease with which the necessary excess stock on heat treated tool steel can be removed using standard grinding wheels. The grinding ratio (grindability index) is the volume of metal removed per volume of wheel wear. The higher the grindability index the easier the metal is to grind. The index is valid only for specific sets of grinding conditions.

It should be noted that the grindability index does not indicate susceptibility to cracking during or after grinding, ability to produce the required surface (and subsurface) stress distribution, or ease of obtaining the required surface smoothness.

Weldability. The ability to construct, alter or repair tools by welding without causing the material to crack may be an important factor in selection of a tool material, especially if the tool is large. It is only rarely of importance in selecting materials for small tools. Weldability is largely a function of composition, but welding method and procedure also influence weld soundness. Generally, tool steels that are deep hardening and that are classified as having relatively high safety in hardening are among the more readily welded tool steel compositions. Never-

Table 2 (continued)

AISI designation	Resistance to decarburization	Hardening response	Amount of distortion(a)	Resistance to cracking	Approximate hardness(b), HRC	Machinability	Toughness	Resistance to softening	Resistance to wear
Low-carbon mold steels (continued)									
P5	High	...	W, high; O, low	High	58-64(c)	Medium	High	Low	Medium
P6	High	...	A, very low; O, low	High	58-61(c)	Medium	High	Low	Medium
P20	High	Medium	Low	High	28-37	Medium to high	High	Low	Low to medium
P21	High	Deep	Lowest	Highest	30-40(d)	Medium	Medium	Medium	Medium
Water-hardening steels									
W1	Highest	Shallow	High	Medium	50-64	Highest	High(e)	Low	Low to medium
W2	Highest	Shallow	High	Medium	50-64	Highest	High(e)	Low	Low to medium
W5	Highest	Shallow	High	Medium	50-64	Highest	High(e)	Low	Low to medium
Chromium hot work steels									
H10	Medium	Deep	Very low	Highest	39-56	Medium to high	High	High	Medium
H11	Medium	Deep	Very low	Highest	38-54	Medium to high	Very high	High	Medium
H12	Medium	Deep	Very low	Highest	38-55	Medium to high	Very high	High	Medium
H13	Medium	Deep	Very low	Highest	38-53	Medium to high	Very high	High	Medium
H14	Medium	Deep	Low	Highest	40-47	Medium	High	High	Medium
H19	Medium	Deep	A, low; O, medium	High	40-57	Medium	High	High	Medium to high
Tungsten hot work steels									
H21	Medium	Deep	A, low; O, medium	High	36-54	Medium	High	High	Medium to high
H22	Medium	Deep	A, low; O, medium	High	39-52	Medium	High	High	Medium to high
H23	Medium	Deep	Medium	High	34-47	Medium	Medium	Very high	Medium to high
H24	Medium	Deep	A, low; O, medium	High	45-55	Medium	Medium	Very high	High
H25	Medium	Deep	A, low; O, medium	High	35-44	Medium	High	Very High	Medium
H26	Medium	Deep	A or S, low; O, medium	High	43-58	Medium	Medium	Very high	High
Molybdenum hot work steel									
H42	Medium	Deep	A or S, low;	Medium	50-60	Medium	Medium	Very high	High
Air-hardening medium-alloy cold work steels									
A2	Medium	Deep	Lowest	Highest	57-62	Medium	Medium	High	High
A3	Medium	Deep	Lowest	Highest	57-65	Medium	Medium	High	Very high
A4	Medium	Deep	Lowest	Highest	54-62	Low to medium	Medium	Medium	Medium to high
A6	Medium to high	Deep	Lowest	Highest	54-60	Low to medium	Medium	Medium	Medium to high
A7	Medium	Deep	Lowest	Highest	57-67	Low	Low	High	Highest
A8	Medium	Deep	Lowest	Highest	50-60	Medium	High	High	Medium to high
A9	Medium	Deep	Lowest	Highest	35-56	Medium	High	High	Medium to high
A10	Medium to high	Deep	Lowest	Highest	55-62	Medium to high	Medium	Medium	High
High-carbon, high-chromium cold work steels									
D2	Medium	Deep	Lowest	Highest	54-61	Low	Low	High	High to very high
D3	Medium	Deep	Very low	High	54-61	Low	Low	High	Very high
D4	Medium	Deep	Lowest	Highest	54-61	Low	Low	High	Very high
D5	Medium	Deep	Lowest	Highest	54-61	Low	Low	High	High to very high
D7	Medium	Deep	Lowest	Highest	58-65	Low	Low	High	Highest
Oil-hardening cold work steels									
O1	High	Medium	Very low	Very high	57-62	High	Medium	Low	Medium
O2	High	Medium	Very low	Very high	57-62	High	Medium	Low	Medium
O6	High	Medium	Very low	Very high	58-63	Highest	Medium	Low	Medium
O7	High	Medium	W, high; O, very low	W, low; O, very high	58-64	High	Medium	Low	Medium

(a) A, air cool; B, brine quench; O, oil quench; S, salt bath quench; W, water quench. (b) After tempering in temperature range normally recommended for this steel. (c) Carburized case hardness. (d) After aging at 510 to 550 °C (950 to 1025 °F). (e) Toughness decreases with increasing carbon content and depth of hardening.

theless, these "weldable" tool steels require careful preheating and postheating to ensure successful welding.

Hardenability includes both the maximum hardness obtainable when the quenched steel is fully martensitic and the depth of hardening obtained by quenching in a specific manner. In this context, depth of hardening must be defined — generally as a specific value of hardness or a specific microstructural appearance. As a very general rule, maximum hardness of a tool steel increases with increasing carbon content; increasing the austenitic grain size and the amount of alloying elements reduces the cooling rate required to produce maximum hardness (increases the depth of hardening). The Jominy end-quench test, which is used extensively for measurement of hardenability of constructional steels, has limited application for tool steels. This test gives useful information only for oil-hardening grades. Air-hardening grades are so deep hardening that the standard Jominy test is not sufficient to evaluate hardenability.

An air-hardenability test has been developed that is based on the principles involved in the Jominy test, but which uses only still air cooling and a 150-mm (6-in.) diameter end block to produce the very low cooling rates of large sections. Such tests provide useful information for research but are of limited use in devising production heat treatments. By contrast, water-hardening grades of tool steel are so shallow hardening that the Jominy test is not sensitive enough. Special tests, such as the Shepherd PF test, are useful for research and for special applications of water-hardening tool steels.

In the Shepherd PF test, a bar 19 mm ($^3/_4$ in.) in diameter, in the normalized condition, is brine quenched from 740 °C (1450 °F) and fractured; the case depth (penetration, P) is measured in 0.4-mm ($^1/_{64}$-in.) intervals, and the fracture grain size of the case (F) is determined by comparison with standard specimens. A PF value of $6-8$ indicates a case depth of $^6/_{64}$ in. (2.4 mm) and a fracture grain size of 8. Fine-grain water-hardening tool steels are those with fracture grain sizes (F values) of 8 or more. Deep-hardening steels of this type have P values of 12 or more; medium-hardening steels, 9 to 11; and shallow-hardening steels, 6 to 8.

Distortion and Safety in Hardening. Minimal distortion in heat treating is important for tools that must remain within close size limits. For a more detailed discussion of this factor, see the article in this section on Distortion in Tool Steels. In general, the amount of distortion and the tendency toward cracking increase as the severity of quenching increases.

Resistance to decarburization is an important factor in determining whether or not a protective atmosphere is required during heat treating. In a decarburizing atmosphere, the rate of decarburization increases rapidly with increasing austenitizing temperature, and for a given austenitizing temperature the depth of decarburization increases directly with holding time. Some types of tool steels decarburize much more rapidly than others under the same conditions of atmosphere, austenitizing temperature and time.

MACHINING ALLOWANCES

The standard machining allowance is the recommended total amount of stock that the user should remove from the as-supplied mill form to provide a surface free from imperfections that might adversely affect response to heat treatment or the ability of tools to perform properly.

The decarburization resulting from oxidation at the exposed surfaces during forging and rolling of the tool steel is a major factor in determining the amount of stock that should be removed. Although extra care is used in producing tool steels, scale, seams and other surface imperfections may be present, and, if present, must be removed.

Besides the standard machining allowance, sufficient additional stock must be provided to allow for cleanup of any decarburization and distortion that may occur during final heat treatment. The amount of this allowance varies with the type of tool steel, the type of heat treating equipment and the size and shape of the tool.

Group W and group O tool steels are considered highly resistant to decarburization. Group M steels, cobalt-containing group T steels, group D steels and types H42, A2 and S5 are rated as poor in resisting decarburization.

Decarburization during final heat treatment is undesirable because it alters the composition of the surface layer, thereby changing the response to heat treatment of this layer and usually affecting adversely the properties resulting from heat treatment. Decarburization can be controlled or avoided by heat treating in a neutral salt bath or in a controlled atmosphere or vacuum furnace. When heat treating is done in vacuum, a vacuum of 100 to 200 μm Hg is satisfactory for most tools if the furnace is in good operating condition and has a very low leak rate. However, it is recommended that a vacuum of 50 to 100 μm Hg be used wherever possible.

If special heat treating equipment is not available, appreciable decarburization can be avoided by wrapping the tool in stainless steel foil. Type 321 stainless steel foil can be used at austenitizing temperatures up to about 1000 °C (1850 °F); either type 309 or type 310 foil is required at austenitizing temperatures from 1000 to 1200 °C (1850 to 2200 °F).

POWDER METALLURGY STEELS

In recent years, tool steels with improved properties have been produced by the powder metallurgy (P/M) process. In this process, a bath of prealloyed molten metal is gas atomized and quenched to produce a fine powder. The particles of this powder are screened and loaded into a steel container, which is then evacuated, and the particles are hot isostatically pressed to full density. The resulting compact is rolled or forged to size on conventional steel-mill equipment or, in some instances, is used in the as-compacted condition to make tools.

P/M tool steels have two major advantages: complete freedom from macrosegregation and uniform distribution of extremely fine carbides. These characteristics provide deeper hardening and faster response to hardening. The latter is important, particularly for molybdenum high speed steels, which tend to decarburize rapidly at austenitizing temperatures. P/M products also show less out-of-roundness distortion in large-diameter bars.

When sulfur is added to P/M tool steels, they exhibit a very fine homogeneous distribution of sulfides. This uniform sulfide distribution promotes better machinability. After heat treating, the refined microstructure of P/M tool steels promotes superior grindability and greater toughness compared with those of conventionally processed tool steels.

The following AISI types of high speed steels are available as P/M tool steels: M2, M2 with high sulfur and high carbon, M3 class 2 with sulfur, M4, M35 with sulfur, M42 and T15. P/M steels can be substituted for their conventional counterparts in all applications, and are particularly advantageous when heavy sections are required.

The freedom from gross segregation provided by the P/M process makes it possible for new higher-alloy tool steels to be readily fabricated. One type now available, which contains 1.50 C, 3.75 Cr, 3.00 V, 10 W, 5.25 Mo and 9.00 Co, is reported to have the highest hot hardness of any high speed steel. It has been used in cutting tools for critical applications such as machining of certain aerospace alloys and cutting applications where the highest speeds and feeds are required. Another type, CPM 10V, which contains 2.45 C, 5.25 Cr, 9.75 V and 1.30 Mo, is designed for extreme wear resistance in cold and warm work tooling. Its microstructure consists chiefly of a uniform distribution of hard, wear-resistant vanadium carbides in a tool steel matrix. The wear resistance of this steel at ordinary temperatures typically exceeds that of T15.

PRECISION-CAST TOOL STEELS

Precision casting of tools to nearly finished size offers important cost advantages through reductions in waste and machining. Casting is particularly advantageous when patternmaking costs can be distributed over a large number of tools.

Experience with cast forging and extrusion dies has shown that cast tools are more resistant to heat checking; minute cracks do occur, but they grow at much lower rates than in wrought material of the same grade and hardness. Slower propagation of thermal-fatigue cracks generally extends die life significantly. Mechanical testing of cast and wrought H13 indicates that yield and tensile strengths are virtually identical from room temperature to 600 °C (1100 °F), but that ductility is moderately lower in cast material. Hot hardness of cast H13 is higher than that of wrought H13 at temperatures above about 300 °C (about 600 °F); this hardness advantage increases with temperature and measures about eight points on the HRC scale at 650 °C (1200 °F).

Because cast dies exhibit uniform properties in all directions, no problem of directionality (anisotropy) exists. Dimensional control of castings is very consistent after an initial die is made and any necessary corrections are incorporated in the pattern. Reasonable finishing allowances are 0.25 to 0.38 mm (0.010 to 0.015 in.) on the impression faces, 0.8 to 1.6 mm ($\frac{1}{32}$ to $\frac{1}{16}$ in.) at the parting line of the mold, and 1.6 to 3.2 mm ($\frac{1}{16}$ to $\frac{1}{8}$ in.) on the back and outside surfaces. The hot work tool steels most commonly cast include hot work grades H12, H13, H21 and H25 and cold work grades D2, D5, A2 and D7.

SURFACE TREATMENTS

In many applications, service life of high speed steel tools can be increased by surface treatments.

Oxide coatings, provided by treatment of the finish-ground tool in an alkali-nitrate bath or by steam oxidation, prevent or reduce adhesion of the tool to the workpiece. Oxide coatings have doubled tool life — particularly in machining of gummy materials such as soft copper and non-free-cutting low-carbon steels.

Plating of finished high speed steel tools with 0.0025 to 0.0125 mm (0.1 to 0.5 mil) of chromium also prolongs tool life by reducing adhesion of the tool to the workpiece. Chromium plating is relatively expensive, and precautions must be taken to prevent tool failure in service due to hydrogen embrittlement.

Carburizing is not recommended for high speed steel cutting tools because the cases on such tools are extremely brittle. However, carburizing is useful for applications such as cold work dies that require extreme wear resistance and that are not subjected to impact or highly concentrated loading. Carburizing is done at 1035 to 1065 °C (1900 to 1950 °F) for short periods of time (10 to 60 min) to produce a case 0.05 to 0.25 mm (0.002 to 0.010 in.) deep. The carburizing treatment also serves as an austenitizing treatment for the whole tool. A carburized case on a high speed steel has a hardness of 65 to 70 HRC but does not have the high resistance to softening at elevated temperatures exhibited by normally hardened high speed steel.

Nitriding successfully increases life for all types of high speed steel cutting tools. However, gas nitriding in dissociated ammonia produces a case that is too brittle for most applications. Liquid nitriding for about 1 h at 565 °C (1050 °F) provides a light case, increasing both surface hardness and resistance to adhesion. For nitrided high speed steel taps, drills and reamers used in machining annealed steel, five-fold increases in life have been reported, with average increases of 100 to 200%. Obviously, if this nitrided case is removed when the tool is reground, the tool must then be retreated, which reduces the cost advantage of the process.

In addition, special surface-treatment processes, such as aerated nitriding baths, improve resistance to adhesive wear without producing excessive brittleness. Sulfur-containing nitriding baths provide a high-sulfur surface layer for additional resistance to seizing.

Sulfide Treatment. A low-temperature (190 °C; 375 °F) electrolytic process using sodium and potassium thiocyanate provides a seizing-resistant iron sulfide layer. This process can be used as a final treatment for all types of hardened tool steels without much danger of overtempering.

MARAGING STEELS

Certain high-nickel maraging steels are being used for special noncutting tool applications; 18Ni(250) is the type most frequently used. However, for the most demanding applications, the higher-strength 18Ni(300) is often preferred. For applications requiring maximum abrasion resistance, any of the maraging steels can be nitrided.

Maraging steels achieve full hardness — nominally 50 HRC for 18Ni(250), 54 HRC for 18Ni(300) and 58 HRC for 18Ni(350) — by a simple aging treatment, usually 3 h at about 480 °C (900 °F). Because hardening does not depend on cooling rate, full hardness can be developed uniformly in massive sections, with almost no distortion. Decarburization is of no concern in these alloys, because they do not contain carbon as an alloying element. If the long-time service temperature exceeds the aging temperature, maraging steels overage with a significant drop in hardness.

The 18Ni(250) grade is used for aluminum die-casting dies and cores, aluminum hot forging dies, dies for molding plastics, and various support

Source: Metals Handbook Desk Edition, 1985, 18.1-18.36

tooling used in extrusion of aluminum. In die casting of aluminum, maraging steel dies can be used at higher hardness than is possible for dies made of H13 tool steel because maraging steel is not as prone to heat checking. Because the aging process results in very little size change, it is possible to machine the intricate impressions for plastic molding dies to final size prior to final hardening.

For molding extremely abrasive types of plastics, the higher surface hardness provided by 18Ni(300) maraging steel is desirable.

Superhard Tool Materials

SUPERHARD TOOL MATERIALS are exceptionally hard—far harder than any tool steel. Some superhard tool materials, such as hard alloys and cemented carbides, exhibit largely metallic characteristics, whereas others, such as ceramic tool materials, are considered nonmetallic.

Cemented carbides are the most widely used superhard tool materials. Their chief applications are in metalworking tools, mining tools and wear-resistant parts. Steel-bonded carbides have a heat treatable matrix, and are used extensively to make forming and stamping tools. Ceramics, boron nitride and diamond are used chiefly to make cutting tools. Diamond is also used for making small-diameter wiredrawing dies.

To take full advantage of the wear resistance of cemented carbides and other superhard tool materials in metalcutting applications, new types of machine tools having the required power and rigidity have been developed. The degree of improvement that can be obtained by using superhard tools in older existing equipment may be limited by a lack of sufficient power and rigidity.

CLASSIFICATION OF SUPERHARD TOOL MATERIALS

At present there is no universally accepted system for classifying carbides and other superhard tool materials. The systems most often employed by both producers and users to describe superhard tool materials are the SAE J1072 system, the ISO classification, the British Hard Metal Association system, and an informal application-oriented system known as the C-grade system. Close cooperation between user and producer is often the most fruitful means of selecting the proper grade.

CEMENTED CARBIDES

Cemented carbides are made by a powder metallurgy process in which finely divided compounds of refractory metals and carbon are bonded together to form a compacted solid of high strength and hardness. The first cemented carbide to be produced was tungsten carbide with a cobalt binder. Over the years, this original material has been modified in many ways to produce a variety of cemented carbides that can be used in a wide range of applications (see Table 4). These modifications consist mainly of varying the amount of metal used as the binder, varying the structure (grain size) of the carbide and substituting other metallic carbides for part of the tungsten carbide. Titanium carbide and tantalum carbide have been the metallic carbides used most widely in complex grades, but because of the limited world supply of tantalum, niobium carbides are replacing tantalum carbides in increasing amounts. Cemented carbides containing only tungsten carbide are commonly referred to as "straight grades," whereas those containing other metallic carbides in addition to tungsten carbide are called "complex grades."

Straight grades of cemented carbides are used for cutting tools, drawing dies, forming-die inserts, punches and many other types of tools. Complex grades are used chiefly to make cutting tools and cutting-tool inserts for machining plain carbon and low-alloy steels. In these applications, tools made of complex grades exhibit less wear, because the crater formed on the top rake of the tool due to chip adherence is much smaller than that developed on tools made of straight grades. Also, the complex grades have better resistance to deformation at the temperatures developed along the cutting edge.

Cemented carbide is very hard, and holds its high hardness at temperatures well above those at which high speed steels begin to soften. Cemented carbide cutting tools originally were constructed by copper brazing small compacts into heavy steel shanks. These tools had to be sharpened by regrinding frequently for maximum tool efficiency. Although regrinding of brazed-tip tools is still done in many instances, it is much more common for cemented carbide to be used as relatively small indexable inserts that are mechanically held in a steel holder. These inserts are indexed until all available cutting edges have become worn, and then are reground or recycled.

Manufacture by Powder Metallurgy. The conventional means of making cemented carbide tools is a powder metallurgy (P/M) process in which finely divided tungsten carbide powders are blended with cobalt powders, compacted by cold pressing to the desired shape (with allowance for shrinkage during sintering), and sintered in vacuum or under a controlled atmosphere at a temperature high enough to melt, or at least partly melt, the binder. One of the main limitations of this conventional process is the requirement that the parts be of uniform cross section along the direction of pressing, and be of limited length-to-diameter ratio, so that the compacting pressure can be applied relatively uniformly.

Large parts, and parts that cannot be conveniently compacted in dies, generally are made by cold isostatic pressing. The parts may or may not be presintered, and may or may not be machined to shape, before being liquid-phase sintered to final density. Large parts occasionally are compacted by hot pressing in graphite dies.

Hot isostatic pressing is commonly used to increase the soundness of sintered carbide parts. Hot isostatic pressing improves the surface finish and increases the average transverse rupture, compression and impact strengths of carbide tools. It is an essential step in the production of certain tools such as compressor plungers; grinder spindles; anvils and dies for very-high-pressure apparatus; cold extrusion punches; and swaging mandrels. Hot isostatic pressing also is preferred

Table 4. Typical applications of cobalt-bonded cemented carbides

Grade	Grain size	Application
Straight grades		
97WC-3Co	Medium	Machining of cast iron, nonferrous metals and nonmetallic materials; excellent abrasion resistance and low shock resistance; the most wear resistant of the straight WC-Co grades; maintains a sharp cutting edge and makes long finishing cuts to close tolerances possible; also used for fine wire dies and small nozzles
94WC-6Co	Fine	Machining nonferrous and high-temperature alloys
94WC-6Co	Medium	General-purpose machining of work materials other than steel; also used for small and medium-size compacting dies, coating dies, burnishing rings and nozzles
94WC-6Co	Coarse	Machining of cast iron, nonferrous metals and nonmetallic materials; also used for small wiredrawing dies, compacting dies, small drawing dies, and caps and rings. The hardest grade used in mining applications where impact is encountered, as in rotary percussive bits
90WC-10Co	Fine	Machining steel and milling high-temperature metals (including titanium and its alloys) at low feeds and speeds: face mills, end mills, form tools, cutoff tools and screw-machine tools
90WC-10Co	Coarse	Primarily used for mining roller bits and percussive drilling bits
84WC-16Co	Fine	Primarily used for mining and metalforming components
84WC-16Co	Coarse	Metalforming and mining components: medium and large dies where great toughness is required, blanking dies for punch presses, and large mandrels
75WC-25Co	Medium	Metalforming components for heavy impact applications, such as heading dies, cold extrusion dies, and punches and dies for blanking heavy stock
Complex grades		
71-74.5WC-10-12.5TiC-11-12.0TaC-4.5Co	Medium	Finishing, semifinishing and light roughing operations on plain carbon and alloy steels and alloy cast irons
72-73WC-7-8TiC-11.5-12TaC-8-8.5Co	Medium	Tough, wear-resistant grade for heavy-duty roughing cuts. Successfully withstands high temperatures encountered in heavy-duty machining, interrupted turning, scale cuts and milling of plain carbon and alloy steels and alloy cast irons
64TiC-28WC-2TaC-2Cr$_3$C$_2$-4Co	Medium	High-speed finishing of steels and cast irons
57WC-27TaC-16Co	Coarse	Cutting hot flash formed in the manufacture of welded tubing; also used to make dies for hot extrusion of aluminum wirebar and tubing

for parts that require a nearly perfect surface finish. Drawing dies and mandrels; extrusion punches; Sendzimer-mill rolls; strip-mill rolls; wire-flattening rolls; liners and plungers for pumps and compressors; burnishing rolls; and balls for burnishing, sizing and valve applications are among the types of tools and other parts that need exceptionally good surface finishes for optimum performance.

Parts that are long and slender, such as rod stock for circuit-board drills, are difficult to produce by either compaction in dies or isostatic pressing. Such parts are produced by extruding a mixture of carbide powders, metallic binders and a suitable organic vehicle, followed by sintering.

PROPERTIES OF CEMENTED CARBIDES

Specific properties of individual grades of cemented carbides depend not only on the composition of the carbide but also on its particle size and on the amount and type of binder.

The compositions and properties of nine straight grades and four complex grades of cobalt-bonded carbide are given in Table 5. Because properties are influenced by both composition and structure, both characteristics must be specified to define a specific grade.

Microstructure. Varying the structures of cemented carbides can improve tool performance in specific applications.

For straight tungsten carbides of comparable grain size (WC particle size), increasing cobalt content increases transverse strength and toughness but decreases hardness, compressive strength, elastic modulus and abrasion resistance. If we compare, for example, medium-grain carbides having 3, 6 and 25% cobalt, we find the 3% Co grade to have the greatest hardness and abrasion resistance — properties that make it well suited for wiredrawing dies and for cutting tools used in machining of cast iron and other abrasive or gummy materials. The 6% Co grade has moderate values for all properties, and is a good general-purpose carbide material. The 25% Co grade has the greatest toughness, and is used for applications involving heavy impact. Because of its relatively low hardness and abrasion resistance, it is not used for cutting tools. Similar parallels in properties and uses can be drawn both for the fine-grain grades and for the coarse-grain grades containing 6, 10 and 16% cobalt.

Another set of comparisons can be drawn for the grades containing 6% cobalt. All three grades — fine, medium and coarse — are used for cutting tools, but the applications to which they are applied involve different machining conditions and different work materials. The fine-grain material is used for finish to medium-rough machining of ductile, gray and chilled irons and of austenitic stainless steels, high-temperature alloys and nonmetallic materials; the medium-grain material for light to heavy machining of these same wrought work materials; and the coarse-grain grade for heavy to extremely heavy rough machining of such materials. The medium-grain material is widely employed for general-purpose machining because its properties have been found to offer a good practical balance between hardness and toughness. The coarse-grain grade, which has the lowest hardness and abrasion resistance and the best toughness of the three grades, is used where a combination of moderate hardness and high toughness is needed. Similar comparisons

can be made for the grades that contain 10 and 16% Co. In general, decreasing grain size improves abrasion resistance and makes it easier to retain the edge on a cutting tool; increasing grain size improves toughness and makes the cemented carbide more suitable for die applications.

For complex grades, comparisons similar to those drawn for the straight grades are not as readily made. Variations in carbide type, as well as in binder content, affect properties, which in turn influence suitability for specific types of service. In the microstructures the WC particles are angular, whereas those of TiC and TaC are more rounded.

The first two complex grades contain about the same amount of WC, but one contains twice as much binder. The lower-cobalt grade is used for lighter-duty cutting.

The complex grade high in TiC is relatively low in transverse strength and high in resistance to abrasion and cratering. It is used extensively for high-speed, light-duty finishing.

The complex grade highest in cobalt content and in TaC is preferred for hot work tools, in both cutting and shaping of metals.

Hardness. Following the practice of the producers and users of cemented carbides in the United States, Rockwell A hardness is reported in Table 5, and Fig. 5 graphically depicts changes in Rockwell A hardness as temperature is increased. The reduction in hardness between room temperature and about 800 °C (about 1475 °F) for 94WC-6Co with a very fine structure is 7 points on the Rockwell A scale — from 93 to 86 HRA. Lower hot hardness generally signifies lower resistance to deformation at high temperature. Nevertheless, cemented carbides with similar hot-hardness values sometimes show very significant differences in resistance to deformation at high temperature in service.

Abrasion Resistance. Most producers of cemented carbides and other superhard tool materials measure abrasion resistance by subjecting the materials to a test in which a specimen is held against a rotating wheel for a fixed number of revolutions while the specimen and wheel are immersed in a water slurry containing sharp aluminum oxide particles. Comparative rankings are reported, usually in terms of wear ratings based on the reciprocal of volume loss. Two standards apply to abrasive wear testing and the method of reporting results: ASTM B611 and CCPA P112. Not all producers have adopted the ASTM method of abrasive wear testing, so values of abrasion resistance cited by producers vary widely. Because of this variance, it is almost impossible to make valid comparisons among test results reported by different producers. It also is fallacious to use abrasion resistance as a measure of wear resistance of superhard tool materials when they are used for cutting steel or other materials: abrasion resistance in a standard test does *not* correspond directly to wear resistance in machining operations.

Values of comparative abrasion resistance are listed in Table 5. These are relative values only, and are based on a value of 100 for the most abrasion-resistant grade. Comparative abrasion resistance is lowered as cobalt content or grain size is increased. However, abrasion resistance is lower for complex carbides than for straight WC grades having the same cobalt content.

Corrosion resistance of cemented carbides is fairly good, and they may be employed advantageously in certain corrosive environments for applica-

tions where outstanding wear resistance is required. Resistance to water, to oils and other cutting fluids used in machining, and to alkaline attack is excellent; but the cobalt matrix in many cemented carbides is subject to attack by acids. When the cobalt is attacked, accelerated wear develops because of the rapid crumbling of unsupported carbide particles. Corrosion of the cobalt binder also may cause a drastic reduction in strength.

Cemented carbides used as tool materials begin to oxidize when heated in air above about 500 to 600 °C (900 to 1100 °F). However, as measured by weight changes or shape distortion, the rate of oxidation of grades containing large amounts of titanium carbide is much lower than that of straight grades at temperatures as high as 1000 °C (1800 °F).

Toughness. Cemented carbides are brittle materials; usually they will show less than 0.2% elongation in a tensile test.

Values of Charpy impact strength for cemented carbides have little significance and may be misleading. The energy absorbed during impact testing of very hard materials consists mainly of energy absorbed in elastic bending of the specimen and energy absorbed by the testing machine. The portion of total absorbed energy that is a measure of the toughness of a material — namely, the energy of plastic work and the energy necessary to create new surfaces — is only a few percent of the total energy measured.

Typical applications of cobalt-bonded cemented carbides are listed in Table 6.

COATED CEMENTED CARBIDES

Coated cemented carbides are materials consisting of substrates having compositions similar to those of conventional cemented carbides onto which thin coatings of very hard material are deposited, usually by chemical vapor deposition.

Coated cemented carbides have combinations of wear and breakage resistance superior to those of uncoated carbides. Coated cemented carbides became commercially available in 1970. In the United States, about 35% of the cemented carbide tools used for metalcutting in 1979-1980 were coated. Acceptance of coated carbides for metalcutting continues to increase steadily as improvements in the coating itself and in substrate materials are introduced. Coatings now are more uniform in thickness and have less porosity than earlier coatings; adhesion to the substrate has been improved, and undesirable interface reactions have been suppressed. New substrates designed to be coated are being used, which has resulted in improved resistance to breakage and thermal deformation, the two most important substrate properties.

Coated carbides are becoming more widely used because, in high-production machining operations, they permit cutting speeds to be significantly increased. For instance, compared with uncoated carbide tools, the same tool life can be obtained at as much as 50% greater cutting speed with a TiC-coated tool, and as much as 90% greater cutting speed with an Al_2O_3-coated tool. Selection of the proper coating should be based, however, not merely on the cutting speed, but also on the tool-wear mode (edge wear vs flank wear vs cratering). Tool configuration, relief angles and type of cut (roughing or finishing) also influence selection of the proper combination of substrate material and coating.

Source: Metals Handbook Desk Edition, 1985, 18.1-18.36

Table 5. Properties of representative cobalt-bonded cemented carbides

Nominal composition	Grain size	Hardness, HRA	Density Mg/m³	lb/in.³	Transverse strength MPa	ksi	Compressive strength MPa	ksi	Proportional limit, compression MPA	ksi	Modulus of elasticity GPA	10⁶ psi
97WC-3Co	Medium	92.5-93.2	15.3	0.55	1590	230	5860	850	2410	350	641	93
94WC-6Co	Fine	92.5-93.1	15.0	0.54	1790	260	5930	860	2550	370	614	89
	Medium	91.7-92.2	15.0	0.54	2000	290	5450	790	1930	280	648	94
	Coarse	90.5-91.5	15.0	0.54	2210	320	5170	750	1450	210	641	93
90WC-10Co	Fine	90.7-91.3	14.6	0.53	3100	450	5170	750	1590	230	620	90
	Coarse	87.4-88.2	14.5	0.52	2760	400	4000	580	1170	170	552	80
84WC-16Co	Fine	89	13.9	0.50	3380	490	4070	590	970	140	524	76
	Coarse	86.0-87.5	13.9	0.50	2900	420	3860	560	700	100	524	76
75WC-25Co	Medium	83-85	13.0	0.47	2550	370	3100	450	410	60	483	70
71WC-12.5TiC-12TaC-4.5Co	Medium	92.1-92.8	12.0	0.43	1380	200	5790	840	1170	170	565	82
72WC-8TiC-11.5TaC-8.5Co	Medium	90.7-91.5	12.6	0.45	1720	250	5170	750	1720	250	558	81
64TiC-28WC-2TaC-2Cr₃C₂-4.0Co	Medium	94.5-95.2	6.6	0.24	690	100	4340	630	...	...	...	...
57WC-27TaC-16Co	Coarse	84.0-86.0	13.7	0.49	2690	390	3720	540	1170	170	441	64

Nominal composition	Tensile strength MPa	ksi	Impact strength J	in.·lb	Relative abrasion resistance(a)	Thermal expansion μm/m·°C at 200 °C	at 1000 °C	μin./in.·°F at 400 °F	at 1800 °F	Thermal conductivity, W/m·K	Electrical conductivity, % IACS
97WC-3Co	...	...	1.13	10	100	4.0	...	2.2	...	121	5.3
94WC-6Co	...	...	1.02	9	100	4.3	5.9	2.4	3.3	...	...
	1450	210	1.36	12	58	4.3	5.4	2.4	3.0	100	7.8
	1520	220	1.36	12	25	4.3	5.6	2.4	3.1	121	10.0
90WC-10Co	...	...	1.69	15	22	...	...	...	...	...	...
	1340	195	2.03	18	7	5.2	...	2.9	...	112	11.4
84WC-16Co	...	...	3.05	27	5	...	...	...	...	...	...
	1860	270	2.83	25	5	5.8	7.0	3.2	3.9	88	9.2
75WC-25Co	1380	200	3.05	27	3	6.3	...	3.5	...	71	9.8
71WC-12.5TiC-12TaC-4.5Co	...	...	0.79	7	11	5.2	6.5	2.9	3.6	35	4.3
72WC-8TiC-11.5TaC-8.5Co	...	...	0.90	8	13	5.8	6.8	3.2	3.8	50	5.2
64TiC-28WC-2TaC-2Cr₃C₂-4.0Co	...	...	...	...	8	...	...	...	...	...	...
57WC-27TaC-16Co	...	...	2.03	18	3	5.9	7.7	3.3	4.3	...	...

(a) Based on a value of 100 for the most abrasion-resistant grade.

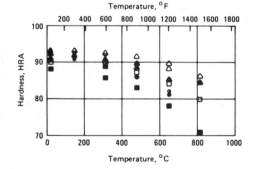

Fig. 5. Hot hardness of cemented carbides

Key:

Symbol	Composition	Grain size
o	94WC-6Co	Fine
●	94WC-6Co	Coarse
△	94WC-6Co	Very fine
▲	90WC-10Co	Very fine
□	85WC-15Co	Very fine
■	85WC-15Co	Coarse
◉	79WC-8TiC-4TaC-9Co	Medium
▲	72WC-8TiC-11.5TaC-8.5Co	Medium

Table 6. C-grade classification system for cemented carbides

C grade	Application category
Machining of cast iron, nonferrous and nonmetallic materials	
C-1	Roughing
C-2	General-purpose machining
C-3	Finishing
C-4	Precision finishing
Machining of carbon and alloy steels	
C-5	Roughing
C-6	General-purpose machining
C-7	Finishing
C-8	Precision finishing
Wear-surface applications	
C-9	No shock
C-10	Light shock
C-11	Heavy shock
Impact applications	
C-12	Light impact
C-13	Medium impact
C-14	Heavy impact
Miscellaneous applications	
C-15	Hot weld-flash removal, light cuts
C-15A	Hot weld-flash removal, heavy cuts
C-16	Rock bits
C-17	Cold header dies
C-18	Wear at elevated temperatures and/or resistance to chemicals
C-19	Radioactive shielding, counterbalances and kinetic-energy devices

Titanium carbide coatings were the first coatings used on cemented carbide and are still the most widely used. TiC usually is deposited to a thickness of about 0.005 mm (0.0002 in.), but commercial TiC-coated tool materials vary greatly in thickness of coating and type of bond between coating and substrate. These variations can cause significant differences in metalcutting performance.

Some manufacturers use several different substrates and may offer as many as three or four grades of TiC-coated carbide that differ significantly in performance characteristics. A photomicrograph of the cross section of a typical cemented carbide coated with TiC is shown in Fig. 6. The TiC layer is at the top.

Commercial grades of TiC-coated carbide do not have the superior breakage resistance of uncoated heavy-duty roughing grades of cemented carbide, such as WC-TaC-8TiC-8.5Co, a roughing grade low in TiC and relatively high in cobalt. Coated grades are not as wear resistant as the most wear-resistant uncoated grades, such as 64TiC-32WC-4Co, a grade high in TiC and low in cobalt. Nevertheless, TiC-coated carbides perform well under a wide variety of machining conditions, often producing two to three times as many parts as can be produced using uncoated finishing or general-purpose grades of comparable breakage resistance. This comparison pre-

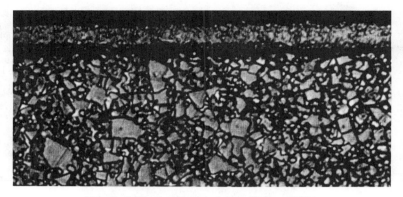

Etched with Murakami's reagent. Magnification, 1500×.
Fig. 6. Cross section of TiC-coated cemented carbide

sumes the use of similar machining conditions and similar criteria for determining the point at which the tool edge no longer can be used.

Aluminum oxide coatings are rapidly gaining acceptance for efficient metal removal at high cutting speeds. Carbide tools coated with aluminum oxide possess an edge strength much higher than that of solid ceramic cutting tools.

Commercial Al_2O_3-coated cemented carbides are available as single-coated or double-coated products. The former consist of an oxide coating 0.005-mm thick that is metallurgically bonded to a specially designed cemented carbide substrate. Double-coated products consist of a thin Al_2O_3 coating over a TiC-coated cemented carbide substrate.

Titanium nitride coatings are claimed to impart superior resistance to crater formation and flank wear in certain metal-cutting applications. TiN-coated products are easily recognized by their gold color.

Multiple coatings consisting of successive thin layers of titanium carbide, titanium carbonitride and titanium nitride are being used in increasing quantities. Multiple coatings are claimed to combine the desirable qualities of both nitride and carbide coatings, and impart to the tools better resistance to edge wear, flank wear and cratering, a combination that cannot be obtained with a single-layer coating.

NICKEL-BONDED TITANIUM CARBIDE

A satisfactory method of cementing titanium carbide using molybdenum carbide and nickel has been developed. The resultant material has good crater resistance, low coefficient of friction and low thermal conductivity. Although penetration hardness is high, abrasion resistance is lower than that of a cobalt-bonded tungsten carbide having equal or lower penetration hardness.

Typically, cemented titanium carbides contain 8 to 25% Ni, 8 to 15% Mo_2C and 60 to 80% TiC. Occasionally, tungsten carbide, cobalt, titanium nitride and other additives may be present in smaller amounts.

STEEL-BONDED CARBIDE

Steel-bonded carbide is a P/M tool material intermediate in wear resistance between tool steels and cemented carbides. Steel-bonded carbide consists of 40 to 55 vol % titanium carbide homogeneously dispersed in a steel matrix.

It is customary to follow tool steel practice and make the entire tool from steel-bonded carbide.

The tool can be joined to a supporting member either of the same material or of steel. This may be done by mechanical fastening, adhesive bonding or brazing. Brazing under vacuum using AWS BNi-8 (a nickel-manganese brazing filler metal) provides a ductile, high-strength joint with a remelt temperature in excess of the temperature recommended for heat treating grade C steel-bonded carbide.

Grade C is used for progressive stamping dies; lamination dies; dies for drawing, bending and curling sheet metal and wire; tube rolls; gages and fixtures. Grade CM is used to make tools for cold and warm forming of heavy-gage stock. Grade SK is used to make hot work rolls and forging dies, including dies for forging hot powder metals. Generally, steel-bonded carbide is not recommended for cutting tools to be used for machining ferrous metals because the hardness drops off too rapidly at the high temperatures developed at the cutting edge. Interface temperatures developed during cutting of nonferrous metals are not too high, on the other hand, and several grades of steel-bonded carbide have been used for machining these materials.

Steel-bonded carbide performs well in severe wear applications involving sliding friction. Its success has been attributed to exceptionally hard, extremely fine, rounded grains of titanium carbide exposed in slight relief at the surface. To provide this condition, the surface preparation of the tool after heat treatment requires lapping with a coarse compound (about 30-μm particle size) to remove any grinding marks and smeared metal, then lapping with a medium compound (15-μm particle size) and finally polishing with a fine abrasive (6-μm particle size) to a mirror finish. Aluminum oxide is satisfactory as the abrasive.

Other grades of steel-bonded carbide are available that have special-alloy matrices to provide corrosion resistance or nonmagnetic properties. Also, some grades can be surface treated to improve the wear resistance of the matrix in applications involving extremely fine abrasives. These surface treatments include nitriding and boriding.

CAST Co-Cr-W-Nb-C ALLOYS

Developed early in the 20th century to provide tools with better capabilities than high speed steel, cast cobalt-chromium-tungsten-niobium-carbon alloys are still produced. In general, as cutting tools, they bridge the gap between high speed steels and cemented carbides. Several cast Co-

Cr-W-Nb-C wear-resistant parts have been used in machinery applications as well.

Use of cast Co-Cr-W-Nb-C cutting tools should be considered:

- Where relatively low surface speeds cause build-up with cemented carbides
- Where machines lack the power or rigidity to use cemented carbides effectively
- Where higher production is desired than is possible with high speed steel tools
- For multiple-tool operations where surface speed of one or more operations falls between the recommended speeds for high speed steel and carbide tools
- For short runs on automatic equipment where form grinding of carbide tools is excessively costly
- For machining rough surfaces of castings where the surfaces contain abrasive materials such as residual sand, surface oxides, slag or refractory particles.

Tools made of cast Co-Cr-W-Nb-C alloys usually are not recommended for light, very fast finishing cuts.

CERAMICS

Ceramic cutting-tool materials use aluminum oxide as the base material. They are available as indexable inserts that can be used in the same holders as those used for cemented-carbide inserts. However, because ceramics are more brittle than carbides, extra care must be used to ensure that the inserts are firmly seated. Also, tool overhang should be kept at 50 to 100% of tool-holder thickness.

The many available ceramic tools fall into three general groups:

- *Group A-1:*
  Al_2O_3 with up to about 10% of other oxides or carbides, primarily those of titanium, magnesium, molybdenum, chromium, nickel or cobalt. The mixtures are cold pressed and sintered.
- *Group A-2:*
  Essentially pure Al_2O_3, hot pressed
- *Group A-3:*
  Al_2O_3 plus 25 to 30% of a refractory carbide such as titanium carbide, hot pressed.

Typical properties of these groups are shown in Table 7. Also, the properties of a specific example of a Group A-1 material have been included, for which the microstructure is shown in Fig. 7.

Because ceramic tools are predominantly oxides, they are not subject to the oxidation that limits the usefulness of cemented carbides at high temperatures in air.

Ceramic tool materials retain their resistance to wear and deformation at much higher temperatures than the best cemented carbides. Consequently, ceramic tools generally can cut for acceptable periods of time at speeds much higher than those possible with cemented carbide tools. Coolants should not be used when cutting with ceramic tools.

POLYCRYSTALLINE CUBIC BORON NITRIDE

In 1973, polycrystalline composite cubic boron nitride (CBN) cutting tools were introduced. By use of high-pressure, high-temperature pro-

Table 7. Typical properties of ceramic tool materials

Property	Group A-1 General	Group A-1 Example(a)	Group A-2	Group A-3
Hardness: HRA	93-94	93-94	93-94	93-94
Density:				
Mg/m^3	3.96-3.98	4.1	4.0	4.24
lb/in.^{3}	0.142-0.143	0.148	0.144	0.153
Transverse strength:				
MPa	480-690	620	640	760
ksi	70-100	90	92.5	110
Compressive strength:				
MPa	3790-4480	2140(b)	4140	3930-4070
ksi	550-650	310(b)	600	570-590
Modulus of elasticity:				
GPa	390	400	390	...
10^6 psi	57	58	57	...
Impact strength:				
J	...	0.23	...	...
in. · lb	...	2	...	...
μm/m · °C	...	6.1(d)	7.2	7.7
Thermal expansion(c): μin./in. · °F	...	3.4(d)	4.0	4.3
Thermal conductivity:				
At room temperature:				
W/m · K	29	...	...	17-21
Btu/ft · °F	17	...	...	10-12
At 100 °C (212 °F):				
W/m · K	22	...	29	...
Btu/ft · h · °F	13	...	17	...
At 450 °C (850 °F):				
W/m · K	11	...	...	...
Btu/ft · h · °F	6.5	...	...	...
At 600 °C (1100 °F):				
W/m · K	...	...	...	14.7
Btu/ft · h · °F	...	...	...	8.4

(a) 89Al$_2$O$_3$-11TiO, cold pressed and sintered. (b) Proportional limit. (c) At 21 to 200 °C (70 to 400 °F). (d) 8.3 μm/m · °C (4.6 μin./in. · °F) at 21 to 980 °C (70 to 1800 °F).

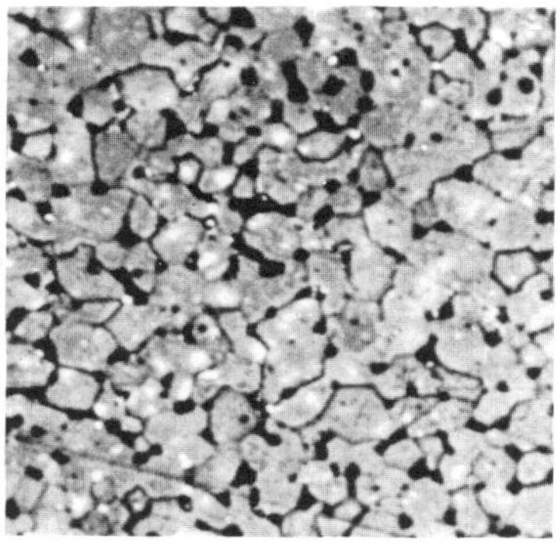

Phosphoric acid etch. Magnification, 750×. For material properties, see A-1 example in Table 7.

Fig. 7. Microstructure of 89Al$_2$O$_3$-11TiO grade A-1 ceramic tool material

cesses, a layer of CBN is bonded to a cemented carbide substrate. The CBN is held together primarily by CBN-CBN intercrystalline bonds.

Most frequently, composite CBN tools are supplied in the same shapes as those used for cemented carbides. In many cases, CBN inserts can be brazed into a steel holder using the same procedures employed for cemented carbides, as long as special care is taken to avoid overheating the structure and to prevent molten flux from contacting the CBN layer.

Because the cost of a CBN tool is several times that of a cemented carbide tool, it is usually necessary to regrind CBN tools with diamond wheels and reuse them to minimize the cost per tool edge.

Because of their high hardness, their ability to hold this hardness and the fact that they resist

oxidation at much higher temperatures than carbides, composite CBN tools are used effectively to cut difficult-to-machine superalloys at speeds several times higher than those possible with cemented carbides. The best conditions for their use are in the self-induced thermal machining mode.

In addition to nickel-base and cobalt-base high-temperature alloys, other ferrous materials — including chilled cast iron, Meehanite cast iron, tool steels (M2, M42, D2, A2, S5 and O1) and other steels (1055, 8620, 52100 and 4140) hardened to 50 to 70 HRC — have been machined successfully with CBN tools.

DIAMOND

Diamond, the tetrahedral form of carbon, is the hardest and most scratch-resistant material known. Its scratch hardness number is 10 on Mohs' scale. It will scratch all other materials and will be scratched by none. Similarly, its penetration hardness is 5000 to 12 000 HK, roughly twice that of the next hardest material.

These characteristics make diamond very attractive as a tool material. However, industrial-grade natural single-crystal diamonds are expensive even in small sizes. In addition, diamonds are very brittle and cleave easily along certain crystallographic planes. Also, diamond starts to oxidize rapidly at 650 °C (1200 °F), and at atmospheric pressure it reverts to graphite above 1500 °C (2700 °F).

These properties and the fact that carbon dissolves rapidly in iron at high temperatures make diamond unsatisfactory for machining ferrous alloys.

However, diamond tools are used effectively in machining high-silicon cast aluminum alloys, copper and its alloys, sintered cemented tungsten carbides, rubber impregnated with silica glass,

glass-fiber/plastic and carbon/plastic composites, and high-alumina ceramics.

Diamonds are used extensively in resin-bonded and metal-bonded grinding wheels. In addition, diamond dies are used for drawing fine wire. Other common industrial uses of diamonds are as tools for dressing abrasive grinding wheels, as diamond cutoff saws and laps (when dispersed in a metal matrix) and as a polishing abrasive when in very finely divided form.

The relative abrasion resistance and Knoop hardness values of laminated diamond, natural diamond and cemented 94WC-6Co tool blanks are compared in the following table:

Tool material	Hardness, HK	Relative abrasion resistance
Laminated diamond/carbide composite	5500-8000	250
Natural diamond single crystal	8000-12 000	96 to 245
Cemented 94WC-6Co	1800-2200	2

The relative abrasion resistance is based on the time, in minutes, required to generate a specific size of wear land on the material when turning a siliceous hard rubber commonly used as a coating for steel rolls in the paper industry.

Distortion in Tool Steels

DISTORTION in tool steel parts includes all irreversible changes in size and shape that result from processing, from heat treatment, and from temperature variations and loading in service.

Changes in size or shape of tool steel parts may be either reversible or irreversible. Reversible changes are those caused by stressing in the elastic range or by temperature variations that neither cause changes in the metallurgical structure nor induce stresses that exceed the elastic range. Under such conditions, the initial dimensional values can be restored by a return to the original state of stress or temperature.

The upper limit of reversible dimensional change in a tool steel is determined by the stress required to initiate deformation (that is, the elastic limit corresponding to a preselected value of plastic strain), the elastic deformation per unit stress (modulus of elasticity), the effect of temperature on these properties, the coefficient of thermal expansion and the temperature-time combinations at which stress relief and phase changes occur.

For practical purposes the modulus of elasticity of all tool steels, regardless of composition or heat treatment, is 210 GPa (30 × 10^6 psi) at room temperature. Therefore, if a tool steel part deforms excessively under service loading but returns to its original dimensions when the load is removed, a change in grade or type of tool steel or in heat treatment will not be useful. To counteract excessive elastic distortion it is necessary to (a) reduce the applied stress by increasing the section size or (b) use a tool material with a higher modulus of elasticity (such as cemented tungsten carbide.)

Irreversible changes in size or shape of tool steel parts are those caused by stresses that exceed the elastic limit or by changes in metallurgical structure (most notably, phase changes). Such irreversible changes sometimes can be corrected by thermal processing (annealing, tem-

374

pering or cold treating) or by mechanical processing to remove excess material or to redistribute residual stresses.

NATURE AND CAUSES OF DISTORTION

Distortion is a general term encompassing all irreversible dimensional changes. There are two main types: size distortion, which involves expansion or contraction in volume or linear dimensions without changes in geometrical form; and shape distortion, which entails changes in curvature or angular relations, as in twisting (warpage) or bending, and nonsymmetrical changes in dimensions. Usually, both size distortion and shape distortion occur during any heat treating operation.

SIZE DISTORTION IN TOOL STEELS

Typical volume percentages of martensite, retained austenite and undissolved carbides are given in Table 8 for four different tool steels quenched from their usual austenitizing temperatures.

Typical changes in linear dimensions for several tool steels are given in Table 9. As shown in this table, some tool steels, such as A10, show very little size change when hardened and tempered over the entire range from 150 to 600 °C (300 to 1100 °F).

Other types, such as the M2 and M41 high speed steels, expand about 0.2% (2 mm/m, or 0.002 in./in.) when hardened and tempered in the usual range of 540 to 595 °C (1000 to 1100 °F) to develop full secondary hardness. Although the information in Table 8 is useful in comparing size distortion in several tool steels, the factor of shape distortion makes it impossible to use these data alone to predict dimensional changes of a particular tool made from any of these steels.

SHAPE DISTORTION IN TOOL STEELS

In considering shape distortion, it is important to recognize that the strength of any tool steel decreases rapidly above about 600 °C (1100 °F). At the austenitizing temperature, the yield strength is so low that plastic deformation often occurs simply from the stresses exerted on the part by its own weight. Therefore, long parts, large parts and parts of complex shape must be properly supported at critical locations to prevent sagging at the hardening temperature.

Rapid heating increases shape distortion, especially in large tools and in complex tools containing both light and heavy sections. If the rate of heating is high, light sections will increase in temperature much faster than heavy sections. Likewise, the outer surfaces will increase in temperature much faster than the interior, especially in moderate to heavy sections. Differences in thermal expansion due to the differences in temperature will be sufficient to set up large stresses in the material. Under these stresses, the hotter regions will deform plastically, thereby relieving the thermally induced stress.

On continued heating, the hotter portions will begin to level off at the furnace temperature, while the cooler portions will continue to increase in temperature. This produces a decrease in thermal differential, which in turn causes at least a partial reversal in thermal stress because of the plastic deformation that took place when the temperature differential was high. This may or may not cause the part to undergo further plastic deformation, but if it does, the deformation will most likely be lesser in extent than the deformation that took place when the temperature differential was high, and will most likely be in a different direction.

Slow heating minimizes distortion by keeping temperature differentials low throughout the heating cycle. Ideally, all heat treatment of tool steel parts should start from a cold furnace to provide the greatest freedom from shape distortion during heating. Starting from a cold furnace is neither very practical nor energy efficient unless heat treating is being done in a vacuum furnace. For heat treating in fused salt or an atmosphere furnace, preheating of parts at an intermediate temperature prior to heating them to the austenitizing temperature provides a useful compromise.

On cooling to form martensite, large temperature differences between surface and interior, and between light and heavy sections, can cause severe shape distortion. This problem is most likely to arise if the hardenability of the steel is so low that a high cooling rate is required to obtain full hardness. In that event, it may be best to substitute a high-hardenability, air-hardening tool steel, especially when a large or complex part is being made.

However, if lower-hardenability steels that require liquid quenching are used, fixturing and pressure die quenching will help minimize distortion. Long, symmetrical parts should be fixtured, and should be quenched in the vertical position and agitated vertically while completely submerged in the quenching medium.

SPECIAL TECHNIQUES FOR CONTROLLING SHAPE DISTORTION

Besides being reduced through control of rates of heating and cooling, as discussed above, shape distortion can be reduced by quenching locally instead of quenching the entire part, or by using flame, induction, electron beam or laser methods to harden only that portion of the tool that must be hardened.

Special hardening procedures such as martempering and austempering may also be useful for controlling distortion.

Controlling out-of-roundness is important for certain precision applications, such as class C and D cutting hobs made of high speed steels. Class C and D hobs must be held to close size limits because they are not ground to size after heat treatment, but rather are used in the unground condition.

Normal size distortion in hardening and tempering can be accommodated by making the tool slightly oversize or slightly undersize, as required, and then heat treating. High speed steel bars, however, have been observed to go out-of-round as much as 0.05 mm (0.002 in.) when conventionally processed. This out-of-roundness problem can be combated by using specially processed wrought bars or bars made from hot isostatically pressed powders, which maintain the best possible symmetry during conventional heat treatment.

Stabilization involves reducing the amount of retained austenite that can slowly transform and thus produce distortion if the material is heated or subjected to stress. Stabilization also reduces internal (residual) stress, which in turn makes distortion in service due to stress relaxation less likely to occur. Stabilization is most important for tools

Table 8. Microconstituents in four tool steels after hardening

Steel	Hardening treatment	As-quenched hardness, HRC	Martensite, vol %	Retained austenite, vol %	Undissolved carbides, vol %
W1	790 °C (1450 °F), 30 min; WQ	67.0	88.5	9	2.5
L3	840 °C (1550 °F), 30 min; OQ	66.5	90	7	3.0
M2	1225 °C (2235 °F), 6 min; OQ	64	71.5	20	8.5
D2	1040 °C (1900 °F), 30 min; AC	62	45	40	15

Table 9. Typical dimensional changes in hardening and tempering

Tool steel	Hardening treatment °C	°F	Quenching medium	Total change in linear dimensions, %, after quenching	°C / °F	150 / 300	205 / 400	260 / 500	315 / 600	370 / 700	425 / 800	480 / 900	510 / 950	540 / 1000	565 / 1050	595 / 1100
O1	816	1500	Oil	0.22		0.17	0.16	0.18	...	...	...	...	...	...	...	...
O1	788	1450	Oil	0.18		0.09	0.12	0.13	...	...	...	...	...	...	...	...
O6	788	1450	Oil	0.12		0.07	0.10	0.14	0.10	0.00	−0.05	−0.06	...	−0.07	...	...
A2	954	1750	Air	0.09		0.06	0.06	0.08	0.07	...	0.05	0.04	...	0.06	...	...
A10	788	1450	Air	0.04		0.00	0.00	0.08	0.08	0.01	0.01	0.02	...	0.01	...	0.02
D2	1010	1850	Air	0.06		0.03	0.03	0.02	0.00	...	−0.01	−0.02	...	0.06	...	...
D3	954	1750	Oil	0.07		0.04	0.02	0.01	−0.02	...	...	...	...	...	...	...
D4	1038	1900	Air	0.07		0.03	0.01	−0.01	−0.03	...	−0.4	−0.03	...	0.05	...	...
D5	1010	1850	Air	0.07		0.03	0.02	0.01	0.00	...	0.3	0.03	...	0.05	...	...
H11	1010	1850	Air	0.11		0.06	0.07	0.08	0.08	...	0.3	0.01	...	0.12	...	...
H13	1010	1850	Air	−0.01		...	...	...	...	...	...	0.00	...	0.06	...	...
M2	1210	2210	Oil	−0.02		...	...	...	...	...	...	...	−0.06	0.10	0.14	0.16
M41	1210	2210	Oil	−0.16		...	...	...	...	...	...	...	−0.17	0.08	0.21	0.23

that must retain their size and shape over long periods of time.

If the tool steel chosen provides the required hardness after tempering at a relatively high temperature, it is possible to reduce the amount of retained austenite and the internal stress by multiple tempering. Initial tempering reduces internal stress and conditions the retained austenite so that it can transform to martensite on cooling from the tempering temperature. Usually, a second or third retempering treatment is necessary to reduce the internal stress set up by the transformation of retained austenite.

Single or repeated cold treatment to a temperature below M_f will cause most of the retained austenite to transform to martensite in plain carbon or low-alloy tool steels that must be tempered at low temperatures to achieve the hardness required for the application. Cold treatment may be applied either before or after the first temper, but if the tools tend to crack because of the additional stress induced by dimensional expansion during cold treatment, it is generally prudent to apply cold treatment after the tools have been tempered the first time. When cold treatment is applied after the first temper, the amount of retained austenite that transforms on cold treatment may be considerably less than would be expected, because some of the austenite may be stabilized by tempering prior to cold treating.

Cold treatment is usually done in a commercial refrigeration unit capable of attaining temperatures of −70 to −95 °C (−100 to −140 °F). Tools must be retempered promptly after returning to room temperature following cold treatment, to reduce internal stress and increase the toughness of the newly formed martensite.

For some tools, a small percentage of retained austenite is desirable to improve toughness and provide a favorable internal stress pattern that will help the tool withstand service stresses. For these tools, little or no stabilization may be preferred, and a full stabilizing treatment may actually result in tools that are unfit for service or only marginally able to perform their required functions.

Tool Materials
for Special Applications

Cutting Tools

CUTTING TOOLS include all of the various styles of cutters used in machining of metals, plastics, woods and other machinable structural materials. The most common styles include single-point tools, drills, reamers, taps, threading dies, milling cutters, end mills, broaches, saws and hobs. For many of these styles, the actual cutting edge is on a detachable portion of the tool called an "insert." Inserts generally are made of materials different from those of the tool holders in which they are affixed.

SINGLE-POINT CUTTING TOOLS

Single-point cutting tools are those types of cutters having essentially only one cutting edge and/or one corner in contact with the workpiece throughout a given cutting cycle. Single-point cutting tools are used for turning, boring, shaping, planing and threading.

A substantial portion of all single-point tools used in the United States consist of superhard tool bits (inserts) mounted in tool holders usually made of carbon steel or alloy steel.

High speed steels are used for single-point cutting tools in the form of both solid tool bits and inserts. Molybdenum types M2 and M4 generally are recommended for solid tool bits used for general-purpose machining of metals whose hardness is below about 250 HB. For machining harder metals, M42 and T15 high speed steels, which have better hot hardness, are generally preferred. Types T4 and T5 are relatively common alternatives for types M2 and M4 in single-point tools for machining cast irons and copper alloys.

Cemented carbides are used largely for inserts. Single-point carbide tools are generally preferred for high-volume production machining, where productivity is significantly enhanced by high machining speeds, relatively deep cuts and relatively high feed rates, and where it is highly desirable to change tool bits only at infrequent intervals. With single-point carbide tools it is most important to establish optimum machining speed, because this factor has the greatest effect on tool life.

Specific types of carbide are preferred for single-point turning of certain metals. Straight grades of tungsten carbide are intended primarily for use in machining cast iron and nonferrous metals. When used for machining steel, the straight grades tend to crater rapidly, and thus it is often better to select a coated carbide or a complex grade containing titanium carbide instead.

Ceramic inserts, such as those of solid aluminum oxide, permit very high machining speeds to be used with no sacrifice in tool life. They also produce finer finishes than those obtained with other insert materials. Ceramics are weaker than carbides or coated carbides, and can be used only for applications where impact loading is low.

Diamond inserts may be either single-crystal natural diamonds or carbide substrates under a layer of randomly oriented fine-grain polycrystalline synthetic diamond. Single-crystal natural diamond inserts have outstanding wear resistance, but they cannot withstand high shock loading. Diamond-carbide composite inserts combine excellent wear resistance with good resistance to shock loading.

Diamond tools are preferred for machining soft, abrasive nonferrous metals, and abrasive nonmetallic materials such as cemented tungsten carbide, unfired ceramics, filled plastics, rubber, carbon and graphite.

Boron nitride bonded to a cemented carbide substrate offers exceptionally high wear resistance and edge life in the machining of high-temperature alloys and hardened ferrous metals. Next to diamond, cubic boron nitride is the hardest known material. This accounts for its cutting properties, and use of a cemented carbide substrate gives the insert satisfactory resistance to shock.

DRILLS

Drilling of holes for assembly of metal parts is performed principally with twist drills made of high speed steel. Carbide drills of several designs are sometimes used for drilling cast iron, high-silicon aluminum alloys and abrasive materials, as well as certain ceramic and plastic materials of high hardness. For drilling holes of large diameter — about 25 mm (1 in.) or greater — in-

dexable carbide insert-style drills are sometimes used. The number of holes drilled with all types of carbide drills is still a small percentage of the total number of holes drilled.

High Speed Steel Drills. High speed steels M1, M2, M7 and M10, which have the highest strength and toughness, are used for most of the drills manufactured in the United States. Each of these steels has a specific range of carbon content normally associated with it. Steels with carbon contents at the upper end of the range normally are used for drills requiring less toughness and more resistance to abrasion. Drills subject to shock during use often are produced from steels with carbon contents at the lower end of the normal range, to provide better toughness.

Cemented Carbides for Drills. Drilling of holes in concrete, glass and various ceramic materials with carbide-tipped "masonry" drills is a well-known application for cemented carbide tool materials. Solid cemented carbide twist drills of special design are being used in large quantities for drilling printed circuit boards made of glass-filled epoxy that is abrasive and comparatively low in strength.

Carbide-tipped die drills have been used successfully on some difficult-to-machine heat-resistant materials, especially where the machine setup is very rigid.

Straight tungsten carbides containing up to 6% cobalt are used successfully in tipped twist drills for drilling cast iron and nonferrous metals.

REAMERS

Reamers are cutting tools used for finishing holes that must meet stringent finish and size-tolerance requirements that cannot be satisfied by simple drilled holes.

High speed steel, with its high resistance to softening at elevated temperatures, has replaced carbon steel for all except hand reamers.

In choosing materials for machine reamers, high hardness and high abrasion resistance are the most important properties. The general-purpose high speed steels (M1, M2, M7, M10 and T1) at high hardness levels, and the high-vanadium types (M3, M4 and T15), have been used successfully. The latter steels have greater resistance to abrasion than the lower-vanadium types. The very high resistance to softening at elevated temperatures provided by cobalt-containing high speed steels such as M33, M42, M6 and T5 is less often required in reaming than in drilling, because heat is more easily controlled in reaming.

Solid cemented carbide reamers and carbide-tipped reamers have been used widely and successfully in reaming almost all metals. General-purpose and harder grades of carbide are used most commonly.

TAPS

Taps are tools used for cutting or forming internal screw threads. Selection of tool materials for taps usually is done by the tap manufacturer. Three families of materials generally are considered: (a) carbon and alloy tool steels, (b) high speed steels and (c) cemented carbides.

Carbon and alloy tool steels are suitable for taps used for hand tapping or other low-speed, light-duty applications.

High speed steels are used for taps that must cut efficiently at high speeds and thus must resist softening under the extreme heat generated at the cutting edge of the tool. Most tap manufacturers use M1, M2, M7 and M10 for general-purpose taps.

For special tap designs that require greater resistance to abrasion, type M3 (class 2) or type M4 is generally used. Where requirements include high abrasion resistance and/or very high resistance to softening at elevated temperatures, types T15 and M42 are the most popular choices. Three common surface treatments — nitriding, oxide coating and hard chromium plating — are often used to improve tap life.

Cemented carbides can be considered for extremely abrasive applications, such as tapping of filled plastics and certain grades of cast iron.

END MILLS

An end mill is a shank-type milling cutter with cutting edges on its periphery and end surface. The shank may be either straight or tapered. These tools are available in diameters ranging from 0.8 to 75 mm ($^1/_{32}$ to 3 in.), and may have one or more cutting teeth (most have 2, 4 or 6 teeth). Larger sizes are made as special items on customer request. End teeth may be of the noncenter type or may be designed to cut to the axis of the tool to permit plunge cutting.

End-Mill Materials. The majority of end mills larger than 16 mm ($^5/_8$ in.) in diameter are made from high speed steel bars, which in the annealed state may be easily machined to shape, heat treated and finished by grinding. End mills smaller than 16 mm in diameter are commonly made from hardened cylindrical blanks of high speed steel into which the flutes are ground.

The majority of workpieces machined with high speed steel end mills have hardnesses below 300 HB. For these applications, end mills made from the more popular and less costly general-purpose tool steels such as M1, M2, M7 and M10 generally have proved satisfactory on the bases of both economy and performance.

For work materials that are difficult to machine and that have hardnesses from 350 to 450 HB, end mills made of cobalt high speed steels such as T15, M33 and M42 are most effective.

Cemented carbide end mills have the ability to withstand higher cutting temperatures and greater abrasion than high speed steel end mills but are somewhat less resistant to shocks caused by interrupted cuts or voids. Carbide end mills are used most often for machining nonferrous alloys; nonmetallic materials; work materials with hardnesses exceeding 450 HB; and materials with highly abrasive scaled surfaces, such as sand castings.

The straight grade of tungsten carbide containing 6% cobalt is used most commonly for solid carbide end mills and for brazed-tip tools, both of which are used for cutting cast iron, titanium alloys, nonferrous metals, stainless steels and plastics. A complex grade containing 72 to 73% WC, 7 to 8% TiC, 11.5 to 12% TaC and 8 to 8.5% cobalt binder is used for end mills in steel-cutting applications.

MILLING CUTTERS

Milling cutters that are relatively small and of complex configuration generally are made of high speed steels. The cutting teeth on tools of this type often have helical cutting edges, deep radial and/or axial gashes, irregular profiles or thin web sections. These complex configurations can be most readily obtained by machining the tool as an integral unit from annealed high speed steel, heat treating it to a suitable hardness, and then grinding it to size.

There are many special applications and operating conditions for which milling cutters are constructed of materials other than solid high speed steel. For economy, large-diameter cutters, and cutters of simple configuration used in high-production applications, often comprise high speed steel blade inserts attached to low-cost alloy steel bodies.

Because of their high cost, cemented carbides are used predominantly in those milling cutters into which carbide tips can be brazed or carbide blades inserted.

For general-purpose cutting of steel, the grades containing complex carbides are widely used in all types of milling cutters. Other types of workpiece materials, such as cast irons, various brasses, aluminum alloys and fiber composites, are most productively cut using the straight tungsten carbide grade containing 6% Co. This straight tungsten carbide also performs well in cutting stainless steels of the 300 series.

Tool steel milling cutters used for cutting plain carbon and low-alloy steels, cast irons and nonferrous alloys, where workpiece hardness does not exceed 30 HRC, normally are made of high speed steels such as M1, M2, M7 and M10. T1 cutters are available from most cutter manufacturers, but they are high in initial cost.

At intermediate hardness levels from 30 to 35 HRC, the high-vanadium grades M3 and M4 provide increased tool performance. As the hardness of the workpiece increases beyond 35 HRC, tool life of the general-purpose grades drops very rapidly. For workpiece hardness levels from 35 to about 45 HRC, cobalt high speed steels such as M42 and T15 are recommended. No high speed steel, however, is recommended for cutting metals with hardnesses much above 45 HRC; for these high-hardness materials, carbide tools must be used. In general, carbide milling cutters can be used at cutting speeds three to six times those permitted with high speed steel cutters.

As a group, nickel and cobalt high-temperature alloys are very difficult to machine, and best results are obtained by using cobalt-bearing high speed steel cutters, low cutting speeds and heavier-than-normal feed rates. Carbide cutters normally are not recommended because they are prone to chipping of the cutting edge.

HOBS

A hob is a type of milling cutter used to generate a repeating form about a center, such as in cutting of gear teeth, spline teeth and serrations. The hob and workpiece must rotate and mesh in a specific timed relationship, and thus the cutting teeth on the hob follow a helical or thread pattern around the periphery of the tool. This is the feature that distinguishes a hob from a milling cutter.

The majority of hobs are made from high speed steels, although hobs for special applications have been manufactured from cemented tungsten carbide, cast cobalt-chromium-tungsten alloys, and even some low-alloy tool steels. Virtually all types of high speed steel have been used, but the most widely used is type M2 in either the standard or the high-carbon version.

General-purpose types like M2 offer good combinations of edge strength and wear resistance, but high speed steels such as M3 and M4, which contain more carbon and vanadium, are used for hobbing harder and more abrasive materials. For applications that require increased resistance to softening at elevated temperatures, cobalt-bearing grades such as M42 and T15 are more suitable.

A recent innovation has been the production of high speed steel by powder metallurgy. A broad range of grades is available, and the same application guidelines apply to them as apply to wrought tool steels.

Shearing and Slitting Tools

MOST SHEAR BLADES are solid, one-piece blades made of tool steel. However, some are composite tools that consist of tool-material inserts in heat treated medium-carbon or low-alloy steel backings.

COLD SHEARING AND SLITTING OF METALS

Blade materials recommended for cold shearing of various metals are presented in Table 1, and blade materials for rotary slitting are given in Table 2.

Tool materials vary in toughness and wear resistance, and the metals being sheared vary in hardness and resistance to shearing. If the material to be sheared is very thin and of relatively low hardness, the shear-blade material can be low in toughness but must have optimum wear resistance. For shearing material of greater thickness and higher hardness, it may be necessary to decrease blade hardness, or change to a less wear-resistant blade material or to a shock-resistant tool steel having a hard case over a tough core, to obtain the toughness needed to resist edge chipping.

HOT SHEARING OF METALS

Shearing is done at elevated temperatures when the work material is thick and resistant to shearing or when hot shearing is otherwise desirable as part of the manufacturing process. The strong

Table 1. Recommended blade materials for cold shearing of flat metals

Material to be sheared	6 mm ($\frac{1}{4}$ in.) or less	6 to 13 mm ($\frac{1}{4}$ to $\frac{1}{2}$ in.)	13 mm ($\frac{1}{2}$ in.) and over
Carbon and low-alloy steels up to 0.35% C	D2, A2	A2, A9	S2, S5, S6, S7
Carbon and low-alloy steels, 0.35% C and over	D2, A2	A9, S5, S7	S2, S5, S6, S7
Stainless steels and heat-resisting alloys	D2, A2	A2, A9, S2	S2, S5, S6, S7
High-silicon electrical steels	D2, T15, cemented carbide inserts(a)	S2, S5, S7	(b)
Copper and aluminum alloys	D2, A2	A2	S2, S5, S6, S7
Titanium alloys	D2	...	...

(a) Carbide inserts usually are brazed to heat treated medium-carbon or low-alloy steel backings. (b) Seldom sheared in these thicknesses.

Table 2. Recommended blade materials for rotary slitting of flat metals

Material to be sheared	4.5 mm ($\frac{3}{16}$ in.) or less	4.5 to 6.5 mm ($\frac{3}{16}$ to $\frac{1}{4}$ in.)	6.5 mm ($\frac{1}{4}$ in.) or more
Carbon, alloy and stainless steels	D2	D2, A2, A9	A9, S5, S6, S7
High-silicon electrical steels	D2, M2	D2	...
Copper and aluminum alloys	A2, D2	A2, D2	A2, S5, S6, S7
Titanium alloys	D2, A2	...	...

secondary hardening of group H tool steels provides sufficient resistance to softening to make them useful for shear blades operating at temperatures up to 425 °C (800 °F).

Hardness of blades for hot shearing varies considerably with conditions such as thickness and temperature of the metal being sheared, and type and condition of available equipment. However, hardness is usually kept within the range from 38 to 48 HRC.

MATERIALS FOR MACHINE KNIVES

Unlike materials for metal-slitting applications, materials for knives used in cutting papers, films and foils usually are selected on the basis of cost and wear resistance, without consideration of toughness.

Score cutters are most commonly made of 52100 steel hardened to 60 to 62 HRC—a hardness level that is adequate for most operations and that will not result in scoring of an opposing platen sleeve or hardened roll. Such sleeves and rolls are made of 52100 or carburized steel, hardened to not less than 60 HRC. Alternative materials for score cutters are O7 and D2 tool steels.

Shear cutters can be produced from a much wider array of alloys, the choice being influenced by such factors as tool design, material to be cut, machine design, and maintenance limitations. The entire range of tool steels (including the popular 52100 and other low-alloy types plus the higher-alloy group D, group M and group T steels), as well as specialty tool steels and cemented carbides, can be considered. For standard applications, 52100 and O1 are selected most often, chiefly because they are readily available as sheet, bar stock or tubing.

Burst cutters are, by design and function, fairly thin knives. Material selection thus is restricted to alloys that are readily available in thin sheet, a group that includes 52100 steel, 1075 steel and razor-blade stock. Except in those applications where resistance to elevated temperature is required (as in core cutting with a stationary blade), use of high speed steels is rarely economical.

Single-knife cutters are used to reduce wide rolls of paper, foam or textile to narrower rolls without unwinding and rewinding. These knives have long, thin, one-sided or two-sided bevels and are kept sharp by means of one or more grinding wheels situated on the machine, which are activated either automatically (for grinding at specific intervals) or manually (for grinding as required).

Under normal circumstances, such knives are made from L2, L6 or D2 tool steel. Certain applications of single knives, including slicing of foam or impregnated fabrics, cause the thin bevel to heat up considerably as it penetrates the material being slit. In these cases, an alloy that has higher resistance to elevated temperatures, such as M2 or T1 tool steel, may be required.

A standard paper-trimmer knife consists of a carbon steel backing and an insert that provides the actual cutting edge. The insert material can range from O1 to M2 to one of the group T high speed steels. For most applications, O1 is ideal with respect to initial cost, ease of maintenance, and adequate performance between resharpenings.

Sheeter and cutoff knives are similar in function to trimmer knives and frequently are made with cutting-edge inserts of M2 high speed steel hardened to 62 to 64 HRC. Under optimum conditions, life of these blades is very long—50 to 100 million cuts between resharpenings.

Some sheeter or cutoff blade designs are too narrow in cross section to allow the use of inserts; such knives are most often made of solid 52100, O1 or D2 tool steel.

Blanking and Piercing Dies

BLANKING AND PIERCING DIES include the punches, dies and related components used to blank, pierce and shape metallic and nonmetallic sheet and plate in a stamping press. The primary measure of the performance of a die material in blanking or piercing service is the number of acceptable parts that can be produced.

Sectional views of the blanking dies and the blanking and piercing punches used for making simple parts are shown in Fig. 1. More complex parts require notching and compound dies.

MATERIALS FOR SPECIFIC TOOLS

Punches and Dies. Typical materials for punches and dies used for blanking parts of different sizes and degrees of severity from several different work materials about 1.3 mm (0.050 in.) thick, in various quantities, are given in Table 3. (Sketches of typical parts are presented in Fig. 2.) Typical materials for the punches and dies used to shave several work materials of this same thickness in various quantities are given in Table 4.

Tables 3 and 4 may be used to select punch and die materials for parts made of sheet thicker or thinner than the 1.3 mm used in the examples. For sheet of greater thickness, use the punch and die material recommended for the next greater production quantity than the quantity actually to be made (the column to the right of the actual production quantity in the table). Similarly, for sheet of lesser thickness, use the punch and die material recommended for the next lower production quantity (shift one column to the left of the actual production quantity).

Typical materials for perforator punches used on several different work materials are given in Table 5. The usual limiting slenderness ratio (punch diameter to sheet thickness) for piercing aluminum, brass and steel is 2.5:1 for unguided punches and 1:1 for guided punches. The limiting slenderness ratio for piercing spring steel and stainless steel ranges from 3:1 to 1.5:1 for unguided punches and from 1:1 to 0.5:1 for accurately guided punches.

Table 6 gives typical materials for perforator bushings of all three types (punch holder, guide or stripper, and perforator or die). These recommendations are particularly applicable to precision bushings—for instance, where the outside diameter is ground to a tolerance of −0, +0.008 mm (−0, +0.0003 in.) and is concentric with the inside diameter within 0.005 mm (0.0002 in.) TIR. The hardness of W1 bushings should be 62 to 64 HRC, and that of D2 bushings, 61 to 63 HRC.

Die plates and die parts that hold inserts normally are made of gray iron, alloy steel or tool steel. For stamping thick sheet or hard materials, either class 50 gray iron, or 4140 steel heat treated to a hardness of 30 to 40 HRC, should be used. For long-run die plates for stamping thick or hard materials, steels such as 4340 and H11 are preferred when inserts are pressed into the die plates, and 4340 is nearly always used when inserts are screwed in. Die plates for stamping thin or soft sheet may be made of class 25 or class 30 gray iron or of mild steel.

Secondary Tooling. Punch holders and die shoes for carbide dies are made of high-strength gray iron or low-carbon steel plate. Yokes for retaining carbide sections usually are made of O1 tool steel hardened to 55 to 60 HRC. Backup plates for carbide tools are preferably made of O1 hardened to 48 to 52 HRC. Strippers ordinarily can be made of low-carbon or medium-carbon steel (1020 or 1035) plate. Where a hardened plate is used for medium-production work, 4140 flame hardened, W1 conventionally hardened or W1 cyanided and oil quenched is often preferred. Hardened strippers for carbide dies and high-production D2, D4 or CPM 10V dies are made of O1 or A2, hardened to 50 to 54 HRC.

Custom-made hardened guides and locator pins usually are made of W1 or W2 for most medium- or long-run dies, or of alloy steels such as 4140 for low-cost short-run dies.

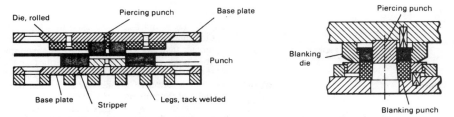

Tools at left are for short-run production of parts similar to parts 1 and 2 in Fig. 2 from relatively thin-gage sheet metal; tools at right are for longer runs. Refer to Table 3 for tooling recommendations.

Fig. 1. Sectional views of typical tools used for blanking and piercing simple shapes

Dimensions are in inches; to find equivalent metric values (mm), multiply listed values by 25. Parts 1 and 2 are relatively simple parts, and require dies similar to those illustrated in Fig. 8. Parts 3 and 4 are more complex, and require notching and compound or progressive dies.

Fig. 2. Typical parts of varying severity commonly produced by blanking and piercing

Table 3. Typical punch and die materials for blanking 1.3-mm (0.050-in.) sheet
For sketches of typical parts, see Fig. 2.

Work material	Tool material for production quantity of:				
	1000	10 000	100 000	1 000 000	10 000 000
Part 1 and similar 75-mm (3-in.) parts					
Aluminum, copper and magnesium alloys	Zn (a), O1, A2	O1, A2	O1, A2	D2	Carbide
Carbon and alloy steel, up to 0.70% C, and ferritic stainless steel	O1, A2	O1, A2	O1, A2	D2	Carbide
Stainless steel, austenitic, all tempers	O1, A2	O1, A2	A2, D2	D4	Carbide
Spring steel, hardened, 52 HRC max	A2	A2, D2	D2	D4	Carbide
Electrical sheet, transformer grade, 0.6 mm (0.025 in.)	A2	A2, D2	A2, D2	D4	Carbide
Paper, gaskets, and similar soft materials	W1 (b)	W1 (b)	W1 (c), A2 (d)	W1 (d), A2 (d)	D2
Plastic sheet, not reinforced	O1	O1	O1, A2	D2	Carbide
Plastic sheet, reinforced	O1 (e), A2	A2 (f)	A2 (f)	D2 (f)	Carbide
Part 2 and similar 300-mm (12-in.) parts					
Aluminum, copper and magnesium alloys	Zn (a), 4140 (g)	4140 (h), A2	A2	A2, D2	Carbide
Carbon and alloy steel, up to 0.70% C, and stainless steels up to quarter hard ...	4140 (h), A2	4140 (h), A2	A2	A2, D2	Carbide
Stainless steel, austenitic, over quarter hard	A2	A2, D2	D2	D2, D4	Carbide
Spring steel, hardened, 52 HRC max	A2	A2, D2	D2	D2, D4	Carbide
Electrical sheet, transformer grade, 0.6 mm (0.025 in.)	A2	A2, D2	A2, D2	D2, D4	Carbide
Paper, gaskets, and similar soft materials	4140 (j)	4140 (j)	A2	A2	D2
Plastic sheet, not reinforced	4140 (j)	4140 (h), A2	A2	D2	Carbide
Plastic sheet, reinforced	A2 (e)	A2 (e)	D2 (e)	D2 (e)	Carbide

(continued on the next page)

Press Forming Dies

PRESS FORMING is a process in which sheet metal is made to conform to the contours of a die and punch—largely by being bent or moderately stretched, or both. The suitability of a tool material for a press forming die is determined by the number of parts that can be produced using that die. This number is influenced by the size and shape of the part, the type and thickness of the metal being formed, lubrication practice, and the allowable variation in dimensions.

Typical Tool Materials. Tooling for the part shown in Fig. 3 consists of a punch and a lower die. In operation, the punch pushes the blank through the lower die, which causes wear of the lower die. The metal closely envelops the punch, with little sliding, and in that event a punch generally produces about ten times as many parts as a lower die made of the same material. However, at areas where the part shrinks against the punch during forming, wear (and possibly galling) of the punch surface will occur, particularly when the forming is done in single-action dies. For a small die and punch, the cost of steel is of minor importance, and type D2 tool steel may be used for production quantities as low as 10 000. If galling occurs during preproduction trials, the tool can be nitrided.

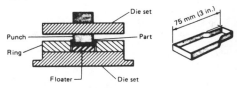

For typical die materials, see Table 7.

Fig. 3. Cross section of die used for small part of mild severity

Tooling for the part in Fig. 4 consists of a punch, an upper die and a lower die. Without the upper die, excessive wrinkling would occur at the shrink flanges. As for the part shown in Fig. 3, a less wear-resistant material is required for the punch and upper die than for the lower die. Under conditions for which the tooling is typically made of tool steel (see Table 8), the tooling is in the form of inserts in a lower die made of cast iron, as indicated in Fig. 4, and the punch is made of a tool steel such as D2.

Typical lower-die materials for press forming large parts similar to that shown in Fig. 4 are given in Table 8. For quantities less than 100 000 pieces, the entire lower die is typically made of the material indicated in the selection table, without inserts. The punch is made of a less wear-resistant material, which usually is the same as the lower-die material in the first column to the left of the quantity being considered.

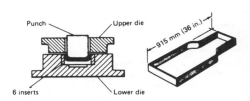

For typical die materials, see Table 8.

Fig. 4. Cross section of die used for large part of mild severity

Source: Metals Handbook Desk Edition, 1985, 18.1-18.36

Table 3. (continued)

Work material	Tool material for production quantity of:				
	1000	10 000	100 000	1 000 000	10 000 000
Part 3 and similar 75-mm (3-in.) parts					
Aluminum, copper and magnesium alloys	O1, A2	O1, A2	O1, A2	A2, D2	Carbide
Carbon and alloy steel, up to 0.70% C, and ferritic stainless steel	O1, A2	O1, A2	O1, A2	A2, D2	Carbide
Stainless steel, austenitic, all tempers	A2	A2, D2	A2, D2	D2, D4	Carbide
Spring steel, hardened, 52 HRC max	A2	A2, D2	D2, D4	D2, D4	Carbide
Electrical sheet, transformer grade, 0.6 mm (0.025 in.)	A2	A2, D2	D2, D4	D2, D4	Carbide
Paper, gaskets and other soft materials	W1 (b)	W1 (b)	W1 (k), A2	W1 (k), A2	D2
Plastic sheet, not reinforced	O1	O1	A2	A2, D2	Carbide
Plastic sheet, reinforced	O1 (m)	A2 (f)	A2 (f)	D2 (f)	Carbide
Part 4 and similar 300-mm (12-in.) parts					
Aluminum, copper and magnesium alloys	A2	A2	A2, D2	A2, D2	Carbide
Carbon and alloy steel, up to 0.70% C, and ferritic stainless steel	A2	A2	A2, D2	A2, D2	Carbide
Stainless steel, austenitic, up to quarter hard	A2	A2	A2, D2	D2, D4	Carbide
Stainless steel, austenitic, over quarter hard	A2	D2	D2	D2, D4	Carbide
Spring steel, hardened, 52 HRC max	A2	A2, D2	D2	D2, D4	Carbide
Electrical sheet, transformer grade, 0.6 mm (0.025 in.)	A2	A2, D2	D2	D2, D4	Carbide
Paper, gaskets, and other soft materials	W1 (b)	W1 (b)	W1 (n)	W1, A2	D2
Plastic sheet, not reinforced	A2	A2	A2	A2, D2	Carbide
Plastic sheet, reinforced	A2 (f)	A2 (f)	D2 (f)	D2 (f)	Carbide

Note: Although carbide is recommended in this table only for 10 million pieces, it should usually be considered also for runs of 1 to 10 million pieces.

(a) Zn refers to a die made of zinc alloy plate and a punch of hardened tool steel. (b) For punching up to 10 000 parts, the W1 punch and die would be left soft and the punch peened to compensate for wear if necessary. (c) For punching 10 000 to 1 000 000 pieces, the W1 punch can be soft so that it can be peened to compensate for wear, or it can be hardened and ground to size. (d) Of the two alternatives listed, A2 tool steel is preferred if compound tooling is to be used for quantities of 10 000 to 1 000 000. (e) This O1 punch may have to be cyanided 0.1 to 0.2 mm (0.004 to 0.008 in.) deep to make even 1000 pieces. (f) For the application indicated, the punch and die should be gas nitrided 12 h at 540 to 565 °C (1000 to 1050 °F). (g) Soft. (h) Working edges are flame hardened in this application. (j) May be soft or flame hardened. (k) For punching 10 000 to 1 000 000 pieces, the punch would be W1, left soft so that it can be peened to compensate for wear, and the die would be O1, hardened. (m) Cyaniding of the punch is advisable, even for 1000 pieces. (n) For punching 10 000 to 1 000 000 pieces, the W1 die would be hardened and the W1 punch would be soft, so that it could be peened to compensate for wear.

Table 4. Typical punch and die materials for shaving 1.3-mm (0.050-in.) sheet

Work material	Tool material for production quantity of:			
	1000	10 000	100 000	1 000 000
Aluminum, copper and magnesium alloys	O1 (a)	A2	A2	D4 (b)
Carbon and alloy steel, up to 0.30% C, and ferritic stainless steel	A2	A2	D2	D4 (b)
Carbon and alloy steel, 0.30 to 0.70% C	A2	D2	D2	D4 (b)
Stainless steel, austenitic, up to quarter hard	A2	D2	D4 (b)	D4 (b)
Stainless steel, austenitic, over quarter hard, and spring steel hardened to 52 HRC max	A2	D2	D4 (b)	M2 (b)

(a) Type O2 is preferred for dies that must be made by broaching. (b) On frail or intricate sections, D2 should be used in preference to D4 or M2. Carbide shaving punches may also be practical for this quantity.

Tables 7 and 8 may be used to select lower-die materials for parts made of sheet thicker or thinner than the 1.3 mm used for the examples, or for parts of greater or lesser severity than those shown in Fig. 3 and 4. For parts of greater severity or sheet of greater thickness, use the die material recommended for the next greater production quantity than the quantity actually to be made (the column to the right of the actual production quantity in the table). Similarly, for parts of lesser severity or sheet of lesser thickness, use the die material recommended for the next lower production quantity (shift to the next column to the left of the actual production quantity).

Deep Drawing Dies

DEEP DRAWING is a process in which sheet metal is formed into round or square cup-shape parts by making it conform to a punch as it is drawn through a die. In conventional deep drawing, successive draws are made in the same direction. The types of dies and other tooling used for conventional deep drawing are illustrated in Fig. 5.

For economy in manufacture, a drawn part should always be produced in the fewest steps possible. Ironing—that is, thinning the walls of the part being drawn by using a reduced clearance between punch and die—is used almost universally in multioperation deep drawing. Ironing helps produce deep draws and uniform wall thickness in the fewest operations. Each operation is designed for maximum practical reduction of the metal being drawn. Accordingly, the information given here is predicated on use of reductions near the maximum of about 35%.

MATERIALS FOR SPECIFIC TOOLS

Draw Rings. Table 9 gives typical materials for draw rings (both dies and backup rings) used in drawing and ironing cups of various diameters and lengths from the three basic types of sheet

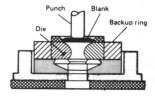

First operation in drawing

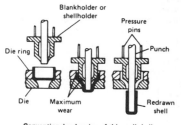

Conventional redrawing of thin-wall shells

Fig. 5. Tooling components used in conventional drawing operations

Table 5. Typical materials for perforator punches

| Work material | Punch material for production quantity of: | | |
	10 000	100 000	1 000 000
Punch diameters up to 6.4 mm (¹/₄ in.)			
Aluminum, brass, carbon steel, paper and plastics	M2	M2	M2
Spring steel, stainless steel, electrical sheet and reinforced plastics . . .	M2	M2	M2
Punch diameters over 6.4 mm (¹/₄ in.)			
Aluminum, brass, carbon steel, paper and plastics	W1	W1	D2
Spring steel, stainless steel, electrical sheet and reinforced plastics . . .	M2	M2	M2

Table 6. Typical materials for perforator bushings

| Work material | Bushing material for production quantity of: | | |
	10 000	100 000	1 000 000
Aluminum, brass, carbon steel, paper and plastics	W1 (a)	W1 (a)	D2
Spring steel, stainless steel, electrical sheet and reinforced plastics	D2	D2	D2 or carbide

(a) When bushings are of a shape that cannot be ground after hardening, an oil-hardening or air-hardening steel is recommended to minimize distortion.

metal listed in Table 10. The data in Table 9 are given for both round and square cups drawn from stock 1.6 mm (0.062 in.) thick in three typical production quantities. Similar data for a large square cup and a large pan are given also. Design dimensions for all seven parts referred to in Table 9 are given in Fig. 6. The square parts have liberal corner radii consistent with favorable die life.

Punches and Blankholders. Typical materials for punches and for blankholders (pressure pads) or shellholders (pressure sleeves) are given in Table 11. The materials listed in Table 11 are for punches and blankholders used in drawing and ironing round and square steel cups similar to parts 2 through 7 in Fig. 6.

More wear-resistant materials are required not only for the tools used in drawing and ironing harder or thicker stock or for those used for longer runs, but also for tools used to achieve greater percent reductions during ironing.

Metalworking Rolls

ROLLS are used to reduce the cross section of metal or to change its shape, or both. Cylindrically shaped, rolls are placed in a mill housing

supported on journal bearings against which screwdown pressure is exerted. Rolls rotate in opposite directions, the metal bar, plate or sheet passing between them longitudinally (Fig. 7a). Cross rolling is used for making seamless tube or for straightening round bar (Fig. 7b). Cross rolling is also used for widening billet (to make wide sheet or plate products), as well as for producing a more homogeneous microstructure.

CAST IRON ROLLS

Cast iron rolls are used in the as-cast condition or after stress relief. Some high-alloy iron rolls are heat treated by holding at high temperature followed by several lower-temperature treatments. Cast irons used for rolls are metastable and may be white or gray depending on composition, inoculation (if any), cooling rate and other factors.

In American practice, cast iron rolls are classified as (a) chilled iron rolls, (b) grain rolls, (c) sand iron rolls, (d) ductile iron rolls and (e) composite rolls.

Chilled iron rolls (hardness, 50 to 90 HSc) have a definitely formed, clear, homogeneous chilled white iron body surface and a fairly sharp line of demarcation between the chilled surface and the gray iron interior portion of the body. Clear chilled iron rolls can be made in unalloyed or alloyed grades.

Alloy chilled iron rolls have hardnesses that range from 60 to 90 HSc and that are controlled by carbon and alloy contents. Customary maximum percentages of alloying elements are 1.25

Table 7. Typical lower-die materials for forming a small part of mild severity from 1.3-mm (0.050-in.) sheet
For die cross section and part shape, see Fig. 3.

Metal being formed	Quality requirements — Finish	Tolerance mm	Tolerance in.	Lubrication(b)	Lower-die materials(a) for total production quantity of: 100	1 000	10 000	100 000	1 000 000
1100 aluminum, brass, copper(c)	None	None	None	Yes	Epoxy-metal, mild steel	Polyester-metal, mild and 4140 steel	Polyester-glass(d), mild and 4140 steel	O1, 4140	A2, D2
1100 aluminum, brass, copper(c)	None	±0.1	±0.005	Yes	Epoxy-metal, mild and 4140 steel	Polyester-metal, mild and 4140 steel	Polyester-glass(d), mild and 4140 steel	4140, O1, A2, D2	A2, D2
1100 aluminum, brass, copper(c)	Best	±0.1	±0.005	Yes	Epoxy-metal, mild steel	Polyester-metal, mild and 4140 steel	Polyester-glass(d), mild and 4140 steel	4140, O1, A2	A2, D2
Magnesium or titanium(e)	Best	±0.1	±0.005	Yes	Mild steel	Mild and 4140 steel	A2	A2	A2, D2
Low-carbon steel, to ¹/₄ hard	None	None	None	Yes	Mild and 4140 steel	Mild and 4140 steel	4140, mild steel chromium plated, D2	A2	D2
Type 300 stainless, to ¹/₄ hard	None	None	None	Yes	Mild and 4140 steel	Mild and 4140 steel	Mild and 4140 steel	A2, D2	D2
Low-carbon steel	Best	±0.1	±0.005	Yes	Mild and 4140 steel	Mild and 4140 steel	Mild and 4140 steel	A2, D2, nitrided D2	D2, nitrided D2
High-strength aluminum or copper alloys	Best	±0.1	±0.005	No(f)	Mild and 4140 steel	Mild and 4140 steel	Mild steel chromium plated and 4140	Cr plated O1; A2	D2, nitrided D2
Type 300 stainless, to ¹/₄ hard	None	±0.1	±0.005	Yes	Mild and 4140 steel	Mild and 4140 steel	Mild and 4140 steel	Cr plated O1; A2	D2
Type 300 stainless, to ¹/₄ hard	Best	±0.1	±0.005	Yes	Mild and 4140 steel	Mild and 4140 steel	Mild steel chromium plated, D2	D2, nitrided D2	D2, nitrided D2
Heat-resisting alloys	Best	±0.1	±0.005	Yes	Mild and 4140 steel	Mild and 4140 steel	Mild steel chromium plated, D2	D2, nitrided D2	D2, nitrided D2
Low-carbon steel	Good	±0.1	±0.005	No(f)	Mild and 4140 steel	Mild and 4140 steel	Mild steel chromium plated	D2, nitrided D2	D2, nitrided D2

(a) When more than one material for the same conditions of tooling is given, the materials are listed in order of increasing cost; however, final choice often depends on availability rather than on small differences in cost or performance. Where mild steel is recommended for forming fewer than 10 000 pieces, the dies are not heat treated. For forming 10 000 pieces and more, such dies should be carburized and hardened. Where 4140 is recommended for fewer than 10 000 pieces, it should be pretreated to a hardness of 28 to 32 HRC. Flame hardening of high-wear areas is recommended for quantities greater than 10 000 pieces. (b) Specially applied lubrication, rather than mill oil. (c) Soft. (d) With inserts. (e) Heated sheet. (f) Use lubrication to make 1 to 100 parts.

Table 8. Typical lower-die materials for forming a large part of mild severity from 1.3-mm (0.050-in.) sheet
For die cross section and part shape, see Fig. 4.

Metal being formed	Quality requirements Finish	Tolerance mm	in.	Lubri-cation(b)	Lower-die materials(a) for total production quantity of: 100	1 000	10 000	100 000	1 000 000
1100 aluminum, brass, copper(c)	None	None	None	Yes	Epoxy-metal, polyester-metal, zinc alloy	Polyester-metal, zinc alloy	Epoxy or polyester-glass(d), zinc alloy	Alloy cast iron	Cast iron or A2(e)
1100 aluminum, brass, copper(c)	None	±0.1	±0.005	Yes	Epoxy-metal, polyester-metal, zinc alloy	Polyester-metal, zinc alloy	Alloy cast iron	Alloy cast iron	Alloy cast iron
1100 aluminum, brass, copper(c)	Best	±0.1	±0.005	Yes	Epoxy-metal, polyester-metal, zinc alloy	Polyester-metal, zinc alloy	Alloy cast iron	Alloy cast iron	Alloy cast iron, A2(e)
Magnesium or titanium(f)	Best	±0.1	±0.005	Yes	Cast iron, zinc alloy	Cast iron, zinc alloy	Cast iron	Alloy cast iron	Alloy cast iron, A2(e)
Low-carbon steel, to 1/4 hard	None	None	None	Yes	Epoxy-metal, polyester-metal, zinc alloy	Epoxy-glass, polyester-glass, zinc alloy	Epoxy or polyester-glass(d), cast iron	Alloy cast iron	
Type 300 stainless, to 1/4 hard	None	None	None	Yes	Epoxy-metal, polyester-metal, zinc alloy	Epoxy-glass, polyester-glass, zinc alloy	Epoxy or polyester-glass(d), alloy cast iron	A2(e)	D2(e)
Low-carbon steel	Best	±0.1	±0.005	Yes	Zinc alloy	Epoxy-glass polyester-glass, zinc alloy	Alloy cast iron	D2, nitrided A2(e)	D2, nitrided D2(e)
High-strength aluminum or copper alloys	Best	±0.1	±0.005	No(g)	Zinc alloy	Polyester-glass, zinc alloy	Alloy cast iron	Alloy cast iron	Nitrided A2(e), nitrided D2(e)
Type 300 stainless, to 1/4 hard	None	±0.1	±0.005	Yes	Zinc alloy	Zinc alloy	Alloy cast iron	D2, nitrided A2(e)	D2(e), nitrided D2(e)
Type 300 stainless, to 1/4 hard	Best	±0.1	±0.005	Yes	Zinc alloy	Zinc alloy	Alloy cast iron	Nitrided D2	Nitrided D2(e)
Heat-resisting alloys	Best	±0.1	±0.005	Yes	Zinc alloy	Zinc alloy	Alloy cast iron	Nitrided D2	Nitrided D2(e)
Low-carbon steel	Good	±0.1	±0.005	No(g)	Zinc alloy	Zinc alloy	Alloy cast iron	Nitrided D2	Nitrided D2(e)

(a) When more than one material for the same conditions of tooling is given, the materials are listed in order of increasing cost; however, final choice often depends on availability rather than on small differences in cost or performance. Where mild steel is recommended for forming fewer than 10 000 pieces, the dies are not heat treated. For forming 10 000 pieces and more, such dies should be carburized and hardened. Where 4140 is recommended for fewer than 10 000 pieces, it should be pretreated to a hardness of 28 to 32 HRC. Flame hardening of high-wear areas is recommended for quantities greater than 10 000 pieces. (b) Specially applied lubrication, rather than mill oil. (c) Soft. (d) With inserts. (e) Use as inserts in cast iron body. (f) Heated sheet. (g) Use lubrication to make 1 to 100 parts.

Mo, 1.00 Cr and 5.5 Ni. Many different combinations are used to produce desired properties. Rolls of this type, particularly in the harder grades, are used chiefly for rolling flat work, both hot and cold. The softer, machinable grades are used for rolling rod and small shapes.

Grain rolls are "indefinite chill" iron rolls (hardness, 40 to 90 HSc) that have an outer chilled face on the body. These rolls have high resistance to wear and good finishing qualities, to considerable depths. The harder grades are used for hot and cold finishing of flat rolled products, and the softer grades are for deep sections (even with small rolls).

Sand iron rolls (no chill; hardness, 35 to 45 HSc) are cast in sand molds, in contrast to chilled iron rolls and grain rolls, the bodies of which are cast directly against chills. Sand iron rolls are used chiefly for intermediate and finishing stands on mills that roll large shapes. They are also used for roughing operations in primary mills.

Ductile iron rolls (hardness, 50 to 65 HSc) are made of iron of restricted composition to which magnesium or rare earth metals are added under controlled conditions to cause the graphite to form, during solidification, as nodules instead of the flakes common to gray iron. The resulting iron has strength and ductility properties between those of gray iron and those of steel.

Composite rolls, sometimes called double-pour rolls (hardness: bodies, 70 to 90 HSc; necks, 40 to 50 HSc), are rolls in which the body surface is made of a richly alloyed, hard cast iron resistant to wear, and the necks, wabblers, and central areas of the body are of a tougher and softer material. The metals are bonded firmly together during casting to form an integral structure. The chief applications of composite rolls in rolling of steel have been work rolls for four-high hot and cold strip mills and for plate mills; in rolling of nonferrous metals, the chief application has been rolls for hot breakdown and cold reduction of sheet and strip.

CAST STEEL ROLLS

Differentiation between cast iron rolls and cast steel rolls cannot be made strictly on the basis of carbon content. Iron rolls usually are of compositions that produce free graphite in unchilled portions; in contrast, steel rolls do not exhibit free graphite.

Cast steel rolls have higher hardness than cast iron rolls, and the superior strength of cast steel rolls often makes them preferable.

Composition. Alloy steel rolls have almost entirely superseded carbon steel rolls. Compositions of most alloy steel rolls are within the following limits: 0.40 to 2.60 C; less than 0.12 S, usually 0.06 max; less than 0.12 P, usually 0.06 max; up to 1.25 Mn; up to 1.50 Cr; up to 1.50 Ni; and up to 0.60 Mo. Higher carbon contents increase hardness and wear resistance. Some rolls have higher alloy contents, but these usually are for special purposes.

Applications. Cast steel rolls are graded according to carbon content. The general applications of these rolls are listed in Table 12.

FORGED STEEL ROLLS

Hardened forged steel rolls are used principally for cold rolling various metals in the form of coiled sheet and strip. Extremely high pressures are used in cold rolling, and forged rolls have sufficient strength, surface quality and wear resistance for cold rolling operations. Forged rolls sometimes are employed in nonferrous hot mills in preference to iron rolls because of their higher bending strength and resistance to metal pickup.

Composition. The most commonly used composition for forged steel rolls, sometimes known as regular roll steel, averages 0.85 C, 0.30 Mn, 0.30 Si, 1.75 Cr and 0.10 V. About 0.25% Mo sometimes is added to this basic composition, and the chromium content may be varied to obtain specific characteristics. For rolling nonferrous metals, a forged steel containing 0.40 C and 3.00 Cr is preferred. In Sendzimir mills, the work rolls and first and second intermediate supporting and drive rolls usually are made from high-carbon high-chromium steel with 1.50 or 2.25% C and 12.00% Cr (D1 or D4). For more severe service, work rolls of M1 molybdenum high speed steel are used. Special P/M alloys produced through

Table 9. Typical materials for draw rings used in drawing and ironing both round and square parts
For part designs and over-all dimensions, see Fig. 6.

Metal to be drawn	Total number of parts to be drawn		
	10 000	100 000	1 000 000

Cups up to 76 mm (3 in.) across, drawn from 1.6-mm (0.062-in.) sheet (parts 1, 2 and 3)

Drawing quality aluminum and copper alloys	W1; O1	O1; A2	A2; D2
Drawing quality steel	W1; O1	O1; A2	A2; D2
300-series Stainless steel	W1 chromium plated; aluminum bronze	Nitrided A2; aluminum bronze	Nitrided D2 or D3; cemented carbide

Cups 305 mm (12 in.) or more across, drawn from 1.6-mm (0.062-in.) sheet (parts 4 and 5)

Drawing quality aluminum and copper alloys	Alloy cast iron(a)	Alloy cast iron(a); A2 inserts(b)	A2 or D2 inserts(b)
Drawing quality steel	Alloy cast iron(a)	Alloy cast iron(c); A2 inserts(b)	A2 or D2 inserts(b)
300-series stainless steel	Alloy cast iron(d); aluminum bronze inserts(b)	A2 or aluminum bronze inserts(b)	Nitrided A2 or D2 inserts(b)

Square cups similar to part 6, drawn from 1.6-mm (0.062-in.) sheet

Drawing quality aluminum and copper alloys(e)	W1	O1; A2	A2; D2
Drawing quality steel(e)	W1	O1; A2	A2; D2; nitrided A2 or D2
300-series stainless steel(f) ...	W1; aluminum bronze	Nitrided A2; aluminum bronze	Nitrided A2 or D2

Large pans similar to part 7; drawn from 0.8-mm (0.031-in.) sheet

Drawing quality aluminum and copper alloys	Alloy cast iron(a)	Alloy cast iron(a); A2 corner inserts(b)	Nitrided A2 or D2 inserts(b)
Drawing quality steel	Alloy cast iron(a)	Alloy cast iron(a); A2 corner inserts(b)	Nitrided A2 or D2 inserts(b)
300-series stainless steel	Alloy cast iron(d); aluminum bronze	Nitrided A2 or aluminum bronze inserts(b)	Nitrided A2 or D2 inserts(b)

(a) Wearing surfaces flame hardened. (b) In flame hardened alloy cast iron. (c) Quenched and tempered for part 4; flame hardened for part 5. (d) Flame hardened on wearing surfaces to not over 420 HB. (e) For drawing aluminum, copper and steel, the tool material would be used as corner inserts. (f) For drawing stainless steel, inserts would be used for all wear surfaces.

Table 10. Sheet metals that require similar drawing-die materials

Type of sheet metal	Maximum hardness	Metals that require similar drawing-die materials
Drawing quality aluminum and copper alloys	64 HRF(a)	All aluminum and clad aluminum alloy sheet, copper and alloys, zinc and alloys, silver, pewter and Monel
Drawing quality steel	70 HRB	Carbon steel, grades 1008 to 1020
	75 HRB	Carbon steel, grades 1021 to 1030
Austenitic stainless steel	95 HRB	301, 302, 304, 305, 308, 310, 316, 317 steel; 410 and 430 carbon steel clad with stainless steel; copper clad with stainless steel; magnesium drawn at 200 to 300 °C (400 to 600 °F) with no ironing of sides; 17-4 PH, 17-7 PH and PH 15-7 Mo stainless steels

(a) Roughly equivalent to 58 HB (500-kg load) or 24 HR30T.

Table 11. Typical materials for punches and blankholders
For part designs and over-all dimensions, see Fig. 6.

Die component	Total number of parts to be drawn		
	10 000	100 000	1 000 000
For round steel cups like part 2			
Punch(a)	Carburized 4140; W1	W1; carburized S1	A2; D2
Blankholder(b)	W1; O1	W1; O1	W1; O1
For square steel cups like part 3			
Punch(a)	Carburized 4140; W1	W1; carburized S1	A2; D2
Blankholder(b)	W1; O1	W1; O1	W1; O1
For round steel cups like parts 4 and 5			
Punch(a)	Alloy cast iron(c)	O1 (d)	A2 (c); D2 (c)
Blankholder(b)	Alloy cast iron(c)	Alloy cast iron(e)	O1; A2
For square steel cups like parts 6 and 7			
Punch(a)	Carburized 4140 (f)	W1; O1 (d)	Nitrided A2; D2 (d)
Blankholder(b)	Alloy cast iron(c)	W1; O1	O1; A2

(a) Chromium plating is optional on punches, to reduce friction between part and punch and thus facilitate removal of the part. Cast iron, however, should not be plated. (b) Also applies to shellholder, pressure pad or pressure sleeve. (c) Flame hardening not necessary. (d) The punch holder is flame hardened alloy cast iron with a nose insert of the indicated tool steel. (e) For part 4, this blankholder is quenched and tempered; for part 5, it is flame hardened. (f) The punch holder is alloy cast iron with a nose insert of the indicated steel.

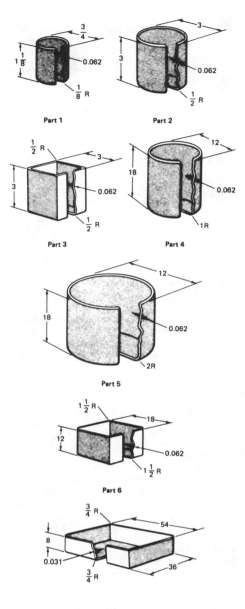

Dimensions are given in inches; to find equivalent metric dimensions (mm), multiply listed dimensions by 25. Corner radii comply with standard commercial practice. For typical deep drawing materials, see Table 9.

Fig. 6. Seven typical deep drawn parts

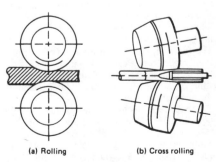

(a) Rolling   (b) Cross rolling

Fig. 7. Typical arrangements of metalworking rolls

Table 12. Applications of cast steel rolls

Carbon, %	Applications
0.50 to 0.65	Applications where strength is the prime and only requirement
0.70 to 0.85	Blooming mills; roughing stands in jobbing, plate and sheet mills; muck mills
0.90 to 1.05	Blooming mills; slab mills; roughing stands in continuous bar mills; backing rolls
1.10 to 1.25	Blooming and slab mills where breakage is not great; piercing mills; roughing stands in billet, bar, rail and structural mills
1.35 to 1.55	Intermediate stands for rail mills; structural, continuous billet and continuous bar mills
1.60 to 1.80	Intermediate stands for continuous bar and billet mills; middle rolls for three-high mills
1.85 to 2.05	Middle rolls for rail and structural mills; finishing mills where housing design is too limited for iron rolls
2.10 to 2.60	Finishing rolls for unusual conditions
2.65 and up	Special applications

application of the P/M process to high-alloy tool steel have wear resistance approaching that of carbide, which makes them attractive for some special forged steel rolls. Other high-wear alloys in the MHO series, or T15, also are applied on these occasions.

Hardness. The hardness range for forged steel rolls varies with the specific application and is developed with the cooperation of mill operators.

Hardness of work rolls for rolling thin strip averages about 95 HSc; lower hardnesses are employed for rolling thicker strip. In temper and finishing mills, work-roll hardness sometimes is higher than 95 HSc, and for special applications such as foil rolls, can be as high as 100 HSc. In nonferrous rolling, especially in aluminum plate mills, work-roll hardness generally ranges from 60 to 80 HSc. Hardness of backing rolls varies from 55 to 95 HSc; values on the high side of this range are specified for rolls in small mills and foil mills.

For Sendzimir mills, customary hardness is 61 to 64 HRC for D1 and D4 steel work rolls and 64 to 66 HRC for high speed steel work rolls. Customary hardness of intermediate rolls is 58 to 62 HRC.

Only the body section of a forged roll is hardened. Journals usually are not hardened, except those for direct-contact roller-bearing designs, for which a minimum hardness of 80 HSc is specified. In normal practice, the journals of forged rolls range in hardness from 30 to 50 HSc.

SLEEVE ROLLS

Use of forged and hardened sleeve-type backing rolls in certain hot strip and cold reduction mills has become a common practice because such rolls are more economical.

Sleeves are forged from high-quality alloy steel. Chromium-molybdenum-vanadium and nickel-chromium-molybdenum-vanadium compositions are generally used. Sleeves are heat treated by liquid quenching in either oil or water and are tempered to hardnesses of 50 to 85 HSc, depending on application.

The mandrel over which the sleeve is slipped may be made from a cast roll that has been worn below its minimum usable diameter, from a new casting made specifically for use as a mandrel, or from an alloy steel forging.

The outside diameter of the mandrel and the inside diameter of the sleeve are accurately machined or ground for a shrink fit. Mounting is accomplished by heating the sleeve to obtain the required expansion and then either slipping the sleeve over the mandrel or inserting the mandrel into the sleeve. This operation is performed with

the mandrel in a vertical position. A locking device prevents lateral movement of the sleeve. Final machining is done after the sleeve is mounted.

MISCELLANEOUS FORGED ROLLS

Auxiliary rolls such as leveler rolls and pinch rolls are employed in processing and handling equipment associated with rolling mills. These rolls are characterized by their long, slender shape. They are made from forged or rolled bars of 52100 steel or carburizing grades, and are processed to a hardness of approximately 95 HSc. Rolls used for various types of straightening machines generally are of sleeve design. One roll may be concave in body shape while its mating roll is straight. Standard compositions for forged steel rolls may be employed, and bodies may be hardened to 85 to 90 HSc.

CEMENTED TUNGSTEN CARBIDE ROLLS

Cemented tungsten carbide rolls have been used for rolling metals under a wide variety of conditions and in many types of rolling mills. They are used for both cold and hot rolling, and are made in all sizes from 6 to 400 mm (^{1}/$_4$ to 16 in.) in diameter. Rolls for slitters and trimmers also have been made of cemented tungsten carbide.

Use of cemented tungsten carbide rolls for applications varying from conventional cold rolling of flat sheet and wire to continuous hot rolling of rod in a wide range of sizes is continuing to expand, and technology of design and composition for these rolls is changing rapidly. Therefore, consultation with experienced carbide manufacturers is advised in selecting materials for specific applications.

————Coining Dies————

IN COLD COINING operations, the surface metal being worked is free to flow only to the small extent required by the coining operation. Flow to a larger extent constitutes cold extrusion.

DECORATIVE COINING

In dies used for decorative coining, materials that can be through hardened to produce a combination of good wear resistance, high hardness and high toughness are preferred.

Typical Die Materials. For dies up to 50 mm (2 in.) in diameter, consumable-electrode vacuum-melted or electroslag remelted 52100 steel pro-

vides the clean microstructure necessary for development of critical polished die surfaces. When heat treated to a hardness of 59 to 61 HRC, 52100 steel provides excellent die life. This steel also is suitable for photochemical etching, a process used in place of mechanical die sinking for engraving many low-relief dies. L6 tool steel at a hardness of 58 to 68 HRC is suitable for dies up to 100 mm (4 in.) in diameter. It can be through hardened, has enough toughness for long-life applications, and is suitable for photochemical etching of low-relief patterns. Air-hardening tool steels are preferred for coining and embossing dies greater than 100 mm (4 in.) in diameter. One of the chief reasons for choosing air-hardening tool steels is their low degree of distortion during heat treatment. Type A6 is a nondeforming, deep-hardening tool steel that often is used for large dies that must be hardened to 59 to 61 HRC. Air-hardening hot work steels such as type H13 are used at a hardness of 52 to 54 HRC for applications requiring especially high toughness.

For dies containing high-relief impressions, lowest die cost is obtained by machining the impressions directly into the dies when the number of pieces to be coined is less than the anticipated die life. For longer runs that require two or more identical dies, it is less expensive to produce the impressions by hubbing.

O1 and A2 tool steels are alternative choices for machined dies in production quantities up to about 100 000 pieces. The small additional cost of A2 is often justified because A2 gives longer life, especially when aluminum alloys, alloy steels, stainless steels or heat-resisting alloys are being coined.

Dies for Coining Silverware. Probably the greatest amount of industrial coining is done with drop hammers in the silverware industry, in producing highly embossed designs on surfaces such as teaspoon handles. Water-hardening steels such as W1 are almost always used for making such coining dies, whether the product is made of silver, a copper alloy or stainless steel. Water-hardening grades are selected because die blocks made of these steels can be reused repeatedly. After a die block fails, the block is annealed, the impression is machined off, and a new impression is hubbed before the die is rehardened. Dies made of deep-hardening tool steels such as O1, A2 and D2 are not reused (as are W1 dies), because they fail by deep cracking.

For ordinary designs requiring close reproduction of dimensions, dies may be made of A2, or of the high-carbon high-chromium steels D2, D3 and D4, to obtain greater compression resistance. For coining of designs with deep configurations and either coarse or sharp details, where dies usually fail by cracking, a deep-hardening carbon tool steel may be used at lower hardness, or O1, or S5 or S6, may be selected. In some instances, it may be desirable to select an air-hardening type such as A2, which would provide improved dimensional stability and wear resistance. A hot work steel such as H11, H12 or H13 may prove to be best where extreme toughness is the predominant requirement. When die failure occurs by rapid wear, a higher-hardness steel, or a more highly alloyed wear-resistant steel such as A2, may solve the problem.

For articles coined on drop hammers from series 300 austenitic stainless steels, it has sometimes been found advantageous to use steels of the S1, S5, S6 and L6 types, oil quenched and tempered to 57 to 59 HRC. Because the carbon contents of these grades are between 0.50 and 0.70%, they are less resistant to wear than W1, A2 and D2, but are tougher and more resistant to chipping and splitting. If necessary, the wear resistance of S5 tool steel dies can be improved slightly by carburizing to a depth of 0.13 to 0.25 mm (0.005 to 0.010 in.).

COINING IN PROGRESSIVE DIES

Tool steels recommended for coining a cup-shape part to final dimensions in the last stages of progressive stamping are shown in Table 13. This press coining operation involves partial confinement of the entire cup within the die. This produces high radial die pressures and thus requires pressed-in inserts on long runs, to prevent die cracking.

The punch material can be the same as the die material, except that O1 should be substituted for W1 in applications where W1 might crack during quenching.

P/M STEELS FOR COINING DIES

The application of hot isostatic processing to P/M production of high speed steels and special high-alloy steels has expanded the range of tool steel grades available for long-run coining dies. Dramatic increases in toughness and grindability have been achieved. Type M4 is an excellent example. When made by P/M processing, M4 has approximately twice the toughness and two to three times the grindability of conventionally processed M4.

____Cold Heading Tools____

THE MANY TYPES of fasteners that are used in large quantities are manufactured in cold heading machines. Most of the tools used in these machines require maximum surface hardness for wear resistance, along with maximum strength and toughness to enable them to withstand high service pressures without breaking.

Materials used for tools in single-die, two-die-three-blow and multistation cold heading machines are listed in Table 14.

TOOL STEEL DIES

Solid cold heading dies made of W1 and W2 tool steels are used for short production runs. These tools are hardened by flush quenching, which provides the desired combination of high hardness, high strength and high toughness. Most long production runs require dies made of the more highly alloyed tool steels M1, M2 and D2, or of cemented carbides. Some of the new P/M alloys have also given excellent performance and are economical alternatives to cemented carbides on long runs. These materials are most commonly used as inserts in H13 tool steel cases; hardness of such cases usually is held to about 48 HRC to provide the best combination of backup strength and freedom from hazardous breakage. However, for applications involving high-interference fits between insert and case, the greater high-strength fracture toughness of maraging steels such as 18Ni(300) makes them very desirable as materials for cases, although at some sacrifice in wear resistance.

Table 13. Typical tool steels for coining a preformed cup to final size on a press

Metal to be coined	Die material(a) for total quantity of:		
	1000	10 000	100 000
Aluminum and copper alloys	W1	W1	D2
Low-carbon steel	W1	O1	D2
Stainless steel, heat-resistant alloys and alloy steels	O1	A2	D2

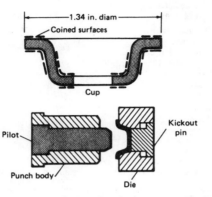

(a) For quantities over 10 000, the materials given are for die inserts. All selections shown are for machined dies. The same material would be used for the punch, except that O1 should be substituted for W1 in applications where W1 might crack during heat treating.

Table 14. Typical materials for tools in cold heading machines

Tool	Material
Cutoff quill (die)	M4, or cemented carbide insert
Cutoff knife	M4, or cemented carbide insert
Upset, cone or spring punch	W1, S1, M1, or cemented carbide insert
Cone-punch knockout pin	M2, or CPM 10V
Backing plug	O1
Finish-punch case	H13
Finish-punch insert	M1, M2, CPM 10V, or cemented carbide
Die case	H13
Die insert	M1, D2, M2, CPM 10V, or cemented carbide
Die knockout pin	M2, or CPM 10V

CEMENTED CARBIDE TOOLS

As a rule of thumb, cemented carbide tools properly designed and utilized should provide roughly ten times the life of steel tools. Although tungsten carbide tooling is higher in initial cost than tool steel tooling and lower in impact resistance, the longer life and superior dimensional integrity of carbide result in far lower cost per thousand pieces produced. For best results, cemented carbide tools should be used in cold heading machines that are in good condition, with tight rams, minimum vibration and accurate alignment.

____Cold Extrusion Tools____

IN COLD EXTRUSION, neither the tooling nor the work is preheated. However, the heat generated by plastic deformation of the work-piece under steady and nearly uniform pressure may be sufficient to require tool steels with relatively high resistance to softening at elevated temperatures.

In the cold extrusion process, backward displacement from a closed die progresses in the direction opposite that of punch travel, as shown at left in Fig. 8. Parts often are cuplike in shape and have wall thicknesses equal to the clearance between punch and die.

In forward extrusion, the metal is forced in the direction of punch travel (Fig. 8, center). One end of the die recess is just large enough to receive the starting slug, and the other end has a small orifice of the shape required for the final part.

Sometimes the two methods of extrusion are combined so that some of the metal flows backward and some flows forward, as shown at right in Fig. 8.

Compressive strength of the punch and tensile strength of the die are among the most important factors influencing the selection of materials for cold extrusion tools. Thus, almost without exception and particularly for extrusion of steel, the primary tools in contact with the work must be made from steels that will through harden in the section sizes involved.

The primary mechanism of tool deterioration in cold extrusion is abrasive and adhesive (galling) wear of both punch and die.

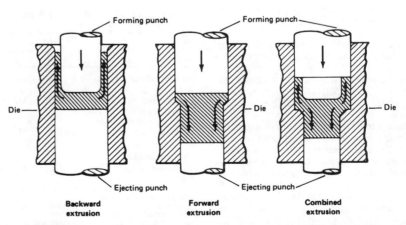

Fig. 8. Backward, forward and combined forward and backward displacement in cold extrusion

Source: Metals Handbook Desk Edition, 1985, 18.1-18.36

Lubrication, at least theoretically, is the prime factor controlling tool wear, because there is no metal-to-metal contact between the tools and the work when lubrication is ideal. In practice, the stresses imposed on tooling can vary by more than 100% with changes in lubrication.

PUNCH AND DIE MATERIALS

Cold extrusion of a part from 1018 or 1021 steel requires about 10% more extrusion pressure than extrusion of the same part from 1010 steel. For low-alloy steels, forming loads are about 20 to 30% greater than those for 1010. Medium-carbon steels such as 1030 and 1040 also require forming loads about 20 to 30% higher than those for 1010 steel.

Dies made of W1 tool steel generally are satisfactory for extruding the softer alloys of aluminum. Steels such as A2 and D2 are preferred for tools used in extrusion of the stronger aluminum alloys, because enough heat sometimes is generated to soften tools made of W1. In extrusion of aluminum, tool wear is roughly proportional to the yield strength of the work metal. Thus, the common impact extrusion alloys 1100, 6061, 2014 and 7075 cause progressively more wear on tools in the order listed.

Tables 15 to 20 list typical tool steels used in tools for cold extrusion of steels and aluminum alloys, in two quantities, for the series of hypothetical parts shown in Fig. 9. These simple parts are seldom encountered in practice; however, the principles described can be related to actual production components of comparable severity.

SECONDARY TOOLING COMPONENTS

Table 21 lists the constructional and tool steels used for the secondary tooling components for cold extrusion of parts 1 to 7 (see Fig. 9) from steel. (Tooling for forward or backward extrusion of aluminum consists of only a die, a die holder, a punch and an ejector.) The steels listed in Table 21 reflect the moderate to high extrusion severity involved in producing parts 1 to 7.

Cemented Carbides. The grades of cemented carbides used most frequently in cold extrusion punches and dies are shown in Table 22. In general, the quantity of parts to be produced and the required dimensional tolerances are more important than tool size in deciding whether or not to use carbide.

As a specific example, it is estimated that carbide tooling could not be justified for extruding part 5 (see Fig. 9) from steel in quantities less than 100 000 pieces. Much larger quantities would be required (perhaps 500 000 or more) to justify use of carbide tooling for extruding part 5 from aluminum.

In addition to its use in cold extrusion dies and die inserts, cemented carbide has been used extensively for cold extrusion punches for many years, producing small parts for automotive, farm-equipment and other high-production industries. Typical of such parts are bearing cups, valve lifters, wrist pins and spark-plug bodies. Larger parts now are being produced by cold extrusion. Primarily, these are cuplike shapes that either are used as is or are subsequently punched out to produce bushings.

Table 15. Typical tool steels for backward extrusion of parts 1 and 2
For designs of parts, see Fig. 9.

Metal to be extruded(a)	Total quantity of parts to be extruded	
	5 000	50 000
Punch material(b)		
Aluminum alloys A2	A2, D2, M4 (c)	
Carbon steel, up	D2, M2 (b),	
to 0.40% C A2	M4 (c)	
Carburizing		
grades of alloy		
steel A2	M2 (d), M4 (c)	
Die material(b)		
Aluminum alloys ......W1 (e)	W1 (e)	
Carbon steel, up		
to 0.40% C O1, A2	A2 (f)	
Carburizing		
grades of alloy		
steel O1, A2	A2 (f)	
Knockout material(b)		
Aluminum alloys A2	D2	
Carbon steel, up		
to 0.40% C, and		
carburizing grades		
of alloy steel A2	A2, D2	

(a) For part 1, starting with a solid slug; for part 2, starting with part 1. In aluminum, part 2 can be made directly from a cylindrical blank. (b) Where two or more tool materials are recommended for the same conditions, they are given in order of cost, with the less or least expensive shown first. (c) First choice in automotive parts processing. (d) Liquid nitrided. (e) The 1.00% C grade is recommended. (f) Gas nitrided on the inside surface only.

Table 16. Typical tool steels for drawing part 3
For design of part, see Fig. 9.

Metal to be drawn(a)	Total quantity of parts to be extruded	
	5 000	50 000(b)
Punch material(c)		
Aluminum alloys A2	D2, M4 (d)	
Carbon and alloy steel, up		
to 0.40% C O1	A2, M4 (d)	
Die material(c)		
Aluminum alloys ....W1 (e)	W1 (e), A2	
Carbon and alloy steel, up		
to 0.40% C O1, A2	A2 (f), D2	

(a) Starting with part 2 (Table 15) for steel. In aluminum, the part would be made in one backward extrusion from a cylindrical slug. (b) For quantities greater than about 100 000 parts in steel, carbide punches and dies should be considered, especially if close tolerances must be maintained. (d) First choice in automotive parts processing. (e) The 1.00% C grade is recommended. (f) Gas nitriding is recommended on the inside surface. F2 tool steel may be used in place of A2.

Table 17. Typical tool steels for drawing part 4
For design of part, see Fig. 9.

Metal to be drawn(a)	Total quantity of parts to be drawn	
	5 000	50 000
Punch material(b)		
Aluminum alloys A2	D2, M4 (c)	
Carbon steel, 1010 ....M2	M2, M4 (c)	
Carbon steel, 1020 to 1040,		
and carburizing grades		
of alloy steel ....M2 (d)	M2 (d), T15, M4 (c)	
Die material(b)		
Aluminum alloys A2	A2, D2	
Carbon and alloy steel A2	A2 (d), D2	

(a) In steel, a part would be made in two operations with an intermediate process anneal (see text). In aluminum, it would be made in one backward extrusion. (b) Where two or more tool materials are recommended for the same conditions, they are given in order of cost, with the less or least expensive shown first. (c) First choice in automotive parts processing. (d) Nitriding treatment is recommended.

Table 18. Typical tool steels for forward extrusion of part 5
For design of part, see Fig. 9.

Metal to be extruded(a)	Total quantity of parts to be extruded	
	5 000	50 000(b)
Punch material(c)		
Aluminum alloys A2	D2, M4 (d)	
Carbon steel, 1010 ....A2	D2, M4 (d)	
Carbon steel, 1020 and 1040,		
and carburizing grades		
of alloy steel A2	M2 (e)	
Die material(c)		
Aluminum alloys ....W1 (f)	A2, D2	
Carbon and alloy steel, up to		
0.40% C A2	A2 (g)	

(a) Starting with part 2 (Table 15) for steel. Aluminum would be extruded from a cylindrical slug. (b) For quantities greater than about 100 000 parts in steel, carbide punches and dies should be considered, especially if close tolerances must be maintained. (c) Where two tool materials are recommended for the same conditions, they are given in order of cost, with the less expensive shown first. (d) First choice in automotive parts processing. (e) Nitrided. (f) The 1.00% C grade is recommended. (g) Liquid nitrided.

Table 19. Typical tool steels for forward extrusion of part 6
For design of part, see Fig. 9.

Metal to be extruded	Total quantity of parts to be extruded	
	5 000	50 000
Punch material(a)		
Aluminum alloys A2	D2, M4 (b)	
Carbon and alloy steels,		
up to 0.40% C A2, D2	M2 (c), M4 (b)	
Die material(a)		
Aluminum alloys A2	A2	
Carbon and alloy steels,		
up to 0.40% C ......(d)	(d)	

(a) Where two tool materials are recommended for the same conditions, they are given in order of cost, with the less expensive shown first. (b) First choice in automotive parts processing. (c) Liquid nitrided. (d) No steel can be recommended without qualification. Medium-carbon alloy tool steels such as H12, H21 and 6F5 have given the best results.

Table 20. Typical tool steels for forward extrusion of part 7
For design of part, see Fig. 9.

Metal to be extruded(a)	Total quantity of parts to be extruded	
	5 000	50 000
Punch material(b)		
Aluminum alloys A2	D2, M4 (c)	
Carbon and alloy steel, up to		
0.40% C, and series 300		
stainless steels A2	D2, M4 (c)	
Die material(b)		
Aluminum alloys A2	D2	
Carbon and alloy steel, up to		
0.40% C, and series 300		
stainless steels ....(d)	(d)	
Knockout material(b)		
Aluminum alloys, steels		
and series 300		
stainless steels ....1020 (e)	1020 (e)	
	O1 (f)	O1 (f)

(a) Starting with a ring-shaped blank. (b) Where two tool materials are recommended for the same conditions, they are given in order of cost, with the less expensive shown first. (c) First choice in automotive parts processing. (d) No tool steel can be recommended without qualification. Medium-carbon alloy tool steels such as H12, H21, 6F5 and 6H2 have given the best results. (e) Or other low-carbon or low-alloy steels for knockout pins. (f) Knockout heads.

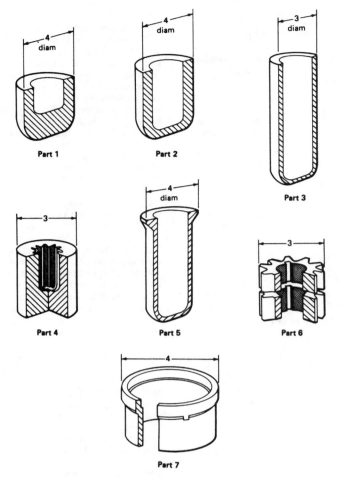

Part 1

Part 2

Part 3

Part 4

Part 5

Part 6

Part 7

Dimensions are in inches; for equivalent metric sizes (mm), multiply by 25.4. For typical die materials used in extruding these parts from low-carbon steel and aluminum, see Tables 15 to 20.

Fig. 9. Seven hypothetical cold extruded parts

Table 21. Typical steels for secondary tooling components used in extruding steel parts

Part number	Related table	Type of operation	Upper die plate	Punch backup plate	Inner shrink ring	Outer shrink ring	Primary support block	Secondary support block	Lower die plate
1 and 2	Table 15	Backward ...1040		S1	S4, S5	1040	W2	W2	1040
3	Table 16	Drawing 1040		S1	S4, S5	1040	W2	W2	1040
4	Table 17	Drawing W2		8620 (a)	4340 (a)	W2	S4, S5	D4	W2
5	Table 18	Forward 1040		S1	S4, S5	1040	W2	W2	1040
6	Table 19	Forward W2		8620 (a)	4340 (a)	W2	S4, S5	D4	W2
7	Table 20	Forward W2		8620 (a)	4340 (a)	W2	S4, S5	D4	W2

Note: Part designs are shown in Fig. 9.
(a) Or other alloy steel having hardenability appropriate for the component.

Table 22. Cemented carbides used most frequently for cold extrusion punches and die inserts

Type of service	Composition, % Tungsten carbide	Cobalt binder	Grain size
Punches			
High impact	84	16	Fine
Medium impact	88	12	Fine
Dies and die inserts			
High impact	75	25	Medium coarse
Medium impact	84	16	Medium fine
Light impact, maximum wear	88	12	Fine

Tools for Drawing Wire, Bar and Tubing

SELECTION of tool materials for cold drawing of metal into continuous forms such as wire, bar and tubing depends primarily on the size, composition, shape, stock tolerance and quantity of the metal being drawn. Cost of the tool material is also important and may be decisive.

Dies and mandrels used for cold drawing are subjected to severe abrasion. For this reason, most wire, bar and tubing is drawn through dies having diamond or cemented tungsten carbide in-

serts, and tube mandrels usually are fitted with carbide nibs. Small quantities, odd shapes and large sizes are more economically drawn through hardened tool steel dies.

WIREDRAWING DIES

Table 23 gives recommended materials for wiredrawing dies. For round wire, dies made of diamond or cemented tungsten carbide are always recommended, without regard to the composition or quantity of the metal being drawn. For short runs or special shapes, hardened tool steel is less costly, although carbide gives superior performance in virtually any application.

Diamond Dies. The use of diamond dies is restricted only by limitations on the sizes of available industrial diamonds and by cost, which is extremely high for diamonds in larger sizes. These tools can outperform cemented tungsten carbide dies by 10 to 200 times, depending on the alloy being drawn, and thus are cost effective despite their high unit cost.

Diamond dies are available in two types: natural and synthetic. Natural diamond dies, which have been used longer, are made from single crystals and produce exceptional surface finishes in the drawn product. Generally, natural diamond is less expensive than synthetic diamond for smaller hole sizes. For a hole size of 0.66 mm (0.0259 in.), the two materials are about the same in cost. For larger sizes, synthetic diamond is more economical.

Cemented tungsten carbide is economical for wiredrawing dies in most applications above the range of size where diamond can be used. The softer cemented carbides, which contain about 8% cobalt, are less brittle and can withstand greater stock reductions without breaking, but wear more rapidly than lower-cobalt grades.

Tool steel used for wiredrawing dies should have near-maximum hardness (62 to 64 HRC) for reductions below about 20%. For greater reductions, because of the possibility of breakage, hardness should be decreased to 58 to 60 HRC, even though the rate of wear will increase.

DRAWING BARS AND TUBING

Table 24 shows die and mandrel materials recommended for drawing bars, tubing and complex shapes. Diamond is virtually never used in larger sizes; cemented tungsten carbide is recommended for three-fourths of all applications. Tool steels are rarely used to make tools for drawing commercial-quality round bars less than 90 mm (3.5 in.) in diameter. Cemented tungsten carbide is used to draw stainless steel tubes as large as 280 mm (11 in.) in outside diameter.

Mandrels. Either carbide or hardened tool steel is satisfactory for mandrels used in drawing tubes, but carbide is more economical for tubes less than 125 mm (5 in.) in diameter. Carbide nibs are available in lengths sufficient for this purpose. Mandrel nibs are available in either a braze-type design or a shell design that permits mechanical attachment to the shank for easy replacement. Carbide tips are also recommended for mandrels used in drawing of shapes, but with reservations on the use of carbide similar to those that apply to the use of carbide in dies for drawing of shapes. Tool steel mandrels are recommended for drawing of tubes over 125 mm (5 in.) in inside diameter.

Table 23. Recommended materials for wiredrawing dies

Metal to be drawn	Wire size mm	in.	Recommended die materials for: Round wire	Special shapes
Carbon and alloy steels	<1.57	<0.062	Diamond, natural or synthetic	M2 or cemented tungsten carbide
	>1.57	>0.062	Cemented tungsten carbide	
Stainless steels; titanium, tungsten, molybdenum and nickel alloys	<1.57	<0.062	Diamond, natural or synthetic	M2 or cemented tungsten carbide
	>1.57	>0.062	Cemented tungsten carbide	
Copper	<2.06	<0.081	Diamond, natural or synthetic	D2 or cemented tungsten carbide
	>2.06	>0.081	Cemented tungsten carbide	
Copper alloys and aluminum alloys	<2.5	<0.100	Diamond, natural or synthetic	D2 or cemented tungsten carbide
	>2.5	>0.100	Cemented tungsten carbide	
Magnesium alloys	<2.06	<0.081	Diamond, natural or synthetic	
	>2.06	>0.081	Cemented tungsten carbide	

Table 24. Recommended tool materials for drawing bars, tubing and complex shapes

Metal to be drawn	Round bars and tubing(a) Common commercial sizes Bar and tube dies	Tube mandrels(b)	Maximum commercial size(c): dies and mandrels	Complex shapes: dies and mandrels(a)(b)
Carbon and alloy steels	Tungsten carbide	W1 or carbide	D2	CPM 10V or carbide
Stainless steels; titanium, tungsten, molybdenum and nickel alloys	Diamond or carbide(d)	D2 or carbide	D2 or M2(a)	F2 or carbide(e)
Copper, aluminum and magnesium alloys	W1 or carbide	W1 or carbide	D2 or CPM 10V	O1, CPM 10V or carbide

(a) Tool steels for both dies and mandrels are usually chromium plated. (b) "Carbide" indicates use of cemented carbide nibs fastened to steel rods. (c) 10-in. OD by ³/₄-in. wall. (d) Under 1.6 mm (0.062 in.), diamond; over 1.6 mm (0.062 in.), tungsten carbide. (e) Recommendations for large tubes or complex shapes apply to stainless steel only.

Closed-Die Hot Forging Tools

THE CLOSED-DIE FORGING TOOLS discussed here are restricted to die blocks, die inserts and trimming tools used for hot forging in vertical presses and hammers. In hammer forging, the hammer—whether a gravity, steam, air-drop or counterblow hammer—strikes a sudden blow, imposing a shock load on the forging dies. In press forging, the working pressure is applied as a fast push rather than a blow, so that the dies are subjected to less shock. However, because a press is much slower acting than a hammer, the dies in a press absorb more heat from the hot blank during the forging cycle.

The size and shape of the part being forged influence the force and energy required to reshape the hot plastic metal—from the initial shape of the forging stock (usually round, square or flat) to that of the finished forging. The force and energy required are further influenced by the composition of the metal being forged. For example, as the alloy content of a steel increases, hot strength and subsequent resistance to flow also increase, and more energy is required to forge the same shape. Similarly, some copper alloys and aluminum alloys are considerably easier to forge than others.

The basic causes of premature die failure are excessive force, abrasion, and excessive temperature. In addition, dies may break in a brittle manner if used cold, and thus preheating to 260 to 300 °C (500 to 600 °F) is recommended. This may be accomplished by placing "warmers" (pieces of hot steel) between the dies or by installing gas-fired or electrical heating devices to maintain temperature during idle periods.

DIE MATERIALS

Prehardened die blocks suitable for making forging dies are available in a range of compositions and hardnesses. Prehardened tool steels also are available in other forms for making small die blocks, die inserts and trimming tools. All steels available in a prehardened condition also are available in the annealed condition for ease of machining; once machined to the desired contours, they can be hardened and finished by methods ordinarily used for standard tool steels.

Prehardened die-block steels usually are purchased on the basis of hardness and proprietary name. Typical tool steels, and their Brinell hardnesses, for die blocks and die inserts used in hammer and press forging are shown in Tables 25A and 25B. In these tables, recommended die materials are listed for the six hypothetical shapes of increasing severity illustrated in Fig. 10. In-

formation is included for two production quantities and four types of work metal.

The recommendations for steel and hardness level in Tables 25A and 25B are based on lowest over-all cost (which includes both material and fabrication costs), and on avoiding breakage. Lower-alloy tool steels at lower hardnesses are acceptable when relatively few parts are to be made from carbon or low-alloy steel. Somewhat higher hardness is required in dies for forging stainless steels and heat-resisting alloys.

The relatively expensive, high-alloy H11 and H12 tool steels are desirable for forging copper alloys, because copper alloys are quite resistant to flow at their maximum forging temperatures. Copper oxide is abrasive, and copper alloy forgings usually are made to closer tolerances than steel forgings; both of these characteristics demand high wear resistance in the die material.

As shown in Tables 25A and 25B, higher hardnesses can be used for press-forging dies than for hammer-forging dies because the former are not subjected to impact. Because of the longer times in contact with hot work metal, press-forging dies must be made of a steel having greater resistance to softening at elevated temperatures.

The compositions of certain nonstandard tool steels used for die blocks are shown in Table 26.

TOOLS FOR TRIMMING

Usually, trimming of metal flash at temperatures below 150 °C (300 °F) is referred to as cold trimming, and trimming at 500 °C (1000 °F) or higher is termed hot trimming. For the purpose of selecting materials for trimming tools, trimming at temperatures between 150 and 500 °C is also considered hot trimming.

Whether trimming is done hot or cold depends chiefly on whether the trim is to be normal or close and on the composition of the metal being trimmed. Table 27 gives typical materials for both hot and cold trimming tools. Carbon and alloy steels may be trimmed either hot or cold, but close trimming is generally done hot. Stainless steels and heat-resisting alloys are usually trimmed hot, whereas nonferrous alloys may be trimmed either hot or cold.

Cold Trimming. Punches for cold trimming carbon and alloy steel forgings are commonly made from discarded die blocks. Because of this practice, prehardened low-alloy die steels are the materials most widely used for punches and are considered satisfactory for many different types of punches. Although high-carbon alloy tool steels such as A2 have been used successfully as blade materials for normal cold trimming of carbon and alloy steel forgings, D2 is usually a better choice because of its longer life.

Hardened and tempered 6150 steel has been used successfully in punches for normal cold trimming of nonferrous forgings. Other alloy steels similar to 6150 in hardenability are also satisfactory for this application. For close trimming, the edge-holding properties of D2 make it more desirable as a punch material. It is also recommended for blades used in close cold trimming of nonferrous forgings.

Hot Trimming. Prehardened die steels are satisfactory punch materials for hot trimming of carbon, alloy and stainless steel forgings. Punch materials for hot trimming of nonferrous forgings are less critical, and 1020 steel, either as rolled or as annealed, is widely used for reasons of economy.

Table 25A. Typical tool steels for die blocks and die inserts
For illustration of forged parts, see Fig. 10.

	Typical die materials and hardness ranges			
	Hammer forging		Press forging	
Work metals	100 to 10 000 parts	10 000 parts or more	100 to 10 000 parts	10 000 parts or more
For making parts of severity no greater than part 1				
Carbon and alloy steels	6F2 or 6G at 341 to 375 HB	6F2 or 6G at 388 to 429 HB	6F2 or 6G at 388 to 429 HB	6F3 at 369 to 388 HB or H11 or H12 at 388 to 405 HB(a)
Stainless steels and heat-resisting alloys	6F2 or 6G at 388 to 429 HB	6F2 or 6G at 388 to 429 HB	6F2 or 6G at 388 to 429 HB	H11 or H12 insert at 477 to 543 HB or H26 at 514 to 577 HB(b)
Aluminum and magnesium alloys	6F2 or 6G at 302 to 331 HB	6F2 or 6G at 341 to 375 HB	6F2 or 6G at 341 to 375 HB	6F3 at 375 to 405 HB or H11 or H12 at 448 to 477 HB(a)
Copper and copper alloys	6F2 or 6G at 341 to 375 HB or H11 or H12 at 405 to 433 HB	H11 or H12 at 405 to 448 HB	6F2 or 6G at 341 to 375 HB or H11 or H12 at 477 to 514 HB	H11 or H12 at 477 to 514 HB
For making parts of severity no greater than part 2				
Carbon and alloy steels	6F2 or 6G at 341 to 375 HB	6F2 or 6G at 341 to 375 HB	6F2 or 6G at 388 to 429 HB	6F3 at 369 to 388 or H11 or H12 at 388 to 405 HB(a)
Stainless steels and heat-resisting alloys	6F2 or 6G at 341 to 375 HB	6F2 or 6G at 341 to 375 HB with 6F3 insert at 405 to 448 HB	. . .	. . .
Aluminum and magnesium alloys	6F2 or 6G at 302 to 331 HB	6F2 or 6G at 341 to 375 HB or H11 or H12 at 405 to 448 HB(a)	6F2 or 6G at 341 to 375 HB	6F2 or 6G at 341 to 375 HB with 6F3 insert at 405 to 448 HB or H11 or H12 at 448 to 477 HB(a)
Copper and copper alloys	6F2 or 6G at 341 to 375 HB	6F2 or 6G at 341 to 375 HB or H11 or H12 at 405 to 448 HB(a)	6F2 or 6G at 341 to 375 HB	H11 or H12 at 477 to 514 HB
For making parts of severity no greater than part 3				
Low-alloy steels, stainless steels and heat-resisting alloys	6F2 or 6G at 302 to 331 HB	6F2 or 6G at 302 to 331 HB with insert of same steel at 341 to 375 HB	. . .	. . .
Aluminum, magnesium and copper alloys	6F2 or 6G at 269 to 293 HB	6F2 or 6G at 302 to 331 HB	. . .	. . .

(a) Recommended for long runs—for example, 50 000 forgings. (b) Recommended for forging higher-alloy heat-resisting materials, such as nickel-base and cobalt-base alloys.

Hard faced carbon steels are often preferred for use as trimming edges (see Table 27). One advantage of hard faced blades is that chipped or broken edges are easily repaired. The cobalt-base alloy indicated in Table 27 (type 4A) is extremely high in resistance to shock, heat and abrasion. Most forging shops prefer to use this alloy (or a similar one) for facing all trimming blades, either hot or cold. Information on alloy 4A and other hard facing alloys may be found elsewhere in this volume.

TOOLS FOR ISOTHERMAL FORGING

The temperatures typically used for isothermal forging of titanium alloys (870 to 980 °C, or 1600 to 1800 °F) and of nickel-base superalloys (925 to 1100 °C, or 1700 to 2000 °F) impose severe hot-strength, creep and oxidation-resistance requirements on the die materials. Dies for isothermal forging of titanium alloys typically are made of cast nickel-base superalloys that were initially developed for blades in gas turbine engines. Of these superalloys, Inconel 713C and IN-100 have been used with great success for isothermal forging dies. However, in the most severe environments of high temperature and stress, excessive creep can be a major cause of failure for IN-100 dies. In severe environments, replacement of IN-100 by alloys of even higher creep strength, such as TRW-NASA VIA, has proved effective.

Isothermal forging of nickel-base superalloys often is performed with dies made of molybdenum alloys. The susceptibility of molybdenum alloy dies to oxidation at working temperatures requires that forging be performed in vacuum.

Hot Upset Forging Tools

THE TOOLS discussed in this article are restricted to header dies, gripper dies and auxiliary tools used in forging machines or upsetters operating in a horizontal plane.

Forgings made by the hot upset method vary in size from 13-mm ($\frac{1}{2}$-in.) bolts to 305-mm (12-in.) flanged pipe sections for the oil industry. Production rate varies with size of forging and amount of automation. Use of automatic feeds allows rates as high as 7200 pieces per hour in such applications as boltmaking, and such rates require tool materials that will serve continuously at high temperatures for long periods of time. For medium-size parts such as automotive forgings, which are produced at rates of 120 to 150 pieces per hour, the dies cool off enough between blows so that die materials with lower hot strength can be used. The high-alloy, high-hardness upsetting tools used in the bolt industry are not suitable for medium-size automotive upset forgings because such high-hardness tools are too susceptible to breakage. Dies for still larger upset forgings made at lower rates may require higher strength at forging temperature and higher alloy content because of the longer sustained contact between the hot workpiece and the tools.

Complexity of die shape also influences selection of die steels for hot upset forging. Sharp corners and edges greatly increase stress concentration, and thin sections may be subject to extreme loads and high thermal stresses. Internal punches and mandrels are subjected to high impact loads and sliding abrasive wear, and often are designed to be replaceable because of their short life. Replaceable inserts may be used for areas of gripper dies subject to short life and for parts requiring close tolerances.

The material to be forged is important in selecting a die material. At forging temperature, carbon and low-alloy steels have lower strength than stainless steels and heat-resisting alloys and can be forged with less costly tool steels. In upset forging of heat-resisting steels and titanium alloys, even the best die steels may have a short life. The compositions and properties of the AISI grades and types of tool steel discussed in this article are presented elsewhere in this volume. The nominal compositions of six nonstandard tool steels used to make tools for hot upset forging are shown in Table 28.

HEADING TOOLS AND GRIPPER DIES

Tools for hot upset forging generally are constructed in the form of insert dies. Table 29 summarizes the typical materials used for hot upset forging tools for making parts of the various degrees of severity shown in Fig. 11. Part 1 in Fig. 11 represents straight upset forging; parts 2 and 3 represent two degrees of severity for pierced and upset forgings. Each of the three shapes is made in one blow. For all three, the original unsupported length of stock required to make the part is 2.5 times the diameter of the starting stock.

In hot upset forging of simple flanged shapes from low-carbon and alloy carburizing steels, the heading tool usually wears faster than the gripper die. For example, in hot upsetting of hexagonal bolt heads on 1020, 1045 or 4140 steel shanks 13 to 25 mm (0.5 to 1 in.) in diameter, H13 tool steel is used for both the heading tool and the gripper die, but the heading tool has greater hardness. Under the same wear conditions, the higher hardness could have been expected to provide a lower rate of wear, but forging of about 28 000 pieces produced the same amount of wear (0.15 mm, or 0.006 in.) in the heading tool as forging of about 40 000 pieces produced in the softer gripper tool.

Table 25B. Typical tool steels for die blocks and die inserts
For illustration of forged parts, see Fig. 10.

Work metals	Hammer forging 100 to 10 000 parts	Hammer forging 10 000 parts or more	Press forging 100 to 10 000 parts	Press forging 10 000 parts or more
For making parts of severity no greater than part 4				
Carbon and alloy steels	6F2 or 6G at 341 to 375 HB, solid or with H11 or H12 plug(a) at 369 to 388 HB	6F2 or 6G at 341 to 375 HB with H11 or H12 plug at 369 to 388 HB or H11 or H12 at 405 to 433 HB	6F2 or 6G at 388 to 429 HB, solid or with H11 or H12 plug(a) at 405 to 433 HB	6F2 or 6G at 388 to 429 HB with H11 or H12 plug at 405 to 433 HB
Stainless steels and heat-resisting alloys	6F2 or 6G at 341 to 375 HB, solid or with H11 or H12 plug(a) at 429 to 448 HB	6F2 or 6G at 341 to 375 HB with H11 or H12 insert at 429 to 448 HB	6F2 or 6G at 388 to 429 HB, solid or with H11 or H12 plug(a) at 429 to 448 HB	6F2 or 6G at 341 to 375 HB with H11 or H12 plug at 429 to 448 HB
Aluminum and magnesium alloys	6F2 or 6G at 341 to 375 HB or H11 or H12 at 405 to 433 HB	6F2 or 6G at 341 to 375 HB with H11 or H12 plug at 405 to 433 HB or H11 or H12 at 405 to 433 HB	6F2 or 6G at 341 to 375 HB or H11 or H12 at 405 to 433 HB	6F2 or 6G at 341 to 375 HB with H11 or H12 plug at 429 to 448 HB
Copper and copper alloys	H11 or H12 at 405 to 433 HB	H11 or H12 at 405 to 433 HB	H11 or H12 at 405 to 433 HB	6F2 or 6G at 341 to 375 HB with H11 or H12 plug at 429 to 448 HB
For making parts of severity no greater than part 5				
Carbon and alloy steels	6F2 or 6G at 302 to 331 HB	6F2 or 6G at 302 to 331 HB, solid or with 6F3 plug(b) at 369 to 388 HB	6F2 or 6G at 341 to 375 HB	6F3 at 369 to 388 HB with H11 or H12 plug at 369 to 388 HB
Stainless steels and heat-resisting alloys	6F2 or 6G at 302 to 331 HB	6F2 or 6G at 302 to 331 HB with H11 or H12 plug at 369 to 388 HB	...	...
Aluminum and magnesium alloys	6F2 or 6G at 269 to 293 HB	6F2 or 6G at 269 to 293 HB with plug of same steel at 302 to 331 HB	6F2 or 6G at 341 to 375 HB	6F2 or 6G at 341 to 375 HB with H11 or H12 plug(c) at 429 to 448 HB
Copper and copper alloys	6F2 or 6G at 302 to 331 HB	6F2 or 6G at 302 to 331 HB with H11 or H12 plug at 405 to 448 HB	6F2 or 6G at 341 to 375 HB	H11 or H12 at 477 to 514 HB
For making parts of severity no greater than part 6				
Low-alloy steels, stainless steels and heat-resisting alloys	6F2 or 6G at 269 to 293 HB(d)	6F2 or 6G at 269 to 293 HB with plug of same steel at 341 to 375 HB	...	...
Aluminum, magnesium and copper alloys	6F2 or 6G at 269 to 293 HB	6F2 or 6G at 269 to 293 HB	...	...

(a) Recommended for 1000 to 10 000 forgings. (b) Recommended for long runs—for example, 50 000 forgings. (c) For long runs—for example, 50 000 forgings—a solid block made from H11 or H12 tool steel at 477 to 514 HB is recommended. (d) For quantities over 1000, a plug of the same material at 341 to 375 HB is recommended.

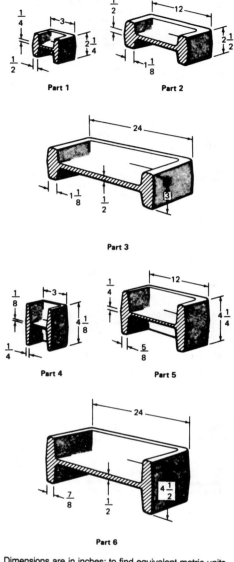

Dimensions are in inches; to find equivalent metric units (mm), multiply listed values by 25. For typical die materials and hardness ranges, see Tables 25A and 25B.

Fig. 10. Six hypothetical parts of progressively increasing severity

Table 26. Nominal compositions of nonstandard tool steels for die blocks

Steel(a)	C	Mn	Si	Cr	Ni	Mo	V
6G	0.55	0.80	0.25	1.00	...	0.45	0.10(b)
6F2	0.55	0.85	0.25	1.00	1.00	0.40	0.10(b)
6F3	0.55	0.60	0.85	1.00	1.80	0.75	0.10(b)

(a) Neither AISI nor SAE has assigned type numbers to these tool steels. (b) Optional.

The reverse may be true under certain severe conditions of forming. For instance, an unsymmetrical steering-gear forging was made with the segment stock all gathered on one side in the gripper die, and this resulted in greater metal movement, with correspondingly heavier die loads and increased abrasion, in that area.

The quantity of water used as a coolant for hot upset forging dies ordinarily does not affect selection of a die steel for forging a given part. In the bolt industry, however, water-cooled tools made of tungsten-bearing tool steels are susceptible to cracking and heat checking at high production rates (on the order of 100 pieces per minute). At high production rates such as these, the hot bolt stock is in nearly constant contact with the dies, allowing them little chance to cool, and under these conditions water cooling imposes severe thermal shock. Low-carbon (0.58 to 0.65% C) M10 high speed steel heat treated to 58 HRC performs well in such applications, as does T1 heat treated to the same hardness.

When a lower-alloy tool steel such as 6G or 6F2 is used for inserts, wear resistance can be improved by adjusting water flow to keep die temperatures below 200 °C (400 °F). At such temperatures, little improvement can be gained by changing to a different low-alloy die steel.

Lubrication of dies impairs speed and is not used extensively for upset forging of steel. Some lubrication may be used in deep punching and piercing, but only enough to prevent the workpiece from sticking to the punch. Resistance to wear and abrasion can be improved by use of die

Table 27. Typical materials for trimming dies

Material to be trimmed	Cold trimming				Hot trimming(a)	
	Normal trim		Close trim			
	Punch	Blade	Punch	Blade	Punch	Blade
Carbon and alloy steels	6F2 or 6G, at 341 to 375 HB	D2 at 54 to 56 HRC	Generally hot trim		6F2 or 6G at 341 to 375 HB	Hard facing alloy 4A on 1035 steel(b); or D2 at 58 to 60 HRC
Stainless steels and heat-resisting alloys	Generally hot trim		Generally hot trim		6F2 or 6G at 388 to 429 HB	D2 at 58 to 60 HRC
Aluminum, magnesium and copper alloys	6150 at 461 to 477 HB	Hard facing alloy 4A on 1020 steel(b); or O1 at 58 to 60 HRC	D2 at 58 to 60 HRC	D2 at 58 to 60 HRC	1020 soft	Hard facing alloy 4A on 1020 steel(b)

(a) Both normal and close trimming. (b) Hard facing alloy 4A has nominal composition as follows: 1 C, 30 Cr, 3 Ni, 4.5 W, 60 Co, rem Fe. For greater detail, refer to the article on Hard Facing Materials in this volume.

Table 28. Nominal compositions of nonstandard tool steels used in tools for hot upset forging

Type(a)	Composition, %						
	C	Mn	Si	Cr	Ni	Mo	V
6F	0.55	0.80	0.25	1.00	...	0.45	0.10(b)
6F2	0.55	0.75	0.25	1.00	1.00	0.30	0.10(b)
6F3	0.55	0.60	0.85	1.00	1.80	0.75	0.10(b)
6F4	0.20	0.70	0.25	...	3.00	3.35	...
6H1	0.55	...	...	4.00	...	0.45	0.85
6H2	0.55	0.40	1.10	5.00	1.50	1.50	1.00

(a) UNS, AISI and SAE have not assigned type numbers to these tool steels. (b) Optional.

Table 29. Typical tool materials for hot upset forging
For part designs and over-all dimensions, see Fig. 11.

Material forged	Tool material types and hardness ranges for total production quantity of:					
	100		1000 to 10 000		50 000 and up	
	Gripper die	Heading tool	Gripper die	Heading tool	Gripper die	Heading tool
For parts of maximum outside upsetting severity (part 1)(a)						
Carbon and low-alloy steels	4150 at 38 to 42 HRC or 4340 insert at 38 to 42 HRC	W1 with 0.70 C at 42 to 46 HRC or 4340 insert at 38 to 42 HRC	6H1 or H11 at 46 to 50 HRC or 4340 insert at 38 to 42 HRC	6H1 or H11 at 44 to 48 HRC or 6G (b) insert at 41 to 45 HRC	6H1 or H11 at 46 to 50 HRC or 4340 insert at 38 to 42 HRC	H11 at 46 to 50 HRC or 6H2 at 52 to 56 HRC or 6G (b) insert at 41 to 45 HRC
Stainless steels and heat-resistant alloys (up to type 310)	6G (b) insert at 38 to 42 HRC	6F3 insert at 42 to 46 HRC	6F3 insert at 42 to 46 HRC	H11 at 46 to 50 HRC or same for insert	6H2 at 52 to 56 HRC or H11 insert at 44 to 48 HRC	6H2 at 52 to 56 HRC or H11 insert at 48 to 52 HRC
For parts requiring both upsetting and piercing (parts 2 and 3)						
Carbon and alloy steels	4340 insert at 38 to 42 HRC or 6G (b) insert at 36 to 40 HRC	4340 insert at 42 to 46 HRC or 6G (b) insert at 41 to 45 HRC	4340 insert at 38 to 42 HRC or 6G (b) insert at 36 to 40 HRC	H11 at 42 to 46 HRC or H11 insert at 46 to 48 HRC	H11 or H11 insert at 42 to 46 HRC	H12 or M10 at 50 to 52 HRC or H11 insert at 46 to 50 HRC
Stainless steels and heat-resistant alloys (up to type 310)	(c)	(c)	(c)	(c)	(c)	(c)

(a) All heads are round and made in one blow with relative dimensions shown. (b) 6F2 die steel may be used interchangeably with 6G. (c) The same tool materials are recommended for upsetting part 2 as are shown for part 1. Part 3 is too severe to be made from a stainless steel or a heat-resistant alloy.

lubricants, but there is no known correlation between lubrication and die performance that would make lubrication a factor in die-steel selection.

Hot Extrusion Tooling

TOOLING for hot extrusion must operate under severe conditions of temperature, pressure and abrasive wear. Fundamentally, the extrusion process consists of forcing material in a plastic condition through a suitable die under high pressure to form a long, continuous shape. Some of the softer metals such as lead and aluminum are sufficiently plastic to be extruded at or near room temperature, but most other metals and alloys are extruded only at elevated temperatures.

The prevailing commercial hot extrusion process, based on tonnage of product, is single-charge direct extrusion with butt discard. The essential parts of the tooling for this process are illustrated in Fig. 12.

With the ram retracted, a hot billet or slug is placed in the container. A dummy block is inserted between the ram and billet; then the hot billet is pushed into the container liner and advanced under high pressure against the die. The metal is squeezed through the die opening, assuming the desired shape, and is severed from the remaining stub by sawing or shearing.

Tool Materials. Table 30 lists typical materials and hardnesses for tools used in hot extrusion. Hot extrusion of aluminum and magnesium is in many respects similar, the major difference being the pressure required. Often, the same tooling materials can be used for extrusion of either aluminum or magnesium.

In addition to the typical materials listed in Table 31, special insert materials and surface treatments have been specified (particularly for tools used in extruding complex shapes) where better resistance to wear at higher temperatures is required. Special insert materials include special grades of cemented tungsten carbide, nickel-bonded titanium carbides and aluminum oxide ceramics. Special surface treatments include nitriding, aluminide coating, and application of proprietary materials by vapor deposition or sputtering.

Die-Casting Dies

MATERIALS for die-casting dies must have good resistance to thermal shock and good resistance to softening at elevated temperatures. Resistance to softening is necessary because dies must be able to withstand the erosive action of molten metal under high injection velocity. Other properties that influence selection of materials for die-casting dies are hardening characteristics, machinability, resistance to heat checking, and weldability.

Performance of die-casting dies is directly related to casting temperature, injection pressure, thermal gradients within the dies, and frequency of exposure to high temperature. These variables are the principal ones used in the selection tables in this article. Tool steels of relatively high alloy content are required for components in direct contact with molten metal, as indicated in Table 32.

Die hardness is less critical for die casting of zinc than for die casting of alloys of higher cast-

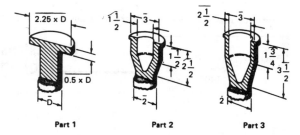

Part 1   Part 2   Part 3

Dimensions for parts 2 and 3 are given in inches; to find equivalent metric dimensions (mm), multiply listed dimensions by 25. Corner radii comply with standard commercial practice. For typical header-tool and gripper-die materials, see Table 29.

Fig. 11. Three typical parts made by hot upset forging

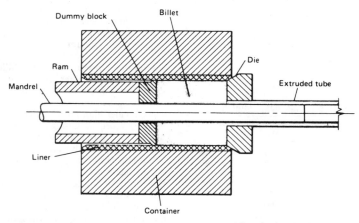

In this arrangement, the mandrel is attached to the press piercer and operates through a hollow ram.

Fig. 12. Typical tooling for extrusion of seamless tube

Table 30. Typical materials and hardnesses for tools used in hot extrusion

Tooling application	Aluminum and magnesium		Copper and brass		Steel	
	Tool material	Hardness, HRC	Tool material	Hardness, HRC	Tool material	Hardness, HRC
Dies, for both shapes and tubing	H11, H12, H13	47-51	H11, H12, H13	42-44	H13	44-48
			H14, H19, H21	34-36	Cast H21 inserts	51-54
Dummy blocks, backers, bolsters, and die rings	H11, H12, H13	46-50	H11, H12, H13	40-44	H11, H12, H13	40-44
			H14, H19	40-42	H19, H21	40-42
			Inconel 718	...	Inconel 718	...
Mandrels	H11, H13	46-50	H11, H13	46-50	H11, H13	46-50
Mandrel tips and inserts	T1, M2	55-60	Inconel 718	...	H11, H12, H13	40-44
					H19, H21	45-50
Liners	H11, H12, H13	42-47	A-286, V-57	...	H11, H12, H13	42-47
Rams	H11, H12, H13	40-44	H11, H12, H13	40-44	H11, H12, H13	40-44
Containers	4140, 4150, 4340	35-40	4140, 4150, 4340	35-40	H13	35-40

Table 31. Typical compositions of superalloys used for extrusion tools

Alloy	C	Mn	Si	Cr	Ni	Mo	Nb	Ti	Al	Fe	V	B
A-286	0.05	1.35	0.50	15.00	26.00	1.25	...	2.00	0.20	53.6	0.30	0.015
V-57	0.08	0.35	0.75	14.80	27.00	1.25	...	3.00	0.25	52.0	0.30	0.010
Inconel 718	0.04	0.20	0.30	18.60	53.00	3.10	5.0	0.90	0.40	18.5	...	...

ing temperature. Consequently, prehardened insert steels (29 to 34 HRC) are often used to make tooling for die casting of zinc.

Hot work tool steels are almost always used to make tooling for casting higher-melting-point alloys, such as aluminum, magnesium and copper. For casting aluminum and magnesium alloys, H13 steels are hardened to about 44 to 48 HRC; for casting copper alloys, H20, H21 and H22 steels generally are used at hardnesses of 38 to 45 HRC.

Injection components for both zinc and aluminum die-casting dies are subjected to considerable contact with molten metal and to severe erosion. Recommended materials and heat treatments for such dies are given in Table 33.

Slides, Guides, Cores and Pins. The moving parts of a die-casting die must have the same general characteristics as those of the stationary parts, and also must provide resistance to wear. Table 34 lists materials recommended for slides, guides, cores and ejector pins.

Die Lubricants. Most commercial die lubricants or mold-release agents reduce soldering or sticking of the casting to the die. The use of an effective lubricant results in easier cleaning of dies, less wear and longer runs between die polishes.

However, some of these die lubricants attack the die steel as well as the particles of the casting alloy adhering to the die surface. If used frequently, such lubricants will shorten die life.

Preheating and Cooling. Water cooling of cores and slides must be considered where significant wear is likely. Die-casting dies used for casting aluminum or copper are water cooled in service and should be preheated to about 175 °C (350 °F) before a run is started. Preheating should be done with water flowing slowly through the die to prevent the serious damage that would result from introduction of cold water into a hot die.

Ejection components are not subjected to high temperatures. Consequently, hot rolled 1020 steel in the as-received condition is used for making ejector boxes, rails, ejector plates, support posts and support blocks for tooling employed in die casting of aluminum and zinc parts.

Trim dies are used in secondary operations and are made of wear-resistant materials, as indicated in Table 35.

Powder-Compacting Tools

THE COMPACTING PROCESS most widely used to convert metallic powders into components is the closed-die process. The basic tooling for this process includes a precision-machined die, an upper and lower punch and, if the part is to be hollow, a core rod. These components are attached to the platens of a mechanical or hydraulic press by means of adapters, which often enclose the basic tooling and give it backup support.

The materials generally used for these tools are (a) cemented carbides and (b) tool steels with or without special surface treatments.

Dies. Die inserts for compaction of carbide, ceramic or ferrite powder most frequently are the medium- or coarse-grain 94WC-6Co grades of cemented carbide. Cemented tungsten carbide containing 12 to 16% cobalt may be used to make inserts for compacting metal powders in medium-to-long production runs.

Cemented carbides are relatively expensive, and shaping of parts to the required form must be done either by electrical discharge machining or by specialized methods of grinding.

Wear-resistant tool steel inserts are sometimes used instead of carbide inserts. Crucible CPM 10V is frequently chosen for medium-to-long production runs. Other wear-resistant tool steels, usually D2 or a high speed steel such as M2 or M4, have been used for short-run applications, as have high-wear P/M alloys. Tool steel inserts generally are heat treated to a working hardness of 62 to 64 HRC. For increased wear resistance, a nitrided case may be specified for dies made of CPM 10V or D2. For certain part designs, a solid die rather than an insert die is a more practical choice; an air-hardening 5%-Cr tool steel such as A2 is generally used for such applications.

Punches. The stresses imposed on punches during service are such that toughness is a much more important material requirement than wear resistance, although wear resistance cannot be ignored. Type A2, and sometimes the shock-resisting type S7, is preferred for punches.

For applications in which A2 or S7 punch faces become severely abraded, a more wear-resistant grade such as D2, D3, high-wear P/M alloys or M2 should be considered. Cemented carbides are too brittle to perform successfully as punches or punch-face inserts.

Core Rods. Both toughness and wear resistance are important criteria in selection of core-rod materials, but generally the primary consideration is wear resistance. For particularly abrasive conditions, CPM 10V has been used successfully, as have D2, M2 and A2 tool steels that have been nitrided or coated with tungsten carbide.

Molds for Plastics and Rubbers

IN MOLDS for parts made of plastic materials (including rubbers), the type of material to be molded is a major factor governing the choice of mold material. For example, some plastics require mold materials that resist abrasion, corrosion, heat, or high compression loads.

This article covers typical mold materials, and methods of producing molds, for injection, blow, transfer and compression molding of the chief groups of plastic materials. The following definitions describe the basic steps in each molding process, and the principal tooling involved.

Injection molding is a method of forming plastic objects from granular or powdered thermoplastics by (a) using heat and pressure to plasticize the material in a chamber and then (b) forcing part of the fluid mass into a cooler chamber, where it solidifies. Injection molding requires pressures of 70 to 140 MPa (10 000 to 20 000 psi).

Blow molding is a method of forming hollow objects in which a thermoplastic in the form of a molten or softened tube (parison) is inserted into a cool mold. The parison is then pressurized at low internal pressure so that it expands against the sides of the mold, where it solidifies. Blow molding requires pressures of only 0.2 to 0.7 MPa (25 to 100 psi). The primary advantage of blow molding is the ability to form re-entrant curves.

Transfer molding is a method of forming plastic objects from granular, powdered or pre-

formed thermosets by softening the material in a heated chamber and then forcing essentially the entire mass into another hot chamber, where it cures.

Compression molding is a method of forming objects from either thermosets or thermoplastics by placing the material in a heated mold cavity open at the top, and then simultaneously applying heat and compressing the material with a "force."

Cavity refers to the female portion of the mold, which forms the outer surface of the molded article. (The cavity frequently is also called the *die*.)

Plunger refers to the ram or piston used to displace the fluid or semifluid material in transfer molding and injection molding. (In compression molding, the ram or piston usually is called the *force*.)

SELECTION OF MOLD MATERIALS

Table 36 lists typical materials used for making machined molds. The choice of mold material depends on the type of plastic to be molded, and on the quantity and shape of the parts to be made. The mold materials given in Table 36 are for several hypothetical parts representing different degrees of molding severity; these hypothetical parts are illustrated in Fig. 13.

Machined Molds. Large cavities usually are produced by machining. For these large cavities, steels that require high-temperature heat treatment and rapid cooling may exhibit an unacceptable amount of distortion; it may be necessary to use prehardened die steels, or steels (such as nitriding steels or maraging steels) that require only a relatively low-temperature final heat treatment to develop acceptable wear resistance.

Hubbed Molds. In many instances, small multiple-cavity molds can be produced at lowest cost by hubbing. This involves pressing a male master plug, known as the hub, into the metal block. (This process is also called "hobbing.") If forming a mold by hubbing is to be economical, the mold steel must be very soft and have a low rate of work hardening. Such steels must be carburized and heat treated to provide the required wear resistance and surface hardness.

Satisfactory hubbing of mold cavities depends not only on the steel to be hubbed but also on selection of the proper steel for the hub. Steels for master hubs should have good machinability and workability in the annealed condition, high compressive strength, and resistance to abrasion in the heat treated condition. Hub steels also should exhibit minimum distortion and size change in heat treatment, minimum scaling, and the ability to be polished to a high finish.

Table 32. Recommended materials for die-casting dies and die inserts

Components	Typical material(s) and hardness
Cavity inserts for Al and Mg castings	H13 hardened to 45 to 48 HRC
Cavity inserts for Zn castings	P20 prehardened to 300 HB
Cavity inserts for long-run Zn castings	H13 hardened to 44 to 46 HRC
Cavity inserts for Cu castings	H20, H21, H22 hardened to 44 to 48 HRC
Holder blocks	4140 prehardened to 300 HB

Table 33. Typical materials for injection components

Component	Metal being cast	Material and condition
Sprue spreader	Zinc	H13 hardened to 250 to 290 HB and nitrided
Sprue bushing	Zinc	Nitralloy hardened to 250 to 300 HB and nitrided
Nozzle and adapter	Zinc	H13 hardened to 46 to 48 HRC and nitrided
Shot sleeve	Aluminum	H13 hardened to 46 to 48 HRC and nitrided
Shot pad	Aluminum	H13 hardened to 46 to 48 HRC and nitrided
Plunger tip	Aluminum	Beryllium copper hardened to 38 to 42 HRC

Table 34. Materials for slides, guides, cores and ejector pins

Component	Material(s)	Condition
Slide carrier	4130, 4140, 6150	Hardened to 46 to 50 HRC
Slide lock	4140, 6150	Hardened to 46 to 50 HRC
Leader pin	1117	Carburized and hardened to 58 to 62 HRC
Guide bushing	1018	Carburized and hardened to 58 to 62 HRC
Guide block	4140	Carburized and hardened to 56 to 60 HRC
Guide plate	4140	Carburized and hardened to 56 to 60 HRC
Ejector pin	H13	Hardened to 34 to 40 HRC and nitrided
Return (surface) pin	H13	Hardened to 34 to 40 HRC and nitrided
Core	H13	Hardened to 48 to 55 HRC

Table 35. Typical materials for trim dies

Component	Material	Condition
Base, holder	1020 (a)	As received
Trim ring, plate	A2	Hardened to 56 to 58 HRC
Punch, pad	1020 (a)	As received
Guide pin, guide bushing	C1117	Carburized and hardened to 58 to 62 HRC
Slide block	A2	Hardened to 56 to 58 HRC
Cam	1020 (a)	Carburized and hardened to 46 to 48 HRC

(a) Hot rolled.

Table 36. Typical materials for machined molds

Material to be molded	Mold materials(a) for making parts shown in Fig. 13 in various production quantities				
	Parts 1, 4 and 7		Parts 2, 5 and 8		Parts 3, 6 and 9
	10 000 to 100 000	1 000 000 to 10 000 000	10 000 to 100 000	1 000 000 to 10 000 000	10 000 or more
Mold materials for thermoplastics					
Group 1: general-purpose plastics and rubbers	P20 or P21, prehardened(b); 414L, prehardened	O1, 53 to 57 HRC; P6 or P20, carburized, 54 to 57 HRC; S7, 51 to 57 HRC; 420 stainless, 45 to 50 HRC	P20 or P21, prehardened(b); 414L, prehardened	P6, carburized, 54 to 58 HRC; P20 or P21, prehardened(b); 414L, prehardened	P20 or P21, prehardened(b); 414L, prehardened
Group 2: fluid plastics	P6 or P20, carburized, 54 to 58 HRC	O1, 53 to 57 HRC; S7, 51 to 57 HRC	P6, carburized, 54 to 58 HRC	P6, carburized, 54 to 58 HRC; H13 (c), 48 to 52 HRC; S7, 51 to 57 HRC; 420 stainless, 45 to 50 HRC	P20 or P21, prehardened(b); 414L, prehardened
Group 3: corrosive and high-temperature plastics	P20 or P21, prehardened(b) and nickel plated(d); 414L, prehardened; 420 stainless, 45 to 50 HRC	O1, 53 to 57 HRC, nickel plated(d); 420 stainless, 45 to 50 HRC; S7, 51 to 57 HRC	P20 or P21, prehardened(b) and nickel plated(d); 414L prehardened; 420 stainless, 45 to 50 HRC	P6, carburized, 54 to 58 HRC, nickel plated(d); 420 stainless, 45 to 50 HRC	P20 or P21, prehardened(b) and nickel plated(d); 414L, prehardened
Mold materials for thermosets					
Group 4: general-purpose plastics	L2, 53 to 57 HRC; P20, carburized, 54 to 58 HRC; S7, 51 to 57 HRC	L2, carburized, 53 to 57 HRC; A2, 53 to 57 HRC; P20, carburized, 54 to 58 HRC; S7, 51 to 57 HRC	P20 or P6, carburized, 54 to 58 HRC	P20 or P6, carburized, 54 to 58 HRC	P20 or P6, carburized, 50 to 55 HRC
Group 5: plastics requiring high-temperature curing	H13, 48 to 52 HRC; S7, 51 to 57 HRC	H13, 48 to 52 HRC; S7, 51 to 57 HRC	P4, carburized, 52 to 56 HRC; H13, 48 to 52 HRC	P4, carburized, 52 to 56 HRC; H13, 48 to 52 HRC; S7, 51 to 57 HRC	P4, carburized, 52 to 56 HRC
Group 6: rubbers	Class 30 gray iron(e)	Class 30 gray iron; 1020, soft, chromium plated(f); A2, 53 to 57 HRC(g)	1020, soft, chromium plated(f)	1020, soft, chromium plated(f)	1020, soft, chromium plated(f)
Group 7: low-pressure and abrasive plastics	P20, carburized, 54 to 58 HRC; L2, 53 to 57 HRC; S7, 51 to 57 HRC	P20, carburized, 54 to 58 HRC; L2, 53 to 57 HRC; S7, 51 to 57 HRC	P20, carburized, 54 to 58 HRC; L2, flame hardened	P20, carburized, 54 to 58 HRC; L2, flame hardened; S7, 51 to 57 HRC	P20 or P6, carburized, 50 to 55 HRC

(a) Where more than one mold material is given for a specific set of conditions, they are arranged in order of preference unless otherwise noted. (b) Hardness of prehardened steels should be 300 HB minimum. (c) Preferred for molding parts 5 and 8. (d) Recommended thickness of plating, 0.005 to 0.025 mm (0.0002 to 0.0010 in.). (e) Cast 356-T6 aluminum recommended for quantities up to 10 000 parts. (f) Recommended thickness of plating, 0.005 to 0.015 mm (0.0002 to 0.0005 in.). (g) Provides increased resistance to handling.

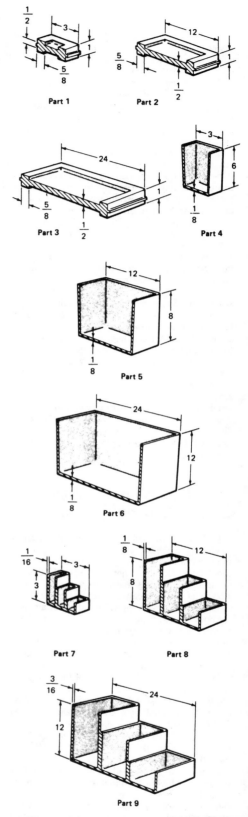

Dimensions are in inches; for equivalent metric dimensions (mm), multiply listed values by 25. See Table 36 for typical mold materials used in producing parts similar in general configuration and degree of severity to those shown above.

Fig. 13. Hypothetical shapes typical of molded plastic products

For relatively simple hubs containing sharp detail but no feather edges (which are susceptible to edge wear, edge breakdown, or loss of detail), S1 is usually recommended. Hardnesses of 59 to 61 HRC are recommended, and slight carburization of the surface during heat treatment is often beneficial. Most hubs fall into this general category. Therefore, S1 is the tool steel most commonly used in hubbing.

Steels O1, A2 and D2, which have high hardness and good abrasion resistance, are used for high-production hubs of simple design for which long life and resistance to bulging are important considerations but for which high toughness is relatively unimportant. A hardness in the range from 60 to 63 HRC is recommended for this type

of hubbing. Hubs of intricate or complex design must be notch tough and often are made of 6F5 tool steel. Such hubs are heat treated to hardnesses of 58 to 59 HRC. However, 6F6 tool steel is more suitable for hubs of complex design that incorporate feather edges, which require exceptional resistance to brittleness and edge failure. Intentional carburization during heat treatment of this type of steel is often helpful.

L6 tool steel is used principally for inexpensive hubs that are pressed into the mold material with relatively low hubbing pressures.

Cast Molds. Many small-cavity and multiple-cavity molds are pressure cast using a hub as the pattern for the die cavity.

Hubs used for pressure casting of beryllium

Table 37. Typical materials and hardnesses for common types of production gages

Types of gage	Gaging tolerance 0.01 mm (0.0005 in.) Part hardness up to 350 HB — Gage material(a)	Hardness, HRC	Part hardness over 350 HB — Gage material(a)	Hardness, HRC	Gaging tolerance 0.05 mm (0.002 in.) Part hardness up to 350 HB — Gage material(a)	Hardness, HRC	Part hardness over 350 HB — Gage material(a)	Hardness, HRC
Occasional gaging of dimensions up to 100 mm (4 in.)								
Gage blocks, gaging pins, anvils, buttons	W1, O1, O2	61-64						
Cylindrical ring and plug gages	1212 (b), W1, O1, O2	61-64						
Threaded ring and plug gages	W1, O1, O2	61-64						
Height and length gages	W1, O1, O2	61-64						
Spline gages	W1, O1, O2	61-64						
Snap gages(c), thread rolls	O1, O2, L7	61-64						
Feeler gages	W1, O1, O2	45-52						
Alignment bars	1212 (b)	...						
Frequent, long-term gaging of dimensions up to 12 mm (¹⁄₂ in.)								
Gage blocks, gaging pins, anvils, buttons	L7 (d)	61-64	D2 (e)	57-64				
	D2 (e)	57-64	M2 (f)	62-65				
			Carbide(g)	...				
Cylindrical ring and plug gages	M2 (f)	62-65	M2 (f)	62-65				
			Carbide(g)	...				
Threaded ring and plug gages	A2 (e)	56-64	M2	62-65				
Height and length gages	M2	62-65	M2	62-65				
Spline gages	O6	61-64	A2 (e)	56-64				
Snap gages(c), thread rolls	L7 (d)	61-64	L7 (d)	61-64				
Feeler gages	L7 (d)	45-50	D2	45-50				
Frequent, long-term gaging of dimensions from 12 to 100 mm (¹⁄₂ to 4 in.)								
Gage blocks, gaging pins, anvils, buttons	A2 (e)	56-64	Carbide(g)	...	A2	62-64	M2 (f)	62-65
	D2 (e)	57-64			D2	62-64		
	M2	62-65			M2	62-65		
Cylindrical ring and plug gages	M2 (f)	62-65	Carbide(g)	...	D2	62-64	M2 (f)	62-65
					M2	62-65		
Threaded ring and plug gages	A2 (e)	56-64	M2 (f)	62-65	A2	62-64	M2	62-65
	D2 (e)	57-64						
Height and length gages	A2 (e)	56-64	M2	62-65	A2	62-64	M2	62-65
	D2 (e)	57-64						
Spline gages	O6	61-64	A2 (e)	56-64	O6	61-64	A2, D2	62-64
			D2 (e)	57-64				
Snap gages(c), thread rolls	L7 (d)	61-64	D2 (e)	57-64	L7	61-64	D2	62-64
			M2	62-65			M2	62-65
Alignment bars	1212 (b)	...	8620 (b)	...	1212 (b)	...	8620 (b)	...
	8620 (b)	...	4140 (h)	...	8620 (b)	...	4140 (h)	...

(a) Where more than one tool material is listed for a specific set of conditions, the last material listed is usually preferred for large sections. (b) Carburized, with a case not more than ¹⁄₅ the section thickness and having a minimum surface hardness equivalent to 61 HRC. (c) Snap-gage bodies generally are made of stress-relieved cast iron, ASTM A48, class 20, 30 or 35. (d) 52100 steel has proved a satisfactory substitute for L7. (e) For close tolerances, this steel must be tempered in the secondary hardening range for maximum stability. (f) Liquid nitriding after full hardening is recommended to produce a surface hardness of about 1100 HV. (g) Cemented tungsten carbide is usually selected, but chromium carbide or boron carbide also can be used. (h) Heat treated to 26 to 30 HRC, then gas nitrided for 24 h.

copper molds must resist softening at the elevated temperatures involved. Therefore, a hot work steel, such as H13, is usually chosen for this application. For some applications, such as gate inserts, machinable steel-bonded carbides can be shaped into the insert, and then heat treated to hardnesses on the order of 68 HRC.

Thread-Rolling Dies

THE PROPERTIES that are most significant in selecting materials for thread-rolling dies are hardness, toughness and wear resistance. Hardness and toughness must be high enough to enable the dies to withstand the forces exerted on them in service; good wear resistance is necessary because the prime cause for removing thread-rolling dies from service is spalling of the crests of the die threads, which is allowed to continue until the contours of the threads being rolled no longer meet dimensional or functional requirements.

The materials most commonly used to make thread-rolling dies are M1 and M2 high speed tool steels; D2 high-carbon, high-chromium tool steel; and A2 medium-alloy cold work tool steel. The steel chosen should be adequately annealed before hardening and should have a sufficiently uniform carbide-particle distribution.

In most applications, D2, M1 and M2 are about equal in performance, whereas the service life of A2 dies is somewhat lower. In general, D2, M1 and M2 should be selected for long production runs and for rolling larger parts, coarser threads and alloys of higher hardness. The hardness ranges given in the following table have been found satisfactory for most applications:

Die material	Recommended hardness, HRC Flat dies	Circular dies
For threading aluminum, copper or soft steel blanks		
A2	57 to 60	56 to 58
D2	60 to 62	58 to 60
M2	58 to 60	58 to 60
For threading ferritic steel (hardness, >95 HRB) or austenitic stainless steel blanks		
A2	57 to 59	56 to 58
D2	59 to 61	58 to 60
M2	59 to 61	58 to 60

Hardness values indicated above should be achieved by double tempering after quenching. Early failure is more likely if double tempering is not used.

If diameter and lead tolerances of the rolled part permit, the dies can be ground or machined before hardening. The recommended steels for that practice, given in descending order of preference, and their average distortion during heat treatment necessary for obtaining the required hardness, are as follows:

Type of steel	Approx average distortion, %
A2	0.04
D2	0.05
M1 or M2	0.11

If tolerances require lower average distortions, dies must be ground after heat treatment. Die life usually is not decreased if grinding is done after heat treatment, as long as proper (nonabusive) grinding techniques are used.

Flat dies are used for producing most standard threaded fasteners and most wood screws. Flat dies made of D2 are usually ground before hardening, because D2 is susceptible to grinding cracks if improperly ground after hardening.

Gages

SELECTION OF MATERIALS for gages depends to a large extent on tolerance to be checked,

Source: Metals Handbook Desk Edition, 1985, 18.1-18.36

number of items to be gaged, composition and hardness of the material being gaged, size and complexity of the item and cost of the gage material. Abrasive wear is the predominant factor in determining the useful life of production gages. Therefore, gage surfaces that contact the workpiece must be hard enough to provide adequate resistance to abrasion. The actual hardness required depends on the hardness and abrasive characteristics of the workpiece surfaces, number of pieces to be gaged and the tolerance to be checked. Production gages that must be used in hostile environments may have to be corrosion resistant. Also, if the gage must operate over a range of temperatures, thermal expansion must be considered when making final material selection.

GAGE MATERIALS

Typical materials and hardnesses for the more common types of production gages are given in Table 37. In this table, it should be noted that

feeler gages, which are thin flexible strips of precisely controlled thickness, must be tempered back to lower hardness so that they will not break during use.

Precision Gages. Making gages to extra-close tolerances often requires special gage materials. In applications that require extremely close gage tolerances, fine surface finishes, exceptional wear characteristics and outstanding dimensional stability, boron carbide and even jewels make excellent gage elements.

Combination Gages. Frames, bodies and bases of combination gages commonly are made of cast iron. The cast iron used for gages is generally a class 20, 30 or 35 gray iron that has been stress relieved at 450 to 480 °C (850 to 900 °F).

Aluminum is also used in combination gages for handles, bodies and bases. Aluminum handles and bodies may be made either of soft grades such as 1100 or, if more strength is required, of harder grades such as 2017 or 2024.

Materials used for wear surfaces in combination gages must have the same properties as those required for the solid gages discussed above. Wear

inserts generally are made from hardened tool steel or cemented carbide. However, when the insert is a relatively simple shape that is not very susceptible to distortion and cracking in heat treatment, it may be possible to use an inexpensive tool steel instead.

Inspection fixtures are gages, but belong to a different category from those discussed so far. They are used mainly for checking dimensions and contours of large stampings such as automobile body components.

Plastics or plastic-faced cast iron composites are frequently used for inspection fixtures (checking fixtures) because these materials are lower in cost and lighter in weight than solid cast iron.

Master gages require high accuracy, good wear resistance and maximum stability. They are used in checking other gages and are expected to hold their accuracy over long periods of time.

No one material has the ideal combination of properties desired for master gages, although high-carbon high-chromium steels such as D2 and D4 have most of the desired characteristics.

———— Selected References on Tool Materials ————

The Metallurgy of Tool Steels, by P. Payson: John Wiley & Sons, Inc., 1962

Metallurgy and Heat Treatment of Tool Steels, by R. Wilson: McGraw-Hill, London, 1975

"Tool Steels" (a Steel Products Manual): American Iron and Steel Institute, March 1978

Tool Steels, by G. A. Roberts and R. A. Cary: American Society for Metals, 1980

Tool Steel Simplified, Revised Ed., by F. R. Palmer *et al.*: Chilton Book Co., Radnor, PA, 1978

ADDITIONAL READING

World Directory and Handbook of Hard Metals, 2nd Ed., by Kenneth J. A. Brooks: Engineers' Digest Limited, London, 1979

"Engineering Properties of Ceramics," by J. F. Lynch, C. G. Rederer and W. H. Duckworth: Technical Report AFML-TR 66-52, Air Force Materials Laboratory, 1966

Some Plain Talk About Carbides, by H. S. Kalish: *Manufacturing Engineering and Management*, July 1973

"A System of Classification of Hard Metal Grades for Machining": Technical Publication No. 1, The British Hard Metal Association, Sheffield, England, 1967

An Analysis of Charpy Impact Testing as Applied to Cemented Carbide, by R. C. Lueth: in *Instrumented Impact Testing*, STP 563, American Society for Testing and Materials, 1974

Where Solid Titanium Carbide Stands, by H. S. Kalish: *American Machinist*, 7 Jan 1974, p 50-52

"Machinable Carbides for High Performance Tooling and Wear Parts," by S. E. Tarkan and M. K. Mal: Technical Paper MR 73-927, Society of Manufacturing Engineers

Tool and Die Failures Source Book, edited by Serope Kalpakjian: American Society for Metals, 1982

"Tantung: The Premiere Cast Alloy": Bulletin 72-1, VR/Wesson Div. of Fansteel, Inc., 2 June 1972

"Ceramic Tools," by E. D. Whitney: Technical Paper TE 73-205, Society of Manufacturing Engineers

Cutting Performance and Practical Merits of Carbide Ceramics, by K. Ogawa, M. Furukawa and Y. Hara: *Nippon Tungsten Review*, Vol C, 1973

"Borazon and Diamond Cutting Tools," by R. E. Hanneman and L. E. Gibbs: Technical Information Series, Report No. 73 CRD 182, General Electric Co., June 1973

Some Experiments to Compare Diamond and Diamond Compact Cutting Tools, by M. Casey and J. Wilks: *Sixteenth International Tool Design and Research Conference*, McMillan Press Ltd., Sept 1975

Distortion in Tool Steels, by B. S. Lement: American Society for Metals, 1959

The Case for TiN-Coated Gear Tools—Part 2

Part 1 of this article (Jan/Feb issue) discussed tool life improvement and increased productivity possible with coated tools. In this concluding installment, the author zeroes in on cost savings and surveys future coatings.

Peter W. Kelly, Gear Tool Product Mgr.
Barber-Colman Co., Machine & Tools Div.

To illustrate the economies and production advantages that can be achieved with TiN-coated gear tools, tests were conducted on the production floor by selected, high volume gear or pinion manufacturers. The results of four separate tests to be described in the following pages are combined in the accompanying table.

Data collected included the specifications of the tool, the work piece being cut and the machine in which the gear cutting operation was performed. The table provides a gear cutting cost analysis for three different tool conditions: 1) The uncoated tool operating under previously established production conditions, 2) The TiN coated tool operating under the same conditions as had been established for the previously applied uncoated tool, and 3) **Assumed** increased speed or depth of cut, the resulting wear and the number of pieces cut when applying the TiN coated tool in order to obtain increased productivity.

To gain the maximum potential available from TiN coated gear tools it is necessary to increase the speed, and/or increase the federate, and/or reduce the number of cuts in a multiple cut operation. For the purpose of illustrating the large savings available by reducing the machining time, it was assumed that the tools tabulated in the tables could be operated at increased speed, and in one case at a reduced number of cuts. It was further assumed that they would produce only half the number of pieces when worn the same amount at the TiN coated tools at the originally established lower speed. These are conservative estimates based on results of other gear tool tests and on experimental work done with other types of high speed steel cutting tools.

Note that the costs tabulated include only the cost of the gear cutting tool and the cost of the actual hobbing or shaping operation itself. The machining costs are based on the calculated cutting time at a conservative estimate of $25.00 per hour. In the simplified analysis presented, the tool sharpening cost and the cost of the gear cutting machine downtime for tool changing have not been included. When these costs are considered, the economic advantages of applying TiN coated tools are increased substantially because the amount of sharpening stock and the frequency of sharpening are greatly reduced.

Half the Wear

The first test consisted of hobbing a 16 tooth automotive transmission pinion with a single thread, unground Class C, hob. Comparing the results of the uncoated and the TiN coated hob under the same operating conditions, the coated hob produced the same number of pieces with only approximately half the hob wear of the uncoated tool. This provided an overall savings per part of 1.7%

Tenfold Savings

However, when operating at a 30% increase of surface speed, the overall savings per part increased over 10 times to 20.7%. This, of course, is due to the fact that of the original uncoated tool application the percentage of the hob cost per part to the total cost per part was only 5.2%. When cutting under the same conditions with a coated hob, that ratio fell to 3.5% and increased to 8.7% when operating at the increased surface speed. Also note that the tool cost at the higher productivity conditions increased over and above that of the uncoated tool, 8.7% vs. 5.2%, but that overall savings per piece when including machining time was 20.7%.

In the second test, for a coarser pitch, double thread Class C accurate unground hob, the surface speed for the TiN-coated tool was increased by 30%. The number of pieces per sharpening and the number of pieces per life of the hob are approximately 2 times and 3 times, respectively, for the coated hob at increased productivity over the uncoated hob. This increase in tool life, combined with an approximate 30% reduction in machining time, yields a substantial 27.1% savings in overall hobbing costs per part.

The third test, involving a shaper cutter for a 51-tooth automotive gear, demonstrates the exceptional wear resistance provided by the TiN coating.

Tool Cost Vs. Machining Cost per piece of Uncoated and TiN Coated Tools

Tool / Work Material	Costs/Savings		Uncoated	TiN Coated	
				Same Speed, Feed & Depth	Increased Speed, Feed &/or Depth
2.75 x 4 15.58 NDP UNG HOB M2 5140H 100 R_B	Tool Cost/PC $		.03	.02	.04*
	Machining Cost/PC $		.55	.55	.42
	Tool Cost of Total %		5.2	3.5	8.7
	Savings/PC %		—	1.7	20.7
3 x 4 10.5 NDP UNG HOB M2 5130H 140 BHN	Tool Cost/PC $		.07	.02	.03*
	Machining Cost/PC $		.52	.52	.40
	Tool Cost of Total %		11.9	3.7	7.0
	Savings/PC %		—	8.5	27.1
4.0886x.975 18.7773 NDP Shaper Cutter M2	Tool Cost/PC $		.27	.03	.06*
	Machining Cost/PC $		1.00	1.00	.81
	Tool Cost of Total %		21.3	2.9	6.9
	Savings/PC %		—	18.9	31.5
4.40x1.25 8.5 NDP Shaper Cutter M2 4027H	Tool Cost/PC $		.45	.05	.09*
	Machining Cost/PC $		1.32	1.32	.93
	Tool Cost of Total %		25.4	3.6	8.8
	Savings/PC %		—	22.6	42.4
					*Assumed

The above summary combined with similar tests from a number of other applications indicates the following values are realistic.

Tool Condition	Production Rates	Tool Cost / Total Cost %	Total Savings %
Uncoated	Existing	5 to 25	—
TiN Coated	Existing	2 to 5	2 to 20
TiN Coated	Increased	5 to 10	20 to 40

When the TiN coated cutter was initially used at previously established production rates, the number of gears produced prior to its first sharpening, that is with TiN coating on the shaper cutter tooth face, was an astonishing 906 pieces or 12 times the number of pieces produced with an uncoated cutter. At that time, the amount of flank wear on the cutter was only .015 compared to .025 for the uncoated cutter.

After sharpening the cutter, thus removing the TiN coating from the tooth face, it was found that 225 pieces could be cut before cratering had progressed to the point that it was feared cutting additional parts would cause tooth breakage. At that time the flank wear on the cutter was only .005.

Continuing to operate under those productivity conditions and running lost of 225 pieces per .005 of sharpening, the cutter produced 22,500 parts during its life which changed the percent of cutter cose per part to the total cost per part from 21.3% for the uncoated cutter to 2.9% for the TiN coated cutter. This resulted in an 18.9% total cost savings per part cut.

However, an even greater savings in total gear shaping cost would be available by increasing the stroke rate by 20% from 750 to 900 strokes per minute. At the assumed increased wear of .010 per 225 pieces cut, the cutter cost per part would increase to $0.06 which would still be $0.21 less than the cost per part of the uncoated cutter. As shown in the table, the resulting machining cost reduction to $0.81 would result in a total savings per part cut of 31.5%.

In the fourth test, the shaping of an 8.5 NPD 20 tooth 22° right hand helical gear, it was intended to utilize as nearly as possible the same speeds and feeds with the coated cutter as were being applied in the existing production line with an uncoated cutter. However, some slight changes in stroke rate and depth of cut were necessary and their values were adjusted in order to approximate the same cycle time. This resulted in an increase of almost 12 times in cutter life and a total savings per part of 22.6%.

During the test of the TiN coated cutter it was suggested that it would be appropriate to reduce the number of cuts in order to increase overall productivity.

Combining the depth for the original first and second cuts and holding the average sharpening stock to .008 would result in being able to cut 112 gears per sharpening. This increase in productivity would reduce the machining cost per part by $0.39 and provide a total savings per part 42.4%.

Summarizing the results of the tests, it can be seen that for 8 to 20 DP high production gear tools the maximum savings potential is available in utilizing TiN coated tools at increased production rates. It is possible to save 20 to 40% of the gear cutting costs.

New Coating Developments

Among the currently anticipated gear tool coating improvements, the following appear to be the most promising:

1. Certain developments are being pursued in titanium nitride coatings with the PVD process to provide reduced processing costs.

2. Titanium Carbo-nitride(TiCN) is most likely the next high speed steel coating which will be developed to provide higher hardness and improved abrasion resistance. The major disadvantage of Titanium Carbo-nitride is its brittleness.

3. Titanium Carbon Oxynitride (TiCON) is generating interest in certain quarters but its total characteristics and advantages are not well known at this time.

4. Hafnium Nitride (HfN) and Zirconium Nitride (ZN) are candidates for future test work. Some basic testing has been done to-date indicating that Hafnium Nitride is not as hard as Titanium Nitride but has higher thermal stability.

5. Titanium Boride (TiB) has a serious disadvantage due to the extreme toxicity of gaseous boron which is used during the coating process. Therefore not much work has been done with this possible coating.

6. Silicon Carbide (SiC), Silicon Nitride (SiN) and Tungsten Carbide (WC) are additional possibilities for future coatings applicable to high speed steel, but currently not much work is being done.

7. There are also possibilities of obtaining enhanced characteristics from combinations of the above coatings through the use of multi-layer coatings.

Until any of the above possibilities of improved coatings are proven realities, the most advantageous tool cost and gear cutting productivity improvements available will be through application of TiN coated tools. △

Source: Cutting Tool Engineering, April 1984, 24-26

INDEX

409